MODERN APPROACHES TO ANIMAL CELL TECHNOLOGY

MODERN APPROACHES TO ANIMAL CELL TECHNOLOGY

Editors

R.E. SPIER

Department of Microbiology, University of Surrey,
Guildford, Surrey, UK

J.B. GRIFFITHS

Porton, Salisbury, Wilts, UK

ESACT

EUROPEAN SOCIETY
FOR ANIMAL CELL TECHNOLOGY
THE 8th MEETING

OHOLO
BIOLOGICAL
CONFERENCE

ISRAEL INSTITUTE
FOR BIOLOGICAL RESEARCH
THE 32nd CONFERENCE

First published 1987

British Library Cataloguing in Publication Data
Esact, Meeting (*8th: 1987*).
Modern approaches to animal cell technology.
1. Biotechnology 2. Cell culture
I. Title II. Oholo Biological Conference
(*32nd: 1987*) II. Spier, R. E.
IV. Griffiths, J. B.
660′.6 TP248.25.C4

ISBN 0-408-02732-0

Library of Congress Cataloging-in-Publication Data

European Society for Animal Cell Technology.
General Meeting (8th: 1987: Tiberias, Israel)
Modern approaches to animal cell technology: ESACT, European Society for Animal Cell Technology, the 8th meeting, Oholo Biological Conference, Irsrael Institute for Biological Research, the 32nd Conference/
[edited by] R. E. Spier, J. B. Griffiths
p. cm.
Bibliography: p.
Includes index.
ISBN 0-408-02732-0
1. Cell culture—Congresses. 2. Cell lines—Congresses.
3. Biotechnology—Congresses.
4. Biological products—Congresses.
I. Spier, R. (Raymond) II. Griffiths, J. B.
III. Oholo Biological Conference (32nd: 1987:
Tiberias, Israel) IV. Title.
QH585.E893 1987
591′.07′24–dc 19 87-30009

Printed and bound by Butler & Tanner Ltd, Frome, Somerset

Organizing Committee

A. Mizrahi, *Meeting Secretary*
S. Cymbalista
A. Kadouri
A. Kohn
A. Lazar
S. Reuveny

The Organizing Committee acknowledges the support given by:

American Cyanamid Company
Biocatalysts & LH Fermentation
Cetus
IRI
Institut Merieux
The Israeli Honey Production & Marketing Board
The Ministry of Science and Development, Israel
The Ministry of Industry and Trade, Israel
Pharmacia
Sandoz
Technad

ESACT Executive Committee 1986–87

B. Griffiths	*Chairman*	PHLS CAMR, Porton, UK
B. Meignier	*Secretary/Treasurer*	Institut Merieux, Marcy L'Etoile, France
A. Mizrahi	*Meeting Secretary*	Israel Institute, Nes Ziona, Israel
R. Crainic		Institut Pasteur, Paris, France
H. Katinger		Univ. of Agriculture, Wien, Austria
R. Mauler		Behringwerke AG, Marburg, F.R. Germany
G. Panina		Inst. Zooprofilattico, Brescia, Italy
R. Spier		University of Surrey, Guildford, UK

8th ESACT Meeting, Tiberias, Israel

Session officials

Chairmen	*Secretaries*
Dr. A. Kadouri	Dr. H. Bahnamann
Dr. B. Meignier	Dr. O. Merten
Dr. O. Schonherr	Dr. W. Scheirer
Dr. W. Tolbert	Dr. S. Reuveny
Dr. E. Lidner	Dr. R. Arathoon
Dr. G. Panina	Dr. M. Butler
Dr. J. B. Griffiths	

Preface

The 59 papers assembled in this book were given as either verbal presentations or posters at the 1987 joint ESACT–OHOLO conference. These original works have been grouped into sections which enable the reader to move from the construction of the animal cell substrate through an understanding of the physiology of those cells in a bioreactor type of environment to the ways in which different products can be made from animal cells in culture and tested. Where a paper was followed by a discussion, or where a group of papers was concluded with a more wide-ranging general debate, the transcribed (and edited) recordings of those proceedings have been added to the text at the appropriate place. It is hoped that by this expedient an additional dimension is given to the work, as many of the verbal discussions not only clarified the material offered in the formal presentations but also brought out points that highlighted the new directions which need to be researched to maintain the already rapid rate of progress in this area.

A further aim of the editors was to have this volume available to the participants and to other interested readers as soon as possible after the termination of the conference. To this end we have accepted a camera-ready version of the manuscripts and we have not required the extensive revision of manuscripts to maintain either a particular style of English or a consistency of terminology. We trust that readers will bear with us on this issue as they will, it is hoped, appreciate the rapidity of the appearance of the publication which will bring readers to the 'work-face' of the subject as it is happening. We also hope that readers will acquire some sense of the international nature of the participants at the conference, many of whom have their own inimitable and idiosyncratic ways of 'putting pen to paper'. We, as editors, firmly believe that the availability of the data presented is of greater importance than the niceties with which it is packaged. We do, however, also recognize that we are required to make every effort to help our colleagues provide us with manuscripts which are as close to perfect as possible. You, the reader, will judge the extent to which we have succeeded in our aims.

As ever at the ESACT sesquiennial meetings there was much to turn the head and the mind. The maintenance of animal cells in small organ cultures and the growth of differentiated cell clumps in a suspension type of growth system presents the animal-cell technologist with challenges yet to be tackled. Furthermore, what are we going to learn about animal cells by the examination of their metabolism in a non-destructive observational system as afforded by the new NMR equipment? Additionally, we seem to be finally coming to grips with the concept of the 'fragility' or otherwise of animal cells grown outside the body. If anything, when the conditions are apposite, they appear to be surprisingly robust!

There is little doubt that animal-cell technologists will have to recognize that the way in which they express the productivity or performance of a system needs to be done in such a way that it becomes clearly understood exactly what is being presented and for what reason. For example:

- yield of product per unit animal cell (or per unit weight or measure of biomass) tells readers about the physiology of the cells and the degree to which those cells are performing in relation to some 'animal-cell biomass ultimate capability'; it does not necessarily relate to economic or system performance,

- yield of product per unit of medium used expresses one of a number of important factors which may be relevant to the overall economics of the process but which may also relate to the physiology of the cell, or when a recycle or make up system is used, it may reflect a more materially efficient process; clearly the two factors of overall raw material cost to the process is contained within this parameter as is some indication of the cost of the downstream process concentration and purification procedures,

- yield of product per unit volume of bioreactor per unit of time is a measure of the efficiency of one of the key elements in the production system; however, this factor can be misleading because in some bioreactors where the cells are held at high cell densities per unit volume (in excess of 10^8 ml) either a large or small volume of medium may be made to flow through the cell containing part of the system; under such circumstances it is more relevant to the determination of the space requirement of the bioreactor system to consider the volume of all the components which have been used to hold and feed the mass of animal cells in the system,

- cost of a unit weight of final product at a defined state of purity is perhaps a valuable determinant of the efficiency of a process, yet is is a parameter which is seldom determined and often difficult to acquire; it is also critically dependent on the performance of the cell and on medium utilization efficiencies, however it does represent an economic absolute which is of great relevance to a manufacturer when deciding on a system for the generation of a particular bioproduct using animal cells as the primary production agent.

In sum, the units chosen to express the productivity or efficiency of a system have to be carefully delineated, and users have to recognize what each of the parameters do, and do not, indicate.

On the product side, whole new areas have to be colonized. Animal cells in culture have, at some time, to become the preferred way of testing the toxicity of materials which are to come into contact with man and his animals. The use of diverse immunomodulators in concert seems to be a way of unlocking the potent curative properties of these materials without incurring the disadvantages of the debilitating side effects. And the exploitation of the properties of anti-idiotyic antibodies has to yield prophylactics and therapeutics of considerable value in the future. These prospects should also include the developing area of the use of autologous, *in vitro* modified, cells to augment the ability of the body to repel an invader to better control its own rogue cells.

It is not unusual in such a review to set out the challenge to further technical developments. We clearly have microcarrier cultures in the 1000 litre or larger scales now, but as yet we have not mastered the latest opportunity to be presented to us in the form of the dense cell cultures. Cells held at local concentrations of greater than 5×10^7 ml can be achieved in a number of manifestations. But can such systems be scaled up and down readily? Can they be operated widely throughout the industry without let or hindrance? Have we an adequate set of principles that will enable us to determine which of such systems is the one which will do for the animal cell technology industry what the stirred tank reactor has done for fungal and bacterial fermentations? Finally, having recognized that the optimization of the bioreactor on its own does not constitute the final stage in the establishment of viable bioprocess how are we going to go about the optimization of the bioprocess as unit about which we subtend our decision making?

R. E. Spier
J. B. Griffiths

Contents

1. The liberation of animal cells: psychology of changing attitudes 1

J. C. Petricciani

2. Cellular DNA biologicals: significance and quantitation 20

F. Horaud

Section 1 New cells and cell lines

3. Human monoclonal antibodies – their potential, problems and prospects 25

K. James

4. Production of cancer specific monoclonal antibodies with human–human hybridomas and their serum-free, high density, perfusion culture 52

H. Murakami and K. Yamada

5. Establishment of a human hybridoma secreting monoclonal antibodies toward rabies virus 77

Y. Gozes, R. Oren, R. Monzain, S. Lerer, O. Galan, H. Ben-Moshe, N. Epstein and D. Kobiler

6. The production of bovine and ovine monoclonal antibodies to gonadal steroids: the use of interspecific hybridomas 89

D. J. Groves, J. Clayton and B. A. Morris

7. The immortalization of differentiated mammalian cells 107

C. MacDonald and D. M. Scott

8. Structural basis of growth factor receptor mediated cellular signals (abstract) 115

A. Ullrich, H. Riedel, E. Livneh and J. Schlessinger

9. Growth control of normal human B lymphocytes 117

E. Lundgren, R. Sanders and G. Roos

10. Anti-proliferative effects and phenotypic alterations induced by methylurea derivatives in B16 mouse melanoma 125

J. Nordenberg, A. Fuchs, L. Wasserman, Z. Malik and A. Novogrodsky

11. Comparative analysis of major biological aspects of human hepatoma in culture 138

P. Bagnarelli, A. Festa, I. Testa, B. Champ, M. Clementi and G. Caroloni

12. Minireview: Essential strategies for the application of herpes simplex virus as a vector for the expression of foreign genes 160

M. Ackermann and U. Kihm

13. The glutamine synthetase gene: a novel marker for gene amplification in mammalian cells (abstract) 179

C. R. Bebbington and C. C. G. Hentschel

14. Recombinant animal cell lines for production of glycoproteins 182

M. Wirth, W. Reiser, G. Zettlmeissl, H. S. Conradt, J. H. Cao, C. Morelle and H. Hauser

15. Foreign gene expression in rat pituitary GH3 cells 199

M. Silberklang, J. Kopchick, S. Munshi, A. Lenny, T. Livelli and R. W. Ellis

16. Use of transfected and amplified *Drosophila* heat shock promoter construction for inducible production of toxic mouse c-myc proteins in CHO cells 215

F. M. Wurm, K. A. Gwinn, O. Papoulas, M. Pallavincini and R. E. Kingston

Section 2 Medium development

17. The use of glutamine-free media for the growth of three cell lines in microcarrier culture 245

T. E. Hassell, I. C. Allen, A. J. Rowley and M. Butler

18. The attachment of MDCK cells to three types of microcarriers in different serum-free media 264

T. E. Sayer, M. Butler and A. J. MacLeod

19. Spreading, growth and function of cells attached to a plasma clot interphase between agarose and the atmosphere 280

E. M. Bey

Section 3 Physiological investigations

20. Investigations of the influence of physical environment on the cultivation of animal cells 297

M. Wudtke and K. Schugerl

21. Shear sensitivity of three hybridoma cell lines in suspension culture 316
C. G. Smith, P. F. Greenfield and D. H. Randerson

22. NMR metabolic studies of human breast cancer cells 328
M. Neeman, E. Rushkin, A. Kadouri and H. Degani

23. Optimization of glass-sphere immobilized bed cultures 342
D. Looby and J. B. Griffiths

24. Examination of a simple model for the diffusion of oxygen into dense masses of animal cells 353
A. D. Murdin, R. Wilson, N. F. Kirkby and R. E. Spier

25. Kinetics of prourokinase production by mammalian cells in culture 365
W. S. Hu, T. Y. Tao, M. A. Bohn and G. Y. Ji

26. Specific kinetic patterns of production of monoclonal antibodies in batch cultures and consequences for fermentation processes 381
O-W. Merten, G. E. Palfi, G. Klement and F. Steindl

27. A model system to study gene expression and scale up of microcarrier cultures of recombinant DNA manipulated cells 397
B. Lundgren and S. Nilsson

Section 4 Bioreactor developments

28. Membrane-stirrer-reactor for bubble free aeration and perfusion 411
H. Büntemeyer, B. G. D. Bödeker and J. Lehmann

29. Immobilization of hybridomas in packed-bed bioreactors 420
A. D. Murdin, J. S. Thorpe and R. E. Spier

30. Production of biologicals by animal cells immobilized in polyurethane foam matrix 437
A. Lazar, S. Reuveny, A. Muzrahi, M. Avtalion, J. P. Whiteside and R. E. Spier

31. Comparison of oxygenation methods in a 39L stirred bioreactor 449
D. Looby and J. B. Griffiths

32. Development of large-scale mammalian cell perfusion culture systems 454
J. Feder

33. Microcarrier DORMACELL does not detectably alter viral or plasmid DNA upon direct exposure 484

I. Keller, U. Brüggemann, H. Mansi-Wothke, D. Schwengers and W. Doerfler

34. Macroporous microcarriers 492

K. Nilsson, S. Birnbaum, F. Buzaky and K. Mosbach

35. Growth of animal cells on microbeads III: Semi continuous *in situ* estimation of numbers 504

D. Thebline, J. Harfield, O. Hanotte, D. Dubois and A. O. A. Miller

36. Evaluation of the continuous perfusion culture system for the production of monoclonal antibodies 513

T. van der Velden-de Groot, W. Witterland, E. C. Beuvery and T. L. van Wezel

37. Recent advances in the control of monitoring of bioreactors 519

D. F. Parsons

38. Automated process control for culturing animal cells 524

J. Rosenquist and C. Wallberg

Section 5 Downstream unit processes

39. Design criteria for the purification of biopolymers for the pharmaceutical industry 535

S. Johansson, H. Johansson and D. Low

40. Use of a dynamic filtration method for separation of animal cells 548

E. Rebsamen, W. Goldinger, W. Scheirer, O-W. Merton and G. E. Pàlfi

41. Continuous culture of murine hybridomas with integrated recovery of monoclonal antibodies 556

J. Rudge, M. A. Desai, S. A. Shojaosadaty and A. Lyddiatt

Sectionh 6 Vaccine products

42. Improved production of Marek's disease vaccine with the multi-tube technology 579

R. Dugré, M. Corbeil and M. Boulay

43. Procedure for the adaption of the new mumps vaccine strain 'rubini' to human diploid cells 587

R. Glück

44. Immuno detection of flaviviruses by monoclonal antibodies directed against West Nile virus specific and flavivirus group epitopes 596

N. Epstein, B. Lachmi, S. Lehrer, D. Katz, S. Cohen and U. Olshevsky

45. A simple ELISA for the estimation of antibodies to equine herpes virus (EHV–1) and its application to the quality control of vaccines 613

M. Ackermann, S. Grüninger, L. Bruckner, H. K. Müller, R. Zutter and U. Kihm

46. The production and application of an oil adjuvant vaccine against foot-and-mouth disease in cattle 628

H. Bahnemann, J. Mesquita, V. Astudillo and F. Dora

Section 7 Immune system products

47. Application of large scale perfusion systems for pharmaceutical production from mammalian cells 643

W. R. Tolbert, W. R. Srigley and C. P. Prior

48. The organotype culture model of human cancer cells: the effect of human recombinant interferons 662

R. Beaupain, M-C. Martyré and E. Falcoff

49. Immune interferon enhances the expression of MHC antigens in human placenta 682

B. Shomer, E. Gazit and V. Toder

50. Gamma interferon and anti-thy-1 antibody induce specific killing activity in memory-like cytolytic hybridomas 687

T. Ozery and Y. Kaufmann

51. Development of a device and a double antibody microradioimmune assay for fast screening and sensitive quantitation of monoclonal antibodies 699

N. Epstein, G. Shavit and Z. Oril

52. Development of monoclonal antibodies directed against carcinoembryonic antigen (CEA) and their application for CEA detection 708

N. Epstein, R. Zandani, I. Inbar, B. Halperin, S. Cohen, S. Reuveny and N. Ariel

53. Carcinoembryonic antigen (CEA) production in low protein growth medium 724

S. Reuveny, A. Lazar, A. Mizrahi, D. Marcus, A. Shahar and N. Epstein

54. Production and purification of a bladder cancer associated membrane protein 738

B. Lundgren, I. Wergeland, S. Braesch-Anderssen, S. Paulie, H. Koho and P. Perlmann

Section 8 Toxicity testing with animal cells

55. The use of animal cells for evaluation of toxicity, carcinogenesis and mutagenesis 747

A. M. Goldberg

56. Development of a model for the study of toxicity and carcinogenicity *in vitro* 000 756

C. B. Seymour, C. Mothersill, A. Cusack and M. McDonnell

Section 9 A glimpse of the future

57. The transformation of science fiction into technical reality by using animal cells in culture 775

R. E. Spier

Postscript

58. Terms and symbols for use in animal cell culture 789

P. P. Gray

Late paper

59. Hybridoma culture for the continuous production of monoclonal antibodies; a fractionalized factorial design experiment for process optimizationI 795

Wergeland, C. Wallberg and R. Ericsson

List of participants *803*

List of exhibitors *825*

1 The liberation of animal cells: psychology of changing attitudes

J. C. Petricciani

INTRODUCTION

The history of cell substrates has been described as a drama (1). In fact, it has been more like a formal dance in which the various partners have fallen prey to seductions of one sort or another during the course of the evening such that at the conclusion of the affair virtually everyone is leaving with someone other than the person with whom they arrived. Seductions come in all sorts of shapes and sizes. Maintaining the *status quo* is one of the most powerful because it offers apparent security, and most of the time it's just plainly comfortable. On the other hand, challenging the order of things as well as the lure of personal recognition for success is strongly seductive, especially if one is convinced he is right. And finally economic considerations of many kinds are clearly powerful forces which play an important role in real life. All of these factors - and more - have to one degree or another led the various participants in the cell substrate dance to take one position or another, and to form alliances with certain other dancers from time to time. The purpose of this paper, however, is not to discuss these aspects of the story, but to review the scientific issues and to summarize the current situation.

BACKGROUND

When the United States Armed Forces Board met in 1954 (1) to consider what cells would be an acceptable substrate to use for the production of an experimental live adenovirus vaccine, they basically had two choices: i) HeLa, a human cancer cell line; or ii) primary monkey kidney cells. Based on very straightforward pragmatic reasoning that there would be less risk in using a cell culture derived from the tissues of normal animals than if cells derived from a human tumor were used, they decided to approve primary monkey kidney cell cultures. The significance of that decision cannot be overstated, because it essentially ruled out as unsafe the use of cells other than those derived from normal tissues of animals in good health, and it set the tone for attitudes towards cell cultures in vaccine production for the following three decades.

It was not until the suggestion by Hayflick and Morehead in 1961 that human diploid cells (HDCs) could serve as a useful and well-standardized cell system for the production of biological products, that any serious reconsideration of the traditional concept of using only primary cells was undertaken. As reviewed by others (1-3), intense debate, disagreement, and dogmatism surrounded the question of safety of products derived from HDCs. Eventually, however, vaccines derived from such cells were used in clinical trials, and ultimately they were licensed by national control authorities in many different countries.

It is interesting to note that during the HDC debates, Dr Robert Hull of the Eli Lilly Company suggested that continuous cell lines (CCLs) could be used for the production of vaccines, and he pointed out that such cells had a significant number of advantages over both primary tissues and HDCs. Although data were presented (4) in support of the safety of vaccines derived from cell lines such as LLC-MK2, no serious attention was paid to that proposal since the focus of controversy and interest was at that time on the HDC issues. The question of heteroploid cell lines (e.g. CCLs) for vaccine production was essentially lost in the turmoil surrounding HDCs. And the very few proponents of CCLs understood very clearly that if HDCs were having such a difficult time gaining acceptance, there was little hope for CCLs.

It was not until the mid-1970s that the next serious proposal to use other than primary or diploid cells was made. At that time there was a great deal of interest in the use of interferon (IFN) in various clinical studies. The major source of IFN at that time was from primary human leukocyte cultures, although some IFN was also available from human fibroblast cultures. In both cases however, the amount of IFN which could be produced was very limited. An alternative source was developed by the Wellcome Research Laboratories using the Namalwa human lymphoblastoid cell line. Rather large amounts of IFN could be produced, and such cells were proposed as a source of IFN for human clinical studies (5). As in the case of the 1954 safety considerations of adenovirus vaccine which could be produced in human cancer cells, IFN from malignant human lymphoid tissue posed serious questions of safety. However, there were two major differences between lymphoblastoid IFN and live adenovirus vaccine: i) interferon was not a replicating agent; and ii) the purification and validation procedures which were available in the mid-1970s were significantly better than those of the mid-1950s.

In an effort to initiate a dialogue within the scientific community on the acceptability of lymphoblastoid cells as a source for IFN, a conference was held at Lake Placid in 1978 to discuss the issues and to open a reconsideration of the entire area of cell substrates (2). The conclusion of that meeting was a very cautious "yes" to the continuing exploration of lymphoblastoid IFN as an experimental biological product.

A decade later it is easy to appreciate how important that research and development effort was because it laid the groundwork for approaches which might be taken to ensure the safety of products derived from a variety of other CCLs including those now being used in biotechnology. Following the Lake Placid meeting, interest began to develop in the use of other CCLs for manufacturing products. Chief among those CCLs was the possibility of using Vero, a continuous monkey kidney cell line, for the production of inactivated polio vaccine.

During the period 1960 to 1980 developments in the veterinary vaccine field were taking place rather independently of those in human biologicals, although the safety of products was also of major concern. Specifically, the use of BHK-21 cells for producing foot and mouth disease vaccine and rabies vaccine was proposed in the 1960s because of the significant advantages that cell system could provide (6). After a period of controversy, and the generation of data in support of the safety of vaccines derived from BHK-21 cells, there was rapid movement towards the use of that cell system such that at the present time over $2x10^8$ doses of vaccine are produced each year in BHK-21.

BIOTECHNOLOGY

The possibility that a wide range of biological products might be produced in bacterial cells through recombinant DNA technology faded when it became evident that bacterial cells were not able to synthesize adequate amounts of certain products. An example is hepatitis B surface antigen which bacterial cells were unable to accumulate in sufficient amounts to make them reasonable as an expression system. In addition, the inability of bacterial cells to glycosolate was thought to be a potential drawback at least for some products in which glycosolation sites might be important for their proper biological functioning.

The use of recombinant yeast and mammalian cells as expression systems for some products was the next step in product development. Among major determinants in the choice of the cell expression system were the following: rapid growth, high level of expression, and the ability to grow to high cell density in suspension culture. Those requirements meant that in the case of mammalian cells the systems of choice would be CCLs. The development of hybridoma technology and the potential usefulness of monoclonal antibodies also led to heightened interest in the use of CCLs.

Once again, however, there was a great deal of concern relating to the safety of products derived from these various abnormal mammalian cells. A number of meetings and discussions tried to address the safety issues, and a consensus began to grow that under certain conditions CCLs would be acceptable. In an effort to develop a more formal consensus on the use of CCLs, the United States Public Health Service sponsored a workshop in July 1984 at the National Institutes of Health in Bethesda (7). The focus of attention at that meeting was on the risks that might be associated with the following cellular contaminants in a biological product: i) transforming proteins; ii) viruses; and iii) DNA. The consensus of that meeting was that CCLs should be considered acceptable as substrates for the production of biologicals if the manufacturing process could be shown to efffectively eliminate causes of concern due to contaminants derived from cells. Table 1 shows some examples of CCLs and their products which are either already approved or at the clinical trial stage. In particular it should be noted that over 10^7 doses of inactivated polio vaccine have already been produced in Vero cells and administered to children in France since 1983, thus establishing a considerable amount of human experience with that CCL as a substrate.

Although, as described above, there has been general movement towards

TABLE 1. EXAMPLES OF CCLs OF CURRENT INTEREST IN BIOLOGICS PRODUCTION

CCL	Origin	Characteristics	Approved (a) or Experimental (e) Uses
Vero	Kidney of a normal adult African green monkey, 1962	Regressing tumors in nude mice at passages < 265; invasive tumors in ATG-treated rats at passages > 146 No evidence of endogenous viruses	Cell substrate for: Inactivated polio vaccine (a) Live polio vaccine (e) Rabies vaccine (human) (a)
BHK-21 (C-13)	Clone of a cell line established from the kidneys of 5, one-day-old Syrian hamsters, 1963	Grows in suspension cultures Tumorigenic in hamsters Virus-like particles present	Cell substrate for: Foot-and-mouth disease vaccine (a) Rabies vaccine (veterinary) (a) Rabies vaccine (human) (e)
CHO-K1	Subclone of a cell line established from an ovary of an adult Chinese hamster, 1957	Grows in suspension cultures Tumorigenic in nude mice Virus-like particles present	Recombinant DNA expression system for: HBsAg (a) Tissue plasminogen activator (a) Factor VIII (a)
PLC/PRF/5 (Alexander cells)	Human hepatoma	Secretes HBsAg Tumorigenic in nude mice	Cell substrate for expression of: HBsAg (e)
Namalwa	Burkitt lymphoma tissue	Human lymphoblastoid cells Tumorigenic in mice Grows in suspension cultures Cell DNA contains integrated EB virus genome	Cell substrate for expression of: Alpha interferon (a)
FRhK-4	Fetal Rhesus kidney	Nontumorigenic at passages < 40 No evidence of endogenous viruses	Cell substrate for: Hepatitis A vaccine (e)

TABLE 2. CELL SUBSTRATES CURRENTLY USED FOR RABIES VACCINE PRODUCTION

CELL	CHARACTERISTICS				
	Growth in Suspension	Endogenous Viruses	Tumorigenic	Virus Yield	
				PFU/ml	LD_{50}/ml
Primary					
Chick embryo	-	+	-	10^6–10^8	
Hamster	-	+	-		$10^{6.3}$–$10^{7.6}$
Diploid					
Human	-	-	-		$10^{6.0}$–$10^{7.5}$
Monkey	-	-	-	$10^{6.5}$–$10^{7.2}$	
Continuous					
Vero	±	-	±		$10^{6.0}$–$10^{7.0}$
BHK-21	+	±	+		$10^{7.3}$–$10^{8.0}$

the acceptance of CCLs for use in the production of various biologicals, a group of consultants recommended that it would be useful for the World Health Organization (WHO) to convene a panel of experts to consider general safety issues associated with the use of cells in the production of biologicals with particular emphasis on contaminating DNA. It was suggested that such a panel could provide international guidance on the acceptability of various cell substrates and production strategies. For that reason, the Director-General of WHO convened the Study Group on Biologicals on 18-19 November 1986 to provide advice on these issues.

STUDY GROUP

The basic conclusions of the Study Group were remarkably straightforward and significant in their implications (8,9). First, and perhaps most important, was their agreement that there is no reason to exclude CCLs from consideration as substrates for the production of biologicals, and that CCLs are in general acceptable when the manufacturing process is shown to yield a product with no detectable risk attributable to the cell substrate. Almost of equal importance was the Study Group's conclusion that when reliable data are provided in support of the safety of a product derived from a CCL, that product should be considered acceptable. The presence of the same product, also approved, but manufactured in another cell system was not considered to be relevant to the acceptance or rejection of a CCL-derived product. Every product should stand on its own merits regarding safety and efficacy. To paraphrase, a product is either safe or it isn't in the same way that pregnancy is a yes or no decision. The Study Group then went on to identify the same three factors from cells, mentioned above, which might carry risk if they were found as contaminants in the product: i) transforming proteins, ii) DNA, and iii) viruses.

Transforming proteins. The apparent risk from oncogene-encoded proteins is limited to growth factors, since they are the only such proteins that can exert their biological effects via the cell surface. Growth factors may be secreted by cells in which biologicals might be produced, but the risks from growth factors are limited, since their growth promoting effects are usually transient and reversible. Because growth factors do not replicate and many of them are rapidly inactivated when administered *in vivo*, their effect will be limited.

Therefore, the Study Group did not consider that the presence of contaminating known growth factors in the concentration at which they are ordinarily to be found constitutes a serious risk in the preparation of biological products from CCLs, and no specific tests for growth factors were recommended.

Heterogeneous Contaminating DNA. The study group concluded that the assessment of the safety of any product with respect to DNA should take into consideration: 1) the elimination of the biological activity of DNA by various steps in the manufacturing process; and 2) the reduction in the amount of DNA during the purification of the product in the manufacturing process. A given product may be considered safe on the basis of reliable data from either of those two elements, or their combination. They also concluded that, based on the experimental data available to date, the probability of risk associated with heterogeneous contaminating DNA in a product derived from a CCL is so small as to be negligible when the amount of such DNA is 100 pg or less in a single dose given parenterally.

The use of special DNA sequences such as viral regulatory sequences in the construction of recombinant cells was considered acceptable by the Study Group because there is no evidence to suggest that such sequences would impose any additional risk beyond that of heterogeneous contaminating cellular DNA in general. Nevertheless, the manufacturing process should be validated to demonstrate that such sequences are not concentrated in the crude harvest or at other steps of the process. In addition, it is expected that recombinant DNA products will be highly purified, and that the amount of DNA/dose will be less than 100 pg and in some cases it may even be undetectable. When DNA that may have important biological activity is involved, its detection should be attempted by appropriate techniques.

Viruses. In considering the potential for infecting man with viruses, cells may be divided into three risk categories with respect to their potential for carrying viral agents pathogenic for man:

High Risk:	Any material derived from either human or nonhuman primate blood or bone marrow cells, caprine or ovine cells, or hybridomas when at least one fusion partner is of human or nonhuman primate origin.
Less Risk:	Any material derived from mammalian non-hematogenous cells such as fibroblasts or epithelial cells.
Low Risk:	Any material derived from either human diploid lines or avian tissue.

Taking into account the above classification of the potential risk for transmitting viruses pathogenic for man, the Study Group agreed that different degrees of concern, and therefore testing, were appropriate for products manufactured from the various types of cells mentioned.

The Study Group stressed the importance of validating the ability of a manufacturing process to eliminate and to inactivate those viruses which may pose a risk to humans when cells or cell lines carrying viruses are proposed for use in the manufacture of human biologicals; a wide margin of safety should be established in any inactivation or purification process. In addition, cells from humans or animals with diseases of unknown origin should not be used, nor should animal cells which may contain "slow viruses" be used to produce biologicals for human administration.

DISCUSSION

The astronomer Johannes Kepler had great difficulty in reconciling his observations that the orbit of Mars was elliptical rather than circular. His difficulty was that for nearly 2,000 years the idea that celestial orbits were circular had become a basic religious and scientific doctrine. And dogmas never die easily or without a fight to save them. It took years for Kepler to finally conclude, first to himself and then publically, that no amount of mathematical juggling could save the dogma of circular orbits, and that the paths of the planets are actually ellipses. Suddenly the entire field of astronomy was released from what has been termed the tyranny of symmetry, and the pathway towards enormous progress was laid open for all those who were willing to travel it.

The story of cell substrates has many analogies to astronomy's release from the constraints of circular orbits, although the timeframe from established doctrine to the initial challenge and finally to a liberated view, has been compressed to about 30 years in the case of cell substrates.

Taken together, the Study Group's conclusions, like those of Kepler, literally liberated animal cells from the dogmatic constraints which had limited the acceptable options to a very narrow range. If the Study Group's conclusions are adopted by manufacturers and National Control

Authorities, then larger amounts of safe biologicals may be produced at lower cost than has been possible in the past because the selection of a cell substrate can now be based on the simple pragmatic determinant of efficiency of production. The manufacturing process then must be shown to reduce DNA to acceptable levels (e.g. 100 pg/dose for parenterally administered products) and/or to eliminate the biological activity of DNA. Similarly, the process must be able to eliminate viruses pathogenic for humans.

One specific immediate result of this new approach is the option of using a CCL to produce rabies vaccine for human administration. In order to replace neural tissue derived rabies vaccine, developing countries need a cell system which: i) gives high yields of immunogenic rabies virus, and ii) can be grown easily in suspension culture. Developed countries could also use the same system to produce more vaccine at much lower cost than current systems allow. The characteristics of the cell culture systems now in use for rabies vaccine production in various countries are shown in Table 2. From a practical point of view it is clear that BHK-21 is the cell system of choice, and plans are now being developed to test such a vaccine in humans.

While the Study Group's conclusions may at first appear to be revolutionary, they are in fact consistent with well-established philosophy and policy of the major regulatory authorities throughout the world. The principle of relying on data to show that a manufacturing process can address specific elements of concern has been basic to the regulation of biologicals for years. There is perhaps no clearer example of this than hepatitis B vaccine derived from human plasma. It took a great deal of courage and confidence by the initial manufacturer and regulatory authority to decide to proceed with the first clinical trial and with eventual licensing of the vaccine because the source material - human plasma derived from persons chronically infected with the hepatitis B virus - could hardly have been of higher risk. But reliance was placed on data to show that the manufacturing process eliminated any infective virus. Now that hepatitis B vaccines derived from recombinant DNA techniques in yeast have been approved in several countries, neither WHO nor National Control Authorities have reconsidered the acceptability of plasma derived vaccines because those vaccines have proved to be safe and effective for humans as currently produced. Those decisions have been based on scientific data rather than emotion. If emotion and anxiety were to be the deciding factors, it is likely that all plasma-derived vaccines would have been removed from the market long ago if for no other reason than because of the fear of AIDS. But a reasoned approach to assessing the safety of these vaccines regarding AIDS has prevailed. This same philosophy is what the Study Group recommended in evaluating the safety of all types of products manufactured in CCLs.

SUMMARY AND CONCLUSIONS

Looking back at the history of issues and decisions that were made regarding the acceptability of various cell substrates leads one to wonder how much more rapidly products might have been developed if different decisions had been made. It is also strikingly evident that primary tissues and primary cells have been the source of whatever problems have existed in the past. This is not to say that everything is known about CCLs or that there are no surprises in store as more experience is gained with their use in biological production. It is difficult to argue, however, that one is not better off with a very well-tested cell system and a manufacturing process which can cope with even theoretically worrisome contaminants.

The prevention and treatment of human diseases will be greatly facilitated by products derived from CCLs in the developing world as well as in technologically advanced countries. As in the case of any evolutionary change in attitudes, the acceptance of CCLs has been slow and difficult; nevertheless, it is now clear that during this past decade there has been a great deal of progress. One might even say that we are approaching the conclusion of the era of controversy and entering into a new stage where there are no longer significant reservations about the use of well-characterized cells of various types to produce biologicals when the manufacturing process can be shown to eliminate and/or inactivate specific contaminants of concern. Indeed, now after 33 years of debate and development, we are back where we started in 1954, but with a different conclusion in very much the way T.S. Elliot described things:

"And the end of all our exploring
Will be to arrive where we started
And know the place for the first time."

As pointed out by the Study Group, even though we've come a long way in our 33 years of exploring, a number of questions remain unanswered about viruses, DNA, and protein as potential contaminants in biological products, and one must be alert to new information and new technology so as to reassess the subject from time to time. Nevertheless, it was also clear to the Study Group that enough is known now to conclude that under certain conditions the use of CCLs is reasonable. Or put another way, there comes a point, as Albert Camus wrote in "The Plague", when someone dares to state that 2 plus 2 equals 4 even though it may be personally risky to say so. But the question is not one of knowing what punishment or reward goes along with that conclusion; rather, it is knowing whether 2 and 2 do really add up to 4.

The next steps in the process which began in 1954 will be for National Control Authorities, industry, and the general scientific community to consider the report of the Study Group as new products are proposed for clinical trials and for approval in various countries, and to see whether they too will conclude that 2 plus 2 equals 4.

REFERENCES

1. Hilleman MR (1979). Cell line saga: an argument in favor of production of biologicals in cancer cells. Adv Exp Biol Med 118: 47-58.

2. Cell Substrates (1979) (JC Petricciani, HE Hopps, and PJ Chapple eds.) Plenum, New York.

3. Hayflick L, Plotkin S, Stevenson RE (1987). History of the acceptance of human diploid cell strains as substrates for human virus vaccine production. J Biol Stand (in press).

4. Hull RN (1968). Immunization of experimental animals with vaccines produced in known oncogenic cell lines. Nat Cancer Inst Monog 29: 503-509.

5. Beale AJ (1979). Choice of cell substrate for biological products. Adv Exp Biol Med 118: 83-97.

6. Beale AJ (1985). The use of abnormal cells for biological products. In A Joyeaux et al, eds. Therapeutic Agents produced by genetic engineering: quo vadis? Sanofi Group, pp 167-178.

7. Abnormal Cells, New Products, and Risk (1985) (HE Hopps and JC Petricciani eds.), Tissue Culture Association, Gaithersburg.

8. Report of the meeting of the Study Group on Biologicals (1987). Wkly Epidem Rec 62: 13-15.

9. Acceptability of cell substrates for production of biologicals (1987). Report of a WHO Study Group. Technical Report Series 747.

Discussion

TOLBERT The amount of DNA detected is 10 pg which is 10 times that mentioned in the meeting og 1984. Is that because of the problems in detecting the lower amounts of DNA.

PETRICCIANI No - it is not based on the problems of detection, the additional data has indicated that the first limit was to conservative and that the higher level would be acceptable.

TOLBERT You mentioned several specific things to look for - oncogenes and viral promotors if one were able to detect DNA at lower concentrations would it not then be necessary for the additional tests.

PETRICCIANI No, one you have to validate the manufacturing process which is different from routine assays. If you were using SV40 in the construct you would look for SV40 sequences to see what the system does with them, which is different from saying that you have to examine each production run for SV40 You need to look at a few runs to establish that such genes do not cause strange effects.

GLUCK You proposed a BHK21 rabies vaccine, Dr Ogo has found it difficult to get less than 100 pg of DNA in such a rabies vaccine - how do you view such a situation.

PETRICCIANI Each product has to be examined individually. In the case of rabies one has to look at how the product is made. Most, but not all of the rabies vaccine incorporate in the production process inactivation with beta-propinolactose which inactivates DNA. If one can show that as used in the manufacture of rabies vaccine the BPL inactivate the DNA I would not worry that there was 100, 1000 or 10,000 ug of DNA in the product because it is biologically unimportant.

ARATHOON Where polio was contaminated with SV40 it would be possible to look at the DNA of the people who had received that vaccine to see if there are any integrated SV40 sequences. Do you think that would be a worthwhile study?

ARATHOON Worldwide there have been millions of doses of rabies and FMD put into animals, into single animals many times and into many generations of animals. This also offers a test system to see happens to contaminating DNA sequences to see if they become integrated.

PETRICCIANI There have been some clinical follow-ups on animals inoculated with BHK derived materials.

It is important in such consideration to determine whether or not it is useful to look at how much DNA is in the FMD type products - 100 - 500 pg/dose, the likelihood of being able to get uptake into the host is remote so that it may not be a worthwhile experiment to try.

SPIER There is some 12 pg of DNA/cell and using 2 ml of 10^6 cells/ml/ dose it gives 24 ug per dose for each of 3 types and then 2 times per year some of the DNA could finish up in say a meal of uncooked steak.

MEIGNIER For some products dot-blot hybridisation is already a routine procedure (polio and Vero cells) Standard values obtained are in the range of 5 pg/dose of human vaccine.

BAHNEMANN All BMK cell produced FMD vaccines are inactivated with first order inactivants. Therefore there is little biological activity in the FMD vaccine.

PETRIE Did the monkey with the tumour on his arm immunosuppressed or healthy.

PETRICCIANI The monkey was specifically immunosuppressed with anti-thymocyte globulin with an untreated monkey you have zero effects.

There are lots of other experiments underway with large amounts of inoculated DNA. Even an injection of large amounts of cellular DNA does not do anything. Using certain viral DNA's in animals which does cause various types of disease and neoplasms. Here we are not talking about high levels of viral DNA we are talking about low levels of cellular DNA.

HORAUD It is important to emphasise each situation. You cannot be dogmatic. Although the amount of DNA in inactivated polio vaccines is low, in rabies vaccine the amount of DNA is higher. This could be due to the fact that polio is purified before inactivation whereas the reverse is the case for rabies.

2 Cell DNA biologicals: significance and quantitation

F. Horaud

We studied in detail the feasability, the reproducibility and the sensitivity of the dot-blot filter hybridization test to detect minute amounts of residual cellular DNA. A comparison was made between ^{32}P radiolabelled probe and a biotine streptavidine probe. In parallel, we developped a dot-blot assay for the detection of DNA by taking advantage of the existence of mouse anti-DNA monoclonal antibodies.

Despite of the fact that both approaches are able to detect DNA on the level of picograms, therefore under the acceptable figure, i.e. 100 pg (WHO - WER, 1987, 4, 13), the test using anti-DNA antibodies presents some advantages being more feasable and standardizable.

Discussion

BEBBINGTON	Did you describe what DNA probes you used.
HORAUD	For mouse DNA we used total DNA extracted from liver
BEBBINGTON	How is that labelled
HORAUD	Nick translation with P^{32}
BEBBINGTON	What is the limit of sensitivity with that method
HORAUD	Up to 0.5 pg for some people we however can

	only detect between 5 -10 pg per dose
WURM	When you use nick translation for getting total genome DNA you are mainly dealing with amplified repetitive sequences. What is the significance of such sequences on the evaluation of product safety
HORAUD	Difficult to exclude risks as we do not know what DNA is in the product.
WURM	Then you agree that you are mainly looking at sequences which have little to do with oncogenes
SPIER	Maybe you should look at the base sequences of the oncogenes and prepare more appropriate probes
J.BROWN	If you detected repeated sequence DNA then you could conclude that unique sequences were also present. To infer the opposite would be unreasonable.
SCHEIRER	You mentioned a sensitivity of 10 pg/dose with monoclonal antibodies, a human dose may be in the gram range. How do you manage to examine a gram of product on a membrane.
HORAUD	You can concentrate the samples and extract the DNA or in a slot-blot technique you can use a large sample and wash out the contaminants leaving the DNA on the filter.
SCHEIRER	Have you proved this method of concentration.
HORAUD	Sometimes, yes.

Section 1
New cells and cell lines

3 Human monoclonal antibodies – their potential, problems and prospects

K. James

INTRODUCTION

The production and application of rodent monoclonal antibodies in science and medicine has proceeded unabated since Kohler and Milstein (18) first showed that specific antibody secreting cell lines could be obtained by the hybridisation of immune mouse spleens and myeloma cells followed by culture in selection medium. This technology has led to new and improved methods for assaying, isolating and characterising a wide variety of molecules of biological importance. In addition it has produced probes for studying complex cellular interactions and reagents for the investigation and treatment of a variety of diseases.

The potential advantages of being able to produce human monoclonal antibodies by similar or alternative procedures was readily appreciated. These included improved serological reagents for diagnosis, prophylaxis and therapy as well as novel approaches for investigating humoral immunity in health and disease. Although immortalised human cell lines secreting specific antibody were described as long ago as 1977 and their potential emphasised (34), progress in this field has been very slow. Nevertheless in spite of the problems encountered to date the potential of human monoclonal antibody technology still continues to generate intensive research interest.

The aim of this brief review is to a) outline the basic procedures used in human monoclonal antibody production; b) summarise the range of human monoclonal antibodies currently available; c) comment on their potential applications; d) emphasise the limitations of the current technology; and e) indicate possible procedures for improving human monoclonal antibody production. For further information in this area the reader should consult a number of recent reviews and books on this topic (7,14,15,20,35).

GENERAL STRATEGIES USED IN HUMAN MONOCLONAL ANTIBODY PRODUCTION

The basic steps involved in human monoclonal antibody production are outlined in Figure 1. The success of the procedure is dependent upon an adequate supply of immune B cells in the appropriate state of differentiation and proliferation and effective methods for immortalising and cloning such cells. Generally speaking these basic requirements pose no difficulty when producing rodent monoclonals. Unfortunately this is not the case with human monoclonal antibody production where there are limitations with respect to immunisation, lymphocyte availability and suitable fusion partners.

Sources of Lymphocytes

To date the immune lymphocytes used in human monoclonal antibody production have come from a variety of sources (Table 1). These include donors who have been immunised by approved schedules or as a result of active disease. Among the former category are antibodies to bacterial and viral antigens while the latter are mainly directed against tumour or autoantigens. Of additional importance are those antibodies derived using lymphocytes from individuals inadvertently exposed to antigens such as red cells and haptens or with no previous record of immunisation or active disease. The latter have provided antibodies reacting with a variety of intra and extracellular components.

For a number of reasons (see later) lymphocytes derived from the

FIGURE 1.

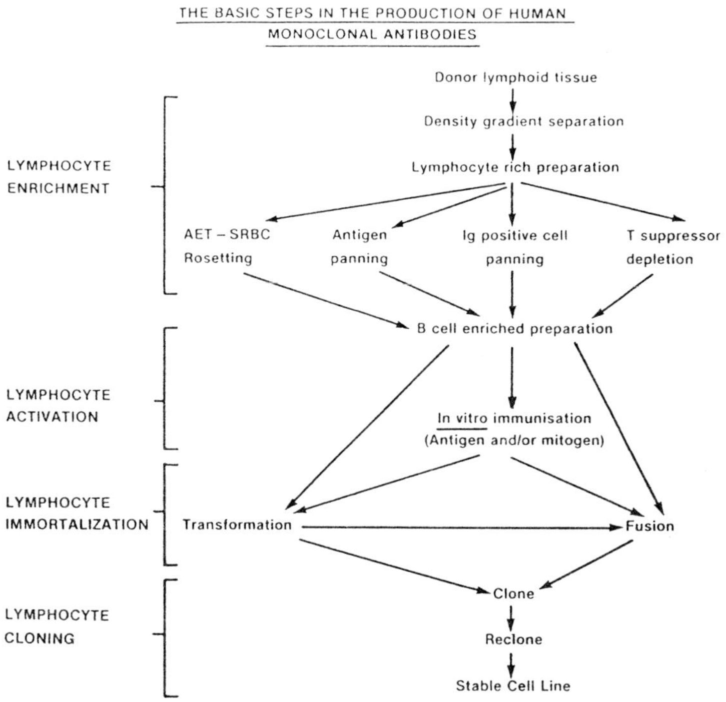

Fig. 1.- The basic steps in the production of human monoclonal antibodies

This flow chart outlines the main strategies which can be used in the production of human monoclonal antibodies. A number of the steps illustrated are desirable though not always essential. They include T cell depletion, B cell enrichment and amplification of antibody forming cell precursors by in vitro incubation with mitogen etc.

TABLE 1

SOURCES OF IMMUNE LYMPHOCYTES FOR HUMAN MONOCLONAL ANTIBODY PRODUCTION

	IMMUNIZATION PROCEDURE	EXAMPLES
1.	Active immunization	Bacterial and viral vaccines
2.	Natural infection	Bacteria and viruses
3.	Active Disease	Cancer and autoimmune disorders
4.	Inadvertent exposure	Red cell antigens and haptens
5.	No obvious exposure	Autoantigens
6.	In vitro sensitization	Bacterial and viral antigens, haptens

For further details see Reference 15.

TABLE 2

PROCESSING OF LYMPHOCYTES FOLLOWING DENSITY GRADIENT SEPARATION

1. T CELL REMOVAL

   Rosetting with AET (2 aminoethylisothiouronium bromide hydrobromide) treated sheep red blood cells.

   Rosetting with anti T* cell coated red blood cells or other particles

   Panning on anti T cell coated plates.

   Lysis with anti T cell antibody and complement.

2. ENRICHMENT OF B CELLS

   Rosetting with anti immunoglobulin or anti B cell coated red blood cells or other particles.

   Panning on anti immunoglobulin or anti B cell coated plates.

3. ENRICHMENT OF SPECIFIC B CELLS

   Rosetting with antigen coated red cells or other particles.

   Panning on antigen coated plates.

   Incubation with fluorescein conjugated antigen followed by fluorescence activated cell sorting.

****

* Antibody based separations usually employ monoclonal antibodies of rodent origin.

above sources are often unsuitable. In addition to the problem of supply and immune status, the success of the immortalisation procedure (especially fusion) is influenced by the state of differentiation and proliferative capacity of the cells used. In an effort to improve the efficiency of this process increasing attention is being paid to *in vitro* methods for sensitising and amplifying specific B cells, an approach which has been successfully used in the development of murine antibodies (5,30).

Unfortunately the *in vitro* immunisation of human B cells has proved more difficult to achieve, a problem attributed to the fact that most human studies have been performed with peripheral blood while the mouse studies have been undertaken with spleen (5). Notwithstanding there are now reports of human monoclonal antibody secreting cell lines which have been obtained by immortalising B cells stimulated *in vitro* with antigens. These include antibodies to bacterial, viral, red cell and other antigens. (Reviewed in 15) It should be noted however that in most cases the lymphocytes used were obtained from seropositive donors, the *in vitro* immunisation merely serving to boost the number and influence the differentiation of the specific B cells.

The basic steps involved in the production of human monoclonals by the *in vitro* technique are portrayed in Figure 2. They involve an initial enrichment of antibody forming cell precursors and the removal of lymphocyte subpopulations which might interfere with the subsequent immunisation and immortalisation process e.g. suppressor and cytotoxic T lymphocytes. The methods used to enrich and deplete various lymphocyte preparations are summarised in Table 2. A number of these steps are also routinely used when immortalising lymphocytes obtained from donors immunised *in vivo*. The success of the *in vitro* immunisation technique

FIGURE 2

BASIC STEPS INVOLVED IN THE PRODUCTION OF HUMAN MONOCLONALS USING THE IN VITRO IMMUNIZATION TECHNIQUE

1. Obtain leukocyte rich suspension from blood or other lymphoid tissue by density gradient separation

2. Remove or inactivate Ts and Tc by treatment with monoclonal antibody or cyclosporine

3. Incubate Ts, Tc depleted leukocyte rich suspension with antigen with/without mitogen, adjuvants or exogenous growth and differentiation factors

4. After 3 – 7 days incubation immortalize cells by fusion or transformation

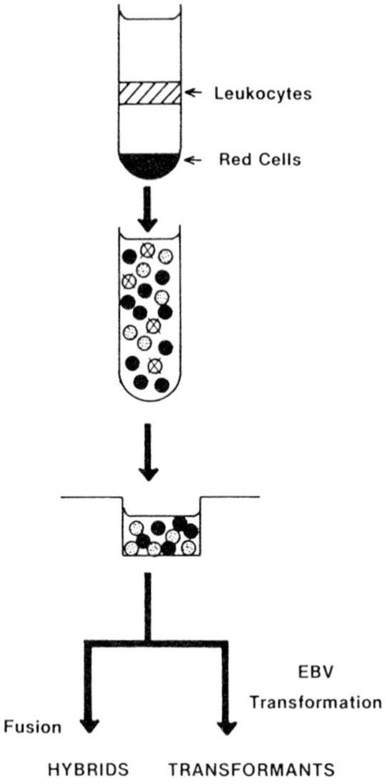

Fig. 2.- Basic steps involved in the production of human monoclonals using the in vitro immunisation technique

This approach is being increasingly employed in human monoclonal antibody production. It is particularly valuable where the antigen in question cannot be safely injected or where the level of immunity achieved *in vivo* is low. The optimal culture conditions for *in vitro* stimulation are dependent upon many factors (see text). The specifically activated B cells obtained are then either directly immortalised by standard procedures or immortalised after further enrichment.

is influenced by many factors including the choice and dose of immunogen, the cellular composition of the culture, the time in culture and the use of mitogens, adjuvants and exogenous growth and differentiation factors. Further studies in this area will be necessary before general guidelines can be proposed.

Procedures for Immortalising Human B Cells

At the present time there are two basic methods for immortalising human B lymphocytes namely fusion with suitable B cell lines or transformation with Epstein Barr Virus (EBV). Both these procedures have been widely employed and are discussed in further detail below.

In view of the success in the development of rodent monoclonals the polyethylene glycol induced fusion procedure has been rigorously pursued. This has stimulated the development of a large number of human fusion partners (see Table 3). Unfortunately few if any of these cell lines perform as well as the fusion partners used in the production of rodent monoclonals. Difficulties with respect to the stability of the hybrids, cloning and antibody secretion have led to the development of hybrid myeloma fusion partners, both human-human and human-mouse, or direct fusion with rodent cell lines.

The alternative immortalisation approach used in human monoclonal antibody production involves transformation of the B lymphocytes with EBV (10,34). This herpes virus is able to bind to the C_3d receptors on the B cell surface, be internalised and its genes incorporated into the lymphocyte genome. This leads to activation of many of the B cells with subsequent transformation ('immortalisation') of a subpopulation of these cells. While this transformation is normally achieved by incubating the lymphocytes with culture supernatant derived from an EBV infected marmoset cell line it may also occur if lymphocytes are

TABLE 3

SOME CELL LINES CURRENTLY USED IN HUMAN MONOCLONAL ANTIBODY PRODUCTION*

CELL TYPE		EXAMPLES
1) HUMAN PLASMACYTOMAS	MYELOMA	SKO-007, RPMI-8226, HFB-1, KMMI
	LYMPHOMA	RH-L4 NAT-30
2) LYMPHOBLASTOID CELL LINES		GM 1500-6TG-OA, KR4, GM467, LICR-LON-HMy2, H35-1-1, W1-L2-727, W1-L2-729 HF2, UC729-6, MC/CAR MC/MNS-2, LTR 228, LSM2-7 GK-5, HO-323.
3) HYBRID MYELOMAS	HUMAN-HUMAN	KR12
	MOUSE-HUMAN	SHM-D3, SBC-H20
	MOUSE-HUMAN-HUMAN	3HL
4) RODENT MYELOMAS		NS.0, NS.1, SP 2/0, P3/HT, P3X63 AgU1, P3X63Ag 8.653

* For further details on these cell lines including their origins, drug markers and immunoglobulin isotypes see Reference 15.

TABLE 4

A SURVEY OF HUMAN MONOCLONAL ANTIBODIES CURRENTLY AVAILABLE

ANTIBODIES TO	EXAMPLES
BACTERIAL ANTIGENS	Diptheria toxin, Gm -ve endotoxin, Exotoxin A, Haemophilus influenzae, Mycobacterium leprae, Neisseria meningitides, Pneumococcus polysaccharides, Pseudomonas aeruginosa, Tetanus toxoid, Chlamydia.
VIRAL ANTIGENS	Cytomegalovirus, Epstein-Barr virus, Hepatitis A,Bs and Bc, Herpes simplex, HTLV-1, Influenzae virus nucleoprotein, Rabies, Rubella, T cell leukaemia virus, Varicella zoster.
OTHER INFECTIOUS AGENTS	Plasmodium falciparum
RED CELL AND WHITE CELL ANTIGENS	Blood groups A and i, Rhesus D and G. Forsmann antigen, chicken and sheep red blood cells, HLA antigens.
TUMOUR ANTIGENS	Bladder, Brain Glioma, Breast, Chronic lymphocytic leukaemia, Colorectal carcinoma, Gastric carcinoma, Lung carcinoma, Lymphoma, Melanoma, Myeloid leukaemia, Renal carcinoma, Vulva carcinoma.
AUTOANTIGENS	Astrocytes, cytoskeletal proteins, ss and ds DNA, Endothelial cells, Golgi, Myelin associated glyco-protein, Nuclear and Nucleolar antigens, Nerve axons, Neurones, Pancreatic islet cells, Pancreatic duct cells, Parathyroid, Pituitary, Platelets, Prostatic acid Phosphatase, Sperm, Sperm coating protein, Stratified squamous epithelium, Thyroglobulin, Thyroid follicle, Vimentin, Rheumatoid factor.
OTHERS	Bombesin, 2-4 dinitrochlorobenzene, 2-4 dinitrophenol, 4 hydroxy 3-5 dinitrophenacetic acid, Keyhole limpet haemocyanin, Phosphorylcholine, Sperm whale myoglobin.

It should be stressed that many of these antibodies have not been adequately characterised and the cell lines are not very stable. For reference to the original articles the reader should consult Reference 15.

outlined in recent reviews (1,3,13) and are addressed elsewhere in these proceedings.

A SURVEY OF HUMAN MONOCLONALS CURRENTLY AVAILABLE AND THEIR APPLICATIONS

Although many problems have been encountered in human monoclonal antibody development the successes achieved and the potential applications have encouraged continuing effort in this area. As a consequence human monoclonal antibodies reacting with a wide variety of targets have been described (Table 5). These include antibodies against bacterial, viral, red cell and tumour antigens developed primarily with diagnosis and therapy in mind. Of additional interest are those derived from normal lymphocytes or individuals with autoimmune diseases; these react with both cell surface and intracellular antigens. It should be appreciated however that many of the antibody secreting cell lines reported have proved unstable. Furthermore a large proportion of the antibodies have not been adequately characterised thus their precise specificity remains unknown.

A wide array of applications have been proposed for monoclonal antibodies and these are outlined in Table 6. They include diagnosis, prophylaxis and therapy and a means of investigating humoral aspects of the immune response in health and disease. The prophylactic potential of one such reagent is illustrated in Fig. 3 where antitetanus antibody of human origin was shown to delay or prevent the development of tetanus toxin induced paralysis in mice. For a detailed appraisal of the therapeutic potential of human monoclonals the readers attention is directed towards a recent review by Larrick and Bourla (22).

In all these applications monoclonal antibodies of human origin offer theoretical advantages over their murine counterparts. In the first place they many interact with important determinants not readily

FIGURE 3

THE PROTECTIVE EFFECT OF A HUMAN ANTI–TETANUS MONOCLONAL IN MICE

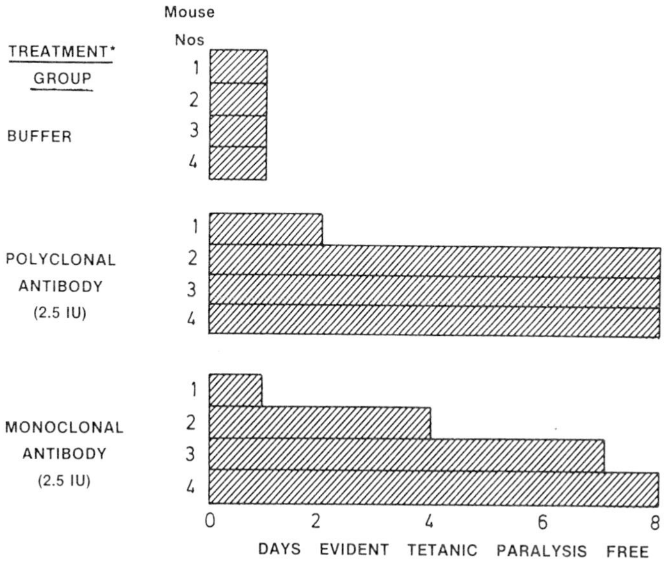

*Antibody injected IP on days 0, 1, 2, 3, 4 and 7
Toxin (4 x Lpl 100 dose) injected SC on day 0

Fig. 3.- The protective effect of a human anti-tetanus monoclonal antibody in mice.

In these studies the antibody used was an IgG3 product secreted by a transformed cell line. Note that while protection was achieved it was not as effective as that observed with a polyclonal preparation. For complete protection a cocktail of 2 or more monoclonals will probably be required.

TABLE 5

POSSIBLE USES OF HUMAN MONOCLONAL ANTIBODIES

DIAGNOSTIC AND MONITORING		Infectious diseases
		Malignancy
		Red cell compatability
		Tissue typing
PROPHYLAXIS AND THERAPY	Infections Diseases and Malignancy	Passive therapy
		Identification and purification of candidate antigens for vaccines.
		As anti-idiotype vaccines
		Targetting drugs
		Imaging lesions
	Other Applications	Autoimmune diseases
		Renal transplantation
		Rhesus incompatability
		Contraception
INVESTIGATING THE IMMUNE SYSTEM IN HEALTH AND DISEASE		Studying normal B cell repertoire
		Studying the molecular biology of the immunoglobulin locus
		Analysing B cell response in infection malignancy, autoimmunity and allergy
		Identifying, purifying and characterising antigens of clinical relevance (see above)

For further details of applications see References 15 and 22.

TABLE 6

REASONS ADVANCED FOR THE INFERIOR PERFORMANCE OF PERIPHERAL BLOOD LYMPHOCYTES IN HUMAN MONOCLONAL ANTIBODY PRODUCTION

1. Insufficient memory cells or antigen specific B cells.
2. Transient appearance of antigen specific B cells in blood following immunisation.
3. Inappropriate state of differentiation of B cells.
4. Low proliferative capacity of B cells.
5. Contains insufficient Ia positive macrophages.
6. Contains too many cytotoxic T cells.
7. Contains suppressor B cells.

incubated with EBV infected lymphocytes themselves, a process known as cell driven transformation.

While the EBV immortalisation procedure has been widely used and resulted in the production of some valuable secreting cell lines there are problems with respect to the cloning and stability of many of the cell lines. This has led to a combination of both transformation and fusion techniques, an approach originally described by Kozbor and colleagues (19). It is believed to significantly increase fusion efficiency, facilitate cloning and result in increased levels of antibody secretion (21).

Cloning and Expansion of Antibody Secreting Cell Lines

Having identified cell lines secreting antibodies of interest the next major hurdle is to obtain stable antibody secreting clones. As already indicated this is much harder to achieve than with mouse hybrids. At the present time the cloning strategies employed are the same, namely dilution cloning or cloning in soft agar with or without the use of feeder layers or media supplements. For precise details on the feeder layers used and other aspects of cloning the reader should consult recent reviews and books (14,15,35).

Another critical stage in the development of monoclonals is the expansion of selected antibody secreting cell lines. As to be expected the approaches used to date have been similar to those adopted for mouse monoclonals; these include a variety of in vitro culture procedures together with growth in nude mice and rats. Some of the problems associated with the latter approach have recently been reviewed by the author (15). It is generally accepted that where large amounts of antibody are required airlift reactors and immobilised cell technologies are the most suitable methods. Further details of these procedures are recognised by mouse monoclonals raised against the same antigen. This can be attributed to the inability of the mouse to recognise the relevant epitopes against a background of immunodominant epitopes. Obvious examples in this are monoclonals to tumour antigens, the Rhesus D antigen and antigens of the major histocompatability complex. Furthermore where the antibodies are to be used to identify and purify candidate antigens for vaccines it cannot be assumed that epitopes recognised by the mouse will evoke protective immunity in man. Finally where prophylaxis and therapy (especially longterm) are envisaged it is obviously desirable to administer antibody of human origin.

While the major impetus in this area has arisen from the desire to develop improved diagnostics and therapeutics the contribution this technology has made to our overall understanding of the B cell repertoire in health and disease should not be overlooked. It has provided valuable information on the host immune response to tumours, infectious agents and normal tissue antigens and has facilitated the isolation and characterisation of the relevant antigens. Knowledge gained in this area should lead to an increase in our understanding of basic immune phenomenon and result in improved therapeutic procedures.

THE LIMITATIONS OF CURRENT TECHNOLOGIES

As emphasised earlier there are a number of major problems which are hindering the advance of human monoclonal antibody technology. These include the supply of suitable immune lymphocytes for fusion and transformation and the poor efficiency of the immortalisation procedures. This is further compounded by the low levels of antibody secreted and the frequent loss of this function.

It is readily apparent that human peripheral blood is not the best source of cells for either *in vitro* immunisation or immortalisation.

The poor performance of such cells has been attributed to a number of factors which are listed in Table 6. In brief it is argued that peripheral blood is deficient in antigen presenting cells, T helper cells and antigen reactive B cells in the appropriate state of differentiation or proliferation (8). It also contains too many suppressor and cytotoxic T cells.

A number of hypothesis have also been advanced to explain the poor secretory capacity and relative instability of immortalised human B cells compared with their murine counterparts and these are summarised in Table 7. They include overgrowth by non-secreting lymphocytes, loss of genes coding for immunoglobulin chains, malfunctioning of regulatory genes and defects in the synthetic and secretory machinery of the cells.

STRATEGIES FOR THE FUTURE

From the above comments it should be apparent that the production of human monoclonal antibodies is a risky affair involving considerable dedication and effort. At the present time there are no simple answers to the problems which limit the advance of this technology. Nevertheless there are a number of areas which are worthy of further investigation and these are outlined in Table 8 and discussed in further detail below.

Supply of Immune Lymphocytes

This still remains one of the major obstacles. More attention will have to be paid to improving both *in vitro* and *in vivo* immunisation schedules and investigating the influence that cellular composition has on the outcome of immortalisation. The selection of lymphocytes will undoubtedly improve as our knowledge of the events associated with lymphocyte triggering, proliferation and differentiation increases.

TABLE 7

POSSIBLE REASONS FOR DEFECTIVE IMMUNOGLOBULIN SYNTHESIS BY HUMAN MONOCLONAL CELL LINES

1) Inhibitory effects of mycoplasma and other microorganisms.
2) Overgrowth of secreting cells by non-secretors.
3) Shortage of relevant growth and differentiation factors.
4) Absence of appropriate receptors for growth and differentiation factors.
5) Loss or inappropriate incorporation of structural genes for immunoglobulin H and L chain.
6) Insufficient copies of structural and regulatory genes.
7) Inappropriate H and L combinations resulting in hybrid molecules.
8) Failure or loss of relevant regulatory genes or other defects in regulation.
9) Defects in the synthetic machinery resulting in impaired transcription, translation and assembly.
10) Defects in the secretory machinery of the cell.

TABLE 8

STRATEGIES FOR IMPROVING HUMAN MONOCLONAL ANTIBODY PRODUCTION

1. Improved immunisation strategies
   a) *In vivo* e.g. optimising use of adjuvants, bleeding times, etc.
   b) *In vitro* e.g. optimising antigen dose, selection of adjuvants, removal of suppressor cells etc.
2. Improved selection of immune B cells
   a) Use of lymphoid tissues other than blood.
   b) Enrichment of specific B cells.
   c) Selection of cells of appropriate differentiation and proliferation status.
3. Improved immortalisation procedures
   a) Optimisation of existing technologies.
   b) Development of better B cell lines.
   c) Application of more efficient fusion procedures e.g. electrofusion.
   d) Use of alternative immortalisation strategies e.g. transfection.
4. Improved cloning procedures
   a) Application of alternative cloning techniques e.g. micromanipulation, fluorescence activated cell sorting.
   b) Use of recombinant growth factors.
5. Studying factors influencing B cell differentiation and proliferation
   a) Expression of receptors for growth and differentiation factors.
   b) Effects of exogenous growth and differentiation factors
6. Studying factors regulating immunoglobulin synthesis and secretion
   a) Expression of immunoglobulin structural genes.
   b) Influence of regulatory genes.
   c) Secretory mechanisms.
7. The application of recombinant techniques
   a) Cloning and expressing human antibody genes.
   b) Cloning and expressing genes for hybrid (mouse/human) antibodies.
   c) Cloning and expressing genes for antibody/non antibody hybrids.
   d) Production of site-directed antibody mutants.

However we would concur with others that the future of human monoclonal antibody technology will be governed largely by advances in *in vitro* immunisation procedures (5,11).

Immortalisation Procedures. Improvements in the immortalisation procedures currently employed are a real possibility. It is obvious that these procedures are far from optimised information being required on the cell populations most susceptible to fusion and transformation (See for example 2,9,10).

The most obvious limitation with respect to the fusion process itself is the lack of suitable non-secreting fusion partners capable of producing stable hybrids secreting high levels of antibody. The development of such lines will inevitably require much effort. In the meantime we, like others feel the use of mouse or mouse/human fusion partners should not be overlooked. A number of laboratories have found these to fuse better and the resultant hybrids clone easier and secrete more antibody than human hybrids; they also grow better in nude mice. Furthermore contrary to popular opinion they have been found to be relatively stable.

Attention will also have to be paid to alternate fusion strategies for as previously emphasised (11) the PEG induced fusion technique is irreproducible and inefficient. Of particular interest in this regard are the antigen directed electrofusion procedures recently described (23,39) which are based on the electrofusion technique originally reported by Zimmerman and his colleagues (4,40). Finally cloning should be made easier by the use of recombinant derived growth factors and the use of alternative cloning procedures including fluorescence activated cell sorting based techniques and micromanipulation of individual cells.

Improving Antibody Production Considerable effort will be required to develop methods for improving the levels of antibody secreted by human cell lines and in ensuring that secretion is maintained. This could conceivably result in part from improvements in immunisation protocols and fusion partners but other strategies will be necessary. The most obvious approach at the present time will be to pursue a rigorous cloning policy and to frequently enrich the specific antibody secreting cell lines, an approach already successfully adopted by some investigators (10,12).

Additional strategies seem possible in view of the observation that cell lines which have stopped secreting immunoglobulin can be 'switched on' by either fusion with normal lymphocytes (32) or stimulation with lipopolysaccharide, a B cell mitogen (29). As previously stressed these observations suggest that the decline in immunoglobulin production by human B cell lines might be a consequence of defects in regulatory gene function of the synthetic and secretory machinery of the cell. In the light of these suggestions techniques for enhancing gene expression would appear worthy of investigation. These include the use of phorbol esters which have been shown to enhance both growth factor expression on B cells and their immunoglobulin secretory capacity (28). These observations are of interest in view of recent reports that the senescence of some human antibody secreting cell lines might be associated with a decline in growth factor receptor expression (24). Alternative approaches would be to transfect cells with specific enhancer sequences (38) or to silence 'inhibitory' genes by demethylation of the genome (27).

Whatever approaches are used it will undoubtedly be some time until they are developed and their effectiveness evaluated. In the meantime Figure 4 illustrates one possible approach to improving human monoclonal antibody production based on existing technology.

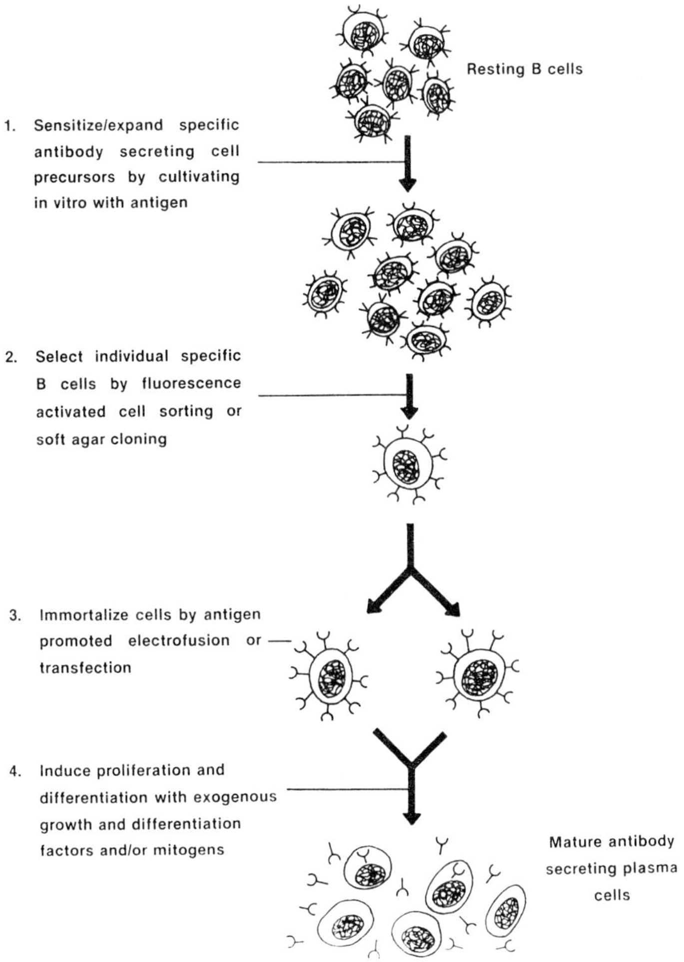

Fig. 4.- A possible future approach to human monoclonal antibody production.

This figure illustrates how existing technology might be employed to improve human monoclonal antibody production. Steps 1 and 2 would provide small numbers of highly purified specific B cells which would then be selectively fused by the antigen promoted electrofusion technique or by transfection with DNA from malignant cell lines. Expansion of the small number of hybrids or transfected cells would be achieved by culture in the presence of growth and differentiation factors produced by recombinant or other procedures.

Gene Cloning Strategies In this review the emphasis has been on conventional approaches to human monoclonal antibody production. Inevitably in the future genetic engineering techniques will begin to play a significant part in the production of both rodent and human monoclonal antibodies. This will involve either the transfection of cell lines with DNA or conventional recombinant procedures using vectors. The transfection approach has already been used to immortalise antibody secreting cells using DNA from permanent cell lines (16) and conversely permanent cell lines with DNA from plasma cells (36).

However of particular interest are the reports describing the production of hybrid molecules by gene cloning techniques. The potential of this approach has recently been reviewed (26) and examples of the types of molecules already developed are shown in Figure 5. The gene constructs have been introduced into mammalian cells by both spheroplast fusion (17,33) and electrophoretic techniques (31). Using these procedures hybrid antibodies have been produced which retain the specificity of the original mouse monoclonal but are intrinsically less immunogenic in man as the constant region is coded for by genes of human origin. These have been directed against haptens (6) tumours (31) and other antigens (37).

Recent developments have further highlighted the potential of the gene cloning approach. These involve replacing the genes coding for the hypervariable regions of human heavy chains with those coding for the hypervariable region of murine monoclonals (17) and changing individual amino acids within combining sites (and hence specificity) by site directed mutagenesis (33). Finally the production by recombinant techniques of antibody/non antibody hybrids with novel effector functions (25) greatly extends the diagnostic and therapeutic potential of these procedures.

FIGURE 5

HYBRID HUMAN MONOCLONALS

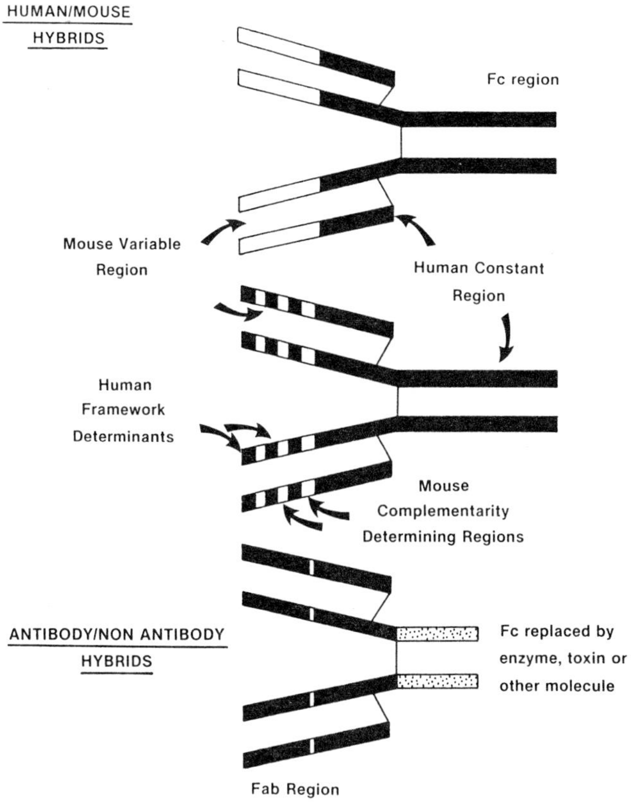

Fig. 5.- Hybrid monoclonal antibodies produced by genetic engineering.

By genetic engineering techniques hybrid monoclonal antibodies can be produced wich offer distinct diagnostic and therapeutic advantages. As illustrated in this figure the changes may be effected in both the variable and constant regions of the immunoglobulin molecule. This strategy enables the construction of molecules which combine the specificity of the mouse monoclonal but which are less immunogenic or alternatively molecules with novel effector functions.

ACKNOWLEDGEMENT

The author wishes to thank the Royal Society of Edinburgh for providing a travelling fellowship enabling him to participate in this meeting. He is also indebted to Drs. J. Boyd and F. Sheffield for performing the studies presented in Figure 3.

REFERENCES

1. Altschuler, G.L., Dziewulski, D.M., Sowek, J.A. and Belfort, G. (1986). Continuous hybridoma growth and monoclonal antibody production in hollow fibre reactors. Biotechnology and Bioengineering, 28, 646-658.
2. Aman, P., Ehlin-Henriksson, B. and Klein, G. (1984). Epstein-Barr Virus susceptibility of normal human B lymphocyte preparations. Journal of Experimental Medicine, 159, 208-220.
3. Birch, J.R., Boraston, R. and Wood, L. (1985). Bulk production of monoclonal antibodies in fermenters. Trends in Biotechnology, 3, 162-166.
4. Birchoff, R., Eisert, I.M., Vienken, J. and Zimmerman, U. (1982). Human hybridoma cells produced by electrofusion. Febs Letters, 147 64-68.
5. Borrebaeck, C.A.K. (1986). In vitro immunisation for the production of murine and human monoclonal antibodies: Present status. Trends in Biotechnology, 4, 147-153.
6. Boulianne, G.L., Hozumi, N. and Shulman, M.J. (1984). Production of functional chimeric mouse/human monoclonal antibodies. Nature, 312, 643-646.

7. Boyd, J.E., James, K. and McClelland, D.B.L. (1984). Human monoclonal antibodies - Production and potential. Trends in Biotechnology, 2, 70-77.

8. Burnett, K.G., Leung, J.P. and Martinis, J. (1985). Human monoclonals to defined antigens: towards clinical applications. Human Hybridomas and Monoclonal Antibodies. (Engleman, E.G., Foung, S.K.H. Larrick, J. and Raubitschek, A., Eds.) pp 113-133. Plenum Press, New York.

9. Chan, M.A., Stein, L.D., Dosch, H.M. and Sigal, N.H. (1986). Heterogeneity of EBV-transformable human B lymphocyte populations. Journal of Immunology, 136, 106-112.

10. Crawford, D.M. (1985). Production of human monoclonal antibodies using Epstein Barr Virus. Human Hybridomas and Monoclonal Antibodies (Engleman, E.G., Foung, S.K.H., Larrick, J. and Raubitschek, A. Eds.) pp 37-53, Plenum Press, New York.

11. Dorfman, N.A. (1985). The optimal technological approach to the development of human hybridomas. Journal of Biological Response Modifiers, 4, 213-239.

12. Doyle, A., Jones, T.J., Bidwell, J.L. and Bradley, B.A. (1985). In vitro development of human monoclonal antibody secreting plasmacytomas. Human Immunology, 13, 199-209.

13. Duff, R.G. (1985). Microencapsulation technology: a novel method for monoclonal antibody production. Trends in Biotechnology, 3, 167-170.

14. Engleman, E.G., Foung, S.K.H., Larrick, J. and Raubitschek, A. (Eds.) (1985). Human Hybridomas and Monoclonal Antibodies. Plenum Press, New York.

15. James, K. and Bell, G.T. (1987). Human monoclonal antibody production: Current status and future prospects. Journal of Immunological Methods. To be published in Volume 100.

16. Jonak, Z.L., Braman, V. and Kennett, R.H. (1984). Production of continuous mouse plasma cell lines by transfection with human leukemia DNA. Hybridoma, 3, 107-118.

17. Jones, P.T., Dear, P.H., Foote, J., Neuberger, M.S. and Winter, G. (1985). Replacing the complementarity determining regions in a human antibody with those from a mouse. Nature 321, 522-525.

18. Kohler, G. and Milstein, C. (1975). Continuous cultures of fused cells secreting antibody of predefined specificity. Nature 256, 495-497..

19. Kozbor, D., Lagarde, A.E. and Roder, J.C. (1982). Human hybridomas constructed with antigen-specific Epstein-barr Virus transformed cell lines. Proceedings of the National Academy of Science, USA, 79, 6651-6655.

20. Kozbor, D. and Roder, J.C. (1983). The production of monoclonal antibodies from human lymphocytes. Immunology Today, 4, 72-79.

21. Kozbor, D. and Roder, J.G. (1984). In vitro stimulated lymphocytes as a source of human hybridomas. European Journal of Immunology, 14, 23-27.

22. Larrick, J.W. and Bourla, J.M. (1986). The prospects for therapeutic use of human monoclonal antibodies. Journal of Biological Response Modifiers, 5, 379-393.

23. Lo, M.S., Tsong, T.Y., Conrad, M.K., Strittmatter, S.M., Hester, L.D. and Snyder, S.H. (1984). Monoclonal antibody production by receptor mediated electrically induced fusion. Nature, 310, 792-794.

24. Melamed, M.D., Gordon, J., Ley, S.J., Edgar, D. and Hughes-Jones, N.C. (1985). Senescence of a human lymphoblastoid clone producing anti-Rhesus D. European Journal of Immunology, 15, 742-746.

25. Neuberger, M.S. (1985). Making novel antibodies by expressing transfected immunoglobulin genes. Trends in Biotechnology, 10, 347-349.

26. Oi, V.T. and Morrison, S.L. (1986). Chimeric antibodies. Biotechniques, 4, 214-221.

27. Olsson, L. and Brams, P. (1985). Human-human hybridoma technology. Five years of technical improvement and its application in cancer biology. Human Hybridomas and Monoclonal Antibodies. (Engleman, E.G., Foung, S.K.M., Larrick, J. and Raubitschek, A. Eds.). pp 227-244, Plenum Press, New York.

28. Polke, C.R., Lowenthal, J.W., Roth, S.A., Rohwer, P., McDonald, R.H. and Kalden, J.R. (1986). Phorbol ester enhances both interleukin 2 receptor expression and immunoglobulin secretion in human Epstein-Barr Virus immortalised B cells. European Journal of Immunology, 16, 146-150.

29. Raison, R.L., Walker, K.Z., Halnan, C.R.E., Brisco, D. and Basten, A. (1982). Loss of secretion in mouse-human hybrids need not be due to the loss of a structural gene. Journal of Experimental Medicine, 156, 1380-1389.

30. Reading, C.L. (1982). Theory and methods for immunisation in culture and monoclonal antibody production. Journal of Immunological Methods, 53, 261-291.

31. Sahagan, B.G., Dorai, H., Saltzgaber-Muller, J., Toneguzzo, F., Guindon, C.A., Lilly, S.P., McDonald, K.W., Morrissey, D.V., Stone, B.A., Davis, G.L., McIntosh, P.K. and Moore, G.P. (1986). A genetically engineered murine/human chimeric antibody retains specificity for human tumour associated antigen. Journal of Immunology, 137, 1066-1074.

32. Schwaber, J.F., Posner, M.R., Schlossman, S.F. and Lazarus, H. (1984) Human-human hybrids secreting Pneumococcal antibodies. Human Immunology, 9, 137-143.

33. Sharon, J., Gefter, M.L., Manser, T. and Ptashne, M. (1986). Site

directed mutagenesis of an invariant amino acid residue at the variable diversity segment junction of an antibody. Proceedings of the National Academy of Science, USA, 83, 2628-2631.

34. Steinitz, M., Klein, G., Koskimies, S. and Mäkelä, O. (1977). EB virus induced B lymphocyte cell lines producing specific antibody. Nature, 269, 420-422.

35. Strelkauskas, A.J. (Ed.) (1987). Human Hybridomas, Diagnostic and Therapeutic Applications. Marcel Dekker, New York.

36. Strelkauskas, A.J., Taylor, C.L., Aldenderfer, P.H. and Warner, G.A. (1987). Transfection of human cells: An alternative method for the establishment of human hybrid clones. Human Hybridomas - Diagnostic and Therapeutic Applications. (Strelkauskas, A.J., Ed.) pp 95-117, Marcel Dekker, New York.

37. Takeda, S., Naito, T., Hama, K., Noma, T. and Henjo, T. (1985). Construction of chimeric processed immunoglobulin gene containing mouse variable and human constant region sequences. Nature, 314, 452-454.

38. Teng, N.N.H., Reyes, G.R., Bieber, M., Fry, K.E., Lam, K.S. and Herbert, J.M. (1985). Strategies for stable human monoclonal antibody production: construction of heteromyelomas, in vitro sensitization and molecular cloning of human immunoglobulin genes. Human Hybridomas and Monoclonal Antibodies. (Engleman, E.G., Foung, S.K.H., Larrick, J. and Raubitscheck, Eds.) pp 71-91, Plenum Press, New York.

39. Wojchowski, D.M. and Sytkowski, A.J. (1986). Hybridoma production by simplified avidin-mediated electrofusion. Journal of Immunological Methods, 90, 173-177.

40. Zimmerman, U. and Vienken, J. (1982). Electric field induced cell to cell fusion. Journal of Membrane Biology, 67, 165-182.

Discussion

ARATHOON — Was the polyclonal used in the tetanus test mouse or human

JAMES — Human polyclonal (standard). We have now to ask ourselves why cells which have been producing antibodies for two months suddenly stop.

ARATHOON — The antibodies produced against plaques from animals which had not been immunised are those altzhiemers plaques or some other plaque.

JAMES — Against altzhiemers. You can get such plaques in Down's syndrome.

MEIGNIER — What are the advantages of using human monoclonal antibodies for diagnostics

JAMES — There would be few advantages for infectious diseases but what about Rhesus D typing and also HLA typing.

KATINGER — Could you estimate how many laboratories world wide have a well established human in-vitro immunisation

JAMES — There are a lot of laboratories working at it - say 100 - 200 of which 20-30 publish papers. It is a question of getting the right dose for in vitro immunisation. If you do not put enough then the cells are not triggered and if you use too much the cells are switched off.

KATINGER — What is the average yield for mouse and human monoclonal antibodies.

JAMES — Mouse hybridomas are successful 90% of the time human hydridomas we have a 1% success rate.

4 Production of cancer specific monoclonal antibodies with human–human hybridomas and their serum-free, high density, perfusion culture

H. Murakami and K. Yamada

The hybridoma technique introduced by Köhler and Milstein (1) has had an enormous impact on biological sciences because it enables us to produce monoclonal antibodies (MoAbs) against almost any existing immunogens. Until 1980, the hybridoma technique was applied almost exclusively to mouse cells. However, human MoAbs are highly desirable for the diangosis and treatment of human diseases(2). To date, the production of human MoAbs has been hindered by the difficulty of obtaining hybridomas which consistently and stably generate human immunogloblins. The production of human MoAbs against various antigens will result in an expansion of the library of MoAbs currently available from murine sources and may lead to the development of clinically useful reagents for the detection of and therapies for human cancers and infections diseases.

It is extremely important to develop the system for obtaining physiologically active bio-macromolecules in large quantity which can be exclusively produced by mammalian cells. We showed here our results on serum-free, high density, perfused culture of human-human hybridomas to obtain cancer specific MoAbs. Discussion was especially focused to compose the culture condition to maintain cells in stationary phase so that cellular product can be obtained continuously, and operation of culture process, recovery and purification can be automated.

I. PREPARATION OF FUSION PARTNER CELL LINES

Preparatipn of a fusion partner, HO-323 was carried out as follows(3):

WIL2-NS (Flow Lab.), a human B-lymphoblastoid cell line, was first cultured in a 2:1:1 mixture of RPMI 1640 (RPMI), Ham's F12 and Dulbercco's modified Eagle's media (RDF), supplemented with 10 µg/ml insulin, 25 µg/mi lactoferrin, 10 µM ethanolamine, and 2.5nM selenite (ILES)(4). Approximately 60% of the cells died by this treatment, but after subculturing for 2 weeks, the selected cells started to grow vigorously in the serum-free medium. These viable cells were cloned on soft agar (0.24%) in RPMI medium supplementd with 10% fetal calf serum (FCS), and non-secretor clones of immunogloublins were selected. The 6-thioguanine-resistant mutants were selected in RPMI/10% FCS medium containing 30 µg/ml of 6-thioguanine, and the resultant mutants were tested for HAT (Hypoxathine, Aminopterin, Thymidine) sensitivity. The mutant clone, HO-323, was selected on the basis of aminopterin sensitivity, fusion efficiencies and growth rates. HO-323 has been subcultured in RPMI/10% FCS medium containing 30 µg/ml 6-thioguanine at 37°C in humidified 5% CO_2, and prior to fusion the cells were cultivated for 3 days without 6-thioguanine.

Methods for preparations of K 562-8AG and NL-6TG were similar to that of HO-323. These were isolated by treating K 562 and NL cells with 100 µg/ml 8-azaguanine or 30 µg/ml 6-thioguanine, respectively. The preparation of NAT-30 was carried out as described elsewhere (5). These HAT sensitive fusion partners were tested for the preparation of hybridomas using lymphocytes from human lymph nodes and peripheral blood.

II. CHARACTERISTICS OF A FUSION PARTNER CELL LINE, HO-323

HO-323, derived from parental WIL2-NS cells, was resistant to 30 µg/ml of 6-thioguanine and died within one week when cultivated in HAT medium (Fig. 1). HO-323 doubles every 15 hr, has a cloning efficiency of 10% and can proliferate well even when plated at low cell density of less than 1 x 10^3 cells/ml. In addition, HO-323 proliferated well in serum-free ILES medium, while most of the other human cell fusion partner, such as LICR-LON-HMy2, did not. HO-323

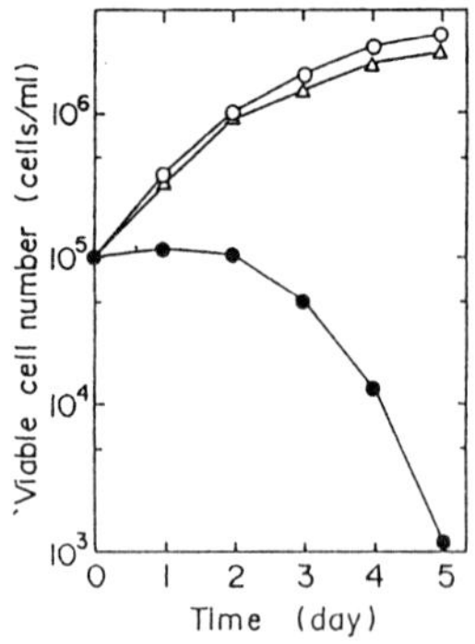

Fig. 1. Growth of HO-323 cells in normal (RPMI/10%FCS, ○), 6-thioguanine-containing (△) or HAT-containing (●) medium. Viability was determined by 0.1% trypan blue exclusion.

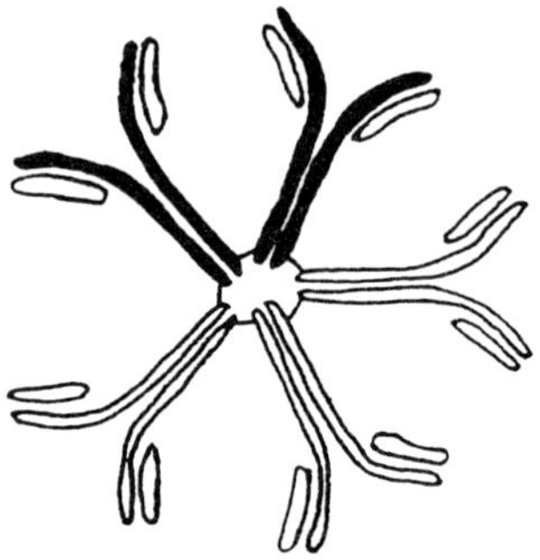

Fig. 2. Chimera antibody Ig (A_2M_3).

Table 1 FUSION EFFICIENCY OF VARIOUS PARENT CELL LINES WITH LYMPHOCYTES FROM LYMPH NODES OF PERIPHERAL BLOOD

Parent cell line	Origin of lymphocytes	Efficiency (clones/10^5 parent cells)
HO-323	Lung cancer	1.8
	Breast cancer	1.1
	SLE	1.3
	Peripheral blood	0.9
NAT-30	Lung cancer	0.3
K562-8AG	Lung cancer	0.0
NL-6TG	Lung cancer	0.0
LICR-LON-HMy2	Lung cancer	0.1
LICR-LON-HMy2[a)]		0.2
SKO-007[a)]		0.1
GM 4672[a)]		0.1
UC 729-6[b)]		0.1-0.6

a):from Houghton et al., 1983; b): Glassy et al., 1983

Table 2 FUSION EFFICIENCY UNDER THE VARIOUS FUSING CONDITIONS

Lymphocytes/HO-323	Efficiency (clones/10^5 HO-323 cells)
1 : 1	0.4
2 : 1	1.8
10 : 1	0.3

This is the same ratio as those of other human partners, such as LICR-LON-HMy2, SKO-007, GM 4672 and UC 729-6 (Houghton et al., 1983; Glassy et al., 1983).

synthesized cytoplasmic μ- and κ-chains (about 10 ng/10^6 cells), but did not secrete any immunoglobulins. Most of the other human cell partners, LICR-LON-HMy2, SKO-007, GM 4672, and LTR 228 (6,7), secreted immunogloblins. The HO-323 was diploid with 46 chromosomes. Thus, HO-323 may be considered to have characteristics suitable for a fusion partner.

The most important point suitable for a fusion partner is the efficiency of the cell fusion. HO-323 has the highest fusion efficiency among fusion partners. As shown in Table 1, the cell line gave more than one hybridoma per 10^5 HO-323 cells. The fusion efficiencies were not greatly variable with regard to the origins of lymph nodes from lung and breast cancers.

Fusion of HO-323 with peripheral lymphocytes of SLE patients yielded about one hybridoma clone per 10^5 HO-323 cells, as well as lymphocytes from those lymph nodes. Lymphocytes in normal peripheral blood also gave a fusion efficiency comparable to those of lung and breast cancers and systemic lupus erythematodes (SLE). IgM, IgG, and IgA secreting hybridomas were obtaind by fusing HO-323 with lymphocytes from any lymph nodes and peripheral blood. The results indicate that HO-323 can fuse with human peripheral lymphocytes sensitized in vivo with various kinds of antigens to make hybridomas secreting corresponding MoAbs. K 562-8AG and NL-6TG were non-secretors of immunogloblins as well as HO-323 but did not give any hybridoma clones. This higher efficiency of HO-323 as a fusion partner may be attributed to its high cloning efficiency and proliferative capacity at low cell density, as described above.

The highest fusion efficiency was obtained when HO-323 was fused with isolated lymphocytes at a ratio of 1:2 (Table 2). This efficiency is almost 10 times higher than those of other human partners, such as LICR-LON-HMy2, SKO-007, GM 4672, and UC 729-6 (6,7).

III. MONOCLONAL ANTIBODIES SPECIFIC TO CANCER CELLS

Monoclonal antibodies were made by fusion of NAT-30 or HO-323 with lymphocytes from a lung or mammary cancer patient. Monoclonal antibodies obtained were tested for binding specificity on the panel of human cancer cells and normal fibroblast cell lines. The specificity was also evaluated by several cancer and normal tissues.

1. Monoclonal Antibodies Specific to Lung Cancer

Fusion of NAT-30 was carried out with human lymphocytes from lymph nodes of lung cancer patients with lung adenocarcinoma. Through repeated fusions, more than 2,000 hybridoma clones were obtained. The proportion of hybridoma clones secreting IgM, IgA, and IgG were 4, 2, and 1, respectively. These MoAbs were tested for reactivity to lung adenocarcinoma cell line PC-8 by indirect immunofluorescent assay. We have found that four clones (HF10B4, HF10F7, HB4G7, and HB4C5) of the tested hybridomas secrete MoAbs reacting against the lung adenocarcinoma. Those MoAbs were all IgM class with lambda light chains. The chromosome number of the HB4C5 clone after 6 months' culture was 56.5 + 5.3, compared with 46.3 + 0.5 for NAT-30 cells, and the chromosome numbers of other clones were also apparently more than that of NAT-30.

IgM in the spent media of these 4 postive hybridoma clones was then examind for specificities on a panel of 29 established human cell lines: 14 lung, 2 bladder, 2 breast, an epidermoid, 2 renal, 2 melanoma, and 2 stomach cancers, and 4 normal fibroblasts (Table 3).

Monoclonal antibodies from 4 clones strongly reacted only with the lung carcinoma lines. Other types of cancer cells except a stomach cancer line, MKN-45 which bound MoAb of HF10B4, HF10F7, and HB4G7, had no reactivity with any of the 4 MoAbs. None of these 4 MoAbs reacted with normal fibroblast lines, including WI-38 and Flow 2000 originating from lung tissue. MoAb of HF10B4 reacted with lung cancer lines more strongly than those of HF10F7, HB4G7 or HB4C5. Monoclonal antibody of HF10B4 reacted with all lung carcinoma cell lines except for PC-9, a differentiated adenocarcinoma. IgM from clone H-7, H-9, or human plasma did not react with any of these cell lines.

Four hybridoma clones producing IgM specific to lung cancer were examined for growth in serum-free culture by using RDF medium supplemented with ITES.

Table 3.

REACTIVITIES OF MONOCLONAL ANTIBODIES TO VARIOUS CELL LINES BY INDIRECT IMMUNOFLUORESCENT ASSAY*

Cell line	HF10B4	HF10F7	HB4G7	HB4C5	H-7	H-9	Human Plasmin
Human cancer lines							
Lung, adenocarcinoma							
PC-8	+++	++	+++	+	-	-	-
PC-3	++	+	+	+	-	-	-
PC-9	-	-	-	-	-	-	-
PC-14	+	-	-	-	-	-	-
11LC-1-109	+	+	+	-	-	-	-
ABC-1	+	-	+	-	-	-	-
Lung, squamous carcinoma							
QG-56	-	-	-	-	-	-	-
PC-1	+	-	-	-	-	-	-
PC-10	-	-	-	-	-	-	-
EBC-1	+	-	+	-	-	-	-
Lung, small cell anaplastic carcinoma							
QG-90	-	-	-	-	-	-	-
SUC-1	++	+	++	+	-	-	-
Lung, large cell anaplastic carcinoma							
PC-13	+	+	+	-	-	-	-
Lung, oat cell carcinoma							
PC-6	+	-	+	-	-	-	-
Bladder							
KU-1	-	-	-	-	-	-	-
NBT-2	-	-	-	-	-	-	-
Breast							
ZR-75-1	-	-	-	-	-	-	-
ZR-75-30	-	-	-	-	-	-	-
Epidermoid							
A431	-	-	-	-	-	-	-
Renal							
KPK-1	-	-	-	-	-	-	-
KPP-1	-	-	-	-	-	-	-
Melanoma							
G361	-	-	-	-	-	-	-
Bowes	-	-	-	-	-	-	-
Stomach, adenocarcinoma							
MKN-20	-	-	-	-	-	-	-
MKN-45	+	+	+	-	-	-	-
Normal human cell lines							
Lung, fibroblast							
W1-38	-	-	-	-	-	-	-
Flow 2000	-	-	-	-	-	-	-
Foreskin, fibroblast							
Flow 7000	-	-	-	-	-	-	-
Intestine, fibroblast							
Flow 11 000	-	-	-	-	-	-	-

* The cells were cultured to confluency, fixed with 0.05% glutaraldehyde and used for the assay. +++; more than 70%, ++; 50-70%, +; 20-50% of the total cells were stained with fluorescent dye-conjugated anti-human IgM after the treatment with each MoAb. -; no anti-human IgM bound to the cells was microscopically detectable. Positiveness was easily discriminated as every positive cell lines, when stained with the dye-conjugated anti-human IgM, exhibited more than 20% of the population reactive with the antibody.

The doubling time of these hybridomas in the serum-free medium was almost twice that in serum-containing medium (Table 4). Long term cultivation of these cells was also successful in the serum-free medium without any loss of MoAb productivity. Specificity of MoAb secreted by the cells showed no difference in the serum-free and serum-containing conditions, indicating that productivity of the IgM by hybridoma cells depended on cell density but not on the presence of serum components. MoAb productivity by HF10F7 cells was the highest among the 4 clones, and reached 16 and 41 μg/ml after 3 day culture in the serum-free and serum-containing media, respectively. There was no difference between the two media in the specificity of secreted MoAb to the cell lines tasted. Purities of the MoAb were 20 and 0.2 % of the total protein in spent media of serum-free and serum-containing conditions, respectively.

2. Monoclonal Antibodies Specific to Breast Cancer

HO-323 cells were fused with lymphocytes obtained from regional lymph nodes of breast cancer. Spent media of the resulting hybridomas were screened for their reactivity to a breast cancer cell line, MCF-7 by enzyme-linked immunosorbent assay (ELISA). Among more than 40 hybridomas which secrete immunoglobulin reactive to MCF-7 cells, so far 11 hybridoma lines have been cloned. Reactivity of these MoAbs were tested with cell lines including 2 other breast cancers, 3 lung cancers, a stomach cancer, a bladder cancer, an epidermoid cancer, and a normal fibroblast line (Table 5). Some MoAb reacted strongly to normal fibroblast. Most MoAbs revealed distinguishable specificity to breast cancer cell lines. It is deduced from the specificity spectra that each MoAb recognizes different antigen determinants on the surface of breast cancer cells.

3. Specific Reactivity of Monoclonal Antibodies to Cancer Tissues

A monoclonal antibody of HF10B4 clone was tested for reactivity with lung adenocarcinoma tissues from patients. As shown in Table 6, this MoAb reacted with all seven adenocarcinoma tissues. Its reactivity against the normal

Table 4

GROWTH OF HYBRIDOMA CLONES IN SERUM-FREE MEDIUM*

Clone	Cell density at day 5 (cells/ml X 10^{-5})		Doubling time in logarithmic phase (hr)	
	serum-free	serum	serum free	serum
HF10B4	12	23	20	13
HF10F7	10	24	31	19
HB4G7	13	29	20	12
HB4C5	10	17	40	21

* Hybridoma clone cells were plated at a density of 10^5 cells per ml of serum-free or serum-containing medium (see text).

Table 5 Reactivity of Human Monoclonal Antibodies against Human Cell Lines by ELISA

Cell lines	Monoclonal antibodies from hybridomas										
	H15A3	H15F2	K15B4	K15C3	K27C3	K15H1	K15F1	H15H2	H15B1	H15F1	H15H1
Breast cancer											
MCF-7	++	++	+	+	++	+++	+++	++	+++	+++	+++
ZR75-1	-	-	-	-	-	+	+	-	+	+	+
HBC-5	+	+	-	-	+	++	++	+	++	+++	++
Lung cancer											
PC-8	-	-	-	-	+	++	+	+	+	++	+
QG-56	-	-	-	-	-	-	-	-	-	+	-
QG-90	-	-	-	-	-	-	-	-	-	-	-
Stomach cancer											
MKN-28	-	-	-	-	-	-	-	-	-	-	-
Bladder cancer											
KU-1	-	-	-	-	-	-	-	-	+	++	-
Epidermoid cancer											
A-431	-	-	-	-	+	+	+	+	-	+	-
Normal fibroblast											
Flow 11,000	-	-	-	-	-	+	+	+	+++	+++	++

tissues adjacent to the corresponding cancer tissues was also tested. Most of the normal tissues were negative. Although one normal tissue (patient 5) was judged positive, inclusion of cancer cells in this tissue was suspected. This MoAb did not react against normal tissues of kidney or liver, or erythrocytes of any blood type. More extensive testings of these MoAbs are under enforcement for the specificities to various cancer tissues.

V. CHARACTERIZATION OF CANCER SPECIFIC ANTIGENS

Any antigen can be easily isolated and purified by affinity chromatography using immobilized specific MoAbs. Molecular biological knowledge on antibody and antigen reaction is indispensable for further understanding of the immune system.

VI. MULTIFUNCTIONAL ANTIBODY

Though HO-323 is a non-secretor of immunogloublin, this cell line synthesizes a small amount of μ-chain inside the cell (3). By fusion of HO-323 and IgA secreting B-lymphocytes, some of the resulting hybridoma clones were analyzed to produce not only IgA but also IgM-like molecules(8). Whether μ-chain secreted by these hybridomas originates from that of HO-323 should be clarified by analyzing the difference of V (variable) region of α- and μ-chains of these cells. We deduce, however, that the μ-chains are derived from HO-323, since preliminary experiments demonstrated that IgM secreting hybridomas were obtained by fusing ouabain resistant HO-323 cells with IgA secreting Epstein-Barr virus (EBV) transformed lymphocytes which are not considered to cause any genomic class switching of immunogloblins. It is not expected that α- and μ-chains having the same V region are simultaneously synthesized from the gene(s) for heavy chains.

Molecular size of the IgM-like immunoglobulin secreted by these hybridomas was estimated gel-chromatographically to be the size corresponding to IgM. These results indicate that the IgM-like immunoglobulin is a pentamer of monomeric IgM's, the heavy chains of which are composed of both α and μ. IgM is composed of five monomeric IgM moieties covalently linked with a J chain peptide. Five monomeric IgM moieties and a J chain produced in the Golgi apparatus join to complete IgM in cytoplasm. These human-human hybridomas have potentiality to make monomeric IgM and IgA which can also combine with the J chain (9). We know that a hybrid molecule between single α- and μ-chain has not been occurred in these hybridomas. Considering that the molar ratio of a α- and μ-chain of the IgM-like immunoglobulin is fairly close to 2:3, the immunoglobulin is profiled as in Fig. 2. In this IgM-like immunoglobulin molecule, 2 of 5 monomeric IgM moieties are substituted for two IgA molecules. Therefore, this IgM-like immunoglobulin is symbolized by Ig (A_2M_3) (10). Molecular size of Ig (A_2M_3) still gel-chromatographically corresponds to that of IgM.

The IgM-like immunoglobulin has two types of light chains κ and λ. The κ-chain was originally made for IgM by HO-323 cells, and λ chain was made for IgA by B lymphocytes. However, we cannot exclude the possibility that the IgM moieties composed of λ-μ combination and IgA moiteties, κ-α in the IgM-like molecule, since there are both κ-and λ-chains in the cytoplasm of hybridomas. This experiment has revealed that IgM molecules with two kinds of monomeric IgM reacting with different antigen determinants can be made by fusion of two kinds of hybridomas secreting IgM and/or IgA whose specificities are different. By a combination of techniques hybridoma and gene technology, it is feasible to process a multifunctional IgM molecule in which each monomeric IgM moiety reacts with a different antigen determinant. These methods are applicable to the creation of macromolecular substances of which biological functions are enhanced or modified from those occurring in nature. Multifunctional IgM composed of human immunoglobulin components is one of examples of macromolecular drugs which neither occur in nature nor show any antigenicity.

VII. SERUM-FREE CULTURE OF HUMAN-HUMAN HYBRIDOMAS

We first developed a hormone- and growth factor-supplemented, serum-free medium for culturing mouse plasmacytomas and hybridoma cell lines (11,12). The medium was composed of an equal volume of Ham's F-12 medium and Dulbecco's modified Eagle's medium (FD) containing insulin, transferrin, ethanolamine, and selenite (ITES). Ethanolamine is the key compound for the successful growth of a variety of murine hybridoma cell lines in serum-free medium. Supplementation of FD containing 5 μg/ml insulin, 2-35 μg/ml transferrin and 10^{-9} M selenite with 20 uM ethanolamine stimulated cell growth of the hybridoma MPC-11 x BL to the same level as supplemenation with 200 μM phosphoethanolamine. This may imply that phosphoethanolamine is converted to ethanolamine before being taken up by the cells or that phosphoethanolamine is taken up by the cells much less efficiently.

This completely defined medium gave a doubling of the murine hybridoma cells every 20 hr in the logarithmic phase, compared to 19 hr in serum-containing medium. The same growth rate was maintained even after 4 weeks of cultivation in the defined medium, a period of more than 30 cell generations. Other hormones such as relaxin, ACTH, growth hormone and prolactin showed some stimulatory effect on the growth of the cells when each of them was added to serum-free FD-ITES. The ITES combination supported the serum-free growth of various types of murine hybridomas (Table 7). Growth rate in the serum-free FD-ITES medium sometimes differed remarkably among clones even if the origin of their malignant fusion partner cells was the same, as typically seen in the case of the MCB lines. Hybridoma cell lines made by the fusion with plasmacytoma NS-1 could not grow well in the serum-free-ITES medium. However, this difficulty has recently been overcome by the addition of 5 mM 2-mercaptoethanol to the serum-free medium. Stirred culture of these murine hybridomas was succesfully achieved in serum-free FD-ITES supplemented with 2-mercaptoethanol, soybean phospholipid, and ascorbic acid (13).
Some human hybridomas, however, did not grow sufficiently in serum-free

FD-ITES medium. Substitution of transferrin in ITES for lactoferrin (ILES) (14) effectuated the growth of a wide variety of human hybridomas. We noticed that growth profiles of mouse myeloma MPC 11 and other hybridomas was sometimes remarkably inhibited when serum-free FD was prepared by the conventional concentration. This hampered experimental reproducibility of cellular growth in serum-free media and was deduced due to an increased susceptibility of cells to osmolality in the serum-free media (15). This difficulty was easily compensated by increase in osmolality by adding such an inorgainc salt as sodium chloride. From the standpoint of view of general convenience, we recommend to prepare the serum-free basal media at a concentration that is 5% higher than the conventional formula (Fig. 3). It should be also realized that the serum-free FD does not contain enough kinds of components which are necessary to grow a wide variety of cells.

The growth rate and maximum cell density of the human hybridomas in ILES or ITES were further increased by substituting FD medium for eRDF medium (16), which is essentially a 1:1 mixture of RPMI 1640 and FD medium but with two-fold concentrations of amino acids and glucose, and an increased osmolality. Human-human hybridoma clones also proliferated and grew indefinitely in the same serum-free medium. MoAb production per cell showed no appreciable difference between serum-free and serum-containing conditions (17). This serum-free condition is a great advantage for the purification of MoAb (12).

VIII. A LARGE SCALE SERUM-FREE STIRRED CULTURE OF HUMAN-HUMAN HYBRIDOMAS

1. High Density Culture.

To increase the efficiency for producing monoclonal antibody in a large scale culture, is should be one of the requisites to cultivate hybridoma cells in high density. Cell density of hybridoma cells usually attains to 1-2 x 10^6/ml in a dish, though it depends on cell lines, media and culture condition.

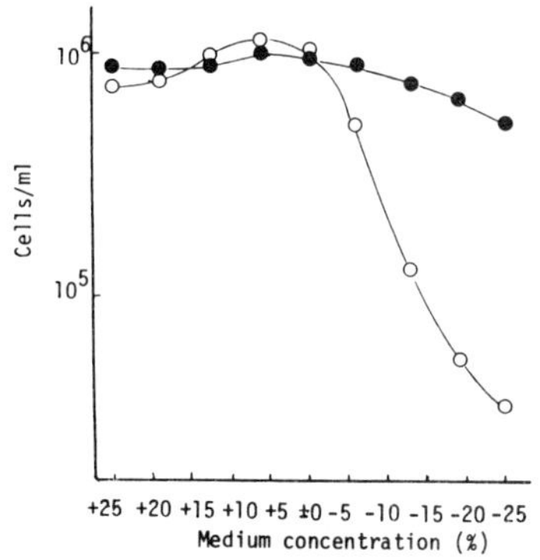

Fig. 3. Growth of MPC-11 cells in varying concentration of serum-free FD supplemented with transferrin, ethanolamine and selenite (○—○) or serum (●—●).

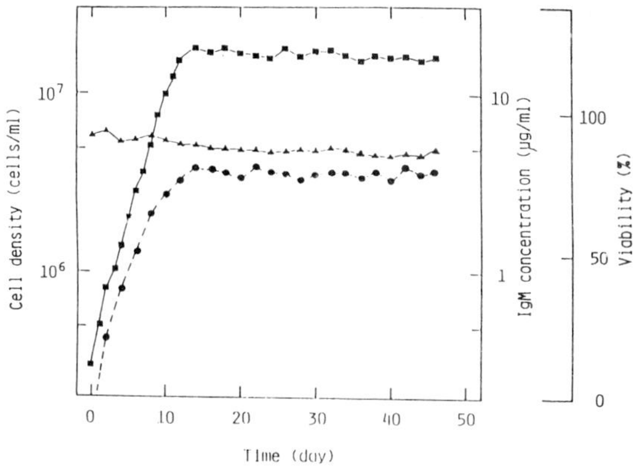

Fig. 4. Stirred culture of human hybridoma HF10B4 cells in ITES-lipoprotein medium. Total cell number: ■—■. Viability: ▲—▲. IgM production a day: ●--●.

To date, some hybridomas have been reported to grow at 2-5 x 10^7/ml in serum-free stirred culture and 10^8/ml in a hollow fiber system. Productivity of monoclonal antibody in these systems will be therefore enormously improved to 10-10^2 times more than that in dish culture, if the productivity of the immunoglobulin per cell is the same in the high density culture as in a culture dish.

Serum-free stirred culture of human hybridomas was performed in eRDF medium supplemented with ITES and lipoprotein. Human hybridoma HF10B4 cells in a culture vessel containing the serum-free medium grew exponentially and attained a stationary phase at which the cell density reached 2 x 10^7/ml. The number of cells which were more than 95% viable cells was maintained consistently for two weeks, and it was obvious that this stationary phase would continue even longer. Generally, when the cell density attained 6 x 10^6/ml, the oxygen gas concentration dissolved in the medium (DO) went down to a zero level even under conduction of oxygen gas into the medium. It has been shown that perfluorotributylamine solubilizes a large amount of oxygen (18). Emulsified perfluorotributylamine, Fluorosol-DA was added to the medium as an oxygen carrier. Addition of emulsified Fluorosol-DA maintained DO concentration at 4 ppm which allowed the cells to increase to 2 x 10^7 /ml (Fig. 4). In the same culture condition, serum medium could not grew the cells to such a high density as 2 x 10^7/ml, only giving 6 x 10^6/ml. Then, the viability was gradually decreased in spite of perfusion of the serum medium. These findings indicate that the serum-free medium is superior to serum medium for high density culture of the hybridoma cells Monoclonal IgM production by the hybridoma cells continued throughout the stationary phase at a rate of 80 μg/ml/day. Productivity of IgM per cell in the stationary phase was identical to that in the exponential phase. A similar relationship was also shown to hold for the producton of IgG by a mouse-mouse hybridoma and a mouse-human hybridoma (19). In the stirred culture, cell density was practically maintained at 10^7/ml throughout the period of culture. It is still unknown why the cells ceased the exponential growth after the cell density reached 10^7/ml in this stirred culture condition. Deposition of such toxic compounds as ammonium ion and peroxides produced by cellular metabolism may inhhibit further proliferation of cells. However, we anticipate that development of a more effective oxygen supply system may be essential for the increase of cell density to more than 10^7/ml. Monoclonal IgM secreted by the human hybridoma cells into the 350 ml spent medium in the vessel was 28 mg/day. From this result, 960 mg IgM could be produced by a liter culture vessel in 12 days, as shown in Table 8. On the other hand, the conventional quiescent cell culture using dishes containing the same volume of medium was calculated to deposit only 60 mg of immunogloblin in the same period, and 10^5/ml cells had to be subcultured 3 times when the cell density attained to 1.5 x 10^6/ml every 4 days. There was no difference in the productivity of MoAbs per cell per day between the stirred culture and the quiescent culture systems. MoAbs in the spent medium continuously provided by the bio-reactor system will be readily purified to an extreme purity by using a column filled with immobilized anti-human immunoglobulin, since the serum-free medium has only a small amount of proteinous macromolecules.

2. Concept for Culture and Control System

Mammalian cells usually double every ca. 20 hr. However, bacterial cells multiply every 20-30 min. This has been thought as a main reason that mammalian cells are inefficient to bacteria for the production of bio-macromolecules derived from mammalian cells. Therefore, as far as we expect to obtain any cellular product in the growth phase of cells, we will not be able to make it efficiently, since average cell density during that phase can not be high enough, even if the final cell density reaches 10^8/ml. It is therefore desirable to make the condition in which the production of bio-macromolecules is not linked to cell multiplication. In broad outline, cells should be maintained in high density without multiplication as long as possible, and keep production. In this case, productivity of bio-macromolecule becomes independent of growth rate of the cells and hence is shown as a

Table 6

REACTIVITIES OF MONOCLONAL ANTIBODY FROM HYBRIDOMA CLONE, HF10B4, TO VARIOUS TISSUES AND ERYTHROCYTES BY INDIRECT IMMUNOFLUORESCENT ASSAY[a]

Sample			Reactivity
Lung adenocarcinoma	Patient	1	+
		2	+
		3	+
		4	+
		5	+
		6	+
		7	+
Normal lung[b]	Patient	1	-
		2	-
		3	-
		4	-
		5	+
		6	-
		7	-
Normal kidney[c]	Patient	8	-
Normal liver[c]		8	-
Erythrocyte blood type			
A	normal Adult	1	-
B		2	-
AB		3	-
O		4	-

a Tissues were sliced with a thickness of 10 um on slide glass, fixed with 10% formaldehyde and used for the assay (5). Erythrocytes were sedimented by centrifugation, fixed with 0.05% glutaraldehyde and used for the assay. +; more than 20% of cells in the tissue was stained with the dye-conjugated anti-human IgM after the treatment with MoAb from HF10B4 cells. -; IgM was detectable on the tissue or cells.

b Histologically normal tissues adjacent to the lung cancer of the same patient whose cancer tissues were assayed in a.

c From patients bearing stomach cancer.

Table 7 Growth of hybridoma lines and hybrid cells between a plasmacytoma and B-lymphoma lines in ITES medium.

Cell lines	Plasmacytoma	Lymphocytes	Cells per dish x 10^{-4} in ITES	in Serum	Growth ratio(%)
Hybridomas					
MPC11xBL	MPC11TG1.7 Oua3	BALB/c (spleen cells)	90	140	64
RDP 45/20	MPC11RG1.7 Oua3	BALB/c (spleen cells)	136	184	74
SPS.6	SP2/0	BALB/c (spleen cells)	66	87	76
AS100-2	SP2/0-Ag 14	BALB/c (spleen cells)	23	16	144
MCB1	MPC11TG1.7 Oua3	(C57B1/6xBALB/c)F (spleen cells)	76	120	63
MCB4	MPC11TG1.7 Oua3	(C57B1/6xBALB/c)F (spleen cells)	18	100	18
MCB6	MPC11TG1.7 Oua3	(C57B1/6xBALB/c)F (spleen cells)	124	144	86
MCB17	MPC11TG1.7 Oua3	(C57B1/6xBALB/c)F (spleen cells)	14	40	35
MCB18	MPC11TG1.7 Oua3	(C57B1/6xBALB/c)F (spleen cells)	18	92	20
MB4	MPC11TG1.7 Oua3	C57B1/6 (spleen cells)	37	68	54
MB8	MPC11TG1.7 Oua3	C57B1/6 (spleen cells)	13	38	34
N-S.1	NS-1	BALB/c (spleen cells)	52	162	32
N-S.3	NS-1	BALB/c (spleen cells)	16	142	11
53-7.313	NS-1	Rat (spleen cells)	13	64	20
T11D7e2	NS-1	Rat (spleen cells)	18	82	12
ROB57.2.1	S194	Rat (spleen cells)	82	48	171
ABM202.5	S194	Rat (spleen cells)	132	79	167
Hybrid cells					
MPC11xW279.1	MPC11TG1.7 Oua3	WEHI 279 (B lymphoma)	44	96	46
MPC11xW279.2	MPC11TG1.7 Oua3	WEHI 279 (B lymphoma)	108	140	77
MPC11xW231.1a	MPC11TG1.7 Oua3	WEHI 231 (B lymphoma)	72	88	82
MPC11xW231-2	MPC11TG1.7 Oua3	WEHI 231 (B lymphoma)	50	72	69

Inoculum of each cell line in a dish was 4 x 10^4.
Cell numbers in a log phase were counted at day 4.
Growth ratio indicates % growth of cells in ITES medium over that in serum medium.

function of productivity per cell. That is to say, in the system to be adopted for the production of bio-macromolecules by mammalian cells in future, we do not expect production in cellular growth phase but in the stationary phase which can be maintained for a long time enough. This concept is depicted in Fig. 5. The cells are cultured in the physiological condition which exclusively stimulates their multiplication in the exponentially growing phase, and then maintained in the physiological environment which is the fittest for the production by shifting the constituents of the serum-free medium and atmospheric gases. In perfusion culture system, the cells grow exponentially under supplementation with nutrients, growth factors and oxygen and then attain to the stationary phase. In the stationary phase, the cell cycle rotates so slowly that dividing time of the cells, which is ca. 20 hr e.g., will be prolonged to 100 hr or even longer. Productivity of biomolecules in a unit time become constant and the situation is maintained for many months. Such toxic metabolites as peroxides and ammonium ion or lactic acid will be kept in a low level. Such high density, serum-free perfusion cultures of hybridoma cell lines have been shown by us (18, 19) and also in Fig. 4.

3. Stimulation of Immunoglobulin Secretion.

By employing the system in which mammalian cells are kept in a high density for long period, secretory proteins can be produced continuously in large quantity. There is another way to increase productivity of the cellular products. That is to disarm the feed-back controle system of cells for the secretion of protein molecule. A mouse myeloma MPC-11 has been known to synthesize 5 pg IgG/min in a cell (20). However, the cells actually deposit only 10-20 μg/ml of the protein when cultured in a dish either in serum or serum-free medium. This indicates that such a high velocity for protein synthesis by the cells will be ceased in a short time and not be kept for sufficient period. Let us suppose if such a rapid protein synthesizing reaction could be kept without any inhibition in the MPC-11 cells cultivated in a one litter vessel at the density of 10^8 cells/ml. Then, the total amount of

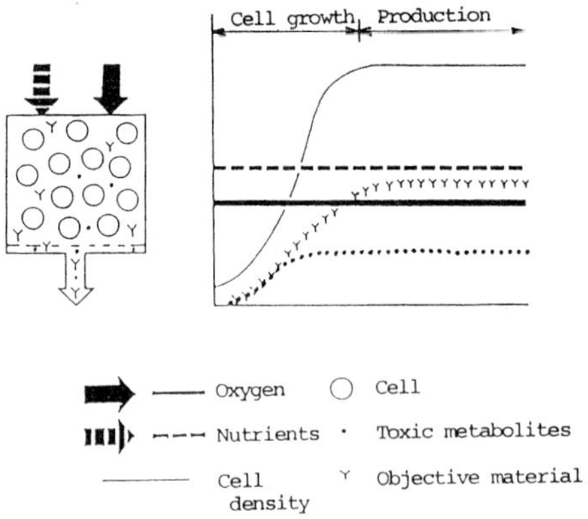

Fig. 5. Concept of perfused high density cell culture.

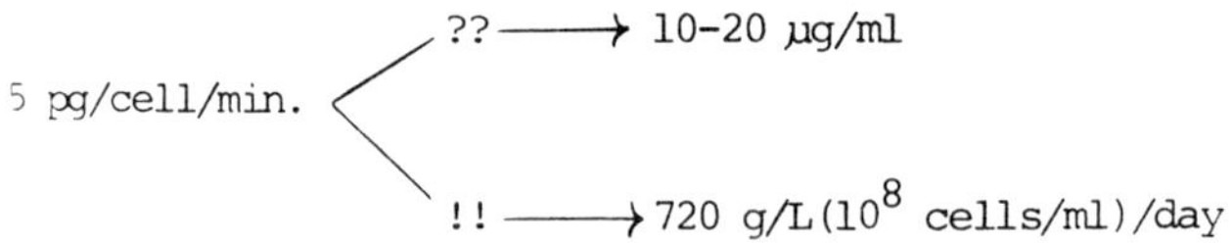

Fig. 6. Putative productivity of IgG by myeloma MPC-11.

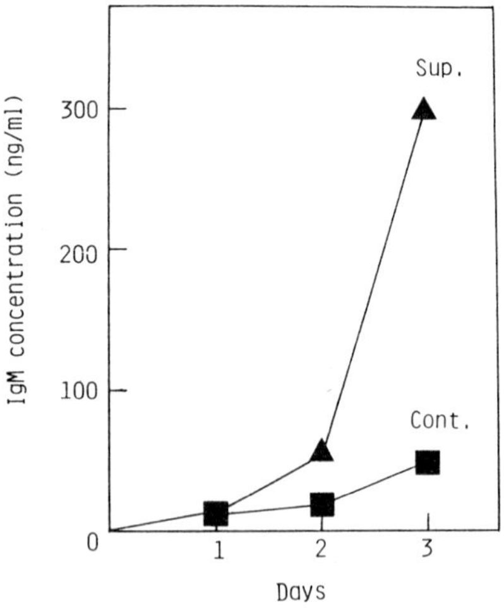

Fig. 7. Culture medium of Namalwa cells contains secretion stimulating fraction (SSF). Secretion of IgM by human hybridoma cells HB4C5 was enhanced 8 times after a 3 day-culture in the presence of SSF.

IgG to be accumulated would be culculated 720 g a day, as shown in Fig. 6. Why such a enormous discrepancy is caused between the actual IgG accumulation in the medium and the calculated value. One of the reason which can explain this discrepancy is that the protein synthesis and secretion by the cells are usually under the feed-back regulations. Cancellation of such regulations will result in the enhancement of production of secretory proteins by the cells.

To this end, we tried to find out secretion stimulating factor (SSF) to enhance the secretion of MoAb from human hybridomas. Proteins synthesized in the cells are secreted by such complex process as shown in Fig. 7, though the mechanism has not been fully understood. A protein extacted from a human lymphoblastoid cell line, and having molecular weight 40 K was so active to stimulate secretion of MoAb by a human hybridoma line. By addition of a few µg of the protein, secretion of a lung cancer specific MoAb by a human hybridoma line HB4C5 was enhanced almost 8 times (Fig. 8).

Mammalian cells may have potentiality to produce enough amount of proteins without applying genetic engineering. Such techniques as introduction of enhancers and gene multiplication have been applied to increase the productivity of secretory proteins by mammalian cells. Even if productivity is accelerated inside the cells by the gene manipulations, it would be soon repressed unless the protein can be secreted. Application of the gene engineering should be rather concentrated to improvement of secretory function of cells. One should remind that Penicillium chrysogenum conventionally used produces several hundred thousand units penicillin into the broth by the improvement on genetic level and of physiological environment, though wild type counterparts made only a few units of the antibiotic. Studies on gene manipulation and physiological environments for mammalian cells will give much outcome to increase the productivity of secretory proteins, and in consequence verify the feasibility of mammalian cell culture for producing wider variety of bio-macromolecules.

Table 8 Productivity of IgM by hybridoma HF10B4 cells in serum-free stirred culture.

Culture system	Final cell density/ml	IgM production mg/vessel(L)	IgM production ug/10^6 cells/day
Stirred	1.0×10^7	960 system(12 days)	8
Quiescent	1.5×10^6	60 3 cycles(12 days)	8

SUMMARY

During the last few years human hybridoma technology has been improved significantly. Production of human MoAbs by this procedure has been reported from several laboratories. However, the production of human hybridomas is still less efficient than rodent hybridoma production. This hampers the effective selection of antigen specific MoAbs among a large number of human hybridoma clones. HO-323, a new fusion partner cell line has fusion efficiency comparable with that of murine counterparts. Therefore, it is now possible to obtain a wide variety of antigen specific MoAbs of human type by fusion of human B-lymphocytes with HO-323 cells. For medical use of the antigen specific MoAbs, the MoAb will be used as a drug carrier which delivers drugs to specific target cells such as cancer cells and infectious pathogens. This drug delivery system will permit not only concentration of a drug on its target but will also utilze some drugs which have not been employed due to the in serious side effects when administered by conventional methods.

Multifunctional MoAbs are artificially engineered antibodies which bind to multiple antigen detreminants and can be drug carriers superior to naturally occurring MoAbs. The characteristics of human MoAbs as drug carries make them very attractive for use for medical purposes.

It is indispensable to establish the efficient culture system providing a large amount of such medically useful bio-macromolecules as human MoAbs. Development of serum-free medium has been highly recomended, since it enables cells to express their specific function and purification of cellular products is far simplified. The serum-free medium we composed has revealed validity for culturing human hybridomas in high density in perfusion culture. Then, we have proposed a new concept for culture system to obtain efficiently bio-macromolecules which were secreted from cells. That is not to expect production of bio-macromolecules during cell growth phase, but to use the stationary phase maintained for long period. Therefore, effort should be concentrated to the researches dealing with maintainance of the stationary phase by delaying cell-cycle and of producibility of bio-macromolecules in high density culture. This concept shows that bio-macromolecules secreted by cells can be continuously obtained in large quantity with automated operation.

This is a new starting point for mammalian cellular physiology in artificial vessels in which we want to modulate cell functions toward our porposes, and this new physiology will be established by fusion of cell physiology and chemical and mechanical engineerings.

REFERENCES

1. Köhler,G. and Milstein,C., Nature, 256, 495 (1975)
2. Olsson,L. and Brams,P. in Human Hybridomas and Monoclonal Antibodies (ed. by Engleman,E.G., Foung,S.K.H., Larrick,J. and Raubitschek,A.) Pp.227. Phenum Press, New York and London (1985)
3. Ohashi,H., Hashizume,S., Murakami,H., Aihara,K., Shinohara, K. and Omura,H., Cell Biol. Intern. Reports, 10, 77 (1986)
4. Hashizume,S., Kuroda,K. and Murakami,H., Biochim. Biophys. Acta, 763, 377 (1983)
5. Murakami,H., in Biotechnology of Mammalian Cells, Development for the production of Biologicals. (ed. by Umeda,H., Koyama,H., Minowada,J. and Oishi,M.) Japan Scientific Societies Press/Springer-Verlag, Tokyo-Berlin, pp.119 (1987).
6. Glassy,M.C., Handley,H.H., Hagiwara,H. and Royston,I. Proc. Natl. Acad. Sci, USA. 86, 6327 (1983)
7. Houghton,A.N., Brooks,H., Cote,R.J., Taormina,M.C., Oettgen,H.F. and Old,L.J. J.Exp.Med, 158, 53 (1983)
8. Shinmoto,H., Murakami,H., Dosako,S., Shinohara,K. and Omura,H., Agric. Biol. Chem. 50, 2217 (1986)
9. Koshland,M.E. Adv. Immud., 20, 41 (1975)
10. Murakami,H., Shinmoto,H., Dosako,S., Yamada,K. and Omura,H., In Vitro Cell.

Developmental Biol. in press.

11. Murakami,H., Masui,H., Sato,G.H. and Raschke,W.C., Anal. Biochem., 114, 422 (1981)

12. Murakami,H., Masui,H., Sato,G,H., Sueoka,N., Chow,T.P. and Kano-Sueoka,T. Proc. Natl. Acad. Sci. USA, 79, 1158 (1982)

13. Murakami,H., Masui,H. and Sato,G.H., in Growth of Cells in Hormonally Defined Media, 9, (ed. by Sato, G.H., Pardee, A.B. and Sirbasku, D.A.) Cold Spring Harbor Laboratory, New York, pp. 711 (1982)

14. Hashizume,S., Murakami,H., Kamei,M., Hirose,S., Shirai,T., Yamada,K. and H.Omura. In Vitro Cell. & Developmental Biol. 23, 53 (1987)

15. Murakami,H., in Methods for Serum-Free Culture of Neuronal and Lymphoid Cells, (ed. by Barnes, D.W., Sirbasku,D.A. and Sato,G.H.) Alan R. Liss, Inc., New York, pp. 197 (1984)

16. Murakami,H., Shimomura,T., Nakamura,T., Ohashi,H., Shinohara,K. and Omura,H. J. Agric. Chem Soc. JAPAN. 58, 575 (1984)

17. Takazawa,Y., Murakami,H., Yamada,K. and Omura,H., Biotechnol. Bioeng. (1987) in press.

18. Murakami,H., Shimomura,T., Ohashi,H., Hashizume,S., Tokashiki,M., Shinohara,K., Yasumoto,K., Nomoto,K. and Omura,H., in Procedings of the Intern Symposium, Growth and Differentiation of Cells in Defined Emviroment pp111, (ed. by Murakami,H., Yamane,I., Barnes,D.W. Mather,J.P., Hayashi,I. and Sato,G.H), Kodansha/Springer-Verlag Tokyo, Berlin (1985)

19. Hamamoto,K., Tokashiki,M., Ichikawa,Y. and Murakami,H. Agric. Biol. Chem. in press.(1987)

20. Laskov,R. and Scharff,M.D., J. Exp. Med., 131, 515 (1970)

Discussion

J. FEDER — Are you harvesting cells every day and what is the perfusion rate?

MURAKAMI — We do harvest spent medium every day

FEDER — So you do have a population which is being maintained in the steady state with an equilibrium between cells that are dying or cells that are growing

MURAKAMI — Yes, in one medium we always find 3-4% of dead cells

FEDER — Do you find the same percentages for the entire 50 day period or do you find that as you are not harvesting cells the percentage of dead cells increases?

MURAKAMI — Cells divide every 100 hours but we still only find 4-5% dead cells.

5 Establishment of a human hybridoma secreting monoclonal antibodies toward rabies virus

Y. Gozes, R. Oren, R. Monzain, S. Lerer, O. Galan, H. Ben-Moshe
N. Epstein and D. Kobiler

ABSTRACT

An hypoxanthine-azaserine (HA) sensitive mutant, derived from the human lymphoblastoid B cell line, Namalva, was selected by growing the cells in the presence of 8-azaguanine (30 μg/ml). The 8 azaguanine resistant cells were cloned, and the appropriate line for hybridoma construction was selected by fusion with stimulated B lymphocytes from peripheral blood. To establish a hybridoma secreting monoclonal antibodies toward rabies virus, 100 ml heparinized blood were drawn from a volunteer immunized against the virus. The lymphocytes used for fusion were separated by Ficoll-Hypaque and cultured in vitro for three days, in the presence of lypopolysaccharide (5 μg/ml) and pokeweed mitogen (1 μg/ml). After incubation, the cells were washed and fused with an equivalent number of cells from the HA-sensitive Namalva-derived mutant. Hybrid colonies, growing in the selective HA-medium, were tested by ELISA for secretion of immunoglobulins and specific antibodies against the rabies virus. Colonies showing positive activity were cloned, and the resulting clones were expanded and injected to produce ascitic fluids in nude mice. The collected ascitic fluids contained human monoclonal antibodies, which were reactive with rabies virus (strain CVS) in ELISA and had neutralizing activity in an in vitro assay.

INTRODUCTION

The technology for the production of murine monoclonal antibodies has been extensively refined and exploited since its introduction in 1975 (1). However, in spite of its obvious advantages, the generation of human monoclonal antibodies (McAbs) is limited in scope, although many attempts have been reported (2). Several factors account for the unsuccessful results: lack of an adequate myeloma cell line suitable for fusion; unsuitable preparation of the B cell prior to the fusion; hybrid instability; mycoplasmma contamination of parent lines; low efficiency in production of McAb (3). The use of other B cell lines, in addition to the human myeloma, also met with limited success. These attempts include the construction of murine-human hybridomas (4,5), establishment of B cell lines by direct Epstein Barr Virus (EBV) transformation of normal human B cells (6,7), use of human lymphoblastoid B cell lines as the tumoral partner for fusion with normal B cells (8,9) and any combination of the above (10).

We have isolated a subclone of a human lymphoblastoid cell line, Namalva, able to fuse efficiently with normal B cells. Herein, we describe the use of this line for the establishment of hybridomas secreting neutralizing human McAbs toward the rabies virus.

MATERIALS AND METHODS

Parental cell line - Namalva is a human lymphoblastoid B cell line established by Klein et al. (11). By culturing these cells in the presence of 8-azaguanine (30 μg/ml), we isolated resistant mutants which are HA sensitive. After cloning, the resulting lines were tested for fusion efficiency. Namalva^R is one of those lines, demonstrating high fusing capacity and a doubling time of 20 hours in cultures grown in RPMI 1640 medium supplemented with 15% fetal bovine serum (FBS).

Preparation of B cells - A volunteer was vaccinated subcutaneously with Merieux inactivated rabies virus (Wistar rabies strain, PM/WI38; Institute Merieux, Lyon, France). After one month interval, the injection was repeated and three weeks later, peripheral blood (100 ml) was drawn in heparin (10U/ml). Lymphocytes were separated by centifugation on a Ficoll-Hypaque density gradient. The cells were washed three times with PBS and cultured at a density of 0.5×10^6 cells/ml in RPMI 1640 supplemented with 15% FBS, 1 μg/ml pokeweed mitogen (PWM; Sigma), 5 μg/ml bacterial lipopolysaccharide (LPS, E. coli 0128: B12; Difco Laboratories, Detroit, Michigan). After 3 days, the cells were washed and used for fusion.

Cell fusion: Cell cultures were harvested and washed three times with RPMI 1640 medium (without serum). The cultured lymphocytes were fused with the Namalva^R cells (at 1:1 ratio) using the 25% polyethylene glycol, 8 minutes method (9). Lymphocytes and parental cells, suspended in serum free RPMI 1640 medium were mixed and sedimented by centrifugation at 350 x G for 5 min at 22°C. The supernatant was decanted leaving about 1 ml in the centrifuge tube. The cells were gently resuspended in this solution. One milliliter 50% of PEG solution was added. The tube was agitated gently, and centrifuged for 3 minutes at 500 x G. Eight minutes after the addition of the PEG (Serva 1500MW), 1 ml of serum free RPMI 1640 was added to the tube for dilution of the PEG. The cell sediment was not dispersed. Immediately thereafter 20 ml of serum free RPMI 1640 were added gently, by dripping it down the side of the tube with gentle agitation without distruption of the sediment. Finally, the tube was again centrifuged for 5 min at 350 x G. After fusion, the cells were suspended carefully in 20 ml of complete tissue culture medium (RPMI 1640, 15% FBS, penicillin 100 μ/ml, streptomycin 100 μg/ml, 2 mM glutamine) and incubated under 92.5% air/7.5 CO_2 in a T-75 tissue culture flask overnight at 37°C. The following day, cells were harvested by centrifugation, resuspended gently in complete tissue culture medium containing hypoxantine (100 μM), azaserine (4 μg/ml) and plated at a density of 10^5 cells per microtiter well for hybrid selection. Cultures were subsequently fed every 3 days with selective medium. Hybrid colonies appeared after 12-16 days. Supernatants of each well were assayed for Immunoglobulins (Ig) secretion towards rabies virus. Positive colonies were cloned by the limiting dilution method and the relevant clones were expanded and further characterized. Hybridoma cells seeded at 2×10^5 cells/ml in complete medium were cultured for 72 hours; and the supernatant, obtained after sedimentation of the cells by centrifugation, was tested for Ig content or anti rabies activity.

Rabies virus ELISA. The secretion of antibodies towards the virus was assayed according to the technique established by Katz et al. (12).

Ig secretion ELISA. Secretion of Ig was estimated according to the technique described previously (13).

Ascitic fluid. Nude mice, 3 months old, were injected with a 0.5 ml pristane intraperitoneally (i.p). Two weeks later 2 x 10^7 hybridoma cells were injected i.p. Ascites developed after 4 weeks. The ascitic fluid obtained was centrifuged and the supernatant was kept for evaluation.

In vitro assay for neutralizing activity of antibodies

Monolayers of Vero cells were prepared by seeding the cells on cover glasses and culturing them for 24 hours. The samples to be tested for content of neutralizing antibodies, were prepared at appropriate dilutions in Dulbecco medium supplemented with 2% FBS. Half of the diluted sample were incubated with 0.1 ml (~10^3 infection units) of the challenging rabies virus (strain CVS) for 1 hour at 37°C.

After incubation, 0.1 ml of the mixtures as well as the virus controls were inoculated on the monolayers of Vero cells on the cover glasses (placed in tissue culture plates) and incubated for 1 hour at 37°C. Two ml of Dulbecco medium supplemented with 5% FBS were added and the infected cultures were incubated for 3 days at 37°C. The detection of viral antigens on the infected cells was done by immuno-flurescence technique. The infected cultures were fixed in acetone, cooled at -20°C for 1 hour and stained with rabbit anti-rabies nucleo-capsid antibodies conjugated with fluorescein. The degree of staining (infection) was evaluated by fluorescent microscopy.

Preparation of Jurkat supernatant - Jurkat cells (10^6 cells/ml) were cultured for 24 hours in complete medium in the presence of 0.2 μ/ml phytohemaglutinin (PHA) (Difco) and 20 ng/ml phorbol 12-myristate-13-

acetate (PMA, Sigma). The cells were sedimented by centrifugation and the supernatant was stored at -20°C until used.

RESULTS

Development of Namalva^R mutant - Namalva cells were grown in complete RPMI medium with 30 μg/ml 8-azaguanine. Most of the cells died within 10 days and the spontaneous resistant mutants could be seen after 4 weeks. The mutants were cloned and the resulting lines were tested for their characteristic growth patterns-8 azaguanine resistance and HA sensitivity. The growth patterns of the Namalva^R cells in comparison with original line (Namalva^S) is seen in Fig. 1.

Screening for anti rabies virus antibodies. Namalva^R cells were fused with human B cells from peripheral blood of a volunteer, vaccinated against rabies virus. Thirty days after fusion, each well was tested for the content of anti rabies virus antibodies. The results of this assay are presented in Fig. 2. Serum of the immunized volunteer was used as positive control. All the colonies showing activity above the negative control value (parental supernatant) were transferred to 24 well plates. After 10 days, the growing colonies were tested for secretion of antibodies in ELISA and the seven cultures showing positive values were cloned by limiting dilution. All the resulting lines were found to secrete IgM. The supernatants of the monoclonal hybridomas were tested for IgM content (Table I).

Enhancement of antibody secretion in vitro. As can be seen in Table I, the rate of secretion of IgM from the established hybridomas was very low, the maximum being 700 ng/ml/10^6 cells/24 h. We tried to increase the IgM secretion by these hybridomas with supernatants of the Jurkat cell line. The Jurkat cell line secretes IL-2 after PHA and PMA activiation (14). It was reported earlier that Interleukin-2 (IL-2) can thus increase antibody secretion by human B cells. We tested the effect of supernatants of activated Jurkat cells on the secretion of antibodies by the two hybridomas A8 and G1 (Table II). We also tested PHA and PMA, which were used to induce IL-2 secretion by Jurkat cells. It can be seen that the IgM secretion was increased 10-50 folds and this increment was probably induced by IL-2.

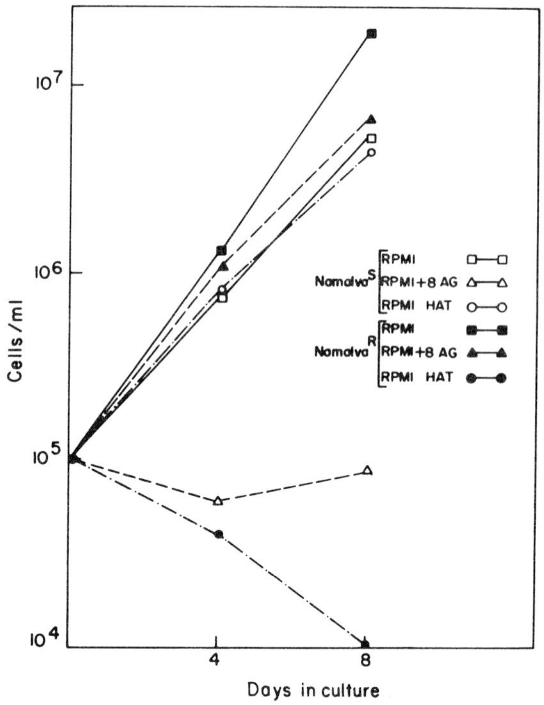

Fig. 1.: Growth of Namalva cells in different media.
Development of a HAT sensitive mutant of the Namalva cell line. The mutant line is obtained by growing the cells in the presence of 8 azaguanine. The differential growth patterns of the parental cell line ($Namalva^S$) and the mutant line ($Namalva^R$) can be seen.

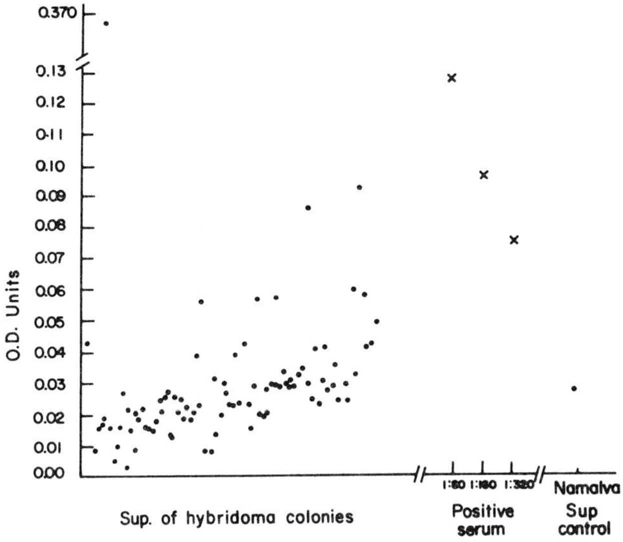

Fig. 2.: Preliminary screening after fusion of colonies secreting Ig towards rabies virus.

Table I

Concentration of IgM in the supernatant of the cloned hybridomas

Hybridoma designation	IgM ng/ml
A8	600
G1	80
F6	550
H10	700
C7	250
A11	450
F10	270

Table II.

Enhancement of IgM secretion* by the addition of Jurkat - derived supernatant** and PMA

Effector added	Hybridoma A8 O.D. units (increment factor)	Hybridoma G1 O.D. units (increment factor)
-	0.120	0.021
Jurkat 5%	1.284 (10.70)	1.166 (55)
" 10%	1.198 (9.99)	1.137 (54)
" 15%	0.947 (7.90)	1.192 (55.7)
" 25%	0.819 (6.80)	0.377 (17.9)
" 35%	0.830 (6.90)	0.654 (31.1)
40%	0.657 (5.47)	0.669 (31.9)
" 50%	0.504 (4.20)	0.502 (23.9)
PHA (0.1 ng/ml)	0.201 (1.60)	0.150 (7.14)
PMA (2 ng/ml)	0.212 (2.35)	0.153 (7.29)
" (10 ng/ml)	0.222 (2.69)	0.201 (9.60)
" (20 ng/ml)	0.510 (4.25)	0.333 (15.90)

* IgM content was measured in supernatants of $2x10^5$ cells/ml, after incubation with the effector for 72 hours. IgM determination was done by ELISA.

** Jurkat supernatant was obtained by activation of these cells with PHA - presumably resulting in the secretion of IL-2.

Table III

McAb towards rabies virus (CVS) determined in ELISA and in an in-vitro neutralization assay

Sample*	ELISA (O.D. Units)	Neutralization** activity
G1 A.F. (1:10;4μg/ml)	1.069	++
G1 A.F. (1:100;0.4μg/ml)	0.312	N.D
G1 Supernatant	0.224	±
A8 Supernatant	0.208	±
Negative control (AF 1:10)	0.092	-
Namalva Supernatant	0.077	-
Human immune serum (1:100)	Over	++

* A.F. - Ascitic fluid (40μg/ml IgM); Supernatant - in-vitro cell culture super natant.

** Neutralization activity: ++ = total neutralization; ± = partial neutralization; - = no activity detected. N.D. - not done.

Ascitic fluids from nude mice. In order to obtain higher production of antibodies, we injected the G1 hybridoma cells into nude mice. The concentration of human IgM was in the ascitic fluid produced in the nude mice was 40μg/ml, showing concordant binding activity in ELISA.

Neutralizing activity. Using an in vitro assay to determine the neutralizing activity of the anti rabies antibodies, we could demonstrate that the G1 hybridoma secretes antibodies which can inhibit the infection of cells by the CVS rabies virus. The degree of neutralizing activity depends on the concentration of antibodies in a given sample (Table III).

DISCUSSION

Several investigators have shown that EBV-transformed B cell lines can be used to produce human monoclonal antibodies (9, 10). Although human myeloma lines are available, they have not been as useful in producing hybridomas as EBV-infected B-cell lines. Human myeloma cell lines fuse with a lower frequency, grow more slowly and are unstable producers of Ig (8).

We have established a lymphoblastoid B cell line - Namalva^R which can serve as an efficient partner in the production of human hybridomas, secreting specific monoclonal antibodies.

Stimulation of peripheral blood mononuclear cells with PWM prior to fusion has been reported to increase the frequency of B lymphocyte fusion by 10 to 100 fold (16). This increase, probably due to the induction of lymphoblasts, facilitates the production of relevant hybridomas even when using a limited number of mononuclear cells derived from 100 ml blood.

The established hybridomas secreted only IgM antibodies. This phenomenon could be explained by the blood donor's stage of immunization or by the nature of the parental cell Namalva.

The concentration of antibodies secreted by the established hybridomas was low both in vitro (maximum 600 ng/ml/10^6 cells/24 hr) and in vivo in nude mice, (40 μg/ml). These concentrations of immunoglobulins are very low compared to the murine system (100 μg in vitro, 10 mg in vivo) respectively. The maximal concentration reported so far for in vitro secretion of antibodies from human hybridomas is 15 μg/ml (17), and in nude mice the maximum concentration is 1.7 mg/ml (18).

The enhancement of antibodies secretion induced by the Jurkat supernatants was remarkable, reaching almost maximal values reported for human hybridomas. It should be noted that in comparison to murine hybridomas, this concentration is still low, and further improvement in secretion efficiency is required for any practical purposes.

The neutralization activity of human monoclonal antibodies against rabies and other viruses and toxins (2,8), will make it possible to use antibodies when murine antibodies are either ineffective or immunogenic. Passively administrated human monoclonal antibodies can be used to treat hemolytic disease of newborns (Rh), tetanus, rabies, hepatitis B, gram negative sepsis, snake bites and tumor antigens, without causing immunogenic priming (2,8).

ACKNOWLEDGEMENTS

We thank Mrs. Nita Ben-David for her assistance in the preparation of this manuscript.

REFERENCES

1. Kohler, G. and Milstein, C. (1975). Nature 256, 495.

2. Carson, D.A. and Freimark, B.D. (1986). Advances in Immunology 38, 275.

3. Sikora, K., Alderson, T., Phillips, J. and Watson, J.V. (1982). Lancet 1, 11.

4. Levy, R. and Dilley, J. (1978). Proc. Natl. Acad. Sci. 75, 2411.

5. Butler, J.L., Lane, H.C. and Fauci, A.S. (1983). J. Immunol. 130, 165.

6. Steinitz, M., Klein, G., Koskimies, S. and Makela, O. (1977). Nature 269, 420.

7. Steinitz, M. and Tamir, S. (1982). Eur. J. Immunol. 12, 126.

8. Larrick, J.W., Kenneth, E.T., Raubitschek, A.A., Senyk, G. and Wang, J.C.N. (1983). Proc. Natl. Acad. Sci. 80, 6376.

9. Strike, L.A., Devens, B.H. and Lundak, R.L. (1984). J. Immunol. 132, 1798.

10. Kozbor, D., Lagarde, A.E. and Roder, J.C. (1982). Proc. Natl. Acad. Sci. 79, 6651.

11. Klein, J., Dombos, L., Bothskar, B. (1972). Int. J. Cancer 10, 44.

12. Katz, D., Ben-Moshe, H., Galan, U., Oren, R., Shafferman, A. and Fuchs, P. (1983). Refuah Veterinarith 40, 120.

13. Gozes, Y., Epstein, N., Halmann, M., Bino, T., Rosenberg, H. and Kobiler, D. (1987). Dev. Biol. Stand. (In press).

14. Kishi, H., Inui, S., Muraguchi, A., Hirano, T., Yamamura, Y. and Kishimoto, T. (1985). J. Immunol. 134, 3104.

15. Gillis, S. and Watson, J. (1980). J. Exp. Med. 152, 1709.

16. Kozbor, D. and Roder, J.C. (1984). Eur. J. Immunol. 14, 23.

17. Isenberg, D.A., Shoenfeld, Y., Madaio, M.P., Rauch, J., Reichlin, M., Stollar, B. and Schwartz, R. S. (1984). Lancet 2, 417.

18. Abrams, D.G., Ochs, J.J., Giandino, S.L., Morgan, A.C., Wilburn, S.B., Witt, A.R., Oldham, R.K. and Foon, K.A. (1984). J. Immunol. 132, 1611.

6 The production of bovine and ovine monoclonal antibodies to gonadal steroids: the use of interspecific hybridomas

D. J. Groves, J. Clayton and B. A. Morris

ABSTRACT

As part of a programme of research into the immunological control of reproduction in sheep and cattle, a straightforward and reproducible protocol for the generation of heterohybridomas secreting ovine and bovine monoclonal antibodies has been developed. Initial attempts at direct fusion of mouse myeloma cells with non-murine lymphocytes tended to result in genetically unstable hybrids. Sensitisation of some of the products of these original fusions to selective media has produced murine x bovine and murine x ovine heteromyeloma lines. These lines can be used in place of the mouse myeloma and fused with activated bovine, or ovine, lymphocytes respectively. The resulting heterohybridomas have proved to be stable lines secreting moderate levels of high affinity antibodies over long periods of time. Further exploitation of this approach could help to realise the potential of non-murine monoclonal antibodies.

INTRODUCTION.

Vertebrate reproduction is controlled by a complex series of interactions between a variety of hormones and their target tissues. A predominant feature of this hormonal system is the existance of gonadal steroid feedback loops exerting both stimulatory and inhibitory effects at all levels of the system. Interference with these feedback controls by immunological neutralization can modulate the reproductive performance of the animal. The degree of negative feedback of gonadal steroids on the hypothalamus and pituitary regulate the release of FSH and LH and so control the number of oocytes developing to maturity and ovulation. Immunoneutralisation of this regulating pathway at an appropriate time can therefore increase the ovulation rate of an animal.

The simplest way to exploit this approach to controlling fecundity is to use the animal's own immune response by vaccinating with protein-steroid conjugates. This method has been shown to work in sheep, and to some extent in cattle (1). The major problem inherent in this method is the variability of the immune response between species and between individuals in the specificity, avidity, and titre of the antiserum produced. An insufficient response will not affect fecundity, whilst an excessive response can make the animal acyclic. Long term effects on successive reproductive cycles can also be very variable.

Passive immunisation, the administration of exogenous

SPECIES	SHEEP	SHEEP	SHEEP	CATTLE	CATTLE
IMMUNOGEN	TESTOSTERONE-3CMO-OVALBUMIN	OESTRONE-6CMO-OVALBUMIN	ANDROSTENEDIONE-3CMO-OVALBUMIN	TESTOSTERONE-3CMO-OVALBUMIN	OESTRADIOL-6CMO-OVALBUMIN
TIME (BOOST TO BLEED)	10 DAYS	5 DAYS	5 DAYS	3 DAYS	5 DAYS
FUSION EFFICIENCY	2×10^{-7}	3×10^{-7}	2×10^{-6}	2×10^{-7}	1×10^{-7}
SPECIFIC F.E.	2×10^{-8}	4×10^{-8}	7×10^{-8}	1×10^{-8}	2×10^{-8}
STABILITY	6 MONTHS *	2 MONTHS + (ONGOING)	2 MONTHS **	15 MONTHS (ONGOING)	12 MONTHS (ONGOING)
Mab SPECIFICITY	TESTOSTERONE	OESTRONE***	ANDROSTENEDIONE***	TESTOSTERONE	OESTRONE/OESTRADIOL
Mab CLASS	OVINE IgG	OVINE IgG	OVINE IgG	BOVINE IgG_1	BOVINE IgG_1
DISSOCIATION CONSTANT (K_D)	7.63×10^{-12}M	NT	NT	2.5×10^{-11}M	5.33×10^{-10}M
SECRETION RATE	2.5 ng/ml	NT	NT	3.4 ug/10^6 cells/ 24 hr	20 ug/10^6 cells/ 24 hr
CHROMOSOME NUMBER	NT	NT	NT	120	NT
FUSION PARTNER	NS1	NS1	O/MF	B/MF-2	B/MF-10

Notes: * NOT RECOVERED AFTER FREEZING
** LOST THROUGH CONTAMINATION
*** TESTED ONLY IN SOLID PHASE TO DATE
NT NOT TESTED

antisera, to neutralise circulating steroids has been used successfully in sheep to improve fecundity (1). Although this approach is more readily regulated, it does depend on the production of specific, high titre, high avidity antisera by donor animals.

In cattle, the paucity of the immune response means that both active and passive immunoneutralisation present serious problems. The first because of the weak endogenous response, and the second, because of the large volumes of antisera necessary to achieve the desired effect. The use of xenogeneic sera from sheep has been tried but there is a potential for anaphylaxis, especially on repeated dosing. Due to their low affinities murine monoclonal antibodies are poor immunomodulators on account of the susceptability of the antigen-antibody complexes to dissassociation. The potential for anaphylactic reaction also exists.

For these reasons, it was found necessary to explore the potential for producing bovine and ovine anti-steroid antibodies using hybridoma technology.

The basis of monoclonal antibody production is the fusion of two cell types - the myeloma, which is potentially immortal, but is also deficient in an enzyme necessary for nucleic acid synthesis in the presence of aminopterin, and the sensitised B-lymphocyte, which has a finite life span but contains the gene coding for the enzyme hypoxanthine (guanine) phosphoribosyl transferase (HGPRT) along with genes coding for antibody production and packaging.

Prior to Kohler and Milstein's (2) pioneering work on the immortalisation of antibody-secreting cell lines, cell fusion had been used primarily as a method of investigating the localisation of genetic material. Several research groups had fused cells from different species to see if continued expression of specific cell products could be correlated with the occurence of particular chromosomes in the hybrids (3). Since 1975, interspecific cell hybridization has increasingly been used in attempts to produce monoclonal antibodies from species for which myeloma lines are not readily available. Suitable lines are available from both mouse and rat sources and considerable effort is being made to derive suitable human myeloma cell lines. However, brief perusal of the volume of literature devoted to the investigation of potentially suitable human lines, and consideration of the intensity with which the human model is studied, will make it apparent that the routine development of myeloma lines from less intensively studied species is currently not a realistic consideration.

A large proportion of modern biotechnological processes are based on the production of foreign proteins by cells in culture. Therefore the use of mouse, or rat, myeloma lines to express non-rodent immunoglobulins was a relatively obvious step. Initial cross-species fusions involved mouse myeloma cells and spleen cells from rats immunised against murine antigens (4). However, the paucity of suitable established human myeloma lines, combined with the obvious demand for human monoclonal antibodies, led to early attempts at fusing lymphocytes from immunised individuals with murine myeloma cells (5). Initial success, at least in the short term, has prompted the extension of this approach to the production of monoclonal antibodies from domestic species, especially cattle and sheep.

The earliest reported attempt to immortalise sheep antibody production was by Tucker et al.(6) using a direct fusion between ovine spleen or lymph node lymphocytes and murine NS1 cells. Although the resulting hybrids grew well in culture, they soon lost the capacity to produce immunoglobulin, presumably due to chromosome loss. Tucker and her co-workers (7,8) subsequently used a technique pioneered in human x murine fusions by Ostberg and Pursch (9) and by Teng et al.(10) to improve the stability of their hybrids. Interspecific hybridomas, obtained from previous fusions, were sensitised to aminopterin and then fused with more lymphocytes from the immunised animal to produce second generation heterohybridomas. This approach has overcome some of the problems associated with interspecific fusions, especially the low fusion efficiency and the instability of the resulting hybrids. The use of heteromyeloma fusion partners has been exploited to generate human, chimpanzee (11), cattle (12) and sheep antibodies and the number of generations of refusion are being investigated to see if any of the benefits continue to accrue on repeated fusion. In the current work, first and second generation interspecific fusions have been used to generate heterohybridomas secreting ovine and bovine antibodies.

Previous reports of the isolation of bovine monoclonal antibodies, using first generation fusions, have been primarily concerned with the provision of isotype preparations for sequencing and serology studies (13). Lymphocytes have generally been obtained from the spleen or lymph nodes of these animals.

In the current work first and second generation interspecific fusions have been used to generate heterohybridomas secreting ovine and bovine monoclonal antibodies.

MATERIALS AND METHODS.

Generation of heteromyelomas.

The murine myeloma line used in the first generation fusions was NS1-1/Ag4-1. The second generation fusions necessitated the production of bovine and ovine heteromyeloma fusion partners.

These heteromyeloma lines were isolated from both bovine x murine and ovine x murine fusions. In both cases NS1 mouse myeloma cells were fused to peripheral blood lymphocytes from cattle or sheep, immunised against testosterone-3carboxymethyloxime-ovalbumin. Cell lines secreting antibodies to testosterone were isolated, but several of these lines subsequently ceased secreting. A number of these were cloned by limiting dilution into media containing 8-azaguanine (Sigma). This purine selects for cells which are deficient in the enzyme HGPRT, and therefore sensitive to aminopterin. These cells were tested frequently for aminopterin sensitivity and maintained in azaguanine medium for 60 days. The resulting cell lines can be substituted for the murine myeloma in fusions since they are both aminopterin-sensitive and potentially immortal. Sensitisation of the products of subsequent fusions is currently in progress.

Cell fusion.

Peripheral blood lymphocytes were collected from the jugular vein of animals previously immunised against steroid-protein conjugates.

The lymphocytes were seperated using ficoll density gradient centrifugation (Lymphopaque, Nyegaard, UK) and washed with Hank's buffered salts solution. Lymphocytes were then fused to either mouse myeloma or heteromyeloma cells at a ratio of between 4:1 and 8:1. The cells were mixed, pelleted and drained before being resuspended in 0.8 ml of PEG 1500 (BDH) over 2 minutes. This was immediatly followed by gradual dilution with 10 ml of Leibovitz L15 (Gibco) medium over 6 minutes. The fused cells were washed twice with L15 and resuspended in Dulbecco's modified Eagle's medium

(DMEM) supplemented with 10% Myoclone foetal calf serum (FCS) (Gibco) and hypoxanthine, thymidine and aminopterin (HAT medium). The cells were plated out over a monolayer of mouse thymocyte 'feeder' cells in 24-well tissue culture plates (Nunc). Plates were then incubated in a CO_2 gassed incubator at 37°C and media was replaced, when exhausted, with further HAT media. After 7 days, media was replaced with HT supplemented medium only.

Screening

All cultures were screened for specific antibody production from 7 days after fusion. Most secreting colonies were initially detected in the first three weeks, but screening continued for about 2 months. The screening assays developed were rapid and simple ELISAs.

The PVC flexible plates (Dynatech) were rinsed with an alkaline carbonate/bicarbonate coating buffer (pH 9.6) and coated with the relevant steroid-protein conjugate. These conjugates were produced using the mixed anhydride reaction, except for the oestradiol conjugate which was purchased from Sigma. The carrier proteins used in the preparation of the coating conjugates were tested to ensure that there was no cross reactivity with either antibodies against the carrier protein used for the immunogen (ovalbumin), or with the enzyme-labelled anti-immunoglobulin antibody. The wash buffer at all stages was phosphate buffered saline containing 0.5% gelatin and 0.5% Tween 20 to reduce non specific binding effects.

Supernatant samples were transferred directly from the tissue culture plates to assay plates and were assayed undiluted. On each plate a dilution series of a polyclonal, conspecific antisera against the relevant steroid was included as a quality control. After incubation and washing, the plates were incubated with horseradish peroxidase (HRP) labelled anti-species antibodies. In the bovine assays, this was a rabbit anti-bovine IgG (Dakopatts) which had been cleaned up prior to use on a Sephadex G200 column. For the ovine assays, an affinity purified donkey anti-sheep IgG, obtained from Guildnay Antisera at the University of Surrey, was used. Following incubation and washing, the enzyme was visualised using orthophenylenediamine.

Colonies detected as secreting specific antibodies in two consecutive assays were rapidly cloned by limiting dilution. Antibody secretion was regularly monitored throughout production. Careful scrutiny of secretion rates, followed by prompt cloning if a significant fall was detected, resulted in no secreting lines being lost during culture after the initial cloning. After 3 or 4 clonings, antibody secretion rates remained relatively constant for any given conditions and medium.

Characterization.

After clones were established as stable secretors,

further characterization of the cells and their secreted product was carried out (Table 1). Of the five lines established to date, the fusion efficiency (clones per lymphocyte fused) has been between 1 x 10^{-7} and 2 x 10^{-6}. This compares with typical values for murine x murine system of between 1 x 10^{-7} and1 x 10^{-5}. The specific fusion efficiency was 5 - 10% of the overall fusion efficiency.

The specificity of the antibodies was established in a liquid phase radioimmunoassay using a tritiated tracer, in order to reflect the <u>in vivo</u> situation as closely as possible. Displacement studies using potentially cross-reacting steroids demonstrated that both ovine and bovine antibodies raised against testosterone were primarily specific for that steroid, although they exhibited a high degree of cross-reactivity with DHT. The bovine antibody raised against oestradiol was suprisingly more specific for oestrone than for oestradiol.

The class of the ovine monoclonal antibodies was established by the use of an anti-IgG antiserum in the screening assays and also by checking the cryogenic stability of the antibodies. Bovine antibody subclass was established using a competitive ELISA (12) with rabbit anti-bovine subclass IgG_1, IgG_2, and IgM antibodies (Jackson IR, PA.) and bovine subclass standards from Miles. The assay was completed with HRP-labelled donkey anti-rabbit IgG (Guildhay Antisera, University of Surrey).

The affinities of the antibodies were calculated, using the same RIA system as that for the cross reactivity determinations, from equilibrium binding studies performed at 4°C according to the methods of Scatchard (14) and Rosenthal (15).

Secretion rates which have been determined were comparable to those reported in murine x murine systems.

Karyotyping studies are underway and most of the first generation clones appear to have only a few more chromosomes than the myeloma line. Both bovine x (bovine x murine) heterohybridoma lines, however, have an average chromosome complement of more than 120, compared to the bovine complement of 60 and the murine one of 40.

Both bovine lines have been transferred to a variety of media in order to facilitate purification. The high levels of bovine globulins and serum proteins in FCS (8 mg/ml) made separation of ug/ml levels of specific antibody difficult. The heterohybridomas grew very well in DMEM supplemented with 5, 10 or 20% donor horse serum (Imperial Labs). However, the increased growth rate appeared to be at the expense of antibody secretion.

Two serum-free media were tested. These were high and low protein formulations of Gibco medium. In both cases secretion rates were higher than in DMEM/10% FCS. The albumin-supplemented high protein medium also supported good cell growth and division, although the low protein formulation was better suited as a maintenance medium.

Interspecific hybridomas will not produce ascites in normal BALB/c mice. Batches of the heterohybridoma cells were therefore injected IP into pristane-primed Onu, genetically athymic, mice. Ascitic tumours became established between 30 and 50 days after the initial injection. Subsequent passages of ascitic cells exhibited a shorter and more uniform delay between injection and harvesting. In the initial passage, yields were variable in volume (0.5 to 3 ml/animal) and antibody concentration (5 to 90 ug/ml). The production of antibody in subsequent passages is being investigated.

Discussion.

There are many situations in which murine monoclonal antibodies are not suitable reagents. There are also many potential uses for wider ranges of specific sources of Mabs, especially for diagnostic purposes.

Polyclonal human antisera are used on a wide variety of therapeutic targets, including the neutralisation of red and white cell antigens, viral antigens such as hepatitis, rabies, CMV and varicella zoster, and bacterial antigens such as tetanus and endotoxins. Antisera are available to combat the effects of snake venom, or to eliminate drugs from the circulation, as well as to neutralise components of the reproductive hormone system. In the veterinary field, antisera to bacterial antigens, such as Clostridium, are already in use. Murine monoclonal antibodies are also used in the treatment of calf scours.

All these current applications could be more easily achieved using conspecific monoclonal antibodies. Current technological limitations in the generation of conspecific myeloma lines, together with the generally poor secretion rates achieved with the EBV transformation system, dictate that non-rodent monoclonals must be generated using interspecific fusions. The fusion efficiencies of murine myeloma cells and non-murine lymphocytes can be significantly lower than homogeneic fusions, and the secretion rates achieved are generally lower. However, it has been conclusively demonstrated that interspecific fusions are a viable technique for the generation of non-murine Mabs, provided sufficient attention is paid to early cloning of secreting lines. Refusion techniques to increase the non-murine chromosome complement have been shown to improve the fusion efficiency compared to that of murine myelomas.

In conclusion, monoclonal antibodies derived from sources other than mouse and rat have a great, but relatively unexploited, potential. Current technology effectively limits the production of non-murine Mabs to interspecific fusions. Many of the reported disadvantages of interspecific fusions, such as the low fusion efficiency and the poor stability of the resulting clones, can be overcome by the use of refusion techniques.

This research group have used, and is continuing to use, a combination of direct interspecific fusion and refusion methods to generate a panel of ovine and bovine antibodies for use in improving the fecundity of domestic animals. These Mabs should present a viable alternative to the use of exogenous synthetic hormones in the control of animal production and allow the exploitation of routes of immunological control not currently available with polyclonal or murine antibody preparations.

ACKNOWLEDGEMENTS.

This work was generously funded by AFRC grant AG 90/9 to BAM. DJG would like to acknowledge the financial support of the OHOLO/ESACT conference organisers, the Council of the Journal of Endocrinology, and Guildhay Antisera in attending this conference.

REFERENCES

1. Webb, R., Land, R.B., Pathiraja, N. and Morris, B.A. (1984) Passive immunization in the female. In: D.B. Crighton (Editor), Immunological Aspects of Reproduction in Mammals. Butterworths, London. pp. 475-499.

2. Kohler, G. and Milstein, C. (1975) Continuous cultures of fused cells secreting antibody of predefined specificity. Nature 256, 495-497.

3. Weiss, M.C. and Green, H. (1967) Human-mouse hybrid cell

lines containing partial complements of human chromosomes and functioning human genes. Proceedings of the National Academy of Sciences, USA. 58, 1104-1111.

4. Trowbridge, I.S. (1978) Interspecies spleen-myeloma hybrid producing monoclonal antibodies against mouse lymphocyte surface glycoprotein T200. Journal of Experimental Medicine 148, 313-323.

5. Scwaber, J.F. and Rosen, F.S. (1978) Induction of human immunoglobulin synthesis and secretion in somatic cell hybrids of mouse myeloma and human B lymphocytes from patients with agamma globulinemia. Journal of Experimental Medicine 148, 974.

6. Tucker, E.M., Dain, A.R., Clarke, S.W. and Donker, R.A. (1981) Culture of sheep x mouse hybridoma cells in vitro. Hybridoma, 1, 77-86.

7. Tucker, E.M., Dain, A.R., Clarke, S.W. and Donker, R.A. (1984) Specific bovine monoclonal antibody produced by a re-fused mouse/calf hybridoma. Hybridoma 3, 171-176.

8. Anderson, D.V., Clarke, S.W., Stein, J.M. and Tucker, E.M. (1986) Bovine and ovine monoclonal antibodies to erythrocyte membrane determinants, produced by interspecific (hetero-) myelomas. Biochemical Society Transactions 14, 72-73.

9. Ostberg, L. and Pursch, E. (1983) Human x (mouse x human) hybridomas stably producing human antibodies. Hybridoma 2, 361-367.

10. Teng, N.N.H., Lam, K.S., Riera, F.C. and Kaplan, H.S. (1983) Construction and testing of mouse-human heteromyelomas for human monoclonal antibody production. Proceedings of the National Academy of Sciences, USA. 83, 7308-7312.

11. Van Meurs, G.J.E. and Jonker, M. (1986) Production of primate monoclonal antibodies. Journal of Immunological Methods 95, 123-128.

12. Groves, D.J., Morris, B.A. and Clayton, J. (1987) Preparation of a bovine monoclonal antibody to testosterone by an interspecies fusion. Research in Veterinary Science, In press.

13. Guidry, A.J., Srikumaran, S. and Goldsby, R.A. (1986) Production and characterization of bovine x murine hybridomas. In: J.J. Langonne and H.V. Vunakis (Editors), Methods in Enzymology, 121, Academic Press, New York. pp. 244-265.

14. Scathard, G. (1949) The attractions of proteins for small molecules. Annals of the New York Academy of Sciences. 51, 660-671.

15. Rosenthal, H.E. (1967) A graphic method for the determination and presentation of binding parameters in a complex system. Analytical Biochemistry 20, 525-532.

7 The immortalization of differentiated mammalian cells

C. MacDonald and D. M. Scott

ABSTRACT

Mammalian tissue culture cells offer major advantages over micro organisms for the production of functional proteins with correct post-translational modifications. Cell lines producing useful products, eg. proteins of therapeutic value, can be obtained either by expressing cloned recombinant genes in animals cells, or by immortalizing differentiated primary cells. Immortalization can be either by

1) fusion to an established tumour cell line or

2) the introduction of viral sequences from cloned recombinant plasmids.

Cell fusion techniques have been important in generating both hybridoma cell lines secreting monoclonal antibodies, and nerve cell hybrid lines which continue to express neuronal functions. However, the method is limited by the range of available tumour cell fusion partners. Immortalization with viral sequences is a promising alternative, which we have used successfully to isolate differentiated cell lines which continue to express the characteristics of the original tissue.

INTRODUCTION

Animal cells can offer important advantages over both yeast and bacteria for the commercial production of mammalian proteins - correct post-translational modifications (such as glycosylation, phosphorylation or proteolytic cleavage) and correct folding of the polypeptide including the formation of quaternary structures. Proteins without the correct secondary modifications may lack biological activity, eg. Factor IX without γ carboxylation of some glutamic acid residues or they may be unsuitable for use as vaccines, due to differences in conformation compared with the native protein; or molecules intended for repeated therapeutic use, may elicit undesirable immune responses, due to inappropriate amino acid additions. Incorrectly folded proteins may be insoluble, and therefore not secreted and difficult to recover (3).

A major limitation to the use of mammalian cells in the biotechnology industry is the shortage of suitable cell lines. These can either be cells which efficiently express recombinant genes which have been introduced into the cell by gene transfer (3); or they can be immortalized lines obtained from differentiated primary cells (12) which produce a useful product in vivo eg. antibody.

Two approaches have been used to generate immortalized cell lines: firstly, fusion to an established tumour cell line; and secondly, the introduction of viral sequences, cloned in plasmid vectors, into mammalian cells. The cell fusion technique has been important for generating hybridoma cell lines secreting monoclonal antibodies (7), and nerve cell hybrid lines which continue to express neuronal functions (8). Liver cell hybrids have also been isolated in this laboratory, and are currently being characterized. However, cell fusion is limited by the range of suitable, and available, tumour cell partners for hybridization. For this reason we consider that immortalization with viral sequences is a promising alternative approach. This paper describes its successful use for the isolation of differentiated cell lines from the thick ascending limb of Henle's loop (TALH cells) which continue to express the characteristics of the original rabbit kidney tissue.

MATERIALS AND METHODS

All chemicals were obtained from Sigma Chemicals GmbH, Frankfurt, or Merck AG, Darmstadt, unless otherwise specified. Hyaluronidase and ouabain were purchased from Boehringer Mannheim GmbH, Mannheim, and collagenase, grade II was from Worthington Biochemical Corporation, Freehold, N.J. Ficoll-400 was supplied by Pharmacia Inc., Uppsala, Sweden. Fetal (FCS) or newborn calf serum, Dulbecco's modification of Eagle's Medium (DMEM), G418 and antibiotics were from Gibco Europe GmbH, Eggenstein. Costar tissue culture ware was supplied by Tecnomara Deutschland GmbH, Ruhberg. Protein A-linked horseradish peroxidase (HRP) was from BioRad Laboratories GmbH, Munchen.

Isolation, Culture and Transfection of TALH Cells

Epithelial cells derived from the TALH of rabbit kidneys were prepared as previously described (4,13) and maintained in culture for approx. 3 weeks prior to introduction of foreign DNA.

Cells were subcultured (with Ca^{2+}- and Mg^{2+}-free PBS + 1 mM EGTA) 16 h prior to exposure to plasmid DNA. Cells were reseeded at a density of 2 x 10^4 cells/cm^2 in DMEM plus 10% FCS. Cells were then exposed for 6½ h to plasmid DNA (1 μg/10^6 cells) from either pSVE (4) (containing the early region of SV40 between the Bam H1 and EcoRV sites) or pSVE2 (containing the SV40 early region between the Bam H1 to Hpa II sites) plus pSV2-neo (14) (5μg/10^6 cells), in the presence of carrier DNA (14 μg/10^6 cells) (L-cell DNA or salmon testes DNA carrier), using the calcium phosphate precipitation technique (6). After exposure to DNA, cells were then washed twice with DMEM + 10% FCS and the medium replaced at 48 h with medium containing G418 (600 μg/ml), a concentration found to be extremely toxic to 1° or 2° cultures of TALH cells (data not shown). Medium containing G418 was subsequently replaced at 72h intervals. Estimation of cell growth was determined according to the method of Mosmann (9).

Immunofluorescence Studies

Cells were examined for the presence of SV40 large T antigen using rabbit anti-SV40 large T antigen antibody (primary antibody) and fluorescein conjugated sheep anti-rabbit immunoglobulin as a second antibody. Tamm Horsfall glycoprotein (THP) was demonstrated using as primary antibody sheep anti-THP antiserum and rabbit anti-sheep immunoglobulin conjugated with biotin as secondary antibody. Bound antibody was detected by the addition of FITC-conjugated avidin.

Ion Transport Studies

Uptake experiments using $^{86}Rb^+$ as a tracer for K^+ (which was present at 5mM) over a 5 minute period were carried out as described by Amsler and Kinne (1). Efflux studies were carried out over a 4 minute period following preloading of cells for 3 hours at 37°C in HBSS + 1% FCS containing $^{86}Rb^+$ (1μCi/cm^2).

Na, K, ATPASE Assay

Na, K, ATPase (EC 3.6.1.3) activity was determined as previously described (13).

RESULTS AND DISCUSSION

Three epithelial cell lines derived from rabbit medullary Thick Ascending Limb of Henle's Loop (TALH) are described here. They maintain their differentiated phenotype following the stable integration of recombinant DNA containing viral oncogenes. The cells were cotransfected with genes from the early region of SV40 virus, together with pSV2-*neo* DNA (which confers resistance to the toxic antibiotic G418 (14). SV40 DNA was chosen as it has been previously shown that certain differentiated functions can be maintained after infection and subsequent transformation. pSV2-neo DNA was introduced to provide a dominant selectable marker, enabling the selection of cells without alterations in the normal cellular phenotype, which may accompany loss of differentiation.

Cells isolated in the presence of G418 (600μg/ml) were demonstrated to exhibit characteristics typical of TALH cells *in vivo*, plus acquired characteristics obtained as a result of their stable integration of 'foreign genes'.

(a) Characteristics exhibited as a result of transfected genes

All examined cells showed the ability to grow in the continued presence of G418 at a concentration (200μg/ml) shown to be lethal to normal TALH cells in ca. 2 weeks. These cells were also shown by indirect immunofluorescence studies to express SV40 large T antigen, an early region gene of SV40 virus. In addition these cells acquired the ability to proliferate in low serum, or completely defined serum-free medium, in which primary and secondary TALH cells can survive for a short period only, and cannot proliferate.

(b) TALH characteristics maintained by transfected cells

Examined cell lines were also seen to maintain certain characteristics of the parental cells. These include the continued expression of high levels of activity (comparable to those seen in primary cultures of freshly isolated TALH cells) of Na, K-ATPase (as shown in Table 1), and the Na, K, Cl-cotransporter (see TAble 2a). These cells also continue to express the apically located Ba^{+}-sensitive K^{+}channel (see Table 2b) and are capable of the synthesis of Tamm Horsfall glycoprotein, a protein reported to be specifically synthesised in the kidney by both medullary and cortical cells of the TALH.

Table 1 Na, K-ATPase activity in 2° and transfected TALH cell cultures.

Cells	Na, K, ATPase activity (pmoles phosphate generated/h/mg protein)
2°TALH	10.7
TALH, SVE.1	7.4
TALH, SVE.2	16.8
TALH, SVE2.1	12.2

Table 2 Demonstration of the Na, KCl- cotransport system and Ba^{2+}-sensitive K channel in transfected cells.

(a) Ouabain and bumetanide sensitive $^{86}Rb^{+}$-uptake into transfected TALH cells.

$^{86}Rb^{+}$-uptake (nmoles/mg protein/min)

TALH cells	Ouabain sensitive	Bumetanide sensitive
SVE.1	not detected	3.5
SVE.2	7.0	4.5
SVE2.1	4.1	3.4

(b) $^{86}Rb^{+}$-efflux from transfected TALH cells in the absence and presence of BaCl.

$^{86}Rb^{+}$-efflux (dpm^{-3}/mg protein/min)

TALH Cells	Control	$+Ba^{2+}$	% Inhibition	Ba^{2+}-sensitive K^{+}-channel activity
SVE.1	11.9	9.4	21.0	2.5
SVE.2	2.1	1.5	28.6	0.6
SVE2.1	4.4	3.1	29.5	1.3

The kidney epithelial cells described here contain viral genes which have been stably integrated into the host cell DNA. These transfected cells are capable of extended (and possibly indefinite) growth, and continue to express differentiated characteristics typical of TALH cells, both in vivo and in primary culture.

This approach has been used successfully to generated macrophage-like cells (10,11) and human epidermal cells (2). Rat hepatocytes have been transfected with SV-40 DNA and well-defined proliferating differentiated cell lines have been obtained. Provided a suitable chemically defined medium was chosen, lines which were epithelial in morphology, secreted albumin, transferin, hemopexin, and expressed the enzyme glucose-6-phosphatase were obtained (15).

There is evidence, both from our work and that from other laboratories, that differentiated cell lines can be obtained by transfected primary or secondary cells with .iral sequences. This approach may be an important alternative to cell fusion for generating novel cell lines secreting proteins of commercial interest.

ACKNOWLEDGEMENTS

This work was supported by Travel Grants from The Biochemical Society and The Wellcome Trust which enabled C. MacD to visit the Max Planck Institut fur Systemphysiologie. The hospitality of the host institution is also acknowledged. Work at the University of Strathclyde was supported by a grant from the Humane Research Trust. A fuller account of the detailed characterisation of the differentiated TALH cells has already been published (12).

REFERENCES

1. Amsler, K. and Kinne, R. (1986) Photoinactivation of sodium-potassium-chloride cotransport in LLC-PK_1/Cl_4 cells by bumetanide. Am. J. Physiol., 250, C799-C806.

2. Banks-Schlegel, S.P. and Howley, P.M. (1983) Differentiation of human epidermal cells transformed by SV40. J. Cell Biol., 96, 330-337.

3. Bebbington, C. and Hentschel, C. (1985) The expression of recombinant DNA products in mammalian cells. Tibtech, 3, 314-317.

4. Brzeski, H., Chambers, M., MacDonald, C. and Stimson, W.H. (1985) The

immortalisation of human lymphocytes by spheroplast fusion. Dev. Biol. Standard, 60, 105-109.

5. Eveloff, J., Haase, W. and Kinne, R. (1980) Separation of renal medullary cells : isolation of cells from the thick ascending limb of Henle's loop. J. Cell Biol., 87. 672-681.

6. Graham, F.L. and van der Eb, A.J. (1973) A new technique for the assay of infectivity of human adenovirus 5DNA. Virol., 52, 456-467.

7. Kohler, G. and Milstein, C. (1975) Continuous cultures of fused cells secreting antibody of predefined specificity. Nature (Lond), 256, 495-497.

8. McMorris, F.A. and Ruddle, F.H. (1974) Expression of neuronal phenotypes in neuroblastoma cell hybrids. Dev. Biol., 39, 226-246.

9. Mosmann, T.J. (1983) A rapid colorimetric assay for cellular growth and survival : application to proliferation and cytotoxicity assays. J. Immunol. Meth., 65, 55.

10. Nagata, Y., Diamond, B. and Bloom, B.R. (1983) The generation of human monocyte/macrophage cell lines. Nature (Lond), 306, 597-599.

11. Schwarzbaum, S., Halpern, R. and Diamond, B. (1984) The generation of macrophage-like cell lines by transfection with SV40 origin defective DNA. J. Immunol., 132, 1158-1162.

12. Scott, D.M., MacDonald, C., Brzeski, H. and Kinne, R. (1986) Maintenance of expression of differentiated function of kidney cells following transformation by SV40 early region DNA. Exp. Cell Res., 166, 391-398.

13. Scott, D.M., Zierold, K. and Kinne, R. (1986) Development of differentiated characteristics in cultured kidney (thick ascending loop of Henle) cells. Exp. Cell Res., 162, 521-529.

14. Southern, P.J. and Berg, P. (1982) Transformation of mammalian cells to antibiotic resistance with a bacterial gene under control of the SV40 early region promoter. J. Mol. Appl. Genet., 1, 327-341.

15. Woodworth, C., Secott, T. and Isom, H.C. (1986) Transformation of rat hepatocytes by transfection with Simian virus 40 DNA to yield proliferating differentiated cells. Canc. Res., 46, 4018-4026.

Discussion

ARATHOON	What are the rates of dedifferentiation - have you seen it at all.
MacDONALD	We have seen none
MEIGNIER	What was the number of passages you have experience?
MacDONALD	6 months - doubling every 24 hours
FEDER	Which protein did you identify
MacDONALD	Tarn Horsfall glycoprotein
FEDER	Do you have any other proteins like this
MacDONALD	All we know is that it is being recogised by the antibody.
FEDER	It is important to demonstrate that there have been no changes in glycosylation.

8 Structural basis of growth factor receptor mediated cellular signals (abstract)

A. Ullrich, H. Riedel, E. Livneh and J. Schlessinger

Polypeptide growth factors play a crucial role in the regulation of proliferation of mammalian cells in culture and in growth, development and maintenance of higher organisms. The biological actions of these growth factor molecules are mediated by specific glycoprotein receptors that are inserted in the cellular plasma membrane. A major criterion that characterizes the specialized role of eukaryotic cells is defined by the nature and composition of receptors on their surface. Interaction of a growth factor with the extracellular domain of its specific receptor triggers a series of cellular responses that ultimately leads to cell division. The receptors for epidermal growth factor (EGF), insulin, insulin-like growth factor I (IGF-I) and colony stimulating factor (CSF-I) all employ an intrinsic tyrosine specific protein kinase activity for the generation of diverse biological signals. To investigate in detail the mechanisms involved in ligand-induced growth stimulation, we characterized as a first step the primary structures of these receptor glycoproteins. Furthermore, we characterized two putative receptors, HER2 and c-kit, which also belong to the tyrosine kinase family and are the cellular homologues of polypeptides (neu and v-kit) with oncogenic activities. The overall structural organization of these receptor molecules suggests the existence of three subclasses which may reflect functional differences and possibly independent evolutionary origins. We have undertaken efforts to functionally characterize structural domains of various growth factor receptors, which we were able to identify on the basis of sequence comparisons. Experimental results derived from studies involving receptor mutants will be discussed.

Discussion

F. HOFFMANN: Do antibodies to receptors mimic the effect of the small molecule which normally activate the receptors.

ULLRICH As yet we do not have any small molecules which stimulate our newly found receptor (HEP2 and C-Kit). We do not know what the effect of antibodies to the receptors would have. Although a dutch group has looked at the effect of antibodies on the EGF receptor which stimulates the auto phosphorylation reaction but does not effect a mitogenic response. However in our work we have found that the tyrosine kinase activity is essential for the biological response. So I think that the MCA to the receptor can have a modulated response as there are 4 or 5 auto phosphorylation reactors on tyrosine residues of the receptor it could be that you can get s partial but not complete stimulation.

SILBERKLANG In all those cell lines which over produce receptors and chimeric GF receptors is the rate of cell division increased as opposed to the rate of movement from G1 to S.

ULLRICH The rate is somewhat reduced in some over expressing cell lines.

9 Growth control of normal human B lymphocytes

E. Lundgren, R. Sanders and G. Roos

ABSTRACT

Primary B lymphocytes depleted of T lymphocytes were triggered to growth either by lymphokines after crosslinking of surface Ig or by phorbol ester and Calcium-ionophore. It was ascertained by demonstration of co-expression of a cell cycle-associated antigen and surface Ig that the B cells were proliferating. A real increase in cell counts was only achieved when cells were first stimulated with phorbol ester and ionophore to express Tac-antigen and then exposed to IL-2 and BCGF.

Primary cultures of non-transformed human B-lymphocytes can be stimulated to enter the cell cycle by some lectins or by cross-linking surface immunoglobulins in the presence of lymphokines (for recent review, see ref. 1). However, the cells usually only grew for a limited period, the requirements for prolonged growth on a larger scale being unknown.

For practical purposes, immortalization techniques have been used to date to obtain larger numbers of B lymphocytes, mainly for making human monoclonal antibodies. Apart froom by conventional cell fusion techniques, B lymphocytes can be immortalized by Epstein-Barr virus, which allows long term cultures (reviewed in refs. 2 and 3).

Rapid progress is expected in our understanding of regulation of B cell growth as a consequence of succesful purification of B cell growth and differentiation factors as well as isolation of genes encoding those lymphokines (4,5,6).

In this study data are presented showing that B lymphocytes were triggered by several factors to enter the cell cycle, estimated by several methods. Several of the factors induced expression of the Tac antigen (defining part of the IL-2 receptor) and it turned out that addition of IL-2 to cells preactivated by phorbol ester and calcium-ionophore resulted in a net increase of cell number.

MATERIALS AND METHODS

Cell culture

Human mononuclear cells were prepared from buffy coat cells by standard procedures. T-cells were depleted by two cycles of E-rosetting. Less than 1% of the remaining cells displayed the T cell antigens CD4 and CD8.

B cell growth factor (BCGF) was purchased from Cellular Products (Buffalo, NY) and used at a final concentration of 10%. IL-2 was produced from Jurkat cells and was used after immune affinity chromatography at a concentration of 20 BRMB U/ml.

Cell growth was estimated by thymidine uptake and cell counts. Flow cytometry (FCM) was performed as described previously (7).

Immunostaining

Cytospin preparations were fixed in ice cold acetone and stained for the Ki-67 antigen using the monoclonal antibody DAKO-PC (Dakopatts AB, Stockholm). Bound antibody was demonstrated by FITC- or biotin labelled anti-mouse antibody. The presence of light Ig chains was demonstrated by FITC-coupled polyclonal antibodies (Kallestad Labs. Inc., Chaska, Mn) and CD4/CD8 antigens with Leu-3a and Leu-2a (Simultest TM, Becton & Dickinson, Mountain View, Ca).

RESULTS

Throughout the study freshly prepared human B lymphocytes were used, obtained from healthy blood donors and depleted of T lymphocytes. The cells were activated by several factors as indicated in Table I. Crosslinking of cell surface receptors with anti-IgM antibodies coupled to Sepharose-beads increased thymidine uptake only in the presence of BCGF. BCGF or IL-2 alone had only marginal effects "marginal" effects (not shown). A clear synergistic effect was observed between anti-IgM, BCGF and IL-2. Also the phorbol ester phorbol dibutyrate (PBt_2) and ionomycin, a calcium-ionophore, increased thymidine uptake more than 20-fold over the background.

Another marker for a proliferative response was the number of nuclei, which were positively stained for the nuclear antigen Ki-67, which has been shown to be a cell cycle marker in T lymphocytes (8). The cells were double stained both for Ki-67 as described and for DNA using propidium iodide and the results will be described in detail elsewhere. The results are summarized in table II. Most of the cells in the medium controls remained in G0/G1 , while a slight increase in cycling cells was observed following treatment with anti-IgM or BCGF.

Table I

The effect of growth factors on (^{3}H)-thymidine uptake in human B lymphocytes.

The cells were exposed to the indicated stimulators in the following concentrations: Anti-IgM, 3 microg/ml, BCGF, 10% v/v, Il-2, 20 IU/ml, ionomycin 5 microg/ml and PBt_2 2.5 ng/ml. (^{3}H)-thymidine uptake was monitored on day 4. One representative experiment out of eleven is shown.

Additions	(^{3}H)-thymidine uptake (mean +/- SD)
Medium	214 +/-184
Anti-IgM	237 +/-45
Anti-IgM, BCGF	1 141 +/-64
Anti-IgM, BCGF, Il-2	8 621 +/-197
PBt_2, ionomycin	4 997 +/-331

Table II

Cell cycle distribution and Ki-67 presentation of activated B cells.

The cells were grown for 4 days in the indicated factors at the concentrations given in table I. Ki-67 staining and propidium iodide staining of nuclei was measured by FCM. DNA content is given for cells separated according to whether they expressed Ki-67 or not.

Treatment	G0/G1	S	G2/M
Ki-67 positive cells			
Anti-IgM, BCGF, IL-2	47.8	32.1	20.1
PBt_2, ionomycin	42.0	34.9	23.1
Ki-67 negative cells			
Anti-IgM, BCGF, IL-2	98.0	2.0	0
PBt_2, ionomycin	97.8	2.2	0

Table III

Frequency of Tac-antigen-positive and IL-2 responsive B cells.

Cells were grown in the factors indicated at the concentrations given in table I for 3 days. Tac-antigen expression was estimated, the remaining cells were washed and propagated for a further 3 days in IL-2.

Treatment	Tac-pos (%)	thymidine uptake
Medium	2.1	340
Ionomycin	27.4	ND[1]
PBt_2	3	ND
PBt_2 , ionomycin	89.1	63 980
Anti-IgM, BCGF, IL-2	48.4	2 001

[1]ND=not done.

A more pronounced effect was observed when the cells were given anti-IgM, BCGF and IL-2 or phorbol ester and calcium-ionophore. More than 40% of the cells resided in S or G2/M and the majority of the cells expressed the nuclear antigen. Most importantly, when the cells were gated according to their staining for Ki-67 it was readily seen that in the Ki-67 negative population all the cells were resting and displayed a DNA content representative of G0/G1 cells. The positive cells on the other hand were distributed throughout the cell cycle.

Thus, Ki-67 seemed to be an excellent marker for cell growth and more specifically for cells in the cell cycle. Moreover, as the system comprises primary cultures with potential contamination by other blood cells, the positive staining for the nuclear antigen Ki-67 could be used to demonstrate that the growing cells were all B lymphocytes. Thus, double staining for immunoglobulin light chains and Ki-67 showed that all cells positive for Ki-67 also expressed surface Ig (not shown). Furthermore staining for CD4 and CD8 showed a less than 1% contamination with T-lymphocytes after 4 days. We therefore conclude that we have observed an increased cycling of B cells, without contamination by growing T cells.

It has been reported by several groups that B lymphocytes can respond to IL-2 and present IL-2 receptors (9,10). In the system used, IL-2 could, by itself, induce thymidine uptake only to a limited extent (not shown). IL-2 receptor presentation was measured as the frequency of cells presenting the Tac antigen. Table III shows that Tac antigen expression was induced either by anti-IgM, BCGF, IL-2 or by phorbol ester and ionomycin. The latter treatment produced a higher frequency of IL-2 receptor presentation, and the responsiveness to IL-2 was far higher in cells pretreated for 3 days with PBt_2 and ionomycin than cells pretreated with anti-IgM and BCGF.

In the experiments shown so far, thymidine uptake and DNA content was estimated on day 4. When cell counts were estimated cell numbers were highly variable. The best factor combination in a single treatment seldom increased cell number by more than 50 %, while phorbol ester and ionomycin decreased the cell number slightly compared to the medium control. As this latter treatment increased Tac expression, we reasoned that the cells pretreated in this way would be sensitive IL-2. Table III shows that cells pretreated with PBt_2/ionomycin were able to respond to IL-2 with thymidine uptake. Estimation of cell count showed that pretreated cells responded to BCGF with only a minimal increase in cell number (fig 1.). In contrast, IL-2 achieved 3 cell doublings after 5 days. A striking synergistic effect was observed between BCGF and IL-2 leading to 4 to

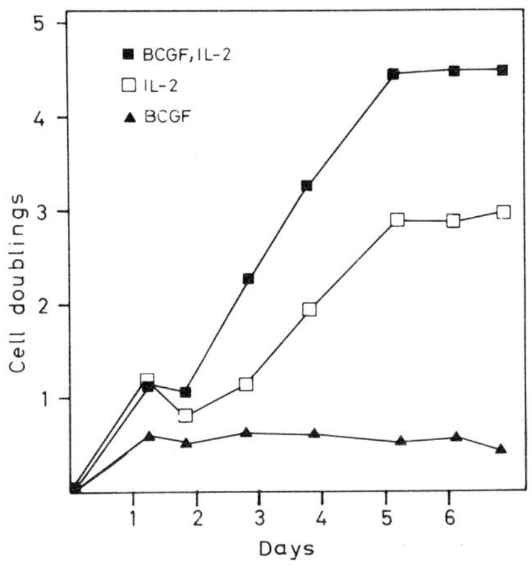

Fig. 1. Growth of B lymphocytes pretreated for 3 days with 5 ng/ml PBt_2 and 2 microg/ml ionomycin. After pretreatment, the cells were washed, suspended in medium (250 000 cells/ml) in BCGF (10%), IL-2 (20 U/ml) or in a combination og both factors.

5 cell doublings during 5 days.

DISCUSSION

In this study it was demonstrated that human peripheral B cells could be stimulated to growth either by lymphokines in the presence of anti-IgM or with phorbol ester and calcium-ionophore. Growth was estimated by thymidine uptake, expression of the nuclear antigen Ki-67 and FCM. Ki-67 expression was a most useful marker as it was only expressed in cycling cells. Thus by double labelling, it could unequivocably be demonstrated that all the cycling cells were B cells. In separate experiments we showed that under the conditions used the contribution to cell growth from T cells was below 1 %. This was true both for the initial 4 day cultures, as well as for the secondary cultures, driven by IL-2 after phorbol ester/calcium-ionophore pretreatment.

Therefore we state that the cell growth observed could be assigned to B cells. However, this does not exclude the possibility that a minor fraction of contaminating T-cells or macrophages might be necessary for the B cell response. Rather this seems problable for two reasons: 1) if macrophages were removed as adherent cells, the reponse to the factors used was always poor and 2) pretreatment with PBt_2 and ionomycin rendered the cells sensitive to IL-2 only after 3 days. Thus, several events could have occurred, before the cells became sensitive to IL-2, with a real possibility that contaminating accessory cells were involved.

Several growth factors considered to be specific for B-cells have been identified and purified in recent years and some of the corresponding genes have been isolated (e.g. 4,5,6). Somewhat unexpectedly these factors are not restriced to B cells only, as both T-helper cells and mast cells can respond. Moreover, they seem to be involved not only in regulation of growth, but also in differentiation. Several authors have shown that B cells can respond to IL-2, originally considered to be strictly a T-cell growth factor, both by growth and differentation (9,10).

There are few reports that these lymphokines will cause a real net increase of cell number. Rather it is a common finding among workers in the field that increase in cell number does not occur using protocols similar to the one used in this study (11).

The finding that phorbol ester together with calcium-ionophore in the condtions used here were so efficient in inducing Tac-antigen, prompted us to propagate the cells in IL-2 after 3 days of pretreatment. Indeed, IL-2 increased cell number, and the activity was highly enhanced by BCGF. Thus, buffy coat cells from one transfusion unit resulted in 2 to 4×10^8 cells, before they went into a resting state. These cells are still viable, and the question as to whether they can be reactivated and the role of protein kinase C will be elucidated in further studies.

In conclusion we confirm that B-cells can be stimulated to incorporate thymidine with lymphokines including IL-2, however seldom does this also lead to an increase in cell number. With the protocol used here we were able to drive the cells in IL-2 synergizing with BCGF to give at least 20-fold more cells within a week, which is ample both for biochemical and molecular studies.

REFERENCES

1. Melchers, F. and Andersson, J.,(1984). B cell activation: three steps and their variations. Cell, 37, 715-720.

2. Shay, J.W. (1985). Human hybridomas and monoclonal antibodies: the biology of cell fusion. In: Human hybridomas and monoclonal antibodies, E.G. Engleman, S.K.H. Foung, J.Larrick, and A. Raubitschek, eds. pp. 5-20, Plenum Press, New York, London.

3. Roder. J.C., Cole, S.P.C., Atlaw, T., Campling, B.C., McGarry, R.C. and Kozbor, D. (1985). The Epstein-Barr virus hybridoma technique. In: Human hybridomas and monoclonal antibodies, E.G. Engleman, S.K.H. Foung, J.Larrick, and A. Raubitschek, eds. pp. 55-70, Plenum Press, New York, London.

4. Mehta, S.R., Conrad, D., Sandler, R., Morgan, J., Montagna, R. and Maizel, A.L. (1985). Purification of human B cell growth factor. J. Immunol., 135, 3298-3302.

5. Yokota, T., Otsuka. R., Mosmann, T., Banchereau, J., DeFrance. R., Blanchard, D., De Vries, J.E., Lee, F. and Arai, K.-I. (1986). Isolation and characterization of a human interleukin cDNA clone, homologous to mouse B-cell stimulatory factor 1, that expresses B-cell and T-cell-stimulating activities. Proc. Nat. Acad. Sci., USA, 83, 5894-5898.

6. Hirano, T., Hasukawa, K., Harada, H., Taga, T.a, Watanabe, Y., Matsuda, T., Kashiwamura, S.-i., Nakajima, K., Koyama, K., Iwamatsu, A., Tsunasawa, S., Sakiyama, F., Matsui, H., Takahara, Y., Taniguchi, T. and Kishimoto, T. (1986). Complementary DNA for a novel human interleukin (BSF-2) that induces B lymphocytes to produce immunoglobulins. Nature, 324, 73-76.

7. Leandersson, T., Hillörn, V., Holmberg, D, Larsson, E.-L., Lundgren, E. (1982). Selective effects of interferon on distinct sites on the T lymphocyte triggering process. J. Immunol., 129, 490-494.

8. Gerdes, J., Lemke, H., Baisch, H., Wacker, H.-H., Schwab, U. and

Stein, H., J. (1984). Cell cycle analysis of a cell proliferation-associated human nuclear antigen defined by the monoclonal antibody Ki-67. J. Immunol., 133, 1710-1715.
9. Jung, L.K.L., Hara. T. and Man Fu, S. (1984). Detection and functional studies on p60-65 (Tac antigen) on activated human B cells. J. Exp. Med., 160, 1597-1602.
10. Pike, B.L., Raubitschek, A. and Nossal, G.J.V. (1984). Human interleukin 2 can promote the growth and differentiation of single hapten-specific B cells in the presence of specific antigen. Proc. Natl., Acad., Sci. USA, 81, 7917-7921.
11. Boyd, A.W., Anderson, K.C., Freedman, A.S., Fisher, D.C., Slaughenhoupt, B., Schlossman, S.F. and Nadler, L.M. (1985). Structural and functional characterization of IL-2 receptors on activated human B cells. J. Immunol., 134, 1516-1523.

10 Anti-proliferative effects and phenotypic alterations induced by methylurea derivatives in B16 mouse melanoma

J. Nordenberg, A. Fuchs, L. Wasserman, Z. Malik and A. Novogrodsky

Abstract.

The methylurea derivatives, dimethylthiourea (DMTU), tetramethylurea (TMU), tetramethylthiourea (TMTU), 1,3 dimethylurea (1,3 DMU) and 1,1 dimethylurea (1,1 DMU) were evaluated for their ability to affect B16 mouse melanoma cell growth and to induce differentiated features in these cells. The order of potency for in vitro growth inhibition was as follows:TMTU>DMTU = TMU>1,3 DMU>1,1DMU. TMTU, however was highly toxic to the melanoma cells. The phenotypic alterations induced by the methylurea derivatives include morphological changes, increase in the activity of the enzyme marker γ-glutamyltranspeptidase, increased melanin content, lipid accumulation and reduction in ATP levels. The mitochondrial damage and decrease in ATP levels induced by DMTU might give a clue to the mode of action of these agents.

Introduction

Dimethylthiourea (DMTU) is a small permeable and relatively

nontoxic scavenger of hydrogen peroxide (H_2O_2) and the H_2O_2-derived product hydroxyl radical (OH)- (1-2). DMTU was reported to prevent free radical-mediated tissue injury in vitro and in vivo (3-4).

Our recent studies have shown that DMTU inhibits B16 mouse melanoma cell growth in vitro and delays tumor appearance in syngeneic mice (5). Growth inhibition was associated with morphological alterations and slight melanogenesis, suggesting that DMTU might belong to the group of chemical inducers of differentiation (5). Other methylurea derivatives, dimethylurea and tetramethylurea were shown to induce differentiation in Friend erythroleukemic cells and to oppose epidermal tumor promotion by retinyl acetate and croton oil in hamster cheek pouch (6,7).

In the present study we evaluated the ability of several methylurea derivatives to affect B16 mouse melanoma cell growth and to induce differentiated features in these cells.

Materials and Methods

Cells:

The B16 F-10 melanoma cell line was kindly donated by Dr. A.Raz, The Weizmann Institute. Cells were cultured in RPMI supplemented with 10% fetal calf serum and antibiotics at 37°C in a humidified atmosphere of 5% CO_2 95% O_2. Cell growth experiments were performed as previously described (5,8).

Morphological studies:

For assessment of cell morphology the cells were prepared as previously described (8,9), stained by the Hematoxyline-Eosin staining and visualized by light microscopy. Transmission electron microscopy was performed as previously described (10).

Biochemical and histochemical studies:

γ glutamyl transpeptidase activity was extracted and determined as previously described (11), according the method of Tate Meister (13). Enzyme activities were expressed per DNA content. DNA was measured according the method of Burton, 1956 (14).

Lipid staining was performed as previously described according to the oil red o method (11,15). Melanin was extracted and measured as previously described (8).

ATP was extracted with perchloric acid 7% (V/V) following rapid freezing of the monolayer. The lysates were neutralized with a mixture of KOH and $KHCO_3$ and the supernatants used for ATP determination. ATP was measured by the fluorometric method of Lowry et al, 1964 (16)

Results and Discussion

B16 mouse melanoma cells were incubated with and without several methylurea derivatives (2.5-10 mM) for 48 and 96 hours. The results depicted in Table I show that the order of potency for cell growth inhibition was as follows: TMTU>DMTU = TMU>1,3DMU>1,1DMU. Only TMTU was found to reduce cell viability. TMTU reduced cell viability even at 2.5 mM and was also toxic to C57Bl mice when administered in vivo. The other methylurea derivatives DMTU, TMU and 1,3DMU inhibited cell growth without affecting cell viability. 1,1DMU up to 10 mM had no

TABLE I: THE EFFECT OF METHYLUREA DERIVATIVES ON B16 F-10 MOUSE MELANOMA CELL GROWTH

Treatment	(Increase in cell number/cm^2)X10^4 48 h	96h
None	5.2 (100)	17.0 (100)
DMTU (2.5 mM)	3.8 (73)	13.0 (76)
DMTU (5.0 mM)	2.9 (55)	9.6 (56)
DMTU (10.0 mM)	2.0 (38)	5.1 (30)
TMU (2.5 mM)	3.9 (76)	12.0 (71)
TMU (5.0 mM)	2.9 (56)	9.6 (56)
TMU (10.0 mM)	1.7 (32)	4.6 (27)
TMTU (2.5 mM)	2.6 (50)	6.7 (39)
TMTU (5.0 mM)	1.5 (31)	4.2 (25)
TMTU (10.0 mM)	1.2 (24)	1.7 (10)
1,1DMU (2.5 mM)	5.6 (108)	17.0 (100)
1,1DMU (5.0 mM)	5.2 (101)	17.0 (100)
1.1DMU (10.0 mM)	4.8 (93)	17.0 (100)
1.3DMU (2.5 mM)	4.7 (91)	14.0 (82)
1,3DMU (5.0 mM)	4.0 (78)	13.0 (76)
1,3DMU (10.0 mM)	3.0 (58)	11.5 (68)

Each number represents the mean of 4 plates. Initial cell number plated $7x10^4$ cells/ml ($1,25x10^4$ cells/cm^2). Numbers in parenthesis represent the percent of control (none).

TABLE II: THE EFFECT OF METHYLUREA DERIVATIVES ON MELANIN CONTENT

Treatment	Melanin (% of control)
None	100
TMU (10 mM)	276
1,1DMU (10 mM)	254
DMTU (10 mM)	163

The melanin content of untreated cells fluctuates in the B16 F-10 melanoma cells. In the present experiments 100% refers to 0.93 μg/10^6 cells. Values are means of 4-6 experiments.

effect on B16 melanoma cell growth.

Morphological alterations, that include cell enlargement and formation of dendrite-like structures, are induced in melanoma cells following treatment with chemical inducers of differentiation (8,17). TMTU, DMTU and TMU induced alterations in cell morphology, whereas 1,3DMU and 1,1DMU did not affect the morphology of the cells (Fig. 1). The effect of the methylurea derivatives on cell morphology seems to correlate with that on cell proliferation.

Determination of melanin content following treatment of the cells with methylurea derivatives reveals that only TMU and 1,1DMU induced a marked increase in melanin content (Table II). We have previously shown that DMTU induced a moderate increase in melanin after prolonged treatment. These results show once again that increased melanogenesis and decreased melanoma cell growth are not necessarily related. Similarly, sodium butyrate that induces many phenotypic alterations in melanoma cells (11) inhibits cell proliferation and decreases melanin content (8). γ glutamyltranspeptidase is a plasma membrane-bound enzyme that has been shown to fluctuate during carcinogenesis and differentiation (12,18,19). Table III shows the effect of the methylurea derivatives on the activity of this enzyme. TMTU, DMTU and TMU induced a marked increase in the activity of γ glutamyltranspeptidase. TMTU was most effective in inducing an increase in the activity of this enzyme. It is unclear whether the increase in the activity of γ glutamyltranspeptidase reflects a differentiated feature. We have recently found that in vitro and in vivo administration of sodium butyrate to leukemic cells induces an increase in the activity of this enzyme. This increase was associated with an increase in differentiation markers (20). DMTU and TMU were

TABLE III: THE EFFECT OF METHYLUREA DERIVATIVES ON γGLUTAMYLTRANSPEPTIDASE ACTIVITY IN B16 F-10 MOUSE MELANOMA CELLS

Treatment	γ glutamyl transpeptidase nmoles/mg DNA/h
None	1500 ± 170
1,1DMU (10 mM)	2400 ± 250
1,3DMU (10 mM)	1800 ± 360
DMTU (10 mM)	4000 ± 700
TMU (10 mM)	3400 ± 1200
TMTU (5 mM)	6450 ± 70

Cells were incubated for 96 h with and without methylurea derivatives. Enzyme was extracted and determined as described in methods. Values are means ± S.D. for 3 experiments performed in duplicate.

TABLE IV: THE EFFECT OF DMTU ON ATP CONCENTRATION IN B16 MOUSE MELANOMA CELLS

Treatment	nmoles/mg DNA
None	218 ± 42
DMTU (72 h)	120 ± 20

Values are means of 4 independent experiments ± S.D.

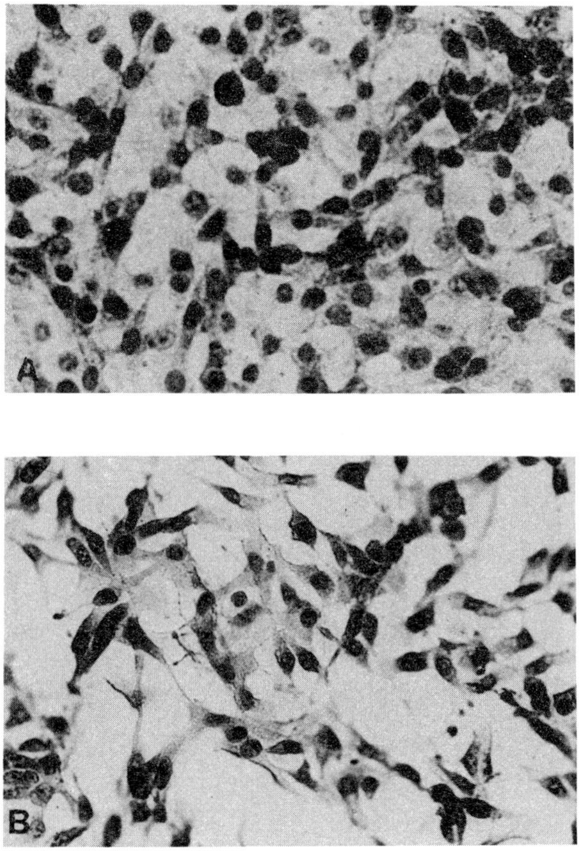

Figure 1: The effect of methylurea derivatives on B16 mouse melanoma cell morphology. Cells were treated for 48 h with 10 mM of the methylurea derivatives.

A. Untreated cells

B. DMTU-treated cells

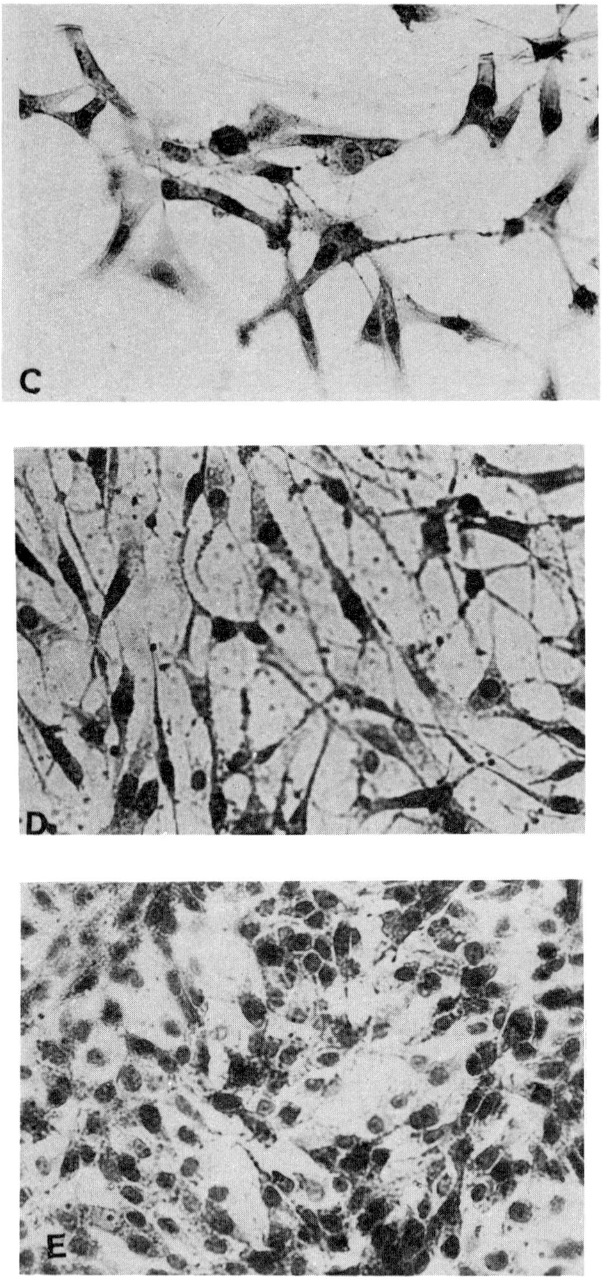

C. TMU-treated cells

D. TMTU-treated cells

E. 1,1 DMU-treated cells (X400)

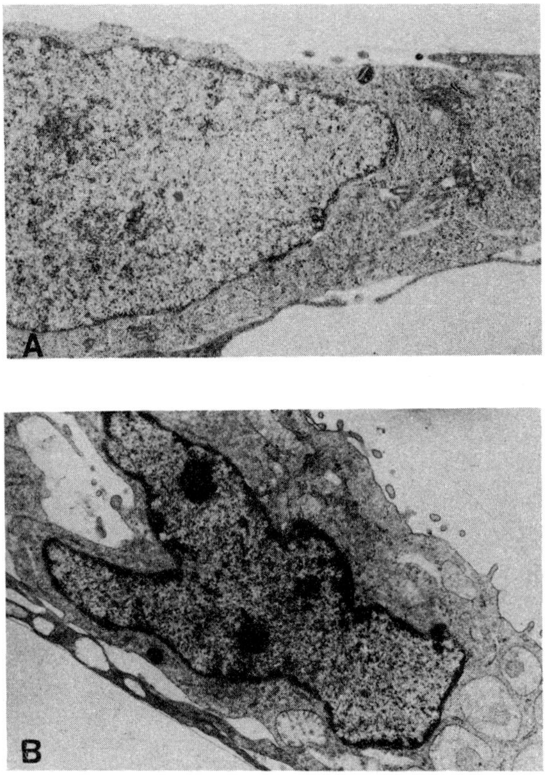

Figure 2: Transmission electronmicroscopy of DMTU-treated cells.

A. Untreated cells (x10,000)

B. DMTU-treated cells. These cells possess swollen mitochondria with disrupted cristae (x13,000).

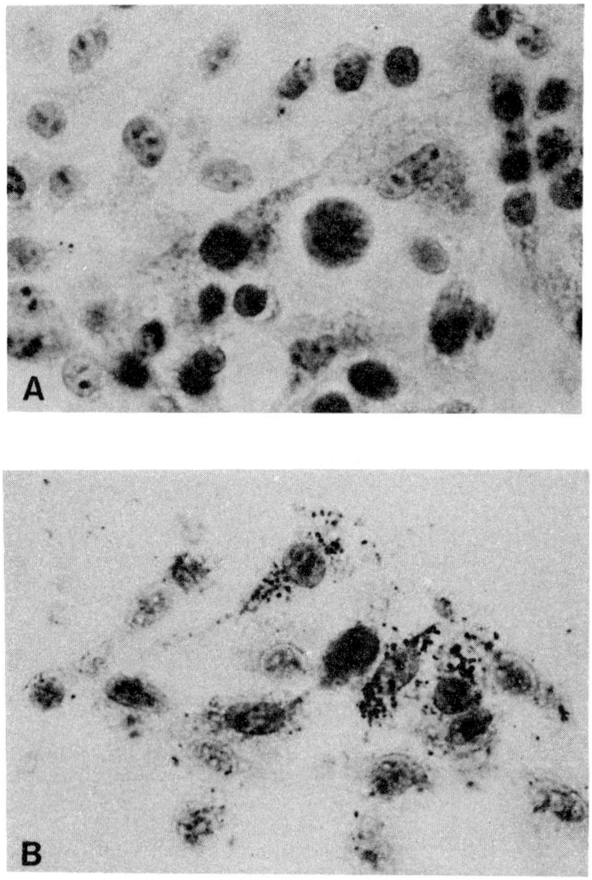

Figure 3: Lipid accumulation following DMTU-treatment of B16 F10 mouse melanoma cells.

Cells were treated for 3 days.

A. Untreated cells

B. DMTU (10 mM)-treated cells. (X400)

recently found to induce an increase in the activity of NADPH cytochrome c reductase, an enzyme marker of the endoplasmic reticulum (Data not shown).

The mechanism(s) by which the methylurea derivatives affect melanoma cells is unclear. One possibility is that these agents cause a metabolic insult, that leads to inhibition of cell growth and to selective gene activation. Transmission electron-microscopy revealed that the mitochondria of DMTU-treated cells were swollen and their cristae were disrupted (Fig. 2). The results depicted in Table IV show that DMTU markedly reduced the intracellular concentration of ATP. DMTU-treated cells accumulated lipid droplets (Fig. 3). It is unclear whether lipid accumulation results from inhibition of the respiratory chain or whether this phenomenon reflects a differentiated feature, resulting from increased lipid biosynthesis, as was shown in fibroblasts (21).

References

1. Fox, R.B. Prevention of granulocyte-mediated oxidant injury in rats by a hydroxyl-radical scavenger, dimethylthiourea. J. Clin. Invest., 74:1456-1464, 1984.

2. Repine, J.E., Fox, R.B. and Berger, E.M. Hydrogen peroxide kills Staphylococcus aureus by reacting with staphylcoccal iron to form hydroxyl radical. J. Biol. Chem., 256:7094-7096, 1981.

3. Linas, S., Parker, N.B. and Repine, J.E. Dimethylthiourea (DMTU) prevents hydrogen peroxide (H_2O_2)-mediated injury and is consumed by H_2O_2 in the isolated perfused kidney (Abstract). Clin. Res., 33:491, 1985.

4. Paller, M.S., Hoidal, J.R. and Ferris, T.F. Oxygen free radicals in ischemic acute renal failure in the rat. J. Clin. Invest., 74:1156-1164, 1984.

5. Nordenberg, J., Aloni, D., Wasserman, L., Beery, E., Stenzel,K.H. and Novogrodsky, A. Dimethylthiourea inhibition of melanoma cell growth in vitro and in vivo.

J. Natl. Cancer Inst., 75:891-895, 1985.

6. McGaughey, C. and Jensen, J.L. Effects of the differentiating agents (inducers) Dimethylacetamide, Di- and Tetramethylurea on epidermal tumor promotion by retimyl (vitamin A) acetate and croton oil in hamster cheek pouch.
Oncology, 37:65-70, 1980.

7. Preisler, H.D. In vitro and preliminary in vivo studies of compounds which induce the differentiation of Friend leukemia cells.
Haematol. Bluttransfus., 19:161-167, 1976.

8. Nordenberg, J., Wasserman, L., Beery, E., Aloni, D., Malik, H., Stenzel, K.H. and Novogrodsky, A. Growth inhibition of murine melanoma by butyric acid and Dimethylsulfoxide,
Exp. Cell. Res., 162:77-85, 1986.

9. Wasserman, L. and Kessler-Icekson, G. A simple method for the preparation of permanent slides from cell cultures.
Stain Technol., 59:353-354, 1984.

10. Glavert, A.M. Fixation dehydration and embedding on biological specimens in: Practical Methods in Electron Microscopy. Glavert, A.M. Ed., North Holland.

11. Nordenberg, J., Wasserman, L., Peled, A., Malik, Z., Stenzel, K.H. and Novogrodsky, A. Biochemical and ultrastructural alterations accompany the anti-proliferative effect of butyrate on melanoma cells.
Br. J. of Cancer, (In Press).

12. Novogrodsky, A., Tate, S.S.R. and Meister, A. -Glutamyl transpeptidase, a lymphoid cell-surface marker. Relationship to blastogenesis differentiation and neoplasia.
Proc. Natl. Acad. Sci. USA, 73:2414-2418, 1976.

13. Tate, S.S. and Meister, A. Interaction of -glutamyl transpeptidase with amino acids, dipeptides and derivatives and analogs of glutathione.
J. Biol. Chem.,249:7593-7602, 1974.

14. Burton, K. A study of the conditions and mechanisms of the diphenylamine reaction for the colorimetric estimation of deoxy-ribonucleic acid.
Biochem.J., 62:315, 1956.

15. Pearse, A.G.E. Histochemistry theoretical and applied, 3rd edition, Vol. 1, Churchill London, pp. 697, 1968.

16. Lowry, O.H.,Passoneau, J.V., Hasselberger, F.Z. and Schultz, D.W. Effect of ischemia on known substrates and cofactors of the glycolytic pathway in brain.
J.Biol.Chem., 239:18-30, 1964.

17. Huberman, E., Heckman, C., Langenbach, R. Stimulation of differentiated functions in human melanoma cells by tumor-promoting agents and dimethyl sulfoxide.
Cancer Res., 39:2618-2624, 1979.

18. Chiba, M. and Jimbow, K. Expression of gamma-glutamyl transpeptidase in normal and neoplatic epithelial cells of human skin: A marker for distinguishing malignant epithelial tumours.
Br. J. Dermatol., 114:459-464, 1986.

19. Fiala, S., Fiala, A.E., Dixon, B. γ-Glutamyl transpeptidase in transplantable chemically induced rat hepatomas and spontaneous mouse hepatomas.
J. Natl. Cancer Inst., 48:1393-1401, 1972.

20. Rephaeli, A., Rabizadah, E., Nordenberg, J., Kluska, A., Prokocimer, M., Rosenberg, I., Novogrodsky, A. and Shaklai, M. Effect of the cell differentiation inducers butyrate, retinoic acid and low dose cytosine arabinoside on leukemic cells of patients with acute non-lymphatic leukemia.
Blood, 68:(suppl.1) 646, 1986.

21. Wawra, E. Long-term effects of sodium butyrate on mouse fibroblasts: A model for differentiation.
Cell. Mol. Biol., 32: 121-129, 1986.

ACKNOWLEDGEMENT: This work is supported by a ICRF U.S. Research Grant.

11 Comparative analysis of major biological aspects of human hepatoma in culture

P. Bagnarelli, A. Festa, I. Testa, B. Champ, M. Clementi and G. Caroloni

Summary

Cell lines established from primary liver cancer of man are currently regarded as major tools in several experimental fields. Since they express and maintain several properties of the hepatic cell stably in vitro, this model system presents a useful opportunity for both elucidating mechanisms of liver biology and studying hepatitis B virus (HBV)/host relationship.

Moreover, since most of these human hepatoma cell lines at present available possess different biological characteristics and a different degree of differentiation, some aspects concerning the role of HBV in hepato-carcinogenesis and the presence of activated cellular oncogenes (c-onc) may be studied comparatively.

Here we show data on a group of seven human hepatoma -derived cell lines representing a wide spectrum in terms of origin and differentiation. Three of them (HepG2, Mahlavu and LI7A) are negative for the presence of specific HBV-DNA sequences using the Southern-blotting technique with a complete HBV-DNA probe; however, the presence of transforming sequences in these HBV-DNA negative lines could be detected by DNA-transfer into NIH 3T3 cell monolayers.

We also studied the effect of Epidermal Growth Factor (EGF) on the serum-free growth of these cell lines, observing a differential response to this mitogen in spite of specific EGF-membrane receptors being present in all the lines tested.

Introduction

Efforts have been made in the past years to establish cell cultures which retain the differentiated functions of the original liver cells; however, normal liver cells grow poorly *in vitro* and a loss of these functions may be observed in culture (1, 2).

The cell lines presently available originate from primary liver cancer (PLC) which stably maintain *in vitro* some properties expressed by the normal liver cells *in vivo* (3), in spite of the above mentioned behaviour of primary liver cultures. This indicates that gene expression is quite stable under *in vitro* culture conditions and a variety of liver-derived cells, showing a different degree of differentiation, may be studied (4).

Furthermore, it has been shown that human hepatoma is closely related to hepatitis B virus (HBV) infection (5) and specific HBV-DNA sequences in integrated or free form have been detected in most of them (6, 7).

Presently, the study of HBV-DNA integration and expression of viral genes appears to be one of the most fruitful and exciting approaches to investigating the mechanisms of liver carcinogenesis (8). Moreover, increased expression of cellular oncogenes (c-*onc*) has been observed in both experimental liver regeneration in rats and chemically-induced rat hepatomas (9) and, similarly, some of these c-*onc* were identified in hepatoma cell lines and in human liver cancer tissues (10, 11).

An additional viewpoint concerning the capability of human hepatoma cell lines to grow under serum-free conditions (12, 13) opens up several possibilities of studying the control of cell growth and its loss in hepatoma transformed phenotype.

The data reported here describe some of the above mentioned features of liver transformed cells *in vitro* :

a) the effect of Epidermal Growth Factor (EGF) on the serum-free growth of human hepatoma cell lines;

b) the ^{125}I-EGF binding to specific membrane receptors of the cell, the affinity of the binding detected and the possible relationship between the EGF-effect and the degree of cell differentiation;

c) the presence of putative activated c-onc and the presence of HBV-DNA in an integrated form.

Materials and methods.

Detection of activated cellular oncogenes.

Transfection assay was performed in NIH 3T3 cell monolayers (14). Briefly, 20 ug of high molecular weight DNA (sup 30 Kbp), precipitated by calcium phosphate treatment (15) was applied to 5 mm petri dishes, seeded at 4×10^5 cells/plate the day before. Five hours later the medium was removed and replaced with Hepes solution 15% glycerole (pH 7.1) for 4 minutes at room temperature. Cells were grown in MEM supplemented with 10% newborn calf serum (NCS).

Foci of morphological transformation (which appeared two weeks later) were cloned using the cloning cylinder procedure and their genomic DNAs were EcoRI digested, migrated into agarose gel, blotted and hybridized with ^{32}P-nick-translated Alu sequences.

The Alu probe, consisting of mainly representative sequences, was prepared according to HoucK et al. (16).

The DNA was transfered to nitrocellulose filters and hybridized with nick-translated probes as described by Maniatis et al. (17).

Analysis of HBV-DNA integrated sequences.

HindIII-digested cell DNA was migrated in 0.8% agarose and transfered to nitrocellulose filters (18); hybridization was carried out after pre-hybridization, using the nick--translated DNA of complete 3.2 kilobase pairs (Kbp) of HBV genomes (HBsAg subtype ayw) (19) inserted into the EcoRI restriction site of the pBR322 plasmid. After hybridization, the unincorporated radioactivity was washed out and the filter was exposed to X-ray film with intesifying screens for 10 days at -80°C.

Cell cultures and the effect of Epidermal Growth Factor.
All the cell lines used in this research (Table I) were cultured in RPMI 1640 medium (M.A. Bioproducts, Walkersville, MD) supplemented with 10% fetal calf serum (FCS) and antibiotics (100 UI/ml penicillin and 100 ug/ml streptomycin). Serum-free cultures were plated using the RPMI 1640 medium with or without 10^{-8}M Na_2SeO_3 (12, 13).
For dose-response experiments cells were plated in medium containing serum; after 24 hours the cells were washed three times and fresh serum-free RPMI 1640 containing different amounts of EGF was added.

Hormone binding and affinity of EGF receptors.
^{125}I-EGF binding to human hepatoma cells was assayed using a previously described procedure with a minor modification (20). Cells were cultured on 4-well tissue culture dishes (Nunc, Roskilde, Denmark) at a density of 1-2 x 10^5 cells/well; the culture medium was discarded and cells were rinsed with Hanks' balanced salt solution. A 0.1 M Hepes buffer containing 0.5 ng/100ul ^{125}I-EGF (2.4 to 3.5 x 10^5 cpm/ug) was used and, after incubation for different periods, cells were washed using phosphate buffered saline (PBS) supplemented with 1 mg/ml bovine serum albumin (BSA) (pH 7.0) at 4°C and lysed by treatment with lysis buffer (0.1 M NaOH, 2% $NaCO_3$ and 1% SDS). Radioactivity was determined in a γ-counter.
All cell lines tested were incubated at 4°C, 24°C and 37°C. 24°C turned out to be the most suitable temperature; binding experiments were similarly performed at different pH values. Non specific binding was determined in the presence of a 200 fold excess of unlabeled EGF. 1 to 600 ng/100ul ulabeled EGF was added to the binding buffer for competition assay; ^{125}I-EGF degradation was assayed by precipitation in the presence

of 10% trichloroacetic acid (TCA).
A monoclonal antibody to the EGF receptor (supplied by Oncogene Science, Mineola, NY) was used to confirm the affinity of the binding detected; two concentrations of monoclonal antibody (containing 6.25 and 0.78 ug/ml mouse IgG respectively) were assayed. ^{125}I-EGF binding of treated and untreated monoclonal cell lines was tested as previously described.

Results.

Integrated HBV-DNA sequences and cellular oncogenes in human hepatoma cell lines.

C-*onc* analogous to retrovirus transforming sequences were identified to be amplificated or activated by single point mutation (22, 23) in a variety of tumors or cell lines deriving from them (24).
Since some of these oncogenes are also overexpressed in some rodent and human cells, we performed transfection assays in NIH 3T3 cells to check which of these putative oncogenes were present in an activated form in liver cancer cell lines. DNAs derived from three cell lines (Mahlavu, LI7A and HepG2) were found to give foci of morphological transformation in the NIH 3T3 recipient monolayers using the transfection assay. The DNA isolated from these representative foci of the first round of transfection was subjected to Southern-blot analysis, thus revealing the presence of human *Alu* sequences and indicating that human transforming DNA sequences derived from human hepatoma cell lines are truly involved in cell transformation (Figure 1). Positive primary *Alu* sequences from rhabdomyosarcoma (RD) cell lines are also indicated, as are positive primary and secondary *Alu* sequences from human bladder carcinoma cells (T24 cell line) containing N-*ras* and Ha-*ras* activated c-*onc* respectively.
Table I summarizes 7 different hepatoma cell lines employed in this study, characterizing them with respect to the presence of integrated HBV-DNA copies, expression of hepatitis B

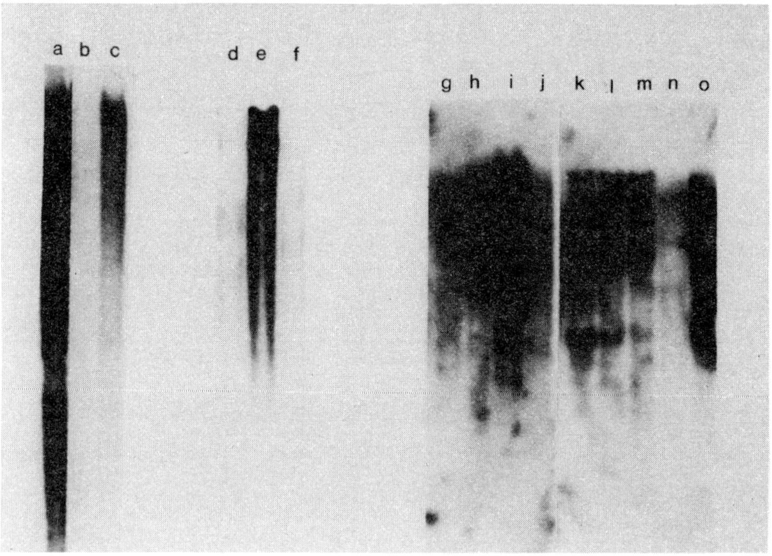

Figure 1

HYBRIDIZATION OF DNA DERIVED FROM LI7A, MAHLAVU AND HEPG2 TRANSFORMANTS WITH *ALU* PROBE.

Primary LI7A (line *a*), Mahlavu (line *e*) and HepG2 (lines *j* and *o*) NIH 3T3 transformants; primary T24 (lines *c*, *g*, *i* and *n*) and secondary T24 (lines *k*, *l* and *m*) NIH 3T3 transformants; primary RD (line *h*) NIH 3T3 transformant; NIH 3T3 control cell DNA (lines *b*, *d* and *f*). DNA samples (20 ug) were digested with *Eco*RI, electrophoresed in 0.8% agarose gel, transfered to nitrocellulose filters and hybridized with ^{32}P-*Alu* probe (2×10^{8} cpm/ug).

Table I

cell line (1) (designation	origin (1)	integrated (2) HBV-DNA copies (number)	HBsAg (2) expression	transforming (3) sequences	references (1)
Hep3B	U.S.A.	+ (2)	+	-	32
HepG2	Argentine	-	-	+	32
PLC/PRF/5	S. Africa	+ (7)	+	-	33
HCC36	China	+ (4)	-	-	34
Mahlavu	S. Africa	-	-	+	35
LI7A	Japan	-	-	+	36
HA22T/VGH	China	+ (3)	-	-	37

(1): designation and origin of human hepatoma cell lines used in this study.
(2): see 'Materials and methods'.
(3): transfection assay was performed in NIH 3T3 cell monolayers as previously reported by Cooper G.M. (24) using 20 ug of genomic cell DNA / plate. DNA derived from foci was EcoRI digested, blotted and hybridized against Alu sequences as described in 'Materials and methods'. Data on both the presence of integrated HBV-DNA sequences and S-gene expression were evaluated for all previously untested cell lines.

Table II

Inhibition of ^{125}I-EGF binding to specific receptors human hepatoma cell lines by monoclonal antibody to EGF-receptors.

cell line	mouse IgG concentration (% of inhibition)	
	6.25ug/ml	0.78ug/ml
HepG2	66	22
Hep3B	54	36
LI7A	3	-
HCC36	54	23

Cells were incubated for 30 minutes in serum-free culture medium containing different amounts of monoclonal antibody to the EGF receptor at 37°C; the degree of binding was then evaluated as reported in 'Materials and methods' for both treated and untreated cells.
Results are reported as the mean of four assays and are expressed as percentage of binding inhibition.

surface antigen (HBsAg) and presence of transforming sequences.

Our data confirm the presence of an activated N-ras gene in HepG2 cell line (25), whereas the transforming sequences detected in LI7A and Mahlavu have yet to be identified. Additionally, these cell lines are negative for free and/or integrated HBV-DNA sequences.

Effect of EGF on serum-free cultured human hepatoma cell lines.

Human hepatoma cell lines can generally grow in fully-defined serum-free media, but a quite different degree of attachment to the solid phase is possible under these culture conditions. In order to test the effect of EGF on these cell lines, we cultured cells for 24 hours after plating in medium supplemented with 10% FCS. After three washings using RPMI 1640, fresh serum-free medium containing EGF was added (Figure 2). Of the cells studied, HepG2 appears to be the most responsive in terms of growth, whereas a slight increase in cell replication within a narrow range can be observed for PLC/PRF/5, Hep3B and HCC36 cell lines. Increasing EGF concentrations inhibited the cell replication only in two cell lines (HA22T/VGH and LI7A). The inhibitory effect observed in HA22T/VGH cells seems to be dependent on the continous presence in the culture medium of the growth factor as shown in Figure 3, because cell replication is reduced in the presence of EGF and they grow normally in its absence.

EGF-membrane receptors.

Figure 4 shows the binding of ^{125}I-EGF detected in human hepatoma cell lines. All cell lines tested possess EGF-receptors, but varying degrees of affinity were observed (Figure 5). Particularly, the fastest binding was achieved at 24°C by HA22T/VGH cells (which reached a maximum within 30 minutes) and by PLC/PRF/5 cells (60 minutes); nevertheless LI7A cells were observed to reach the highest level of binding within the longest period of incubation.

HA22T/VGH and Hep3B membrane receptors for EGF showed the

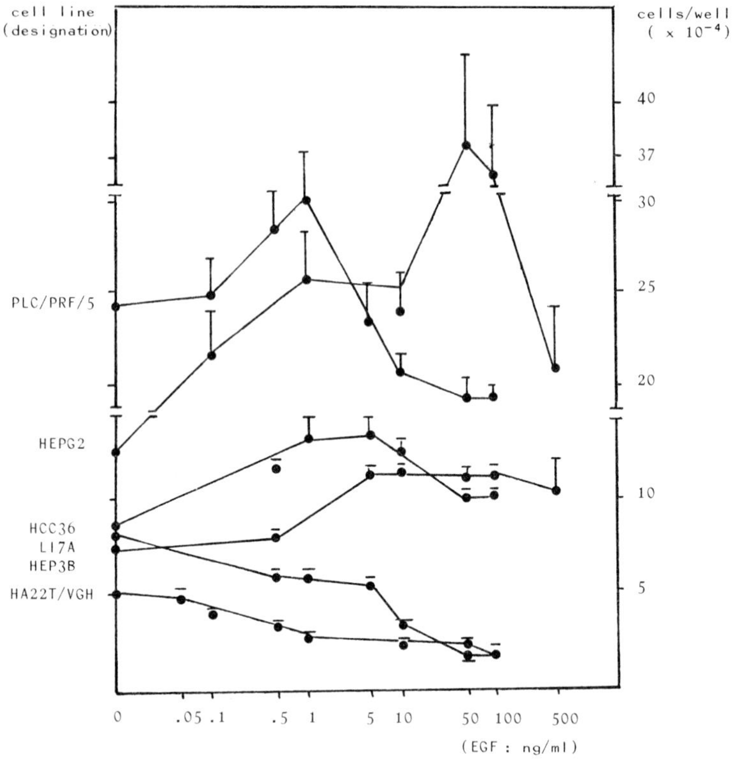

Figure 2

HUMAN HEPATOMA CELL LINES: DOSE/RESPONSE CURVES FOR EGF

Cells were plated ($5x10^4$ cells/well) in medium supplemented with 10% FCS; 24 hours later, after repeated washing, fresh serum-free RPMI 1640 supplemented with different amounts of EGF was added. Cell counts were performed in hemocytometer 92 hours later. Results are given as the mean of four replicates ± S.D.

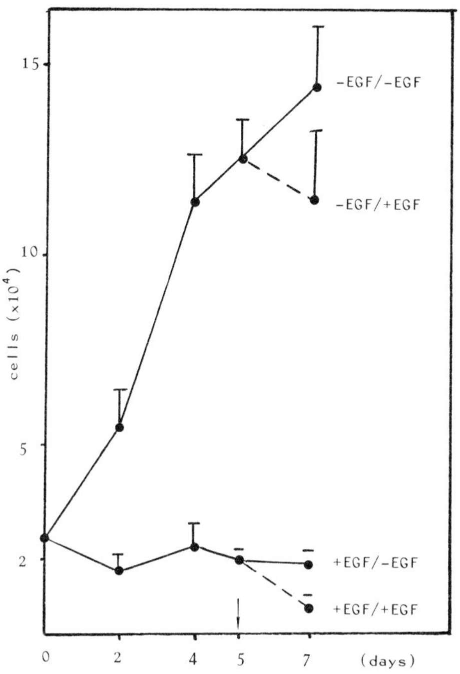

Figure 3

REVERSAL OF EGF INHIBITION OF HA22T/VGH CELL GROWTH

4×10^4 cells/well were plated in RPMI 1640 medium supplemented with 10% FCS and, after repeated washing 24 hours later, the medium was changed to RPMI 1640 supplemented with 10^{-8}M Na_2SeO_3 in the absence or in the presence of 10 ng/ml EGF (day 0). Each experimental group was subdivided on day 5 (arrow) into two groups with the substitutions indicated. Values given are the mean $\pm$ S.D. (n=4).

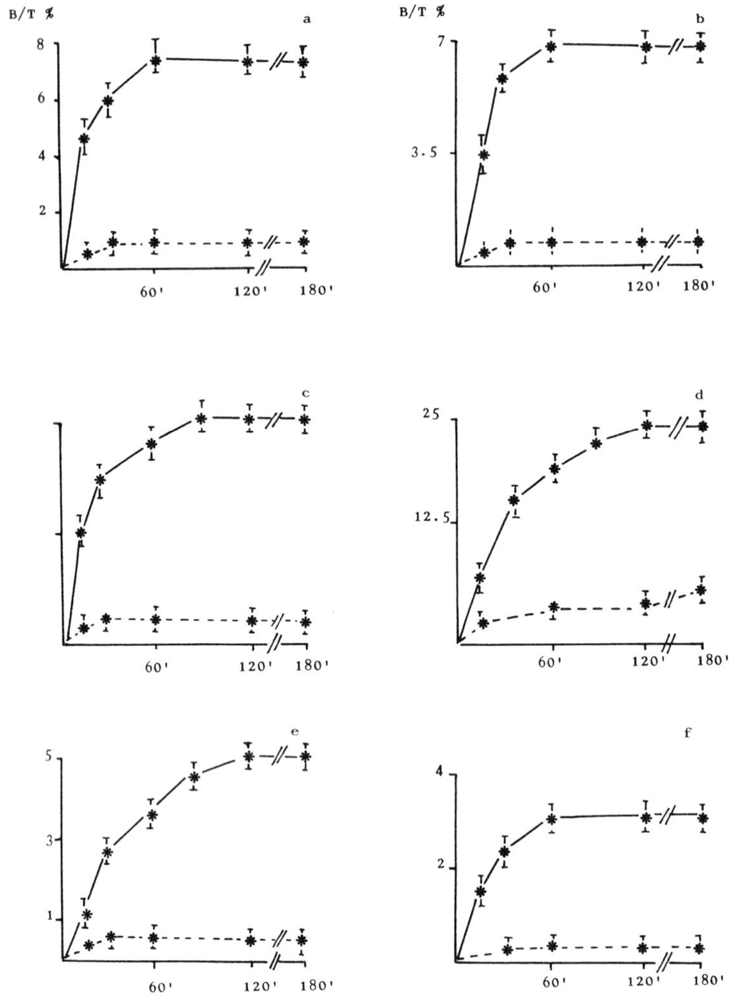

Figure 4

EGF RECEPTORS OF HUMAN HEPATOMA CELLS

^{125}I-EGF binding was performed using 4-well dishes (1-2x10^5 cells/well) as described in 'Materials and methods'. Data are shown as percentage of bound/total ^{125}I-EGF after incubation for different periods (minutes); the non specific activities detected are also reported (dashed lines).

(a= PLC/PRF/5; b= HA22T/VGH; c=Hep3B; d=LI7A; e=HCC36; f= HepG2)

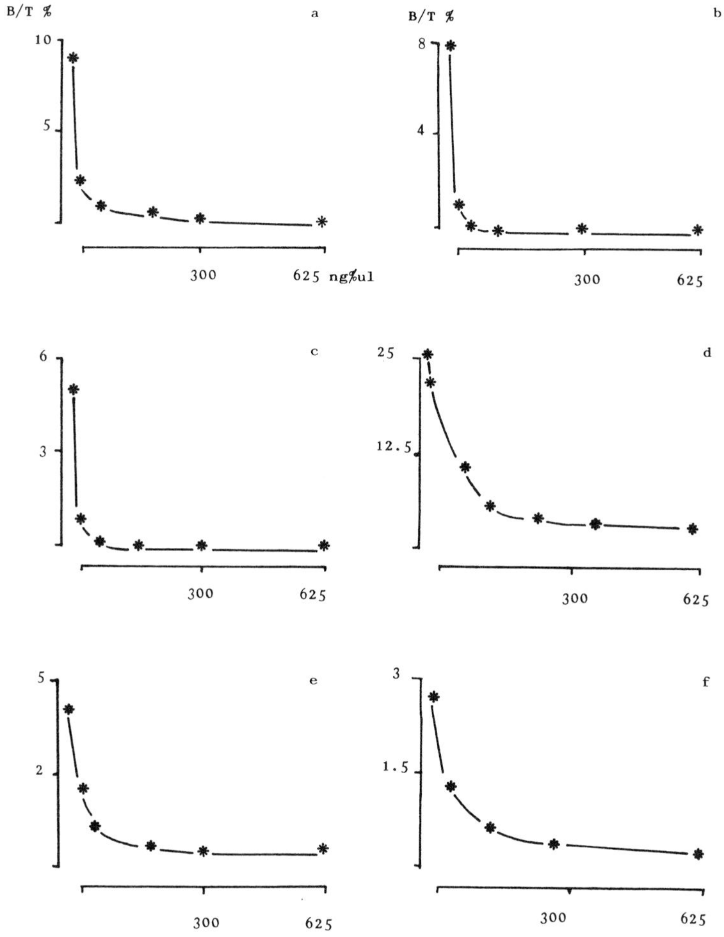

Figure 5

COMPETITION ASSAY OF ^{125}I-EGF BINDING.

0.1 to 600 ng/100 ul of unlabeled EGF were added to the binding buffer for competition assays. Data are given as percentage of bound/total ^{125}I-EGF.
(legend of single figures is that reported forFigure 4)

greatest affinity, whereas a particularly low degree of competition was observed in LI7A receptors.
All the experiments described were carried out at various temperatures (4°C, 24°C and 37°C) and differing binding buffer pHs: 24°C and pH 7.5 turned out to be the most suitable incubation conditions (Figure 6a and 6b).
We also tested whether a monoclonal antibody to the EGF receptor of A431 epidermoid carcinoma cell line could inhibit EGF-binding; Table II reports the degree of inhibition observed for HepG2, Hep3B, LI7A and HCC36 cell lines. Poor inhibition was observed for LI7A cells, thus underlining that these cells probably possess a heterogeneous population of EGF-membrane receptors.

Discussion

Cell lines are now available, derived from primary liver tumors, which express many of the functions of normal liver cells.

The principal aim of this report has been to improve the understanding of several main biological aspects of a group of human hepatoma derived cell lines. These aspects could be involved in mechanisms of carcinogenesis and/or could also be relevant to the biotechnological exploitation of these lines.

The epidemiological correlation between the incidence of hepatocellular carcinoma of man and the prevalence of chronic carrier status (5) and the evidence of HBV-DNA integration in most hepatomas (26, 27, 28) seems to support an etiological role of HBV in PLC. However, since HBV-DNA is not found in all human hepatomas, some other agents must be implicated, such as aflatoxin B1 and chemical carcinogens (29).

In order to study several biologic features of PLC, including the analysis of HBV-DNA integrated sequences, we decided to carry out experiments using hepatoma cell lines rather than bioptical specimens, because different methods can be standardised more rigorously under these experimental conditions.

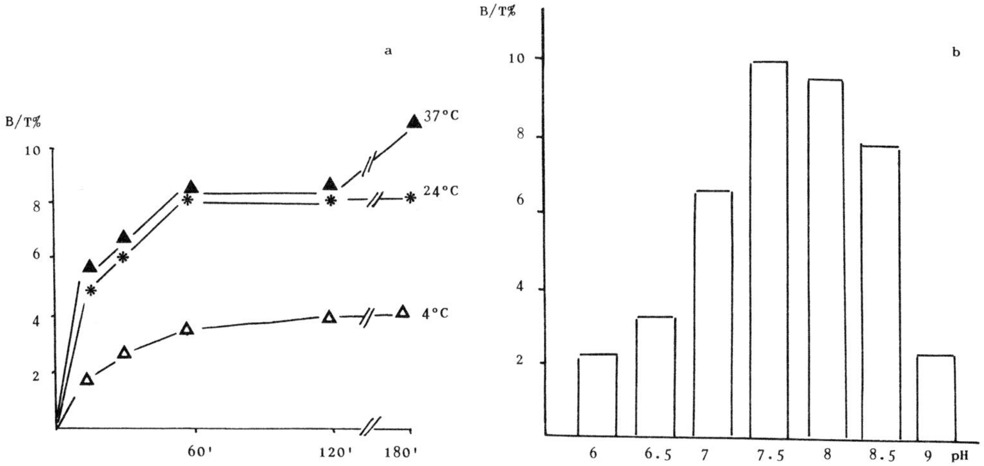

Figure 6

EFFECT OF TEMPERATURE AND pH on EGF BINDING.

The effect of the temperature and pH on the ^{125}I-EGF binding to PLC/PRF/5 cell membrane receptors was investigated.

a= temperatures; b= pHs.

Our data revealed the presence of transforming sequences only in cell lines devoid of HBV-DNA copies, thus supporting the hypothesis that HBV could not be directly involved in all the liver tumors.

In addition, since c-onc are structurally and functionally related to the autocrine growth capability of cancer cells (30, 31), we studied the transformed phenotype of these cell lines in terms of both effect of EGF on cell growth (observed under chemically controlled conditions) and the presence of specific membrane receptors for EGF. The replication of two cell lines was inhibited by the presence of EGF, while a different degree of stimulation may be observed for the other cell lines studied.

EGF was seen to enhance DNA synthesis in cell culture (38), and additionally it increases the uptake of glucose, amino-acids and uridine (39, 40), thus indicating that it can regulate a variety of cell functions. On the other hand, EGF was seen to inhibit the growth of A431 human epidermoid carcinoma cells *in vitro* (43, 44, 45), but was not seen to inhibit normal cell growth. More recently, the growth of the human hepatoma cell line LI7A was seen to be inhibited by the presence of EGF (46). It has been proposed that these inhibitory effects might be due to the presence of a large number of high affinity membrane receptors, causing an excessive kinase stimulation (47).

Our data pricipally show up the lack of transforming sequences in all the HBV-DNA positive human hepatoma cell lines tested, while all the HBV-DNA negative lines exhibited positive *Alu* sequences. Moreover, we observed a differential response to EGF, which does not appear to be linked to the presence of specific high affinity receptors, while the inhibition of growth observed in two cell lines seems principally related to the degree of differentiation shown by single cell lines, thus involving other mechanisms not as yet fully clarified.

Overall, taking a global view of the data, the major biological aspects of human hepatoma-derived cell lines in culture may be correctly defined. It may actually represent a complement necessary for both the understanding of liver carcinogenesis and biotechnological purposes.

Acknowledgements.

We wish to thank Dr. C. Chang and Dr. B.B. Knowles for providing part of human hepatoma cell lines, Dr. F. Galibert for HBV-DNA probe, Dr. P. Nardeux for preparing Alu sequences and Dr. R. Cassingena for hospitality to G.C.

References

1. Tokiwa T. et al. : In Vitro 15, 393-397, 1979
2. Auberger P. et al. : Biochem. J. 210, 361-367, 1983
3. Knowels B.B et al. : in Advances in hepatitis research. F.U.Chisari ed, Masson Publishing USA, Inc, New York, 1984
5. Maupas P. and Melnick J.L. : Progr. Med. Virol.27, 1-10,1981
6. Brèchot C. et al.:Proc. Natl. Acad. Sci. USA 78, 3906-3910,1981
7. Koshy R. et al. : J. Gen. Virol. 57, 95-102, 1983
8. Tiollais P. and Wain-Hobson S.: in Advances in hepatitis research. F.U. Chisari ed, Masson Publishing USA, Inc, New York, 1984
9. Makino R. et al.: Nature 310, 697-698, 1984
10. Jianren G. et al. : Scientia Sinica (series B) 29, 173-180, 1986
11. Ochiya T. et al.: Proc. Natl. Acad. Sci. USA 83, 4993-4997,1986
12. Nakabayashi H. et al.: Cancer Res. 42, 3858-3863, 1982
13. Bagnarelli P. et al. : Develop. Biol. Standard 60, 45-53, 1985
14. Jainchill J.S. et al.: J. Virol. 4, 549-553, 1969
15. Wigler M. et al.: Cell 16, 777-785, 1979
16. Houck C.M. et al.: J. Mol. Biol. 132, 289-306, 1979
17. Maniatis T., Fritsch E.F., Sambrook J.: Molecular cloning: a laboratory manual. C.S.H. ed, New York, Cold Spring Harbor Laboratory. 1982
18. Southern E.M. : J. Mol. Biol. 98, 503-517, 1975
19. Galibert F. et al.: Nature 281, 646-650, 1979
20. Engstrom W. et al.: J. Cell.Sci. 73, 361-363, 1985

21. Kawamoto T. et al.: Proc. Natl. Acad. Sci. USA 80, 1337-1341,1983
22. Tabin C.J. et al.: Nature 300, 143-149, 1982
23. Reddy E.P. et al.: Nature 300, 149-152, 1982
24. Cooper G.M.: Br. J. Cancer 50, 137-142, 1984
25. Notario V. et al.: in Cancer cells, oncogenes and viral genes. Vande Woude G.F., Levine A.J.,Topp W.C. & Watson J.D. eds, Cold Spring Harbor Laboratory, Cold Spring Harbor, New York 1984, pp.425-432.
26. Brèchot C. et al.: Nature 286, 533-535, 1980
27. Edman J.C. et al.: Nature 286, 535-538, 1980
28. De Jean A. et al.: Proc. Natl. Acad. Sci. USA 50, 2505-2509 1983
29. Lin J.J. et al.: Lab. Invest. 30, 267-278, 1974
30. Sporn M.B. and Todaro G.J. N. Engl. J. Med. 303, 878-880, 1980
31. Heldin C.H. and Westermark B. Cell 37, 9-20, 1984
32. Aden D.P. et al.: Nature 282, 615-617 , 1979
33. Alexander J.J. et al.: S. Afr. med. J. 50, 2124-2129 , 1976
34. Sung C.H. et al.: Abstracts' volume of the 1986 Meeting on Molecular Biology of hepatitis B Viruses C.S.H. ed, Cold Spring Harbor , New York.1986
35. Prozesky O.W. et al. in :'Liver'. Saunders S.J. and Terblanche J. Eds., London, Pitman Medical, 1973.
36. Hirohashi S. et al. Cancer Res. 39, 1819-1828, 1979
37. Chang C. et al. Mol. Cell. Biol. 3, 1133-1137, 1983.
38. Carpenter G. and Cohen S. Ann. Rev. Biochem. 48, 193-207, 1979
39. Hollemberg M.D. Vitam. Horm. 37, 69-74, 1979.
40. Hollemberg M.D. and Cuatrecasas P. Proc. Natl. Acad. Sci. USA 70, 2964-2969, 1973.
41. Barnes D.W. J.Cell. Biol. 93, 1-4, 1982.
42. Gill G.N. et al. J. Cell. Biochem. 19, 249-253, 1982.
43. Johnson L.K. et al. Proc. Natl. Acad. Sci. USA 77, 394-398, 1980.
44. Imai Y. et al. Cancer Res. 42, 4394-4399,1982.
45. Kore M. et al. J. Clin. Endocrinol. Metab. 62, 874-880, 1986.

46. Knowles B.B. *et al.* Biochem. Biophis. Res. Commun. 126, 8-14, 1985.
47. Kawamoto T. *et al.* Proc. Natl. Acad. Sci. USA 80, 1337--1341, 1983.

Discussion

R. E. SPIER: Can we ask Dr. Murakami about his human - human hybridomas? If they are used therapeutically what kind of response is obtained. Why should one use Hu-Hu hybridomas? What not use murine hybridomas in humans? Let us examine these issues because Hu-Hu hybridomas are not very productive and few investigators have been able to achieve yields in excess of 10ug?ml in batch culture. Also you quoted a productivity of 720 gms per litre of culture per day is that correct.

MURAKAMI: A murine monoclonal is know to produce IgG or 5pg/cell/min in the initial state for a short time, so for cells at 10^8/ml in a producing at this initial rate you could expect the productivity I mentioned.

We want to use antibodies to cure diseases in the body, if you use murine monoclonal antibodies, 50% of the patients make anti-antibodies, so human monoclonal antibodies.

SPIER: It is a debatable point for you will get a reaction against the idiotope of the injected material, or if you use the 'missile' approach then the warhead is normally a foreign protein which would produce an immune reaction,

	therefore the problem may be more focused on the effects of the proteins present as impurities which would either be from a foreign cell (mouse) or homologous source, human - human cell line.
	Also have you tried to use any Hu-Hu monoclonal antibodies therapeutically
MURAKAMI	It is under investigation
SPIER	Some of your monoclonal antibodies have been injected into humans
MURAKAMI	Yes
SPIER	What happens
MURAKAMI	Antibody specifically binds to the target in an imaging system
SPIER	Did you identify any anti-antibodies?
MURAKAMI	No, not yet
K. JAMES	What do you add to your culture to increase HMCA secretion by 8-10 fold.
MURAKAMI	They have this theoretical possibility which is not yet realised in practice so I added substances derived from a lymphoblastoid cell line derived with a molecular weight of 20KD
JAMES	Is this a differentiation or growth promoting factor
MURAKAMI	A secretion promoting factor
JAMES	You would make a lot of money from this
KADOURI	I agree that parotic cells can produce a 30% of their weight of enzyme per day
TOLBERT	What has been your experience with the stability of HMAB secreting cell lines.
MURAKAMI	Our hybridoma cell has been in culture for 2 years with no change in productivity. The

secret is that the chromosome number of the fusion partner is 46 which is the normal number so when you fuse the lymph...... with this t.... cell the fusion partner has exactly double the normal number of chromosomes (92), This enables all the P-cell chromosomes to be maintained in the hybridoma thus leading to stable expression.

MAIORELLA Having returned from the Kiova Island Conference on MCAb it would seem that the general levels of expression of 10ug as mentioned by Ray Spier are somewhat low as there are a number of groups indicating levels of 25 ug/ml for hu-hu hybridomas and up to 100 ug/ml for hu-mouse heterohybridomas.

Regarding magic factors for EB human Ab producing cells 1L2 (recombinant) has been used by Jim Lowick at CRC and also a group at the University of Oklahoma and they have seen a factor of 5 to 10 increase in expression levels.

FEDER Do we need a cell which produced 5pg/cell/sec if you had 100 pg/cell/day in the steady state then a reactor with 10^{12} cell would produce 100 gms/day of product. We tend to overlook the problems of recovery of product from the system, the DSP problems. The scale of the production system for such products is of this order and is much smaller than that generally associated with microbial technology systems. Dr Murakami seems to have solved some of the problems, but as an enzymologist I am uncomfortable is using initial production rates; I am more comfortable in the real world with data on what the cells are doing over a month at a steady state rate. There are certainly people in this room with hybridomas (Human, Ed) which can produce 10-30ug/10^6 cells/day, and if you have a good system with good recovery you should be able to obtain adequate amounts of material. Note if you increase productivity by a factor of 10 noone will wish to do it because the rewards will be inadequate.

MURAKAMI Potential of the cell is higher than that which is generally obtained

SPIER In a recent book (Molecular Cell Biology (Darnell, Lodishe & Baltimore) it was noted that animal embryo cells can divide every 20 minutes. You can, therefore, back calculate the rate of protein a cell can produce per unit time. This is the kind of figure that accountants and line-managers will hold the scientist-technicians against. For while, as Dr. Feder says, we don't need to produce at theoretical maximal rates now, in the future it is not unlikely that only those Companies producing close to the theoretical maxima will survive economically.

FEDER We can use the secretion capability of the hybridoma to provide us with a model (as a tool) to further develop other recombinant cells (BMK) which have not got the same inherent secreting potential

SPIER We should take a cell which normally secretes materials (pancreas, gastric, mucosal cell) and engineer that to make recombinant products.

FEDER Harvy Lodish would like a liver cell line.

ARATHOON You have made a lot of antibodies and you have screened them with different cell lines, did you see any effects of those different antibodies on the growth and secretary activity of those different cells

MURAKAMI The secretion factor is specific for certain cell lines

ARATHOON Did it stimulate and inhibit growth

MURAKAMI Neither

KADOURI In our competition with nature, nature is still ahead. The metabolising cell in the egg embryo uses fat rather than carbohydrate as the basis of its metabolism.

12 Minireview: Essential strategies for the application of herpes simplex virus as a vector for the expression of foreign genes

M. Ackermann and U. Kihm

INTRODUCTION

Viral surface proteins and viral proteins that are expressed on the surface of infected cells are important elicitors of the host immune response and of viral pathogenicity (13,18). The amount of these proteins is very often crucial for the efficacy of vaccines (16). The expression of high levels of foreign genes in Herpes Simplex Viruses (HSV) has been reported (15,17). Since herpes viruses are widely applied as live virus vaccines, they might as well express foreign viral surface antigens to induce simultaneously protection against other viral infections. In contrast to prokaryotic vectors, herpes viruses can be engineered to produce foreign viral proteins in a manner similar to their synthesis during a natural infection (9,17).

In order to use HSV in recombinant vaccines as vector for the expression of foreign genes, the following has to be regarded:

i) Undesirable features, e.g. genes responsible for pathogenicity, have to be inactivated and/or replaced by foreign genes (6,7,12).

ii) The location of insertion of the foreign gene has to be chosen carefully such as not to interfere with the replication of the vector (15).

iii) Deletions as well as insertions are made possible by recombination.

iv) Foreign genes fused to ALPHA promoters of HSV are expressed at high levels (11,17).

v) Two important strategies exist for the selection of recombinant herpes viruses:

a.) Using appropriate substrate conditions, it is possible to select for and against virus expressing thymidine kinase (TK)(15).

b.) The identification of surface antigens with monoclonal antibodies allows the selection of recombinants directly out of a mixed population (1).

The purpose of this paper is to elucidate the experimental design used for the construction of recombinant herpes viruses, to give examples for the selection of recombinants, to show how selected genes can be inactivated by recombination, and to explain the most elegant way to determine sites within the viral genome that are suitable for insertions.

EXPERIMENTAL DESIGN

The pathway of experiments for the construction of recombinant viruses is shown in Fig. 1. The decisions to be made depend on the availability of data concerning the parent viruses. For the construction of a recombinant herpes virus expressing a foreign gene, the following steps are carried out (1,15,17):

i) The foreign gene is fused to an HSV ALPHA promoter.

ii) The foreign DNA fragment is engineered to be flanked by sequences homologous to the DNA of the vector at the site of the desired introduction of the foreign gene.

iii) This new DNA fragment is cotransfected with infectious HSV DNA into susceptible cells. A mixed population of recombinant- and parent viruses is grown.

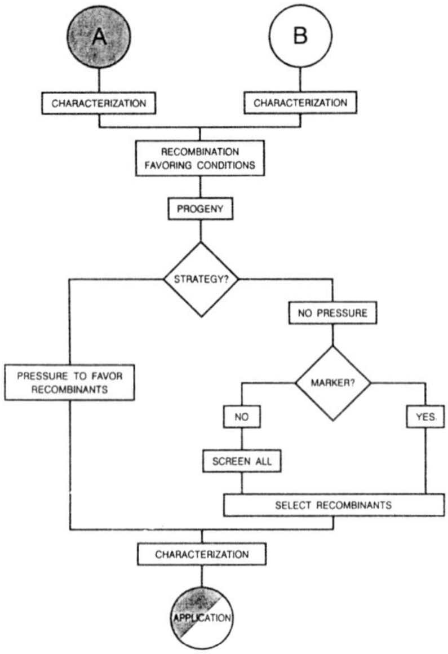

FIG.1

PATHWAY FOR THE CONSTRUCTION OF RECOMBINANT VIRUSES

In order to recognize recombinants between parent A and parent B, the viruses have to be characterized.
Characteristic features of the parents serve as markers for the recognition of an AxB recombinant virus.
The two parents are mixed and grown under conditions favoring recombination events. The mixed population of progeny virus may be subjected to selective pressure, depending on the selection strategy applied. Conditions that allow only the replication of recombinants but not the replication of the parents are applied whereever possible. If pressure is not applicable, the recombinants may be recognized according to a phenotypical marker. Such recombinants are selected directly out of a mixed population. If no phenotypical marker is available, progeny virus has to be plaque purified and screened for recombination by DNA analysis.
Recombinants have to be characterized before their application.

iv) Desired recombinant progeny virus is selected by the application of selective pressure, e.g. selection for or against viral TK, or by the nondestructive identification of a phenotypical marker expressed by the desired recombinants.

The genomes, proteins, and functions of parent- and recombinant viruses are characterized at the beginning and at the end of these experiments in order to optimize the experimental design and to reveil clues for the recognition of the recombinants.

SELECTION OF DESIRED RECOMBINANTS

The specific selection of the recombinants and their separation from the parents is crucial. Commonly used is the selection on the basis of the expression of viral enzymes, e.g. the selection of TK+ or TK- virus. More recently, the selection based on the nondestructive recognition of proteins expressed on the surface of infected cells has been introduced (1,5).

a.) SELECTION FOR RECOMBINANTS WITH THE LOST (TK-) OR AQUIRED (TK+) CAPACITY TO SPECIFY VIRAL THYMIDINE KINASE.

Herpes virus genomes encode several enzymes involved in the DNA metabolism (14,15). One of these enzymes is thymidine kinase (TK: phosphorylation of thymidine). The selection of TK- virus is based on the difference of the substrate specificity of the viral and cellular TK, respectively. For example arabinosylthymine (Ara T) in the growth medium will not affect uninfected TK+ cells and will permit the multiplication of TK- viruses but not TK+ viruses in these cells. The selection of TK+ progeny virus is more difficult. TK- cells are maintained in a medium containing hypoxanthine, aminopterin, and thymidine (HAT). The TK- cell line is required in order to prevent phosphorylation of thymidine by cellular TK. The selective medium is used to block the main pathway of DNA synthesis and to provide the metabolic products used in the TK mediated pathway of DNA synthesis. For this reason only TK+ virus is able to replicate (7,8,10,12).

These selection procedures are used for the construction and selection of recombinant viruses as follows:
The wild type (wt) viral TK gene is regulated as a BETA gene (2,3). In a first step, the TK coding sequences can be deleted and TK- virus can be selected. In a second step, the cloned coding sequences of the TK gene are fused to an ALPHA promoter (ALPHA TK, chimeric gene). Upon cotransfection with TK- virus DNA, ALPHA TK is not likely to recombine back into the original site of BETA TK because of the absence of TK coding sequences in the TK- virus and because the ALPHA promoter of the cloned chimeric gene lacks sequence homology to the BETA promoter upstream from the original TK gene (4). Depending on the sequences flanking the chimeric ALPHA TK gene, it is possible to insert ALPHA TK into any site of the viral genome, provided that the insertion does not interfere with viral replication. These TK insertions serve not only for the selection of TK+ virus, but also as markers for positions in the viral genome, where insertions can be introduced (4,12,15). These TK sequences can be replaced in a following recombination step. Recombinants containing an insertion in the chimeric TK gene are selected by the selection for TK- virus.

b.) SELECTION OF RECOMBINANTS BY THE NONDESTRUCTIVE RECOGNITION OF PROTEINS EXPRESSED ON THE SURFACE OF INFECTED CELLS.

An example of this technique is illustrated in Fig. 2. In this selection no pressure is applied on nonrecombinant viruses. The identification of the recombinants is based on the lost or acquired capacity of viral surface antigens to react with specific monoclonal antibodies (1,5):

i) The specific reaction of a monoclonal antibody to the

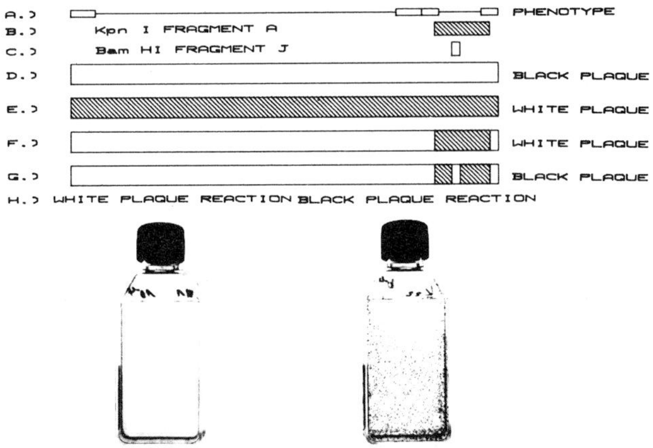

FIG.2

EXAMPLE FOR THE CONSTRUCTION AND NONDESTRUCTIVE IDENTIFICATION OF RECOMBINANTS ACCORDING TO THE ACQUIRED OR LOST CAPACITY TO EXPRESS SPECIFIC SURFACE ANTIGENS (1).

Cells were transfected with viral genomic DNA and cotransfected with cloned DNA fragments as indicated below.

Progeny virus was subjected to the nondestructive immune staining of surface proteins (black plaque assay).

Plaques phenotypically differing from those of the parent virus were selected and plaque purified. Deproteinized DNA of these recombinants was used as parent in further cotransfections as indicated below.

Line A.) HSV DNA sequence arrangements.

Line B.) KpnI fragment A of HSV-2.

Line C.) Bam HI fragment J of HSV-1.

Line D.) Genomic DNA of HSV-1 (black plaque phenotype with monoclonal antibody H1379) used for cotransfection with KpnI A.

Line E.) Genomic DNA of HSV-2 (phenotypically white plaque with H1379) used for isolation of KpnI fragment A.

Line F.) DNA composition of phenotypically white plaque HSV-1x HSV-2 recombinant progeny virus. This recombinant was selected after cotransfection of HSV-1 DNA with KpnI fragment A of HSV-2. DNA of this recombinant was co-transfected with the Bam HI fragment J of HSV-1. As indicated in Line G.

Line G.) This cotransfection resulted in the selection of phenotypically black plaque recombinants.

Line H.) shows dishes with white and black plaque reactions respectively.

The black plaque reaction with H1379 was obviously knocked out with the recombination of KpnI fragment A of HSV-2 into the HSV-1 genome. On the other hand, the black plaque phenotype could be brought back by the Bam HI fragment J of HSV-1.

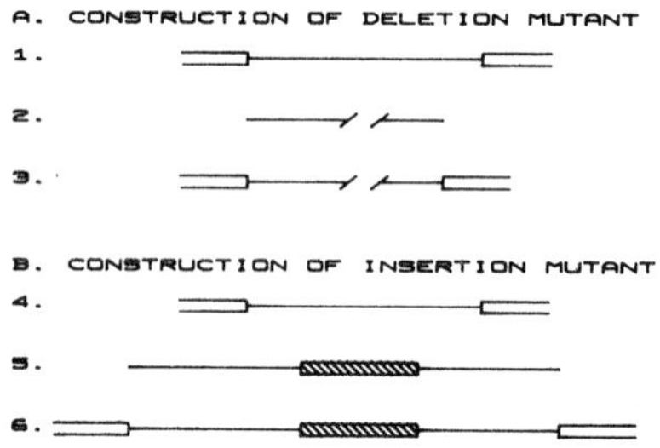

FIG.3

DESIGN FOR THE CONSTRUCTION OF: A. DELETION- AND
B. INSERTION MUTANTS.

1. Homologous sequences flanked by any viral DNA.
2. Cloned homologous sequences containing deletion.
3. Upon cotransfection of viral DNA shown in 1. and plasmid DNA shown in 2., the homologous sequences containing the deletion may recombine into the viral genome.
4. same as 1.

5. Foreign DNA sequences flanked by homologous viral DNA cloned in plasmids.
6. Upon cotransfection of 4. and 5. the foreign DNA sequences may recombine into the viral genome.

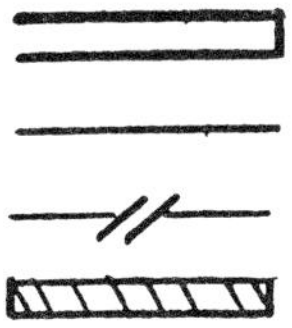

= viral DNA
= homologous viral DNA sequences
= deletion
= foreign DNA sequences

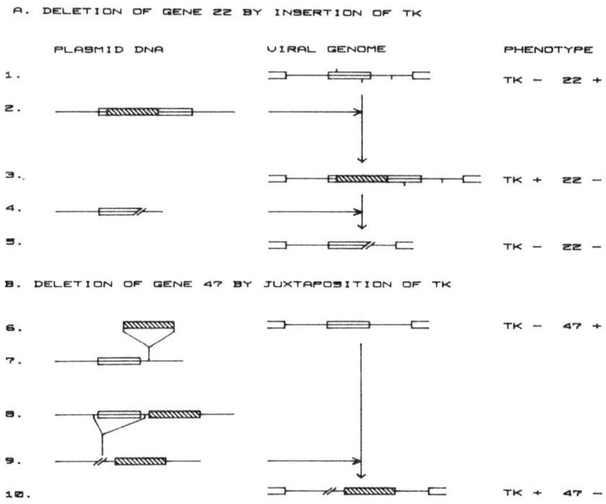

FIG.4

INACTIVATION OF GENES BY RECOMBINATION OF ALTERED DNA SEQUENCES

Prokaryotic events are shown on the left. Events in the viral genome and the resulting phenotypes are shown on the right.

A. CONSTRUCTION OF DELTA 22 DELETION MUTANT SPECIFYING TRUNCATED ALPHA 22 PROTEIN (12).

1. The ALPHA 22 gene and its flanking sequences in the genome of TK- virus.
2. Plasmid containing ALPHA TK in the body of the ALPHA 22 gene.
3. Cotransfection of TK- viral DNA in 1. with plasmid 2. followed by TK+ selection led to recombination of TK into the body of the ALPHA 22 gene and to TK+, 22- progeny virus.
4. Truncated ALPHA 22 gene (DELTA 22) in plasmid.
5. Cotransfection of viral DNA in 3. with plasmid DNA 4. followed by TK- selection may lead to recombination of DELTA 22 sequences into the viral genome and to the selection of 22-, TK- virus.

B. CONSTRUCTION OF 47- DELETION MUTANT (6).

6. (left) ALPHA TK to be inserted in the plasmid containing the ALPHA 47 gene.
6. (right) The ALPHA 47 gene and its flanking sequences in the genome of TK- virus.
7. Cloned ALPHA 47 sequences and site for the juxtaposition of ALPHA TK in the 47 flanking sequences.
8. Plasmid containing the ALPHA 47 gene, ALPHA TK next to it, and the sites for the collapse deletion of the ALPHA 47 gene.
9. Plasmid containing ALPHA TK flanked by the ALPHA 47 flanking sequences.
10. Cotransfection of TK- viral DNA in 6. (right) and the plasmid shown in 9. followed by TK+ selection may lead to the replacement of the ALPHA 47 gene by ALPHA TK and to the selection of 47-, TK+ viral progeny.

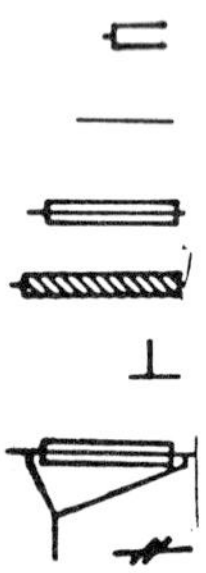

= viral DNA
= homologous (flanking) viral DNA
= specific viral gene
= ALPHA TK
= site of insertion
= site of deletion
= deletion

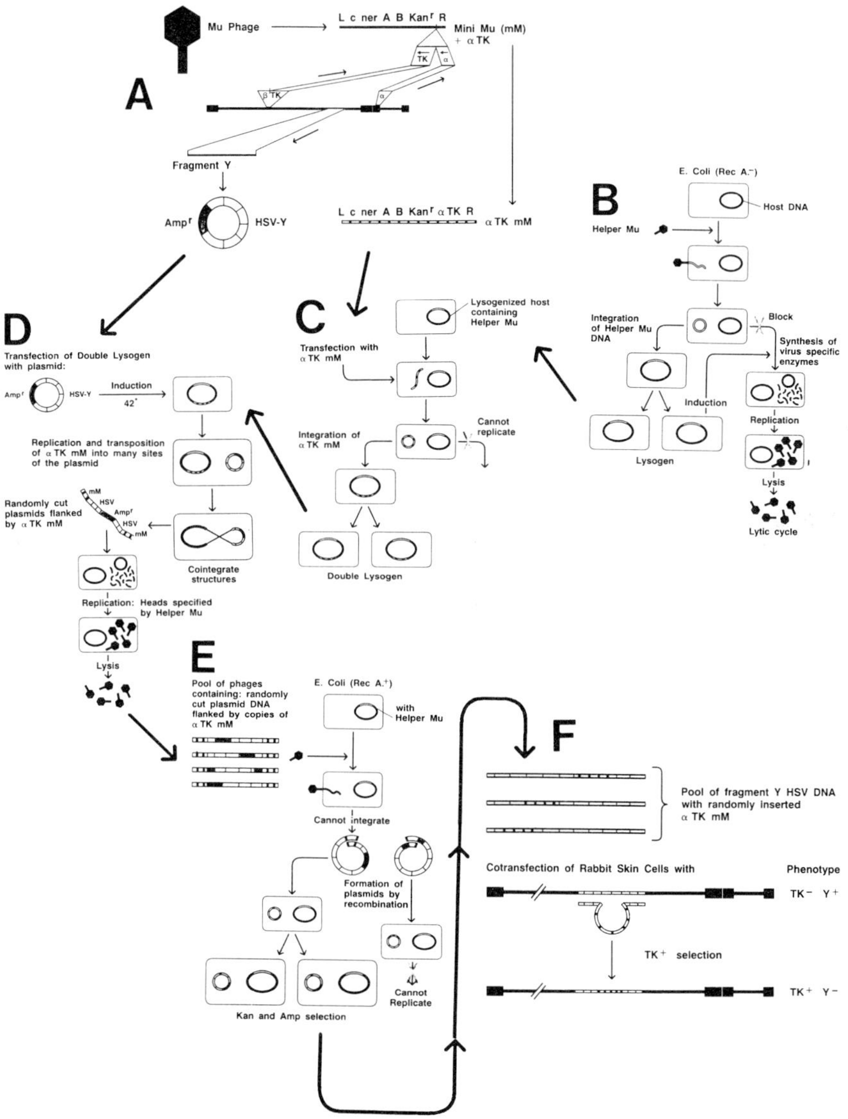

FIG.5

RANDOM INSERTIONS OF ALPHA TK INTO SELECTED DNA FRAGMENTS OF HSV USING THE MINI MU (mM) SYSTEM AND THE TK SELECTION (4,15).

A. DNA of the Mu Phage is engineered to consist of the minimal amount of genes necessary for integration, transposition and selection (mini Mu = mM).

The cloned coding region for HSV thymidine kinase (TK) is fused to an ALPHA promoter (ALPHA TK) and is inserted into the mM genome between the right end (R) of the mM genome and the selectable gene specifying resistance to Kanamycin (Kan*r). This construction is designated the ALPHA TK mM.

Furthermore a randomly selected fragment Y (HSV-Y) is cloned into a plasmid specifying resistance to Ampicillin (Amp*r).

B. E. Coli bacteria are infected with Helper Mu phages. The lytic cycle of the Helper Mu can be blocked at low temperatures because of a temperature sensitive gene. The Helper Mu DNA is thus integrated into the host DNA (Lysogen). The block of the lytic cycle can be released, however, by the induction of the lysogen at 42°C.

C. The lysogenized host containing Helper Mu DNA is transfected with the ALPHA TK mM and incubated at low temperatures where replication and transposition are blocked. ALPHA TK mM DNA is also integrated into the host DNA (Double Lysogen).

D. The plasmid containing fragment Y of HSV is transfected into the Double Lysogen. At the same time replication of the Helper Mu and transposition of the ALPHA TK mM is induced by a temperature shift to 42°C. Thus, ALPHA TK mM starts transposition and jumps randomly into many sites of the plasmid.

ALPHA TK mM sequences are packaged into heads specified by the Helper Mu. Since the heads are always packed full, it depends on the length of fragment Y whether one or two copies of the ALPHA TK mM flanking the plasmid sequences are packed into the heads.

E. The lysed hosts release among others a pool of phages containing randomly cut plasmid DNA flanked by copies of ALPHA TK mM. These phages are used to infect hosts where replication as well as integration of the phages is blocked at low temperatures.

Under Ampicillin and Kanamycin selection, bacterial cells can replicate only if:

i) Two copies of ALPHA TK mM are able to recombine thus forming episomal DNA: a new plasmid.

ii) ALPHA TK mM specifies resistance to Kanamycin.

iii) ALPHA TK mM is inserted in fragment Y of HSV but not into the gene specifying resistance to Ampicillin.

F. Plasmids purified from these bacteria contain a pool of fragment Y DNA with random insertions of ALPHA TK mM sequences. These DNAs are cotransfected with viral DNA purified from TK- HSV (Phenotype TK-, Y+). Selection for TK+ leads to the replication of Y- recombinants only where the TK gene is inserted into a site where viral replication is not affected.

= ALPHA TK mM sequences

= HSV-Y sequences

 = Amp*r sequences

surface protein of interest is established using native plaques in a monolayer of cells as antigen and the nondestructive immunestaining technique.

ii) Infectious DNA of nonreactive (vector) virus is cotransfected with cloned DNA fragments containing the gene of interest flanked by appropriate sequences.

iii) Recombinant progeny virus is detected by the immune-staining method. Plaques specifying the desired antigen are readily detected because of their black colour.

The cells have not to be fixed for this procedure. For this reason, black or white (colourless) plaques can be picked and the desired recombinants can be selected directly from a mixed population.

INACTIVATION OF GENES BY DELETION AND BY INSERTION

Homologous sequences may replace sequences in the viral genome by recombination. The DNA between homologous flanking sequences may be deleted or may consist of a foreign gene (Fig. 3). The frequency of recombination varies, however, with the size and location of nonhomologous sequences. Large fragments recombine more efficiently than small fragments. A nonhomologous insert at the end of a fragment is less likely to recombine than the same piece of DNA inserted in the middle of the same fragment. The following three examples elucidate three different strategies for the inactivation of HSV genes:

i) Mocarsky et al (7) cloned a DNA fragment, spanning the junction of the long and the short component of the HSV genome (L-S junction), into the body of the TK gene.
This plasmid was cotransfected with wild type (wt)

HSV DNA (phenotypically TK+). Recombination of the junction fragment into the body of the wt TK gene was mediated by the homologous TK sequences of the plasmid and the viral DNA respectively. The growth of TK- recombinants was favoured by selection against TK by supplementing the medium with Ara T.

ii) Post et al (12) introduced a procedure for the deletion of specific genes which is illustrated in Fig. 4A: ALPHA TK was first inserted into the cloned ICP 22 gene. Cotransfection of this plasmid with TK- viral DNA followed by selection for TK+ virus led to the growth of TK+, 22- recombinant virus. DNA of this new virus was cotransfected with a cloned DNA fragment containing the truncated 22 gene. Selection against TK led to the growth of TK-, 22- recombinants.

iii) The above technique was improved by Mavromara et al (6). The procedure is illustrated in Fig. 4B: The ALPHA 47 gene and its flanking sequences were cloned. ALPHA TK was inserted in this plasmid next to the 47 gene. The 47 gene was then removed from the plasmid by restriction enzyme digestion followed by religation (collapse). ALPHA TK was thus flanked in this new plasmid by DNA sequences homologous to those flanking the 47 gene in the viral genome. Upon cotransfection of this plasmid with genomic DNA of TK- virus, the 47 gene was deleted where ALPHA TK recombined into the viral genome. Thus, ALPHA 47 deletion mutants were selected by the selection of TK+ virus.

These examples show that genes may be inactivated as well by deletions as by insertions. In recombinant vaccines, undesired genes might be inactivated by the insertion of the desired foreign genes. Thymidine Kinase is used for selection purposes and the TK gene for providing homologous sequences that allow the introduction of foreign sequences by recombination.

EVALUATION OF THE SITE OF INSERTION

In order to get viable recombinants, DNA sequences that are not essential for replication have to be determined within the vector genome. These sequences are most probably suitable for insertions.

The most elegant technique to screen the HSV genome for sequences and genes nonessential for growth in cell culture was reviewed recently by Roizman and Jenkins (15). The simplified pathway of the procedure is shown in Fig. 5. The technique involves the TK selection and depends on the ability of a derivated bacterophage Mu (ALPHA TK mM) to introduce random insertions into a DNA by transposition within an E.Coli cell. Cotransfections of these randomly engineered fragments and the DNA of TK- virus, result in TK positive viable progeny only when the ALPHA TK mM sequences are inserted into nonessetial parts of the genome.

Aspects of pathogenicity and of stability of the resulting recombinant have also to be regarded in the evaluation of the site of insertion. Most preferable are sites where insertions do not impair viral replication and render the least pathogen and most stable recombinants.

CONCLUSIONS

A most promising application of recombinant herpes viruses may be their use as vectors for the expression of foreign genes in recombinant vaccines. Herpes viruses are able to express foreign genes at high levels. Furthermore, they specify foreign proteins with similar post translational modifications as the authentic gene product (15,17).

Foreign genes are fused to herpes virus promoters and are

introduced into the vector genome by recombination. The growth of recombinant progney virus is favored by selection for TK+ or TK- virus. In certain instances, recombinants may be picked directly from a mixed population by the nondestructive immunstaining of surface antigens.

Undesirable genes are inactivated by the introduction of deletions or insertions.

Sites suitable for the insertion of foreign genes are determined with the ALPHA TK mM system.

REFERENCES

1. Ackermann, M., R. Longnecker, B. Roizman, and L. Pereira. Identification, properties, and gene location of a novel glycoprotein specified by herpes simplex virus 1. Virology 150: 207-220. 1986.

2. Honess, R.W. and B. Roizman. Regulation of herpesvirus macromolecular synthesis. I. Cascade regulation of the synthesis of three groups of viral proteins. J. Virol. 14: 8-19. 1974.

3. Honess, R.W. and B. Roizman. Regulation of herpesvirus macromolecular synthesis: sequential transition of polypeptide synthesis requires functional viral polypeptides. Proc. Natl. Acad. Sci. 72: 1276-1280. 1975.

4. Jenkins, F.J., M.J. Casadaban, and B. Roizman. Application of the mini-Mu-phage for target-sequence-specific insertional mutagenesis of the herpes simplex virus genome. Proc. Natl. Acad. Sci. 82: 4773-4777. 1985.

5. Kousoulas, K.G., P.E. Pellet, L. Pereira, and B. Roizman.

Mutations affecting conformation or sequence of neutralizing epitopes identified by reactivity of viable plaques segregate from syn and ts domains of HSV-1 (F) gB gene.
Virology 135: 379-394. 1984.

6. Mavromara-Nazos, P., M. Ackermann, and B. Roizman.
Construction and properties of a viable herpes simplex virus 1 recombinant lacking coding sequences of the ALPHA 47 gene.
J. Virol. 60: 807-812. 1986.

7. Mocarski, E.S., L.E. Post, and B. Roizman.
Molecular engineering of the herpes simplex virus genome: Insertion of a second L-S junction into the genome causes additional genome inversions.
Cell 22: 243-255. 1980.

8. Mocarski, E.S. and B. Roizman.
Herpesvirus-dependent amplification and inversion of a cell-associated viral thymidine kinase gene flanked by viral "a" sequences and linked to an origin of viral DNA replication.
Proc. Natl. Acad. Sci. 79: 5626-5630. 1982.

9. Patzer, E.J. and J.F. Obijeski.
Vaccines prepared from translation products of cloned viral genes. in Immunochemistry of viruses. The basis for serodiagnosis and vaccines. Eds. M.H.V. van Regenmortel and A.R. Neurath.
Chapter 11: 153-170. 1985 Elsevier Science Publishers B.V.

10. Poffenberger, K.L., E. Tabares, and B. Roizman.
Characterization of a viable, noninverting herpes simplex virus 1 genome derived by insertion and deletion of sequences at the junction of components L and S.
Proc. Natl. Acad. Sci. 80: 2690-2694. 1983.

11. Post, L.E., S. Mackem, and B. Roizman.
Regulation of ALPHA genes of herpes simplex virus: Expression of chimeric genes produced by fusion of thymidine kinase with ALPHA gene promoters.
Cell 24: 555-565. 1981.

12. Post, L.E. and B. Roizman.
A generalized technique for deletion of specific genes in large genomes: ALPHA gene 22 of herpes simplex virus 1 is not essential for growth.
Cell 25: 227-232. 1981.

13. Roizman, B. J. Warren, C.A. Thuning, M.S. Fanshaw, B. Norrild, and B. Meigner.
Application of molecular genetics to the design of live herpes simplex virus vaccines.
Develop. Biol. Standard. 52: 287-304. 1982.

14. Roizman, B. and W. Batterson.
Herpesviruses and their replication.
in Virology. Eds. B.N. Fields et al.
Chapter 25: 497-526. 1985 Raven Press, New York.

15. Roizman, B. and F.J. Jenkins.
Genetic engineering of novel genomes of large DNA viruses.
Science 229: 1208- 1214. 1985.

16. Seligmann,Jr, E.B.
The NIH test for potency
in Laboratory techniques in rabies. Eds. M.M. Kaplan and H. Koprowski.
Chapter 33: 279-286. 1973 World Health Organization, Geneva.

17. Shih, M.F., M. Arsenakis, P. Tiollais, and B. Roizman.

Expression of hepatitis B virus S gene by herpes simplex virus type 1 vectors carrying ALPHA- and BETA-regulated gene chimeras. Proc. Natl. Acad. Sci. 81: 5867-5870. 1984.

18. Spear, P.G.
Antigenic structure of herpes simplex viruses.
in Immunochemistry of viruses. The basis for serodiagnosis and vaccines. Eds. M.H.V. van Regenmortel and A.R. Neurath.
Chapter 23: 425-446. 1985 Elsevier Science Publishers B.V.

13 The glutamine synthetase gene: a novel marker for gene amplification in mammalian cells (abstract)

C. R. Bebbington and C. C. G. Hentschel

Efficient expression of recombinant DNA products in mammalian cells has frequently been achieved using DHFR gene-mediated amplification of vector sequences integrated into host cell chromosomes. This system suffers from a number of drawbacks including the requirement for Methotrexate selection and the fact that its use is largely restricted to dhfr$^-$ CHO cells. We have developed a new system for vector amplification based on the glutamine synthetase (GS) gene. A GS cDNA and a GS minigene containing a single GS intron have been shown to act as dominant selectable markers in CHO-K1 cells when expressed under the control of SV40 promoters. The introduced genes can be amplified by selection with methionine sulphoximine and very high copy numbers (>10 000 per cell) have been achieved. We have also used this system to express tissue plasminogen activator and have attained levels of secreted protein comparable to those obtained with DHFR co-amplification. The GS vector developed therefore provides a versatile alternative to DHFR for overexpressing protein products.

Discussion

ULLRICH: In your results with the two different vector systems, did you look at the expression of your marker protein which with the stronger promotor should be 1500 times more expression than with the weaker promotor

BEBBINGTON: In addition to the different promotors in the more efficient constructs we used the 5V40

	polyadenylatio signal in the other we used GS at the 3' end, there is also some sequence which is an antefact of C-DNA cloning put upstream beyond the 5' end of the GD-CDNA which could have significantly affected the expression level.
ULLRICH	But in both cases, to obtain the same resistance you should have the same amount of protein.
BEBBINGTON	We did not measure the GS activities independently, but since the MSX levels were the same this would be reflected in the amount of protein produced.
ULLRICH	Your description of the disadvantages of the DHFR system were incorrect in that you do not need a $DHFR^-$ cell line for expression and amplification. You can amplify the DHFR vector in any cell line for for reasons which are not completely clear but have to do with the fact that tranfected DNA is very labile and therefore expanded advantageously over the endogenous DHFR gene.
BEBBINGTON	You can amplify DHFR in other cell lines by using neomycin for selection and then amplify subsequently but no as simple or convenient and not all the clones are going to express DHFR.
WURM	Do you know anything about the integration f the DNA into the chromosomes.
BEBBINGTON	No
WURM	How long did it take you to select and amplify the expression system?
BEBBINGTON	About three weeks to pick colonies then two

rounds of amplification of 4 weeks per round.

WURM That is a protocol which is similar to the methotrexate selection.

BEBBINGTON Yes

WURM Do you know anything about the stability of expression after releasing the selective pressure.

BEBBINGTON Not enough data but some clones are stable, others are unstable.

14 Recombinant animal cell lines for production of glycoproteins

M. Wirth, W. Reiser, G. Zettlmeissl, H. S. Conradt, J. H. Cao, C. Morelle and H. Hauser

Abstract

Eukaryotic expression plasmids for constitutive or inducible expression with different transcription control and RNA processing elements were constructed. The plasmids contain human structural genes coding for Antithrombin III (ATIII), Interleukin-2 (IL-2) and Interferon-ß (IFN-ß). The plasmids were stably transferred into three different mammalian cell lines, hamster CHO, BHK21 and mouse L cells. The expression of the gene products reflects predominantly the activity of the transcriptional control elements in different cell lines and the influences of the respective structural gene on the posttranscriptional level. Selected transformants were screened for high expression by clonal selection before and after gene amplification. The productivity is up to 10 $\mu g/10^6$ cells/24 h. The products were characterized by biochemical, immunological and biological methods. In all cases biological activities are indistinguishable from the authentic gene products. Analysis of the carbohydrate moieties indicates that composition and extent of glycosylation are cell type specific.

Introduction

Recombinant DNA techniques have enabled the synthesis of therapeutically or industrially relevant proteins in nonmammalian cell hosts in large quantities. However, in many cases proteins of mammalian origin are functional only if they are properly modified (e.g. 1,2). Such co- and posttranslational modifications (disulfide bridging, processing and trimming of glycosylated structures, sulfation, attachment of lipids) are a specific feature of higher eukaryotic cells. The development of gene transfer techniques enabled researchers to introduce foreign genes into animal cells and to express them under the control of strong convenient regulatory DNA elements. Recombinant animal cells should provide a means of producing large quantities of appropriately modified and biologically active proteins for industrial or medical purpose.

Human Antithrombin III is a 60-63 kDa glycoprotein containing 4 N-glycosydically linked carbohydrate chains of the biantennary complex type. The protein is a member of the serpin family of plasma proteinase inhibitors (3). ATIII can inactivate a series of blood plasma proteins that are involved in the coagulation cascade, whereby its inhibitory activity is markedly increased by heparin (4,5). A deficiency of this glycoprotein- either acquired or hereditary - is closely correlated with an increased risc of thromboembolic disease (6). Presently, ATIII is isolated from human plasma in high quantities.

Human Interferon-ß is a biologically active polypeptide which is therapeutically used as an antitumor and antiviral agent. Synthesis is induced mainly in fibroblasts by exposition of cells to virus in vivo and in vitro. The yield of IFN-ß from human primary fibroblasts is poor and production is expensive. Although it can be produced in E. coli in high quantities, the bacterial product is not glycosylated and needs extensive refolding. The natural protein contains one N-linked carbohydrate chain at a single N-glycosylation site.

Human Interleukin-2 is a 133 amino acid protein which is secreted by T-lymphocytes upon stimulation with antigen or lectins. It is a T cell growth factor which plays an essential role in the network of interactions between signal factors and cells in the immune system. The protein produced by human peripheral blood lymphocytes (PBL) *in vitro* is a mixture of glycosylated and unglycosylated species, the carbohydrates being O-linked via Thr_3 at the NH_2-terminus of the polypeptide chain. Its possible application as a therapeutic agent has been discussed (7).

Methods

Construction of plasmids:

Cloning was done according to established procedures (8). Most plasmid constructs are described elsewhere in detail (9,10,11). The plasmid DNA was propagated either in strain DH1 or JM103 of *E. coli*.

Cell culture:

Ltk^- and BHK cells were grown in Dulbecco's modified Eagle medium (DMEM) supplemented with penicillin/streptomycin, glutamine and 10 % newborn calf serum. CHO cells deficient in the DHFR-gene additionally were supplemented with deoxyadenosine, adenosine and thymidine. For production the confluent cells were grown in successive alternating cycles for 24 hours each in medium with or without serum. Adaption of BHK cells to serum free medium was done as described (12).

Transfection, selection of cell clones and gene amplification:

The expression plasmid constructs were introduced into the

respective cell lines as supercoiled DNA by the calcium phosphate co-precipitation method (13) together with a plasmid carrying neomycin resistance in a molar ratio of 10:1 and with a plasmid containing the mouse dihydrofolate reductase gene (dhfr). Upon selection with G418 the clones were pooled and grown to confluency. The amount of recombinant protein was determined separately in the mixture after a 24 h production period in DMEM without serum. Expression values are normalized and given in ng/10^6 cells/24 h. For subcloning of cell pools the limiting dilution method was used. For gene amplification single clones were subjected to stepwise selection (twofold increases) with methotrexate starting with a concentration of 20 nM. Amplification was monitored by Southern Blot analysis.

Biochemical assays:

Western Blots were done according to Burnette et al. (14). Radioimmunoprecipitation and ELISA were performed using procedures adapted for each given gene product.

For ATIII-detection polyclonal antibodies originated from immunized rabbits (Behringwerke) were used. Western blotting of IFN-ß protein was performed using a goat anti-Interferon-ß antiserum (A. Billiau, Rega Inst., Leuven, Belgium). For metabolically labelling cells were starved for 2 h in methionine deficient medium followed by the additon of 100 μCi ^{35}S-methionine (800 mCi/mMol) for a 16h period. For IL-2 detection monoclonal antibodies (GBF) and a sheep anti mouse serum was used. IL-2 was purified from the culture medium as described (15).

Determination of biological activity:

The antiviral activity of human IFN-ß was determined by the

microassay method using african green monkey kidney cells (Vero) challenged with vesicular stomatitis virus (VSV) (16).

IL-2 activity was determined using a cloned murine IL-2 dependent growing cytotoxic T-cell line as target cell (17).

The heparin cofactor and progressive inhibitor activity of ATIII were measured according to Schrader et al. (18) and Hensen et al. (19), respectively.

Results and Discussion

We were interested in isolating high producer cell clones in different recombinant mammalian cell lines. In eukaryotic cells transcriptional activity is a predominant parameter for gene expression. However, the use of a certain transcriptional control region including enhancer and promoter sequences is dependent on the differentiation state as well as on the species where the respective cell line is derived from. Therefore, we evaluated the influence of different promoters in three continuous rodent cell lines: mouse L thymidine-kinase-negative-cells (Ltk$^-$), chinese hamster ovary cells (CHO) and baby hamster kidney cells (BHK21).

We have chosen different kinds of promoters. All of them should guarantee constitutive transcription of the respective gene. The mouse metallothionein I promoter as well as the human metallothionein II_A promoter can be further induced by addition of heavy metal ions like Cd^{2+} and Zn^{2+} to the culture medium.

Figure 1 shows the comparison of expression levels of different constructs in the three recipient cell lines. In general the expression in CHO cells is less efficient, whereas the expression in BHK cells directly after stable transfer is significantly higher. Transfer into the Ltk$^-$ cells gives rise to

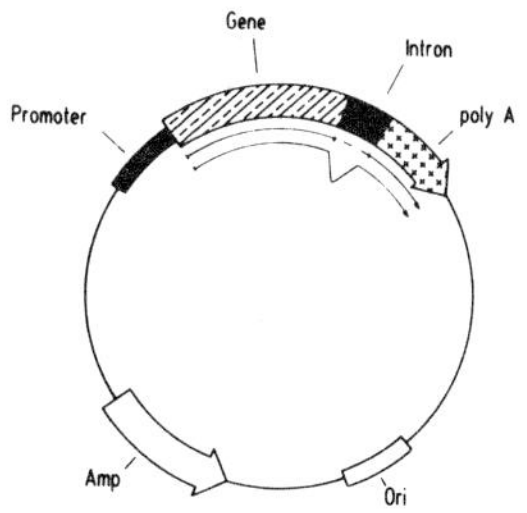

Gene	Promoter	Intron	Polyadenylation signal	Expression in CHO	LTK⁻	BHK21
IFN-ß, human	SV40 early	SV40 small t	SV40 early	1.5	22.5	100
	MTI mouse	SV40 small t	SV40 early	0.75	50	100
	HSV-TK	-	endogeneous	n.d.	15	5
	Adeno-ML	SV40 small t	endogeneous	3	10	11
IL-2, human	SV40 early	SV40 small t	SV40 early	20	100	300
	SV40 early	endogeneous	endogeneous	20	160	300
AT III, human	SV40 early/ MMTV	-	SV40 early	200	18	1500
	SV40 early	-	SV40 early	210	13	1000
	HSP Drosophila	-	SV40 early	11	11	12
	CMV	SV40 small t	SV40 early	78	4	22
	$MTII_A$human	-	SV40 late	5	250	45
	MPSV	-	SV40 early	-	25	1600

Fig. 1 Expression of human ATIII, IL-2 and IFN-ß in recombinant cell lines.

A schematical representation of the expression plasmids is given in the upper part.

The indicated expression plasmids were introduced into the respective cell line. Upon selection the clones were pooled and grown to confluency. The concentration of the respective gene products was determined in the supernatant after a 24 h production period. The values are given as ng/10^6 cells/24 h.

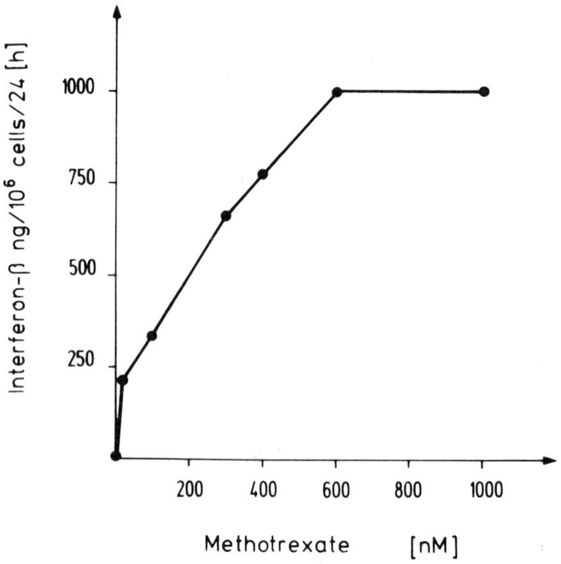

Fig. 2 Expression of human IFN in CHO-cells upon gene amplification.

A selected IFN-ß producing CHO cell clone was subjected to stepwise selection in methotrexate (MTX). Resistant cells were maintained for several days at the respective concentration. After determination of human interferon production the MTX concentration was increased.

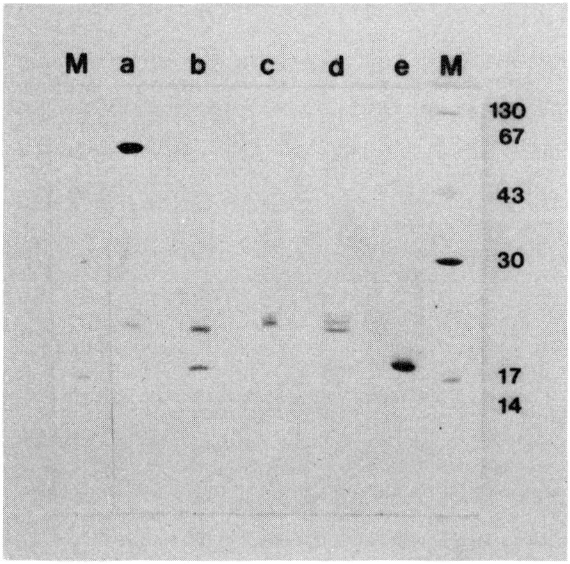

Fig. 3 Western blot of human IFN-ß protein from different cell types.

50 ng of IFN-ß from different cell lines FS-4 (a), BHK (b), CHO (c) and L (d). 0.5 μg of human IFN-ß from E. coli (e) was used as a standard for nonglycosylated IFN-ß.

intermediate expression levels. Southern blot experiments reveal a low copy number of the transferred genes in primary selected CHO•cells and a relatively high copy number in Ltk$^-$ and BHK21 cells (data not shown). Therefore, the difference in expression may be explained at least partially by the gene dosage effect.

In BHK cells the SV40 early promoter and the promoter from myeloproliferative sarcoma virus (MPSV) (20) appear to be the most effective ones. In CHO cells the Adenovirus m.l. promoter is as strong as the SV40 early promoter. Induced metallothionein promoters from both human and mouse are the best in mouse L cells, although the basal expression is enhanced only 2-3-fold. Their low induction ratio is partially due to the concentration dependent toxicity of the inducers Cd2+ and Zn2+. Higher ratios are prevented by extensive cell death at higher concentrations of the heavy metal ions. This effect is even more pronounced in serum free medium.

In contrast to most constitutively expressed mRNAs the mRNAs of many regulated genes like those coding for human IFN-ß or IL-2, respectively, have short half life times. Stability of mRNA which is dependent on primary structure and specific sequences are parameters for gene expression (21). As shown in Table 1 all 3 genes are expressed to a different extent. This can be directly compared in plasmid constructs in which the expression control regions are identical. We interprete this gene specific expression as a function of posttranscriptional events, most probably mRNA stability or translational control.

Due to the chromosomal integration site integration of the transgenomic DNA and the number of integrated functional copies expression variability in single clones is observed. Because it is presently not possible to achieve high expression by localizing the transgenome in a specific way, we decided to perform screenings for single cell clones with respect to high expression. Single cell clones from transfected L cells obtained by screening produced 0.3 $\mu g/10^6$ cell/24 h of human IFN-ß. In BHK cells clones producing $1\mu g/10^6$ cells/24 h of IFN-ß and $10\mu g/10^6$ cells/24 h of ATIII could be isolated.

The expression of CHO cells can be improved following established protocols involving selection for MTX resistance which is accompanied by amplification of the respective DHFR gene (22). Coamplification of the cotransferred DNA in most cases leads to higher yields of expression. This protocol was used for the screening of amplified IFN-ß and ATIII in CHO cells. The result of IFN-ß gene amplification is shown in Fig. 2. It is conceivable that successful amplification in BHK and Ltk^- cells would lead to much higher expression.

For characterization of the specific gene products we have produced serum free supernatants by growing the recombinant cells in mass culture plates or roller bottles. BHK cells were adapted to serum free medium. CHO, L and BHK cells could be grown in fermenters with bubble-free aeration to cell densities up to $7x10^6$ cells/ml (see Büntemeier, H. and Lehmann, J., this volume). The productivity of the respective recombinant cells was not changed significantly under these conditions. Purification of the recombinant proteins from serum free medium was done according to established protocols.

We have characterized the proteins produced by the three cell lines with respect to biochemical and immunological properties. The molecular weight of the gene product tested on SDS PAGE for IL-2 is identical with the authentic protein from natural sources. However, the ratio of glycosylated and non-glycosylated species is different depending on the producer cell line. The electrophoretic mobility of ATIII and IFN-ß clearly depends on the producer cell line. As shown by treatment with tunicamycin, an inhibitior of N-glycosylation, this difference in electrophoretic mobility can be related to heterogeneities at the level of the carbohydrates attached to the polypeptide. All recombinant proteins are typically processed containing the N-terminal amino acid sequence as determined by Edman degradation.

The carbohydrate moieties of IL-2 from PBL and recombinant cell lines have been analyzed in detail (Fig. 5, c.f. 10,15,23). The composition of the carbohydrate side chains from these cells is indistinguishable. The only difference between IL-2 from these cell lines is the ratio of the nonglycosylated form (B) and the glycosylated form (A). In addition, we have found that the extent of glycosylation is not dependent on the synthesis rate of the recombinant protein, but is an inherent property of the individual cell line.

Recombinant ATIII from BHK and L cells is heterogeneous in its electrophoretic mobility if compared to ATIII isolated from human plasma (data not shown). For a rough analysis of the carbohydrate structure we used several inhibitors of the glycosylation pathway (Fig. 4). The results show, that the oligosaccharide chains are of complex type and therefore resemble those of plasma ATIII.

As shown in Fig. 3, human IFN-ß's secreted by the three cell lines all have different mobilities in SDS-PAGE, in neither case a recombinant protein migrates identical with IFN-ß from human foreskin cells indicating most strikingly host cell dependent glycosylation of this secretory polypeptide.

The immunological properties of the recombinant proteins were studied by Western Blot analysis, ELISA and radioimmuno-precipitation and were related to the biological activity. In these tests the proteins from the recombinant cell lines are

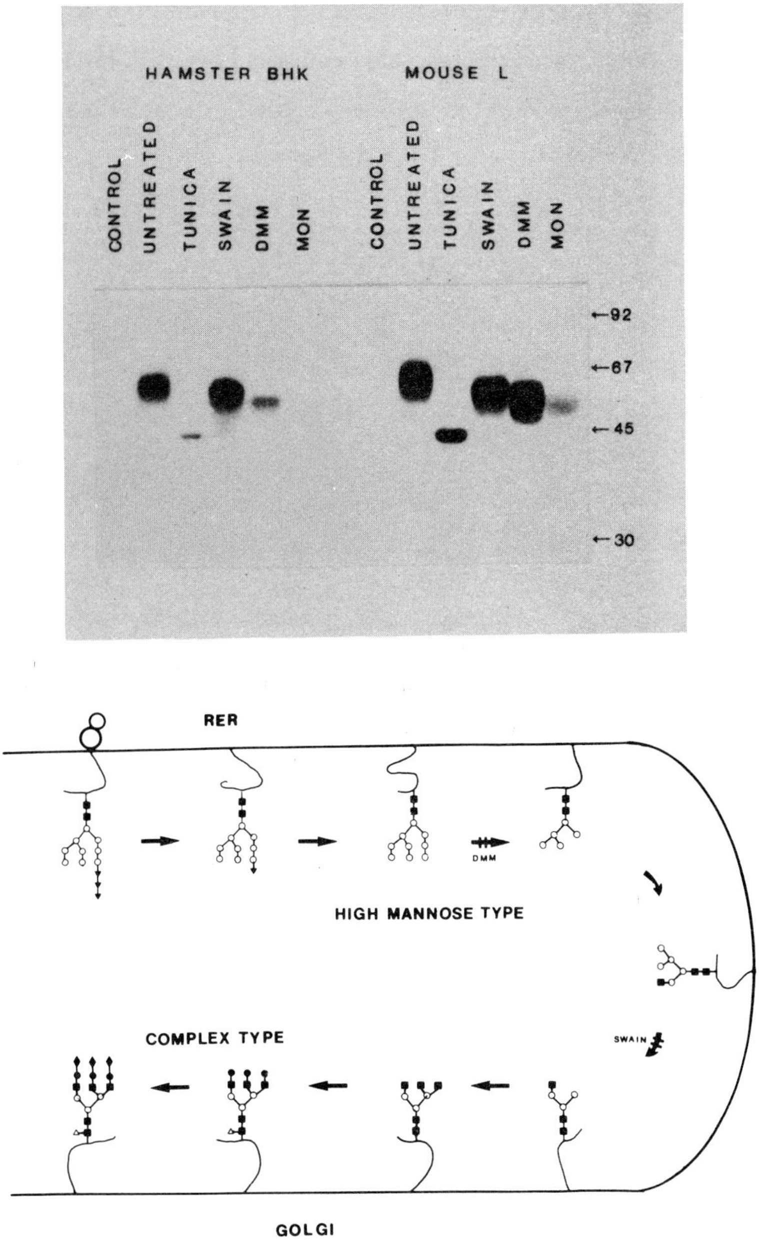

<u>Fig. 4</u> Studies on the glycosylation of ATIII in BHK and L-cells.

The upper part shows a schematic drawing of the N-glycosylation pathway in the Rough Endoplasmatic Reticulum (RER) and the Golgi apparatus (■ N-Acetylglycosamin, O Mannose, ● Galactose, △ Fucose, ♦ Neu5Ac, ▼ Glucose). Intrinsic glycosylation inhibitors deoxy-mannojirimycin (DMM), swainsonine (Swain), monensin (Mon), tunicamycin (Tunica) were used in metabolical labelling experiments. Secreted radiolabelled ATIII was immunoprecipitated, separated by SDS-PAGE and visualized by fluorography.

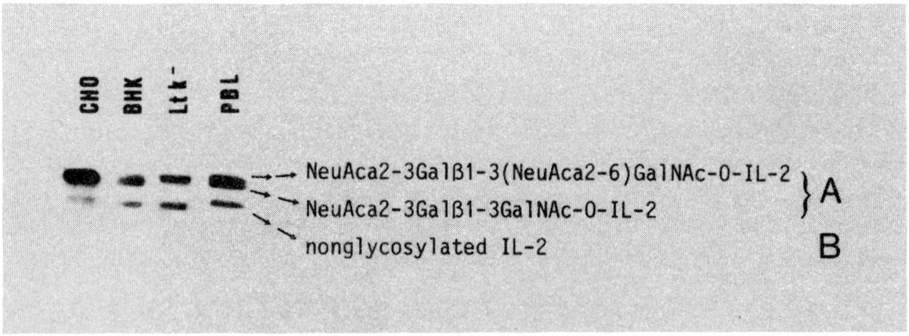

Fig. 5 SDS-PAGE pattern of purified recombinant human IL-2 of different mammalian cell lines.

Purified recombinant human IL-2 preparations were subjected to SDS-PAGE. For comparison the pattern of IL-2 of normal human lymphocytes is shown (PBL). Arrows indicate the position of the differently glycosylated IL-2 forms.

indistinguishable from their natural counterparts.

The biological activity of the recombinant proteins was determined. After purification and separation of both IL-2 forms, A and B, (Fig. 5) have the same specific activity in the proliferation activation test on murine CTL (10,15,23). The specific activity of purified ATIII from BHK and CHO cells was compared with ATIII from human plasma. In the progressive inhibitor test as well as in the heparin cofactor test no differences were detected. The antiviral activity on different cell lines of IFN-ß from the recombinant cells and human FS-4 cells was compared with ELISA and Western Blot analysis data. No significant deviations could be seen. We therefore conclude that although IL-2, IFN-ß and ATIII derived from different cells have distinct glycosylation patterns, the specific activities are at least similar if not identical.

Acknowledgements

We wish to thank M. Ausmeier, M. Cieplik, M. Fitzek, S. Heiser, B. Kornak and K. Schaper for excellent technical assistance and I. Dortmund for preparation of the manuscript. We are indebted to Dr. J. Hoppe for protein sequencing and Dr. J. Lehmann for fermentation.

This work was partly supported by grant PBE 0387 069 from the BMFT.

References

1. Calvo, F.O. and Ryan, R.J. (1985). Inhibition of adenylyl cyclase activity in rat corpora luteal tissue by glycopeptides of human chorionic gonadotropin and the α-

subunit of human chorionic gonadotropin. Biochemistry **24**, 1953-1959.

2. Morrison, S.L. (1985). Transfectomas provide novel chimeric antibodies. Science **229**, 1202-1207.

3. Carrell, R. and Travis, J. (1985). α1-Antitrypsin and the serpins: variation and countervariation. Trends. Biochem. Sci. **10**, 21-24.

4. Rosenberg, R.D. and Damus, P.S. (1973). The purification and mechanism of action of human antithrombin-heparin cofactor. J. Biol. Chem. **248**, 6490-6505.

5. Rosenberg, R.D. and Lam, L.H. (1979). Correlation between structure and function of heparin. Proc. Natl. Acad. Sci. U.S.A. **76**, 1218-1222.

6. Egeberg, O. (1965). Inherited antithrombin deficiency causing thrombophilia. Thromb. Diatl. Haemorrh. **13**, 516-520.

7. Platzer, E., Gramatzki, M., Röllinghoff, M. and Kalden, J.R. (1986). Lymphokines and monokines in the clinic. Immunol. Today **7**, 185-187.

8. Maniatis, T., Fritsch, E. and Sambrook, J. (1982). Molecular cloning (A Laboratory Manual), Cold Spring Harbor, New York.

9. Reiser, W. and Hauser, H. (1987). Recombinant human interferon ß from mammalian cell lines. Drug Res., in press.

10. Conradt, H.S., Ausmeier, M., Dittmar, K.E., Hauser, H., and Lindenmaier, W. (1986). Secretion of glycosylated human interleukin-2 by recombinant mammalian cell lines. Carbohydr. Res. **149**, 443-450.

11. Zettlmeißl, G., Ragg, H. and Karges, H.E. (1987). Expression of biologically active human ATIII in chinese hamster ovary cells. Bio/Technology, in press.

12. Boedeker, B.G.D., Berg, G.J., Hewlett, G. and Schlumberger, H.D. (1984). Screening method to develop serum-free culture medium for adherent cell lines. Develop. biol. Standard **60**, 93-99.

13. Wigler, M. Sweet, R., Sim, G.K., Wold, B., Pellicer, A., Lacy, E., Maniatis, T., Silverstein, S. and Axel, R. (1979). Transformation of mammalian cells with genes from procaryotes and eucaryotes. Cell **16**, 777-785.

14. Burnette, W.N. (1981). "Western Blotting": Electrophoretic transfer of proteins from sodium dodecylsulfate-polyacryl-amide gels to unmodified nitrocellulose and radiographic detection with antibody and radionated protein A. Anal. Biochem. **112**, 195-203.

15. Conradt, H.S., Geyer, R., Hoppe, J., Grotjahn, L., Plessing, A. and Mohr, H. (1985). Structures of the major carbohydrates of natural human interleukin-2. Eur. J. Biochem. **153**, 255-261.

16. Finter, N.B. (1969). Dye uptake methods for assessing viral

cytopathogenicity and their application to interferon assays. J. Gen. Virol. **5**, 419-427.

17. Gilles, S., Ferm, M.M., Ou, W. and Smith, K.A. (1978). T cell growth factor: Parameters of production and a quantitative microassay for activity. J. Immunol. **120**, 2027-2032.

18. Schrader, J., Züchner, C., Köstering, H. and Scheler, F. (1986). Chromogenic substrates for the determination of Antithrombin III. Ärztl. Lab. **32**, 14-114.

19. Hensen, A. and Loeliger, E.A. (1963). Antithrombin III assay. Thromb. Diatl. Haemorrh. **9** (Supplement 1) 8-29.

20. Stacey, A., Abuthnott, C., Kollek, R., Coggins, L. and Ostertag, W. (1984). Comparison of myeloproliferative sarcoma virus with moloney murine sarcoma virus variants by nucleotide sequencing and heteroduplex analysis. J. Virol. **50**, 725-732.

21. Shaw, G. and Kamen, R. (1986). A conserved AU sequence from the 3' untranslated region of GM-CSF mRNA mediates selective mRNA degradation. Cell **46**, 659-667.

22. Kaufman, R.J. and Sharp, P.A. (1982). Amplification and expression of sequences cotransfected with a modular dihydrofolate reductase cDNA gene. J. Mol. Biol. **159**, 601-621.

23. Conradt, H.S., Hauser, H., Dittmar, K.E., Lindenmaier, W., manuscript in preparation.

15 Foreign gene expression in rat pituitary GH3 cells

M. Silberklang, J. Kopchick, S. Munshi, A. Lenny, T. Livelli and R. W. Ellis

ABSTRACT

We have found the rat pituitary cell line, GH3 to be an efficient host for expression of foreign genes, and especially productive with secreted proteins, such as bovine Growth Hormone (bGH). A more adherent cell subpopulation was first derived, which can be grown as a stably, albeit loosely, attached monolayer culture. Under the appropriate growth conditions, this subpopulation is amenable to both transient DNA transfection with DEAE-dextran, and stable DNA transfection with calcium phosphate or by electroporation. Using bGH vectors, transient expression levels of bGH of 200-800 ng/ml/24 hr are routinely reached by day 4 post-transfection (5,11). We have used this assay to optimize relative gene orientation in expression vectors containing a selectable marker gene. We have also used it to compare transcriptional promoters as to efficiency in these cells, and have found the human cytomegalovirus immediate early promoter to be particularly strong (5). Using either the calcium phosphate or electroporation methods, we have developed stable bGH-secreting GH3 clones expressing as much as 10-20 ug/10^6 cells/24 hr. In other experiments, we have extended these observations to expression of truncated, "anchor-less" viral glycoprotein antigen genes, such as the Epstein-Barr Virus membrane antigen. For this particular protein, GH3 cells were the only line of several tested that gave useful levels of expression (1). When the complete, membrane-anchored form of EBV gp350/220 was expressed in these cells, however, levels of expression were much less striking and resembled results obtained with other cells (1). We attribute this difference to the particularly efficient secretory capabilities of these cells.

INTRODUCTION

In selecting a prokaryotic or eukaryotic cell substrate for expression of recombinant DNA derived mammalian genes, the importance of post-translational processing and modification to the biological or antigenic activity of the particular protein of interest often indicates the choice of eukaryotic and especially mammalian cells. However, attempts at establishment of stable, genetically engineered expressor cell lines can sometimes be defeated by a degree of autotoxicity attendant to high-level expression of certain genes (1). Where the protein of interest is secreted, which is often the case, it might be hoped that a cell substrate most efficient at secretion would also be least subject to autotoxicity, due to the short residence time of the protein in the cytoplasm. In this report, we describe some of our work with one cell line that may meet these criteria, the rat pituitary tumor-derived line GH3 (2,3).

We have previously explored the use of these cells for expression of homologous, non-toxic proteins, such as bovine Growth Hormone (bGH) (4,5). Recently, as we developed better and more general plasmid shuttle vectors for gene expression in these cells, we have also used GH3 cells for efficient expression of heterologous proteins, such as viral glycoproteins. In one case, the Epstein-Barr virus (EBV) membrane antigen (1), GH3 cells in fact proved to be the only cell line of several tested that yielded high-level expression of a genetically engineered, truncated, secreted form of the protein.

MATERIALS AND METHODS

Adaptation of GH3 Cells to Monolayer Culture

GH3 cells in plastic T-flasks containing Dulbecco's Modified Eagle's Medium (DMEM), supplemented with 10% Nu-Serum (Collaborative Research) were incubated at 37°C with constant shaking. The culture fluid, containing floating cells, was removed daily, while those cells remaining attached to the flasks were re-fed and incubated further. After 2 weeks of this routine, adherent cells predominated in the population, and, following passage, could grow as a monolayer. This selected subpopulation maintained this phenotype upon further passage.

Plasmid Constructions

The construction of the bGH expression vector pBGH-4 has been described (4). Briefly, pBGH-4 contains, in a pBR322 background, a 1.5 kb segment from the cloned Schmidt-Ruppin strain of Rous Sarcoma Virus (RSV), which has been ligated to the genomic bGH sequence such that the viral Long Terminal Repeat (LTR) acts as a promoter for bGH transcription. A series of derivative vectors was constructed by excising the RSVLTR-bGH fusion, modifying it with a SalI linker so as to have a SalI site at each end, and inserting it into the SalI site of the plasmid pBRd(7) in both orientations. In addition, the Tn5 aminoglycoside phosphotransferase (neo) gene under Herpes Simplex Virus type I thymidine kinase promoter regulation, comprising a ca. 2kb PvuII segment (8), was modified with BamHI linkers and inserted into the BamHI site of pBRd in 2 orientations (Fig. 1). Four gene orientation combinations were thus generated (Fig. 1).

PCMVIE-AK1-DHFR (Fig. 2) was constructed as follows. Plasmid pBRd (7) containing the neo gene (previously deleted of an internal BglII site by blunt-ending and religation) inserted into its BamHI site, as above (cf. Fig.

TABLE I

Orientation (Fig. 1)	Relative BGH Expression*
A	0.50
B	1.3
C	0.47
D	0.72

*Normalized to pBGH-4 (4,5,11) expression

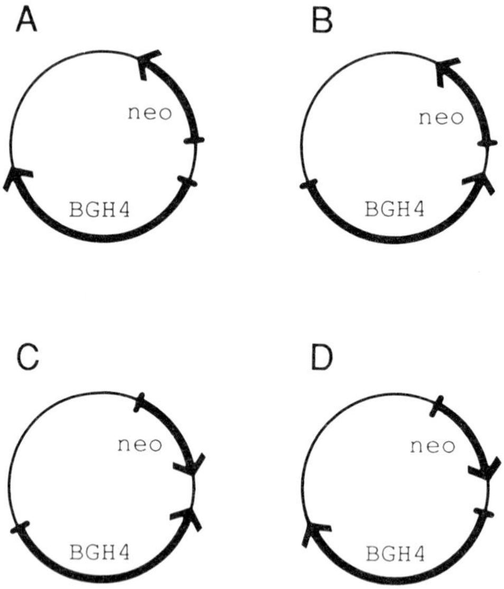

Figure 1: RSVLTR-bGH promoter-gene fusion, BGH4 (4), and neo marker gene in 4 relative orientation combinations in the SalI and BamHI sites of pBRd (7), respectively. Arrows indicate direction of transcription.

1B, the "counterclockwise" orientation), was digested with EcoRV and the ends were modified with 8-mer XhoI linkers. A PvuII-EcoRI segment comprising the translation termination and poly(A) signal region of the bGH gene (9) was modified at the PvuII end with an 8-mer BglII linker, and the resulting BglII-EcoRI fragment was ligated to the ClaI-BglII segment comprising the major immediate early promoter of cytomegalovirus (CMVIE) (5). The resulting ClaI-EcoRI fragment was blunt-ended, modified at both ends with XhoI 8-mer linkers, and inserted into the SalI site of the above-modified pBRD containing the neo gene. Finally, a blunt-ended PvuII-BamHI fragment comprising the mouse dihydrofolate reductase (DHFR) cDNA gene under SV40 early regulation (10) (previously modified by sequential deletion of internal HindIII and BglII sites by blunt-ending and rejoining of the correspondingly restricted DNA) was modified at both ends with SalI linkers and inserted into the XhoI site described above. The resulting 3-cassette vector, designated PCMVIE-AKI-DHFR (Fig.2), has a unique BglII site for insertion of genes to be expressed between the CMVIE promoter and the bGH poly(A) signal.

The insertion of the Epstein-Barr Virus membrane antigen gene into the BglII site of this vector, either as a complete gene or a truncated gene, to yield pCMVIE-EBMA or pCMVIE-MA-delta, respectively (Fig. 3), has been described (1).

Transient Expression Assays

Our standard protocol for DEAE-dextran mediated DNA transfection of GH3 cells (12) was developed to minimize the toxic effects of exposure to DEAE-dextran. 5 x 10^5 GH3 cells were plated per 35mm culture dish and transfected 24 hr later. After a 1-2 ml wash with serum-free DMEM, each plate received 1 ml of a solution containing 500ng/ml supercoiled plasmid DNA and 200 mg/ml DEAE-dextran in serum-free DMEM. The cells were incubated 45-60 min at 37°C in this solution in a 5% CO_2 atmosphere, after which the solution was

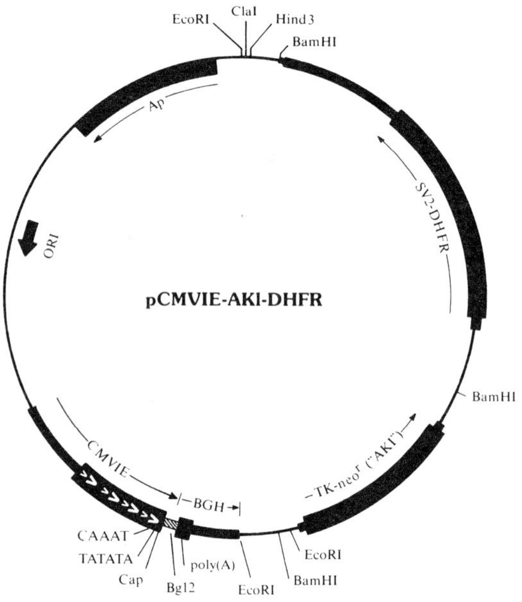

Figure 2: Expression Vector PCMVIE-AK1-DHFR. The three gene cassettes are, CMVIE promoter-bGH poly(A) signal, neo selectable marker gene (8) and DHFR selectable marker gene (10). The BglII site is for insertion of the gene to be expressed.

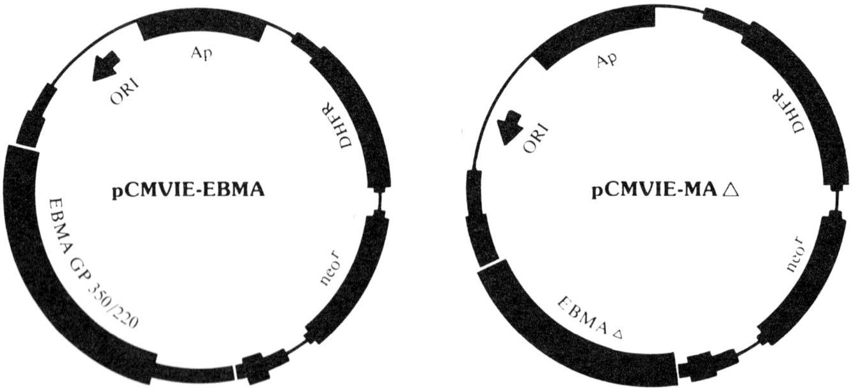

Figure 3: Vectors for EBV gp350/220 expression, consisting of two forms of the gp350/220 coding region inserted into the BglII site of PCMVIE-AK1-DHFR (1) (Methods). EBMA designates the complete gene,

including the membrane anchor coding region, while EBMA-delta designates the anchor-minus gene, lacking 44 codons and terminated by insertion of termination codons via a synthetic DNA linker (1).

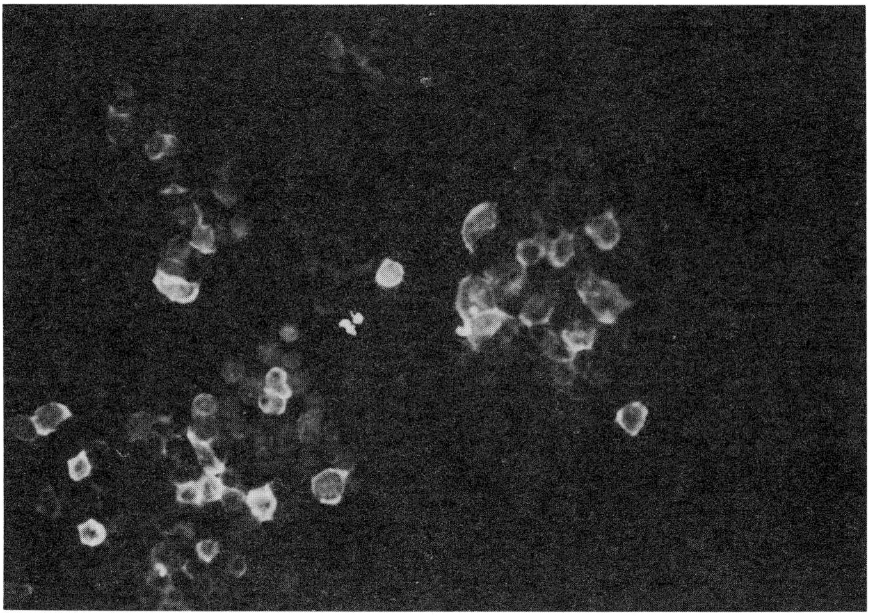

Figure 4: EBV gp350/220 expression in a GH3 clone, examined by fixed-cell indirect anti-gp350/220 immunofluorescent staining; note the relatively low percentage of positive cells.

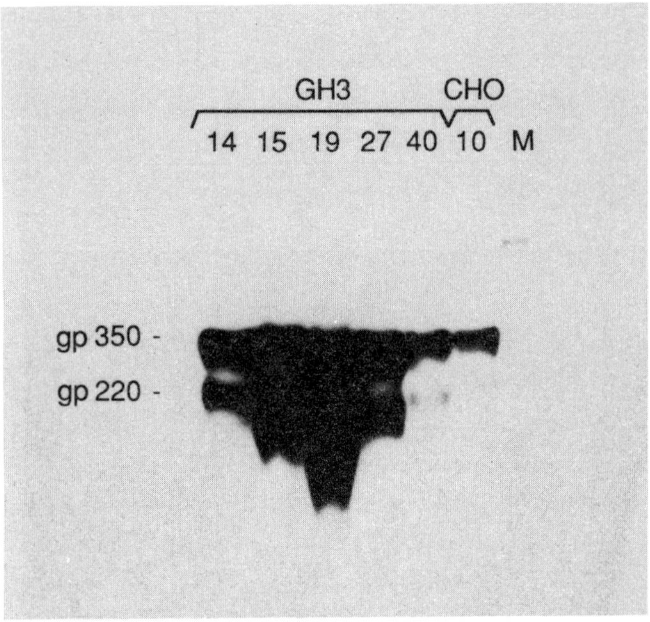

Figure 5: Western blot of media samples harvested from GH3 clones stably transfected with PCMVIE-MA-delta (Fig. 3), including, for reference, one of the best corresponding CHO clones obtained in a separate experiment (1). M, protein marker lane.

replaced with fresh DMEM containing 10% Nu-Serum. Medium was replaced daily, and bGH in the harvested medium was assayed by specific RIA (similar to that described previously for chicken GH (12)). For vector comparison assays, day 3 or 4 media harvests were generally used.

Stable Transfection

GH3 cells were transfected with 5ug of supercoiled plasmid DNA and 15ug of high molecular weight carrier DNA per 100mm plate by the calcium phosphate method (13), and selective medium containing 200 ug/ml G418 antibiotic was applied 24 hr after removal of DNA. Resistant colonies were picked after 12-14 days by cylinder-cloning, expanded, and assayed for expression -- bGH by RIA and EBV membrane antigen by dot immunoblot (1).

Transfection by electroporation was essentially as published (15), with all steps subsequent to DNA exposure exactly as above.

EBV Membrane Antigen Detection

Western blotting, dot immunoblotting and fixed-cell indirect immunofluorescence were all as described (1).

RESULTS

Adherent Cell Population

The parental GH3 line (ATCC CCL82.1) is epthelial-like in morphology and has a low plating efficiency (3); it can be grown in suspension culture (6). We found that the serum substitute, Nu-Serum (Collaborative Research) promoted somewhat better attachment than foetal bovine serum. In addition, by selecting cells that remained adherent to a plastic cultureware surface under conditions of continuous shaking for 2 weeks (Methods), a subpopulation was derived that grew essentially as an attached monolayer. The attachment phenotype persisted with passage, and allowed direct application of DNA transfection and colony selection protocols originally developed for fibroblasts (13).

Expression Vectors For GH3 Cells

We have previously described the optimization of the DEAE-dextran transient expression assay system for GH3 cells (Methods) using bGH as a reporter gene (11). Expression of bGH generally reaches 200-800 ng/ml/24 hr. by day 4 post-transfection (5,11).

In order to develop a more general expression vector for stable transfection, it was necessary to add a dominant selectable marker, such as the *neo* drug-resistance gene (8). To confirm the best relative orientation of the two gene cassettes, expression gene and marker gene, we constructed plasmids containing four relative orientations of the bGH gene (under RSVLTR regulation) and the *neo* gene (Fig. 1), and compared for levels of bGH secretion by transient expression assay. The 4 vectors yielded the results shown in Table I; values are normalized to expression from pBGH-4 (4). As the counterclockwise head-to-tail orientation (Fig. 1B) gave the highest expression, this design was adopted for further constructions.

Vector B (Fig. 1) was also used to obtain stable transfectant clones by the calcium phosphate co-precipitation method (13) or by electroporation (15). In both cases, levels of expression of bGH in the best clones were similar to those previously reported for GH3 clones derived from co-transfection with unlinked marker gene and bGH vectors (5). In some cases, the level of expression in the primary transfected clones exceeded 20 ug/ml/24 hr from a confluent monolayer culture.

In experiments reported elsewhere (5), we found that the CMV immediate-early (CMVIE) promoter surpassed all other promoters tested for constitutive expression of bGH in GH3 cells. We, therefore, substituted it for the RSVLTR in subsequent constructions. First, the CMVIE promoter was joined to the termination codon and poly(A) signal region of the bGH gene so as to leave a unique BglII insertion site, derived from a synthetic linker, in between (Methods). Then this entire cassette was substituted for BGH4 in a construction like that of Fig. 1B. Finally, the SV40 early promoter-driven mouse dihydrofolate reductase cDNA gene from pSV2-DHFR (10) was subcloned into the EcoRV site of this vector (Methods) as an additional selectable marker to yield the general expression vector PCMVIE-AKI-DHFR (Fig. 2).

EBV Membrane Antigen (gp350/220) Expression

The major envelope glycoprotein of Epstein-Barr Virus consists of 2 species with molecular weights of 250kd and 220kd, respectively, both resulting from the same primary transcription unit via differential mRNA splicing (14). Two cloned forms of this gene, a complete open reading frame and a truncated open reading frame generated by synthetic DNA insertion, were spliced into the BglII site of PCMVIE-AKI-DHFR as described (1) (Fig. 3). The truncated gene is deleted for the region coding for the stretch of hydrophobic amino acids at the carboxy terminus of the protein that ordinarily acts as a membrane anchor; the resulting truncated protein is therefore secreted (1).

Transfection of GH3 cells with the complete gp350/220 gene vector, PCMVIE-EBMA (Fig. 3), yielded expressor clones among the neo+ colonies with a phenotype similar to that described for other cell lines used to express this gene (1). Basically, gp350/220 gene expression was unstable and rapidly lost from the population. This phenomenon can be visualized by fixed-cell anti-gp350 indirect immunofluorescent staining (Fig. 4), whereby the percent brightly fluorescent cells can be seen to decline rapidly with passage of the cells (not shown).

When GH3 cells were transfected with PCMVIE-MA-delta, however, the majority of neo+ clones were found, by dot immunoblot, to express secreted gp350/220 relatively stably. In fact, as judged by dot blot or Western blot, many primary GH3 clones, were found to secrete larger amounts of the protein than primary transfectant clones from other cell lines (Fig. 5), such as CHO (1). When examined by fixed-cell immunofluorescent staining, these clones exhibit no detectable gp350/220 fluorescence above background whatsoever (not shown), implying very efficient secretion of newly-synthesized protein and a relatively short cytoplasmic residence time. We have maintained one of the GH3 clones indicated in Fig. 5, GH3-delta-19, in continuous passage for up to 6 weeks without significant loss of expression.

DISCUSSION

We have shown that stably transfected GH3 cells can efficiently express naturally secreted proteins, such as bGH, as well as genetically engineered secreted proteins, such as EBV envelope glycoprotein gp350/220. In the latter case, expression of the truncated, secreted protein was maintained much more stably, upon continuous passage of the cells, than expression of the complete membrane-anchored protein.

In parallel work with EBV gp350/220 expression in CHO cells, we have speculated that rapid loss of expression upon passage of the expressor clones was due to toxic effects on the host cell of high-level expression of this gene (1). In CHO clones, it was noted that the vast majority of

anti-gp350/220 immunofluorescently labelled live cells selected by fluorescence-activated cell sorting either failed to divide upon plating, or divided very slowly, implying that high-level gp350/220 expression inhibits cell growth (1). Non-expressing variant cells in the population presumably therefore outgrow expressors. This phenomenon also appears to take place in CHO cells expressing the secreted form of EBV gp350/220, as loss of expression upon passage of these clones is also relatively rapid (1). Interestingly, fixed-cell immunofluorescent staining for gp350/220 of CHO clones that secrete the truncated protein does reveal a slight staining above background (not shown), in contrast to the total lack of staining of secretor GH3 clones. We would therefore further speculate that the residual toxicity of gp350/220 in its secreted form may be proportional to its finite residence time in the host cell cytoplasm, and that the relative resistance of GH3 cells to this toxicity may be due to their more efficient secretory capabilities.
Pulse-chase experiments with radioactive tracers to directly measure the processing and secretion rate of newly synthesized gp350/220 are in progress.

Preliminary investigations into larger scale culture of gp350/220 expressor GH3 cells indicate that both on plastic cultureware surfaces as well as on microcarrier beads, the secretion rate of gp350/220 remains relatively stable, even in serum-free medium (data not shown). It has also been reported that the secretory phenotype of GH3 cells is retained in suspension culture (6). Scale-up potential should therefore present no barrier to the wider use of these cells for recombinant-derived gene expression. One caveat that must be mentioned, however, is that the cells also continue to secrete other pituitary proteins, such as rat growth hormone, which may prove undesirable for expression of low molecular weight foreign proteins with which these might copurify. Nevertheless, for many applications, such as EBV gp350/220 expression, GH3 cells may prove to be an excellent cell substrate.

ACKNOWLEDGEMENTS

The authors wish to thank J.G. Taylor for performing bGH RIA's, M. Armstrong and E. Emini for help in assaying gp350/220 expression, and S. Pols for typing the manuscript.

REFERENCES

1. Whang, Y., Silberklang, M., Morgan, A., Munshi, S., Lenny, A.B., Ellis R.W. and Kieff, E. (1987). Expression of the Epstein-Barr Virus gp350/220 Gene in Rodent and Primate Cells. J. Virol., in press.

2. Tashjian, A.H. Jr., Yasumura, Y., Levine, L., Sato, G.H. and Parker, M.L. (1968). Establishment of a Clonal Strain of Rat Pituitary Tumor Cells that Secrete Growth Hormone. Endocrinology 82, 342-352.

3. Tashjian, A.H. Jr., (1979). Clonal Strains of Hormone-Producing Pituitary Cells. Methods Enzymol. 58, 527-535.

4. Kopchick, J.J., Malavarca, R.H., Livelli, T.J. and Leung, F.C. (1985). Use of Avian Retroviral-Bovine Growth Hormone DNA Recombinants to Direct Expression of Biologically Active Growth Hormone by Cultured Fibroblasts. DNA 4, 23-31.

5. Pasleau, F., Tocci, M.J., Leung, F. and Kopchick, J.J. (1985). Growth Hormone Gene Expression in Eukaryotic Cells Directed by the Rous Sarcoma Virus Long Terminal Repeat or Cytomegalovirus Immediate-Early Promoter. Gene 38, 227-232.

6. Bancroft, F.C. and Tashjian, A.H. Jr. (1971). Growth in Suspension

Culture of Rat Pituitary Cells Which Produce Growth Hormone and Prolactin. Exp. Cell Res. 64, 125-128.

7. DiMaio, D., Treisman, R. and Maniatis, T. (1982). Bovine Papilloma Virus Vector that Propagates as a Plasmid in Both Mouse and Bacterial Cells. Proc. Nat. Acad. Sci. U.S.A. 79, 4030-4034.

8. Colbere-Garapin, F., Horodnidanu, F., Kourilsky, P. and Garapin, A. (1982). A New Dominant Hybrid Selective Marker for Higher Eukaryotic Cells. J. Molec. Biol. 150, 1-14.

9. Woychik, R. Camper, S., Lyons, R., Horowitz, S., Goodwin E. and Rottman, F. (1982). Cloning and Nucleotide Sequencing of the Bovine Growth Hormone Gene. Nucl. Acids. Res. 10, 7191-7210.

10. Subramani, S., Mulligan, R. and Berg, P. (1981). Expression of the Mouse Dihydrofolate Reductase Complementary Deoxyribonucleic Acid in Simian Virus 40 Vectors. Molec. Cell. Biol. 1, 854-864.

11. Kopchick, J.J., Pasleau, F. and Leung, F.C. (1986). Expression of the Bovine Growth Hormone Gene in Cultured Rodent Cells, in Genetic Engineering of Animals (Eds. Evans, W.J. and Hollander, A.), Plenum, New York, pp. 19-37.

12. Leung, F.C., Taylor, J.G., Steelman, S.L., Bennett, C.D., Rodkey, J.A., Long, R.A., Serio, R., Weppelman, R.M. and Olson, G. (1984). Purification and Properties of Chicken Growth Hormone and the Development of a Homologous Radioimmunoassay. Gen. Comp. Endocrinol. 56, 389-400.

13. Wigler, M., Pellicer, A., Silverstein, S., Axwl, R., Urlaub, G. and Chasin, L. (1979). DNA-Mediated Transfer of the Adenine Phosphoribosyltransferase Locus into Mammalian Cells. Proc. Nat. Acad. Sci. U.S.A. 76, 1373-1376.

14. Beisel, C., Tanner, J., Matsuo, D., Thorley-Lawson, D., Kezdy, F. and Kieff, E. (1985). Two Major Envelope Glycoproteins of Epstein-Barr Virus are Encoded by the Same Gene. J. Virol. 54, 665-674.

15. Potter, H. Weir, L. and Leder, P. (1984). Enhancer-Dependent Expression of Human K Immunoglobulin Genes Introduced Into Mouse Pre-B Lymphocytes by Electroporation. Proc. Nat. Acad. Sci. U.S.A. 81, 7161-7165.

16 Use of transfected and amplified Drosophila heat shock promoter construction for inducible production of toxic mouse c-myc proteins in CHO cells

F. M. Wurm, K. A. Gwinn, O. Papoulas, M. Pallavincini and R. E. Kingston

ABSTRACT

After transfection and selection with methotrexate, CHO cell lines were established which contained up to 2000 copies of an expression vector for c-myc protein. The vector contained the Drosophila heat shock protein 70 (hsp70) promoter fused with the coding region of the mouse c-myc gene. Incubation of cells for up to 3 hours at $43^{o}C$ resulted in at least a 100-fold induction of recombinant c-myc mRNA. When cells were shifted back to $37^{o}C$, within 1 to 4 hours this RNA was translated into protein to yield about 250 μg per 10^9 cells. Cells died a few hours later, suggesting that high concentrations of intracellular c-myc are cytotoxic.

The gene of interest was introduced and amplified to high copy numbers in a silent form. The presence of so many Drosophila hsp70 promoters in the cells did not impair their induction. The method described here may provide a general tool for expression of proteins, especially those unsuitable for constitutive overproduction due to toxicity or other factors.

INTRODUCTION

A number of mammalian or viral proteins of pharmaceutical interest have been produced in _Escherichia coli_ and yeast using inducible expression vectors (1,2). These vectors were chosen because the proteins, when expressed constitutively, proved to be toxic to the microbial host or inhibited its' growth rate.

In most transfected mammalian cells, production of proteins of interest has been accomplished by use of constitutive expression vectors, often having viral components (3-10). Inducible expresion systems, however, have certain advantages: Firstly, noninduced cells are expected to have higher growth rates because more of their metabolic activity can go into housekeeping functions. Secondly, there might be a substantial class of mammalian proteins whose high level expression is toxic to host cells. Thirdly, it might be desireable to initiate the formation of the product at will, especially in an industrial situation with hardware and working schedule constraints. Although of increasing interest, only a few inducible vectors have been used so far in mammalian cells (11-16). We describe here the _Drosophila_ heat shock protein 70 promoter (hsp70 promoter) and its molecular biology in stably transfected cell lines. We believe it is the most effective inducible promoter so far evaluated in mammalian cells. Its potential as a general expression vector in mammalian cells is discussed.

RESULTS AND DISCUSSION

General remarks

The Drosophila hsp70 promoter originates from a genomic fragment containing one of the multiple heat shock loci in _Drosophila melanogaster_ (17,18). When used to drive the expression of a marker gene in mammalian cells, this promoter has an extremely low level of transcription at $37^{o}C$ (19). Even in cells containing up to $4x10^{4}$ copies of a heat shock promoter, a basal level of transcription could not be detected (20). When cells were shifted to $43^{o}C$ for

60 minutes or more, heat shock mRNA is induced at a rate equal to or greater than that of the strong SV40 promoter (19,20).

The product of the protooncogene c-myc is suspected to be toxic to mammalian host cells when overexpressed constitutively. This is based on the assumption that the expression of a protein with regulatory functions is controlled in a strict and sensitive way - so constitutive high level expression would create an imbalance in the cells' regulatory mechanisms. C-myc is highly conserved in vertebrates; several lines of evidence suggest that the gene and its products play an important role in mechanisms involved in control of cell growth and differentiation (21,22). In many instances the gene is responsible for the generation of malignant tumors after chromosomal rearrangements or when activated by retroviruses (23-26). The nature and function of the protein itself remains obscure due to its extremely low expression level in cells. Attempts to overexpress c-myc in mammalian cells using constitutive promoters led either to very poorly growing refractile cells or to cell lines in which the gene was partially or totally deleted (27). In order to inducibly express the mouse c-myc protein we chose to transfect and stably integrate the c-myc gene under the control of the Dr. hsp70 promoter into dihydrofolate reductase (DHFR) deficient Chinese Hamster Ovary cells (CHO) (28). These cells are widely used for production of exogeneous mammalian or viral proteins (3-7). This is due to two major advantages of these cells over other systems. First, transfected DNA usually becomes integrated into the genome, thus enabling the establishment of stable cell lines (30). Second, using the gene for the enzyme dihydrofolate reductase as a selectable marker and treating cell populations with the DHFR antagonistic drug methotrexate, the integrated DNA can be amplified many times, generally leading to elevated expression levels (29, 30, 3-6).

Construction of expression vectors

To produce high yields of the c-myc protein, we transfected CHO-DUKX cells (28) with the vector pHS-Myc. This vector (see fig. 1) was made with a promoter containing a DNA fragment of 850 bp whose essential elements are - from 5' to 3': four "heat shock boxes", conserved sequences of 25 bp which have been found to be crucial for heat inducibility (19,33,34), the TATA box, the transciption start site and a stretch of 88 bp of the nontranslated leader sequence of the hsp70 gene. This part of the heat shock leader contains one of two "consensus sequences", which are conserved in various heat shock genes. The function of these sequences has not yet been elucidated (32). A 4.5 kb genomic mouse c-myc fragment consisting of exon 2 and 3 together with an intron and nontranslated 3'-sequences was linked to the promoter. The first exon of mammalian and avian c-myc genes contains multiple stop condons in all reading frames and is regarded as noncoding (35, 36). In our construction, the first translation start codon following the transcription start site of the heat shock promoter is that of the c-myc gene (fig. 1). The selectable marker was provided by a second vector in which the expression of the c-DNA of a mouse DHFR gene was driven by the constitutive Adeno EII promoter (pCVSVEII-DHFR, 37). The plasmid pHS-Myc and pCVSVEII-DHFR were mixed in a ratio of 5:1 and transfected. Recombinant clones were picked after two weeks of growth in selective medium, propagated separately into 100 mm petri dishes, and then subjected to increasing concentrations of methotrexate. After six months 4 cell lines resistant to 320 μ molar concentrations of methotrexate were established.

Copy number and chromosomal location of transfected heat shock-c-myc construction

To test whether the transfected c-myc DNA had been integrated into the genome and amplified during methotrexate selection a Southern experiment was performed. Genomic DNAs were isolated, restricted, electrophoretically

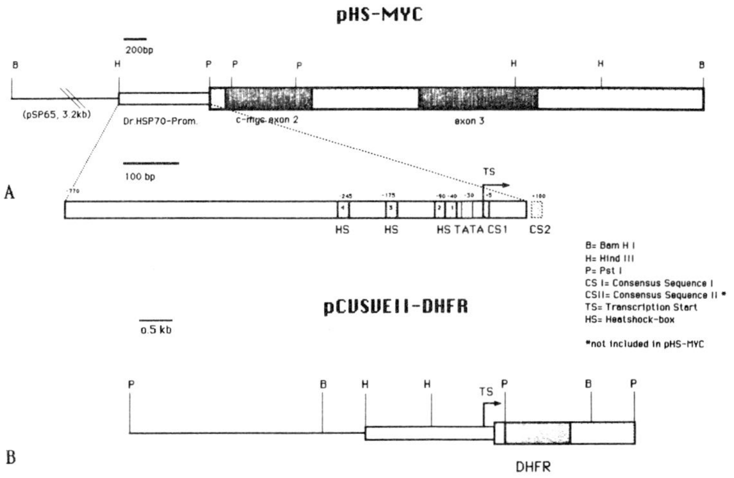

FIGURE 1

Plasmids used to establish recombinant myc lines:

A. Expression vector pHS-Myc: This plasmid contains a genomic mouse c-myc fragment from a unique XbaI site 48 bp upstream of the translation start codon in exon 2 to the unique BamHI site about 2.5 kb downstream of exon 3. The filled boxes denote the c-myc coding region and the light dotted boxes denote the remainder of mouse c-myc genomic sequences. The open box denotes the <u>Drosophila</u> hsp70 promoter region, and the arrow denotes the start site of transcription. The fusion between the promoter region and the c-myc gene is at +88 of the 5' untranslated region of the hsp70 promoter.

B. DHFR selection vector pCVSVEII-DHFR (37): The DHFR coding region is denoted by the solid box, splice and polyadenylation signals are denoted by the hatched box, and the adenovirus EII promoter and SV40 enhancer region are contained in the area denoted by the open box.

separated, and transferred to nitrocellulose. The filters were probed with *in vitro* transcribed P32-labelled c-myc specific RNA. As can be seen in fig 2, each of the 4 lines contained c-myc sequences, though to differing degrees. By comparing the intensities of the signals with those generated by dilutions of the plasmid pHS-Myc, the copy number of the construction in each of the four lines was estimated. The line 4HSMyc contained about 900 copies, 5AHSMyc 2000 copies, 5BHSMyc about 100 copies and 6HSMyc about 2500 copies. The additional bands generated by the latter line indicate DNA rearrangements which occurred probably early in the selective procedure and which were amplified together with the intact construction.

In the line 5AHSMyc the chromosomal location of the hsp70/c-myc construction has been investigated by using a biotinylated c-myc probe in *in-situ* hybridizations of metaphase chromosomes. The amplified heat shock c-myc construction was found to be located on the long arm of chromosome 2. (data not shown).

Inducibility and maintenance of recombinant c-myc transcripts

Cells were induced by feeding them with 43°C prewarmed medium and keeping them at that temperature for 1 to 2 hours. To measure the extent of the induction, RNA from induced and uninduced cells was isolated and hybridized to *in vitro* transcribed c-myc specific RNA. The hybrid molecules were digested with RNase to remove single stranded regions and subjected to polyacrylamide gel electrophoresis. The result of this experiment is shown in fig. 3a. Each of the cell lines exhibited a clear induction of the c-myc mRNA (lanes 1-10). It is important to note here that not a trace of a signal could be detected from cells kept at 37°C - even after prolonged exposures of the film. The induced concentrations of recombinant mRNA were proportional to the copy number of the introduced construction. Therefore there did not appear to be an insufficiency of cellular factors responsible for the induction of the hsp70

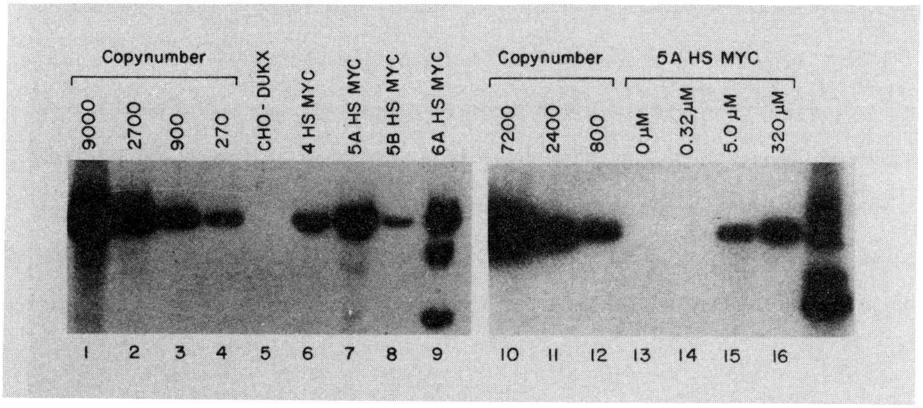

FIGURE 2

Copy number of c-myc constructs in recombinant CHO cell lines:

Genomic DNA isolated from four different recombinant cell lines growing at 320 μM methotrexate (lanes 6-9, 2 μg each), or from the cell line 5A HS MYC growing at different levels of methotrexate (lanes 13-16, 0.5 μg each) was restricted to completion with HindIII and together with HindIII restricted dilutions of the plasmid pHSmyc (lanes 1-4): 20ng, 6ng, 2ng, 0.6ng and lanes 10-12: 4ng, 1.3ng, 0.45ng) subjected to agarose gel electrophoresis and Southern transfer to Genescreen Plus membranes (NEN). The membranes were then hybridized to SP6 polymerase in-vitro synthesized ^{32}P-RNA consisting of the sequences of the PstI-HindIII fragment containing c-myc exonxx 2, intron 2 and exon 3. The last lane on the right contains radiolabelled 1kb-ladder DNA (BRL).

promoter region, even in the most highly amplified lines. This is shown in Fig. 3a, lanes 11-16, where the RNA from the line 5AHSMyc was analyzed using cells at different levels of methotrexate selection. We estimate that the degree of induction of the Drosophila hsp70 promoter in the amplified lines is at least 100 fold, because we could not detect a basal level of expression in uninduced lines. A similar kind of experiment was done to investigate how long after heat shock recombinant RNA remained detectable inside the cells.

The cells were induced at 43^{0}C for 2 hours and then shifted back to 37^{0}C for up to 3 hours. RNA isolated after 1,2, and 3 hours was hybridized and analyzed as described before. Recombinant mRNA was found even 3 hours after shifting the cells back to 37^{0}C (Fig. 3b). We have not determined whether this was due to high stability of the recombinant c-myc RNA, continued high promoter activity or a combination of these effects. Recently, Simcox et al. (38) found that a deletion of the 3'end of hsp70 mRNA increased the stability of the mRNA during the recovery period after heat shock. Our construction did not have the hsp70 3'end either, hence the maintenance of the recombinant hsp70/cmyc mRNA may be due to the fact that it did not contain sequences responsible for rapid degradation of normal heat shock messages during recovery.

c-myc protein synthesis in induced recombinant cell lines

Initially, cell lines were investigated for c-myc protein synthesis immediately after a 2 hour heat shock at 43^{0}C. At that time point no c-myc protein could be detected as judged by labelling with (35S) methionine or by immunoblotting. When cells were allowed to recover at 37^{0}C for extended periods, however, c-myc protein was detected by both methods. The level of protein increased rapidly during the first 2-3 hours of recovery (fig. 4), but continued to increase only slightly over the next 6 hours. Usually cells were harvested for detection and isolation of c-myc after 3 hours of recovery.

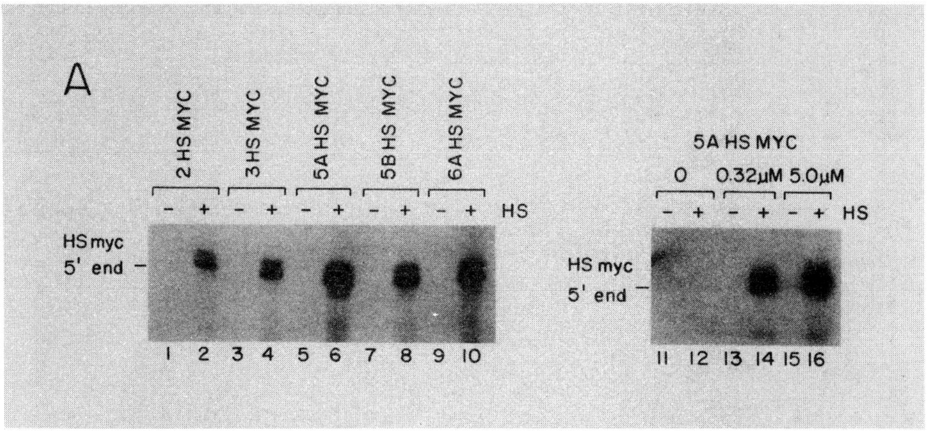

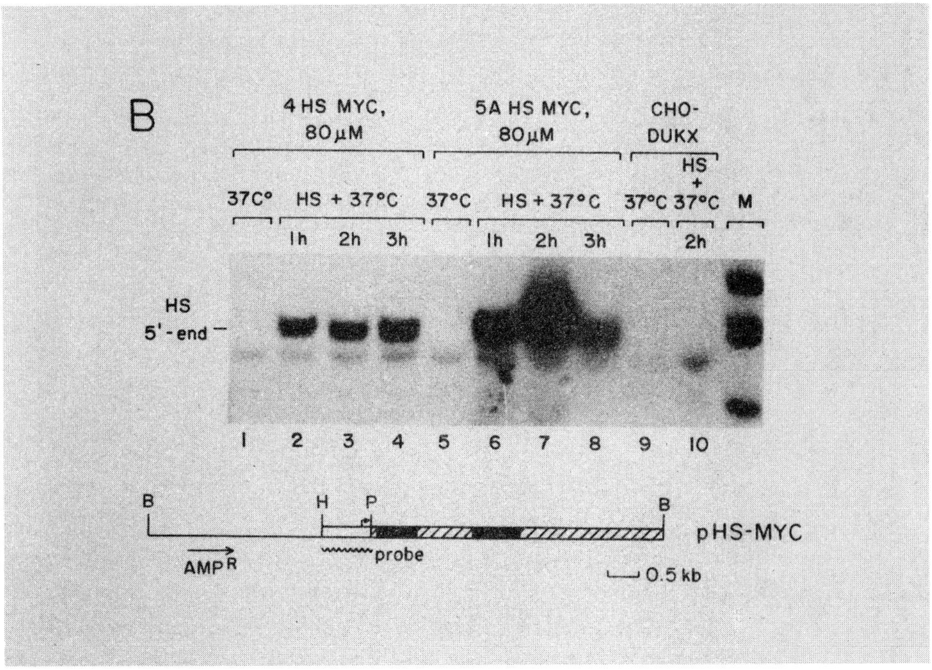

FIGURE 3

A and B: Presence, inducibility and stability of c-myc mRNA in recombinant CHO cells.

Cytoplasmic RNA from uninduced (-) and heat shock induced (2 hours at 43

C, +) recombinant cells was analyzed using an internally labelled RNA probe (shown in B) on a 10% denaturing polyacrylamide gel (lanes 1-10) or on 6% non-denaturing polyacrylamide gels.

A. Lanes 1-10; RNA samples (10 μg) from recombinant cell lines (as indicated) in 0 methotrexate were analyzed. Lanes 11-16; RNA samples (5 μg) from the line 5A HS MYC growing at different levels of methotrexate were analyzed.

B. RNA samples (3 μg) from line 4 HS-MYC (lanes 1-4), 5A HS-MYC (lanes 5-8) and CHO-DUKX (lanes 9 and 10) were analyzed as in A. RNA was isolated from cells growing at 37°C (lanes 1,5, and 9), or from cells that had been incubated at 43°C for 2 hours and allowed to recover at 37°C for 1 hour (lanes 2 and 6), 2 hours (lanes 3, 7 and 10) or 3 hours (lanes 4 and 8). An SP6 RNA polymerase generated transcript containing the sequences of the hsp 70 promoter from PstI (+88) to HindIII (-700) was used as probe. The bands labelled 5' end migrate at 90 bases in denaturing gels.

Three protein bands with apparent molecular weights of 64,000, 66,000 and 74,000 were recognized by monoclonal and polyclonal antibodies raised against a portion of the human c-myc protein that is conserved in mouse c-myc (42).

The smaller two of these were seen in previous studies (39, 40). It is not clear whether the third species with a molecular weight of 74,000 is a form of c-myc present in normal cells or is an aberrant form. It is possible, for example, that overproduction of c-myc results in some modification, such as ubiquitination (43), not relevant to the normal biology of the c-myc protein. The cell line with the higher copy number (5AHSMyc) produced at least twice the amount of c-myc protein as the line 4HSMyc. In experiments using metabolic labelling of cells with 35S-methionine, it was found that protein synthesis in the recombinant lines during recovery after heat shock was reduced to about 20% of the values seen in the nonrecombinant parental line which was induced in the same way. This may have been due to the suspected c-myc toxicity on the cells metabolic pathways. In spite of this finding, the intensity of the c-myc band in SDS/polyacrylamide gels revealed that this protein was one of the most actively synthesized during the recovery period (fig. 4b, lane 4). In cells labelled with orthophosphate (32P), all three recombinant c-myc species were phosphorylated (fig 4b), indicating proper modification of these molecules. All avian and mammalian c-myc proteins analyzed so far have been found to be phosphorylated (41,42,44). It should be noted that the pure genetic information of the c-myc gene would produce an unmodified protein with a molecular weight of 48,000 D. It is not known why the c-myc proteins migrate in SDS/polyacrylamide gels at the molecular weight range of 62000 D to 68000 D (45). Whatever the basis of this anomalous migration, the CHO cells provided a means to produce a protein which by all known criteria appeared identical to authentic c-myc protein.

In immunoblot experiments we estimated the yields of recombinant c-myc produced by these cells relative to a standard of purified c-myc protein

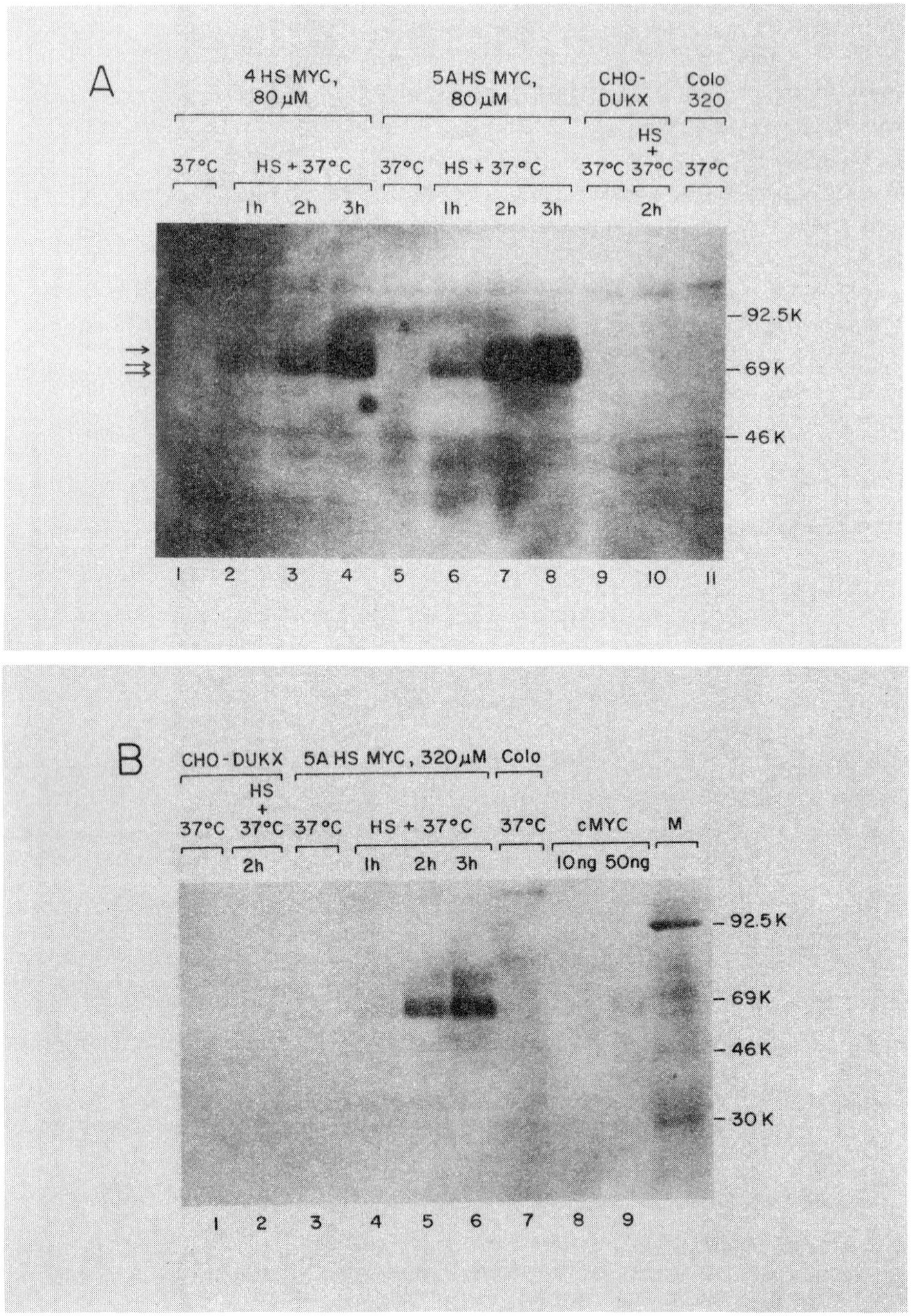

FIGURE 4

A and B: Time course of induction of c-myc protein and analysis of protein levels in induced cells.

Levels of c-myc protein in cell lines growing at 37^{0}C (A: lanes 1, 5, 9, 11; B: lanes 1, 3, 7,) and after heat shock induction for 2 h at 43^{0}C and varying time periods of recovery at 37^{0}C (A: lanes 2 and 6, 1 hour recovery; lanes 3, 7, and 10, 2 hours recovery; lanes 4 and 8, 3 hours recovery. B: lane 4, 1 hour recovery; lanes 2 and 5, 2 hours recovery; lane 6, 3 hours recovery). RIPA buffer lysates (2 ml) were prepared from $2x10^7$ cells. Samples (20 μl) were analyzed by immunoblotting using anti c-myc monoclonal antibodies (R. Chizzonite, Hoffmann-La Roche) and ^{125}Iodine labelled Protein A (NEN). In B, lanes 8 and 9 contain purified c-myc (lane 8;10 ng lane 9; 50 ng) dervied from recombinant insect cells (G. Ju, Hoffmann-La Roche).

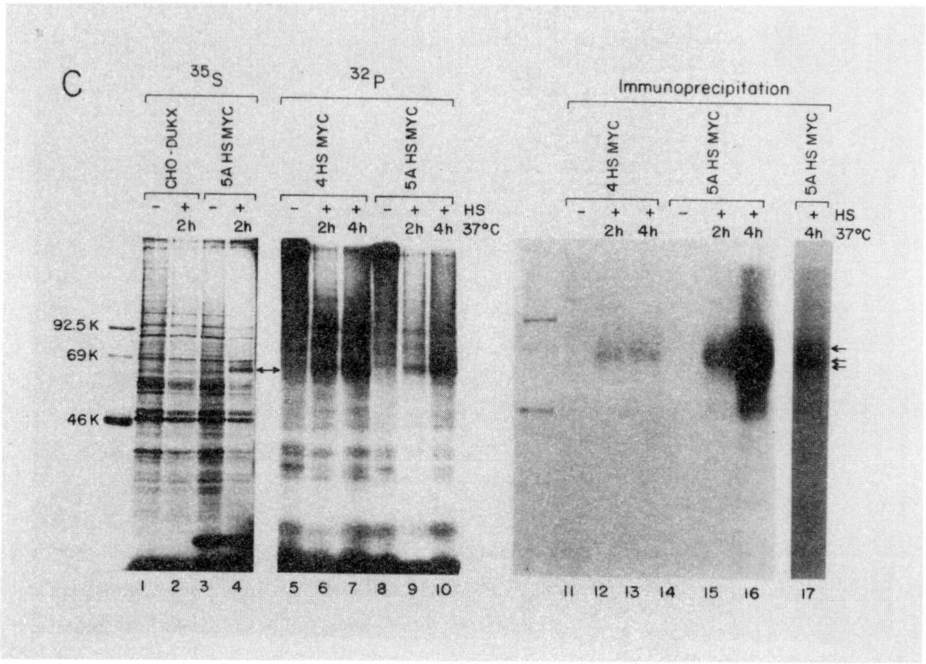

C. The c-myc protein is one of the major proteins synthesized in the recombinant lines during recovery and is made as a phosphoprotein.

RIPA buffer lysates of ^{35}S methionine (lanes 1-4:400 000cm/lane) and ^{35}phosphate labelled proteins (lanes 5-10): 400 000cpm/lane, immunoprecipitiation: lanes 11-17) were separated on 12% SDS-PAGE. Protein isolates were from the indicated cell line either growing at 37^{0}C (-) or induced for 2 hours at 43^{0}C (+) and allowed to recover at 37^{0}C for either 2 hours (lanes 2, 4, 6, 9, 12 and 15) or hours (lanes 7, 10, 13, 16, and 17). Lane 17 is a light exposure of lane 16.

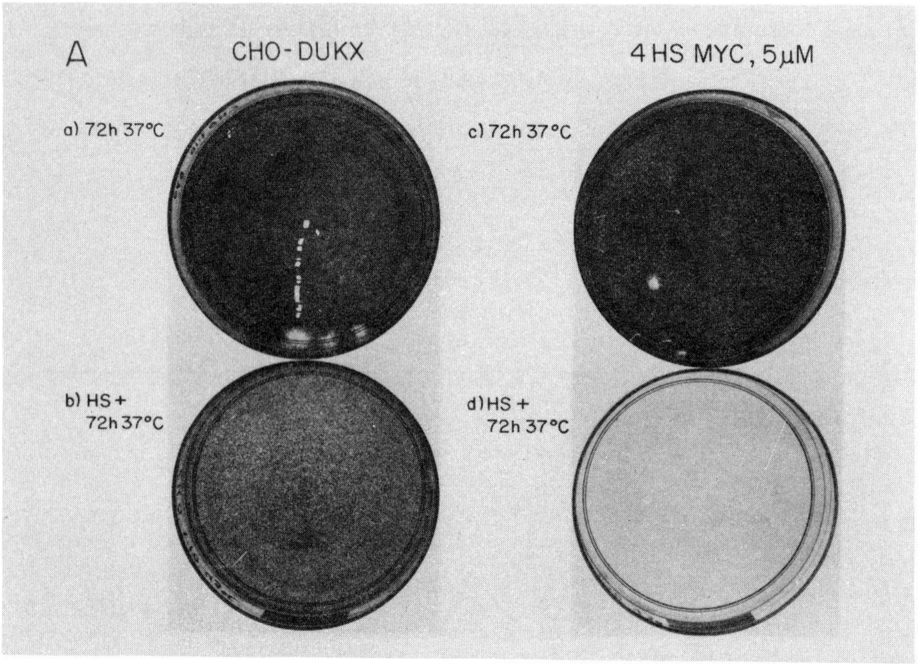

FIGURE 5

Overexpression of c-myc is cytotoxic.

A) 100 mm petridishes seeded with parental or recombinant CHO cells at about 10% confluency were heat shocked for 2 hours at 43^{0}C and recovered for 3 daysat 37^{0}C (b,d) or left at 37^{0}C all the time (a,c). Petridishes were stained with alcoholic cristalviolett solution (1%).

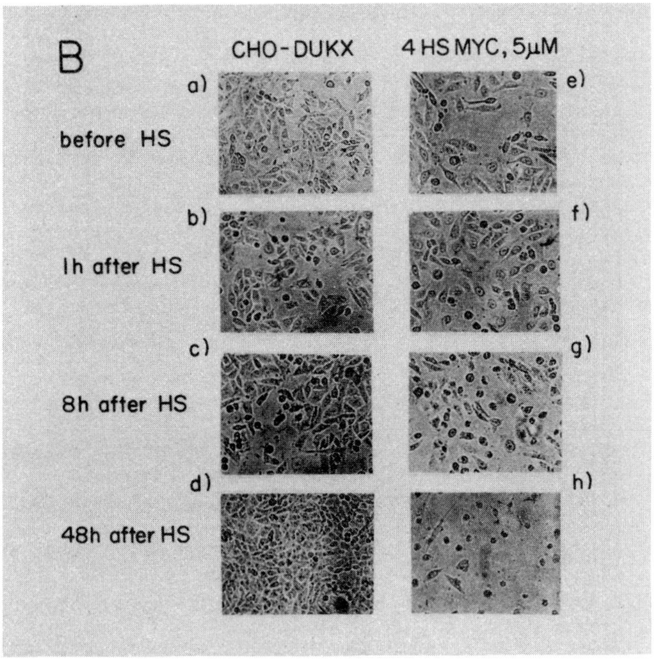

B. Micrographs from approximately 50% confluent CHO-DUKH B1 cells (a-d) and from recombinant cell line 4 HS MYC (e-h) before heat shock and after different periods of recovery after heat shock as indicated.

produced in insect cells (kindly donated by R. Chizzonite at Hoffman LaRoche). We estimated the amount of c-myc protein to be 200 to 300 μg per 10^9 cells. Theoretically, this corresponds with a specific productivity of about 2 fg per cell per day. We have developed a purification scheme for c-myc protein and are in the process of isolating larger amounts for further analysis.

Overproduction of c-myc protein is cytotoxic.

It was suspected that constitutive overproduction of c-myc is toxic. We tested, therefore, the effect of the induced overproduction of this protein on the viability of the cells. Parental and recombinant lines were plated at about 10% confluency, kept at 37^{0}C for 3 hours, then either heat shocked at 43^{0}C for 2 hours and then returned to 37^{0} for a period of 3 days, or, kept at 37^{0}C all the time. No living cells were left on heat shocked dishes of any of the recombinant lines, as well as both the heat shock and control dishes of the parental line, were confluent. An example of this experiment is shown in fig. 5a.

Microscopic examination of the recombinant cell lines during the recovery period showed detachment and slow deterioration of cells and no further cell divisions (fig. 5b). This may have been the result of a general cytotoxic effect of the high c-myc protein levels. Alternatively, the recombinant hsp70/c-myc constructions may block or slow the normal cellular response to heat. Two observations favor the former possibility: Firstly, in SDS-polyacrylamide gels containing samples from the recombinant cells, the normal heat shock proteins were observed at their usual intensity relative to the other proteins (fig. 4). Therefore, there was a proper induction of the endogeneous heat shock genes despite the presence of so many recombinant heat shock genes. Secondly, as mentioned earlier, recombinant lines containing high constitutive levels of c-myc are extremely difficult to establish in culture and have an abnormal and labile phenotype. Hence there may be a continuous selection against cells which express the c-myc protein at high levels, due to the toxicity of the protein itself.

A general method for inducible expression in mammalian cells?

We chose to overexpress the mouse c-myc protein in CHO cells for two reasons: Firstly, we believed that this approach would enable us to obtain this protein in an authentic form in sufficient quantities to allow detailed analysis. Secondly, it was hoped the experiments would show whether the *Drosophila* hsp70 promoter could be used as a general tool for obtaining proteins which interact adversely with the mammalian cell host. We have demonstrated that the *Drosophila* hsp70 promoter is highly inducible even when amplified. The control exerted by this promoter was so tight that even after a 2000 fold amplification no protein was detected without an inducing heat shock. The expression levels of other mammalian inducible promoters, like MMTV-LTR, human metallothionein and interferon promoter are lower than those of the *Drosophila* hsp70 promoter when induced and they have higher levels of basal transcription in the range of 1% to 30% of the induced activity (11-16).

This may present an obstacle for the expression of toxic proteins, especially after amplification. Amplification methods presently used can often lead to at least 50 copies of the introduced genes. Under these conditions, a promoter with only 1% of its maximal activity would, without induction, give moderately high mRNA levels constitutively, which could prohibit expression of toxic proteins of interest.

A third reason to favor the *Drosophila* hsp70 promoter was the observation that the growth of the CHO cells was not impaired by the introduction and amplification of the DNA constructs. This again is probably due to the tightness of the promoter at 37°C preventing diversion of cellular metabolic and housekeeping functions thus permitting maximal growth rates. In large scale situations, where mammalian cells are used for production of proteins of pharmaceutical interest (46), cell growth and medium consumption rates are very important factors and economic advantages will accrue where they can be influenced positively. Thus the Drosophila hsp70 promoter might also be valuable for non-toxic proteins. It should be emphasized that the use of this promoter is not restricted to CHO cells but is functional in various other mammalian substrates as previously shown (19, 33, 47).

In summary, we have described a promoter and its use for the induced production of a rare protein in mammalian cells. This protein is thought to play a key role in the metabolism of cells and seems to exert a cytotoxic effect when expressed to very high levels for prolonged periods. We think the tightness of this promoter at 37^{0}C and its degree of inducibility in highly amplified lines as shown will prove applicable for inducible expression of a variety of other mammalian proteins as well.

LITERATURE

1. Remaut, E., Stannsens, P. and Fiers, W. (1981). Plasmid vectors for high-efficiency expression controlled by the PL-promoter of coliphage lambda. Gene 15, 81-94.

2. Tuite, M.F., Dobson, M.J., Roberts, N.A., King, R.M., Burke, D.C., Kingsman, S.M., and Kingsman, A.J. (1982). Regulated high efficiency expression of human interferon-alpha in S. cerevisiae. EMBO J. 1, 603-608.

3. Kaufman, R.J., Wasley, L.C., Spilitiotes, A.J., Gossels, S.D., Latt, S.A., Larsen, G.F., and Kay, R.M.: Coamplification and coexpression of human tissue-type plasminogen activator and murine dihydrofolate reductase sequences in Chinese hamster ovary cells. Mol. Cell. Biol. 5,7,1750-1759 (1985).

4. Lau, Y-F., Lin, C.C., and Wai Kan, Y.: Amplification and expression of human alpha globin genes in Chinese hamster ovary cells. Mol. Cell. Biol. 4,8, 1469-1475 (1984).

5. Scahill, S.J., Devos, A., Van der Heyden, J. and Fiers, W.: Expression and characterization of the product of a human immune interferon cDNA gene in Chinese hamster ovary cells. Proc. Natl. Acad. Sci. USA 80. 4654-4658, (1983).

6. Lasky, L.A., Dowbenko, D., Simonsen, C., and Berman, (1984): Production of an HSV Subunit Vaccine by genetically engineered mammalian cell lines. In: CSH "New approaches to vaccines, Vol. p. 189-194.

7. Moriarty, A.M., Hoyer, B.H., Shih, I.W., Gerin, J.L., and Hamer, D.H. (1981). Expression of the hepatitis B virus surface antigen gene in cell culture by using simian virus 40 vector. Proc. Natl. Acad. Sci. USA 78, 2606-2610.

8. Mulligan, R.C., Howard, B.H., and Berg, . (1979). Synthesis of rabbit beta- globin in cultured monkey kidney cells following transfection with a SV40-beta-globin recombinant genome. Nature 277, 108-114.

9. Sarver, N., Gruss, P., Law, M.F., Khoury, G., and Howley, P.M. (1981) Bowine Papilloma Virus Deoxyribonucleic Acid: a novel eucaryotic cloning vector. Mol. Cell. Biol. 1, 486-496.

10. Tabin, C.J., Hoffmann, J.W., Goff, S.P., and Weinberg, R.A. (1982) Adaptation of a retrovirus as a eucaryotic vector transmitting the Herpes Simplex Virus Thymidine Kinase Gene, Mol Cell. Biol. 2, 426-4367.

11. McCormick F., Trahey, M, Innis, M., Dieckmann, B., and Ringold, G.: Inducible expression of amplified human beta interferon genes in CHO cells. Mol. Cell. Biology 4,1, 166-172, (1984).

12. Page, M.J.: Expression of amplified human beta interferon genes using heavy metal induction in Chinese hamster ovary cells. Gene 37, 139-144 (1985).

13. Karin, M, Cathala, G., and Nguyen-Huu, M.C.: Expression and regulation of a human metallothionein gene carried on an autonomously replication shuttle vector. Proc. Natl. Acad. Sci. USA 80, 4040-4044, (1983).

14. Lee, F. Mulligan, R., Berg, P., and Ringold, G.: Glucocorticoids regulate expression of dihydrofolate reductase cDNA in mouse mammary tumor virus chimaeric plasmids. Nature, 294, 228-232, (1981).

15. Papkoff, J. and Ringold, G.M.: Use of the mouse mammary tumor virus long terminal repeat to promote steroid inducible expression of -mos. J. Virology, 52, 2, 420-430 (1984).

16. Hamer, D.H. and Walling, M.J. (1982). Regulation in vivo of a cloned mammalian gene: Cadmium induces the transcription of a mouse metallothionein gene in SV40 vectors. J.Mol. Appl. Genets. 1, 273-288.

17. Livak, K.F. Freund, R., Schweber, M., Wensink, P.C. and Meselson, M. (1978): Sequence organization and transcription of two heat shock loci in Drosophila. Proc. Natl. Acad. Sci. USA 75, 5613-5617.

18. Holmgren, R., Livak, K., Morimoto, R., Freund, R. and Meselson, M.(1979): Studies of cloned sequences from four Drosophila heat shock loci. Cell, 18, 1359-1370.

19. Kingston, R.A. and Gwinn, K. (1985) unpublished observation.

20. Mirault, M.E., Southgate, R. and Delwart, E. (1982): Regulation of heat-shock genes: a DNA sequence upstream of Drosophila hsp 70 genes is essential for their induction in monkey cells. EMBO J. 1, 10, 1279-1285.

21. Weinberg, R.A. (1985): The action of oncogenes in the cytoplasm and nucleus. Science 230, 770-776.

22. Dean, M., Levine, R.A., and Campisi, J. (1986): c-myc regulation during retinoic acid-induced differentiation of F9 cells is post-transcriptional and associated with growth arrest. Mol. Cell. Biol. 6,2, 518-524.

23. Adams, J.M., Harris, A.W., Pinkert, C.A., Corcoran, L.M., Alexander, W.S., Cory, S, Palmiter, R.D., and Brinster, R.L. (1985): The c-myc oncogene driven by immunoglobulin enhancers induces lymphoid malignancy in transgenic mice. Nature 318, 533-539.

24. Erikson, J. Miller, D.A., Miller, O.J., Abcarian, P.W., Skuria, R.M., Mushinski, J.F., and Croce, C.M. (1985): The c-myc oncogene is translocated to the involved chromosome 12 in mouse plasmacytoma. Proc. Natl. Acad. Sci. 82, 4212-4216.

25. Schwab, M., Ramsay, G. Alitalo, K. Varmus, H.E., Bishop, J.M., Martinson, T., Levan, G. and Levan, A. (1985): Amplification and enhanced expression

of the c-myc oncogene in mouse SEWA tumor cells. Nature 315, 345-347.

26. Selten, G., Cuypers, H.T., Zijlstra, M. Melief, C., and Berns, A. (1984): involvement of c-myc in MuLV-induced T. cell lymphomas in mice: frequence and mechanisms of activation. Embo J. 3, 13, 3215-3222.

27. Unpublished observtion: Bob Kingston, Phil Sharp.

28. Urlaub, G. and Chasin, L.: Isolation of Chinese hamster cell mutants deficient in dihydrofolate reductase activity, Proc. Natl. Acad. Sci. USA 77, 4216-4220 (1980).

29. Kaufman, R.J. and Sharp, P.A.: Amplification and expression of sequences cotransfected with a modular dihydrofolate reductase complementary DNA gene. J. Mol. Biol. 159, 601-602 (1982).

30. Schimke, R.T.: Gene amplification in cultured animal cells. Cell 37, 705-713 (1984).

31. McGarry, T.J., and Lindquist, S. (1985): The preferential translation of Drosophila hsp 70 mRNA requires sequences in the untranslated leader. Cell 42, 903-911.

32. Amin, J., Mestril, R., Lawson, R., Klapper, J., and Voellmy, R. (1985): The heat shock consensus sequence is not sufficient for hsp 70 gene expression in Drosophila melanogaster. Mol. Cell. Biol. 5,1, 197-203.

33. Corces, U., and Pellicer, A. (1984) Identification of sequences involved in transcriptional control of a Drosophila heat shock gene. J. Biol. Chem. 259, 14812-14817.

34. Topol, J, Ruden, D.M., and Parker, C.S. (1985): Sequences required for in vitro transcriptional activation of a Drosophila hsp 70 gene. Cell 42, 527-537.

35. Stanton, L.W., Watt, R. and Marcu, K.B.: (1983) Translocation, breakage and truncated transcripts of c-myc oncogene in murine plasmacytomas, Nature 303, 401-406.

36. Stanton, L.W., Fahrlander, P.D., Tesser, P.M., and Marcu, K.B. (1984): Nucleotide sequence comparison of normal and translocated murine c-myc

genes. Nature 310, 423-425.

37. Kingston, R.E., Kaufman, R.J. and Sharp, P.A. (1984) Regulation of transcription of the Adenovirus EII promoter by Ela gene proucts: Absence of sequence specifity. Mol. Cel. Biol. 4, 1970-1978,

38. Simcox, A.A., Cheney, C.M., Hoffman, E.P., and Shearn, A. (1985): A deletion of the 3' end of the Drosophila melanogaster hsp 70 gene increases stability of mutant mRNA during recovery from heatshock. Mol. Cell. Biol. 5, 12, 3397-3402.

39. Hann, S.R. and Eisenman, R.N. (1984): Proteins encoded by the human c-myc oncogene: Differential expression in neoplastic cells. Mol. Cell. Biol. 4, 11, 2486-2497.

40. Persson, H. Gray, H.E., and Godeau, F. (1985): Growth dependent synthesis of c-myc encoded proteins: Early stimulation by serum factors in synchronized mouse 3T3 cells. Mol. Cell. Biol. 5, 1, 2903-2912.

41. Alitalo, K., Ramsay, G Bishop, J.M., Pfeifer, S.O., Colby, W.W., and A.D. Levinson (1983): Identification of nuclear proteins encoded by viral and cellular myc oncogenes. Nature 306, 274-277.

42. Hann, S.R., Abrams, H.D., Rohrschneider, L.R. and Eisenman, R.N. (1983): Proteins encoded by u-myc and c-myc oncogenes: Identification and localization in Acute Leukemia Virus transformants and bursal lymphoma cell lines. Cell 34, 789-798.

43. Parag, H.A. Raboy, B. and Kulka, R.G. (1987): Effect of heat shock on protein degradation in mammalian cells: involvement of the ubiquitin system. Embo J. 6,1, 55-61.

44. Ramsay, G. Evan, G.I., and Bishop, J.M. (1984): The protein encoded by the human proto-oncogen c-myc. Proc. Natl. Acad. Sci. USA 81, 7742-7746.

45. Miyamoto, C., Chizzonite, R., Crowl, R., Rupprecht, K., Kramer, R., Schaber, M., Kumar, G., Poonian, M., and Ju, G. (1985): Molecular cloning and regulated expression of the human c-myc gene in Escherichia coli and Saccharomyces cerevisiae: Comarison of the protein roducts. Proc. Natl. Acad. Sci. USA 82, 7232-7236.

46. Arathoon, W.R. and Birch, J.R. (1986): Large scale cell culure in biotechnology, Science 232, 1390-1395.

47. Pelham, H.R.B. (1982): A regulatory upstream promoter element in the Drosophila hsp70 heat-shock gene. Cell 30, 517-528.

Discussion

SILBERKLANG When you put the Drosophila heat shock promotor into a mammalian cell can you use any of the induction systems that work in Drosophilla like cooling the cell down and bringing them up to 37° or oxygen tension or inhibition of oxidative phosphorylation. Might there not be some inductive mechanism less damaging to the cells than heat shock per se.

WURM I would think that the mechanisms that work in Drosophilla would work in mammalian cells. Because the main trigger for the heat shock response is denatured proteins and whatever induces those proteins would also be implicated in other induction systems.

OFFENBURG Is the induction of the promotor a critical step - you said you induced the promotor for 2 hours at 37°C. Have you tried other times.

WURM I looked at different times between 10 min and 3 hrs but 2 hours was the best but each product would be different but it is not critical. 5 min at 45°C has also been used.

General Discussion

SILBERKLANG — Is expression proportional to copy number. There is something about the integration of the vector which is a function of copy number which results in RNA being transcribed from both directions leading to the neutralisation of RNA translation possibilities.

BEBBINGTON — As amplification proceeds you get to a plateau of EPA secretion. There is no correlation in initial tranfectants between copy number and expression levels, sites of integration and rearrangements of the vector might explain that. Yet as you have inteprotein the level of secreted product parallels the degree of amplification.

SILBERKLANG — In DHFR amplification you start from 1 or 2 copies and build from that, and in the early amplification it is relatively linear. When you have 270 copies you do not have an expression level 270 times the original so I wondered if it would not have been better to use the PSV2 vector, starting with a low copy number and amplifying more laboriously but then reach a higher (expression Ed.) plateau, or the same expression with fewer copies.

BEBBINGTON — With PSV2-GS you are getting to as high an MSX concentration as possible and still only getting 10 copies.

SILBERKLANG — Why cannot you go to higher MSX concentrations.

BEBBINGTON — Its not very soluble.

CLARKE — What happens when you try to maintain cells for

long periods of time in the amplified state.

BEBBINGTON You can maintain them with MSX without problems.

ACKERMANN The advantages of the Thymidine Kinase System is that you can select for or against the expression of the enzyme. Is there any selective system which could be used against the glutamate synthetase.

BEBBINGTON No

WURM While one is looking for constitutive production in industrial situations you end up with a batch system for filling etc. Therefore product induction is useful as in the heat shock system where you send the cells through a pipe heated to 42°C for a while then you transfer to a vessel kept at 37°c and leave it for product formation and harvest.

MEIGNIER Combining the discipline of Chemical Engineering, Biochemistry and Microbiology, there is something in me which rejects the concept of going for batch processes. If you have a continuous process you can establish control, if you have control you can get productivity, you do not have the interbatch downtime now the irreproducibilities of setting up a steady state which can be difficult. I agree with my colleagues in the pharmaceutical industries and regulatory agencies that it would be nice and convenient if all products come in identifiable batches with secured labels etc. Yet it is poor from an efficiency and control standpoint that the continuous production has advantages. Batched can be created by collecting material between defined times. I would like to see culture vessels running for years at a time

WURM — I won't agree to that but any system has the advantage that you can grow cells for years and you have another line fore removing cells from the biomass generator which on induction will generate a batch of product so you have the advantage of both a continuous system and a batch process.

SPIER — The big gain from using a continuous product formation system is that you do not have to waste raw materials building biomass. Because the ultimate goal is to use a cell which merely converts raw materials to find products by doing a biotransformation.

GREY — Ideally you need to produce product in a 'clean' system so you need to be able to do it in an inducible system outside the growth system.

FEDER — Using regulatory considerations would be a poor excuse to make batch processes as long as you could define a batch in a continuous system. The advantages of continuous systems are obvious, you have expression by the cells in a maintenance state (you do not need growth factors but you do need maintenance factors). I do not think anything can compete with a pot that contains 10^{12} - 10^{14} cells maintained in a defined steady state. The idea would be to maintain the induced state long term.

SPIER Would Dr. Beaupain like to indicate if there are any tricks required to establish his organ cultures.

NORDENBERG Have you tried to adapt the (organ cultures Ed.) system to grow primary tumour cell cultures on Millipore membranes.

BEAUPAIN It is difficult to keep neuroblastoma cells alive for long term culture -3 months.

MEIGNIER How do you see the future of human monoclonals.

JAMES We have first to ask the question why do so many of these cell lines stop making antibody. Is it secretion, assembly or synthesis then we would be half way to success.

I am optimistic that solutions can be found but I am not optimistic about the attitude of those who fund this kind of work who fail to appreciate that there is yet a problem.

The basic cell biology has to be done concentrating on in vitro immunisation finding out the correct time for fusion or transformation what is the correct differentiation state of the cell, we work up methods of enhancing expression or secretion of the immunoglobulin. Were we to do this we would get there.

Section 2
Medium development

17 The use of glutamine-free media for the growth of three cell lines in microcarrier culture

T. E. Hassell, I. C. Allen, A. J. Rowley and M. Butler

ABSTRACT

The phenomenon of ammonia related growth inhibition has been widely reported. The accumulation of ammonia in culture is a direct result of the chemical and metabolic deamination of glutamine which is normally present in growth medium at concentrations between 2-4mM. We have investigated the effects of ammonia on three cell lines; Baby Hamster Kidney cells (BHK21 C13), McCoy, and Vero cells grown on serum supplemented GMEM in T-flasks or microcarrier cultures. Concentration dependent inhibition of cell growth, by ammonia, was shown to vary between cell lines. In order to reduce ammonia accumulation in these cultures, we have developed a protocol which allows adaptation of cells to glutamine free media. Glutamate or 2-oxoglutarate were substituted for glutamine in these media formulations and resulted in the reduction of accumulated ammonia by >50%. Other effects of adaptation of cells to glutamine free media include altered glucose consumption and lactate production. The implications of this adaptation protocol to the achievement of higher cell productivities in microcarrier cultures will be discussed.

INTRODUCTION

The three major factors thought to limit population density in anchorage dependent cell culture systems are: (i) availability of surface area for growth; (ii) the exhaustion of essential nutrient(s); or (iii) the accumulation of inhibitor(s). The use of microcarriers (23) in animal cell culture systems can remove the cell productivity limitation of surface area. However, the composition of the culture medium is critical in obtaining high yields. During cell growth the medium deteriorates in two ways; firstly by depletion of nutrients and secondly by the accumulation of waste products which may prove growth inhibitory. This study has considered changes which occur in the medium during growth of three anchorage dependent cell lines; Baby Hamster Kidney (BHK21 C13), Vero, and McCoy.

Under culture conditions, glucose has been assumed to be the major energy source for cultured cells owing to its rapid metabolism (11). However, recent observations suggest that a large proportion of glutamine is utilised in aerobic energy metabolism in the cell(1,16,20,25). Glutamine although essential for some anabolic processes in the cell, is principally used for aerobic energy metabolism, in which deamination results in the excessive production of ammonia (4). The term ammonia is used throughout to cover the sum of the ammonium ion, NH_4^+, and ammonia NH_3. In this study the possibility of reducing the final concentration of ammonia by the substitution of glutamine with either glutamate or 2-oxoglutarate was investigated. Adaptation protocols were developed which permitted the omission of glutamine from the medium, and the cells cultured for over 20 passages in these glutamine free media.

This study details the effects of this adaptation process on McCoy cells providing information on amino acid utilisation and ammonia production, glucose consumption, and lactate production rates and concomitant changes in cell yields.

MATERIALS AND METHODS

Cells.

Baby Hamster Kidney (BHK21 C13), McCoy, and Vero cells were grown in $125cm^2$ T-flasks. The cells were harvested by trypsinisation for 5 minutes with 0.25% trypsin in phosphate buffered saline (PBS) and immediately inoculated into experimental T-flasks or spinner flasks.

Medium.

Glasgow's modification of Eagles medium (GMEM)(21) supplemented with 10% bovine serum was used for growth. This standard medium contained 4mM glutamine and 25mM glucose.

Cultures

(i) T-flasks: Cells were seeded into 10ml of culture medium contained in $25cm^2$ T-flasks. Growth was determined by harvesting cells into trypsin (10ml) and counting in trypan blue (18) using a haemocytometer.

(ii) Microcarrier cultures: Cells were grown in 100ml cultures at 5mg ml^{-1} microcariers (Cytodex 1). This offered a surface area of almost $2800mm^2$ ml^{-1} of culture medium. Spinner flasks (Bellco) were placed on a stirrer platform and maintained at a constant stirring rate of 50 r.p.m. Surface area to volume ratio was the same in all microcarrier cultures. All cultures were incubated at 37°C with a CO_2 overlay of 10%.

Ammonia assay.

Ammonia is mainly in the ionic form, NH_4^+, in medium at pH 7.4. Media samples were treated with 10M sodium hydroxide to convert the ammonia to the gaseous form before analysis by an ammonia specific electrode (Orion 95-10). The electrode gave a characteristic 56mV change per decade change in concentration.

Glutamine and Glutamate analysis.

Medium samples were derivatised with phenylisothiocyanate then analysed by High Performance Liquid Chromatography as previously described (5).

2-Oxoglutarate analysis.

2-oxoglutarate was measured by the oxidation of NADH in the presence of glutamate dehydrogenase. This spectrophotometric assay was modified from the method described by Bergmeyer and Bernt (2).

Glucose and Lactate analysis.

These were measured with a substrate specific analyser which is based on enzymes immobilised on membranes. (YSI model 27 industrial analyser)

Adaptation protocol for replacement of glutamine with glutamate.

Cells were grown to about 50% of confluence in T-flasks in standard GMEM with glutamine at 4mM. The medium was removed, the cell monolayer washed with PBS, and then fresh medium added in which the glutamine was substituted by glutamic acid at 4mM.

Adaptation protocol for replacement of glutamine with 2-oxoglutarate.

Cells were grown to about 80% of confluence in T-flasks in standard GMEM with glutamine at 4mM. This medium was removed, the cell monolayer washed with PBS, and glutamine-free fresh medium added containing 2-oxoglutarate at 20mM. At each of four subsequent sub-cultures the 2-oxoglutarate concentration was gradually lowered to a final concentration of 4mM.

RESULTS

Ammonia related growth inhibition.

Cell growth was examined in medium containing added concentrations of ammonium chloride (Figure 1). Cell counts were determined after 3 days growth in T-flasks. In the cases of both BHK21 C13 and McCoy cells, significant growth inhibition was observed in cultures containing 1mM ammonium chloride. Almost complete inhibition was observed at 3mM. Growth inhibition of Vero cells was not observed at these concentrations. In control cultures, without added ammonia, the final concentration of ammonia reached 2.25mM for BHK21 C13, 2.31mM for McCoy and 2.90mM for Vero cells.

Cell adaptation.

(i) *Growth rate.* Initial studies, not detailed here, determined the mean doubling times of the three cell lines when grown in glutamine containing or glutamine-free media. These studies showed initially longer mean doubling times of glutamine free grown cells, but demonstrated no significant differences after 5 passages in the different media.

(ii) Ammonia production. Each of the three cell lines; BHK21C13, McCoy and Vero produced significantly less ammonia when grown in glutamine free media. Ammonia concentrations reached between 30% and 50% of their counterparts grown in glutamine containing media. However, no significant difference, in terms of final ammonia concentration, could be detected between those cells grown in glutamate and those grown in 2-oxoglutarate.

(iii) *Glucose utilisation/Lactate production.* Glucose was present at 25mM in GMEM used for growth of cells. The glucose consumption and lactate production of McCoy cells, grown in T-flasks, in the presence or absence of glutamine is shown in Figure 2. The specific glucose consumption rate was 5-10 fold greater during the lag phase compared to any other period in

Figure 1. Inhibition of BHK, McCoy and Vero cell growth in T-flasks by ammonium chloride.

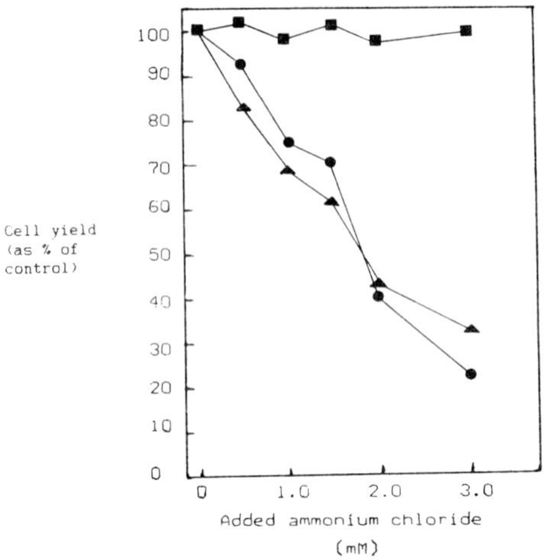

Cells (BHK ●, McCoy ▲, and Vero ■) were inoculated into 10ml of medium held in T-flasks with varying concentrations of added ammonium chloride. Cell yields are expressed as a percentage of that obtained in medium without added ammonia.
All cultures were incubated at 37°C, with a 10% CO_2 overlay, for 3 days before harvest. Growth medium was GMEM with 4mM glutamine supplement.
n = 5

Figure 2. Glucose consumption and lactate production by McCoy cells grown in T-flasks, in GMEM supplemented with; glutamine, glutamate or 2-oxoglutarate.

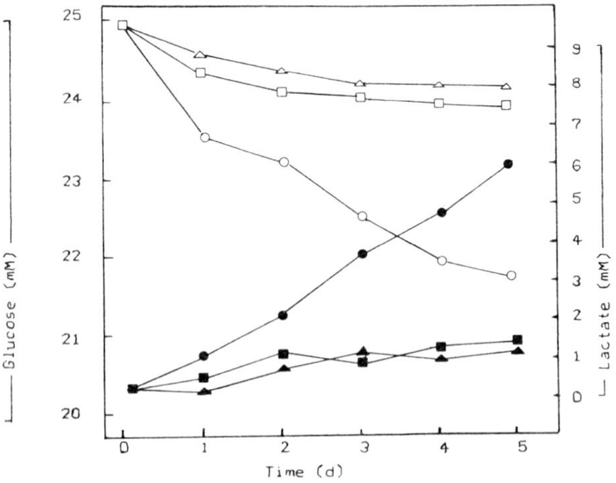

McCoy cells were inoculated, 0.5×10^4 cells ml^{-1}, into T-flasks in GMEM supplemented with 4mM glutamine ○, 4mM glutamic acid □, and 4mM 2-oxoglutarate △. They were harvested daily and assayed for glucose (open symbols) and lactate (closed symbols). All cultures were supplemented with 10% bovine serum, incubated at 37°C, with a 10% CO_2 overlay.
n = 3

glutamine supplemented cultures. These high glucose consumption figures were accompanied by high specific lactate production values. This contrasted with lower and relatively constant specific rates of glucose consumption and lactate production by both the glutamate and 2-oxoglutarate suplemented cultures.

Cell growth on microcarriers.

In order to further investigate cell growth in glutamine free media, McCoy cells were selected for more detailed study. After 10 passages in either glutamate or 2-oxoglutarate supplemented GMEM these cells were grown in microcarrier culture and compared with their glutamine grown counterparts at the same passage. Of particular interest were the changes in ammonia production and maximum cell yields (Figure 3). Cell yield for glutamine supplemented cells peaked at about 1.7 x 10^6 cells ml^{-1} with glutamate grown cells reaching higher concentrations at about 2 x 10^6 cells ml^{-1}. In the glutamine supplemented cultures ammonia concentration was >3mM at maximum cell density, whereas at the equivalent stage for the glutamate supplemented cultures ammonia concentration was <1mM. The 2-oxoglutarate supplemented cultures reached maximum cell densitites of about 6 x 10^5 cells ml^{-1} with an ammonia concentration <1.05mM.

Substrate utilisation.

Glutamine, glutamate, and 2-oxoglutarate were at an initial concentration of 4mM in the media used for the growth of the McCoy cells which had been adapted to their respective substrates.

(i) *Glutamine.* Was reduced to zero during cell growth over a 5 day period (Figure 4). Specific utilisation rates were highest during the lag phase. The high rate of glutamine utilisation is responsible for the rise in ammonia concentration of the medium, although a small percentage of this rise is due to spontaneous chemical decomposition of glutamine to ammonia

Figure 3. McCoy cell growth (a) and ammonia production (b) in microcarrier cultures supplemented with glutamine, glutamate , or 2-oxoglutarate.

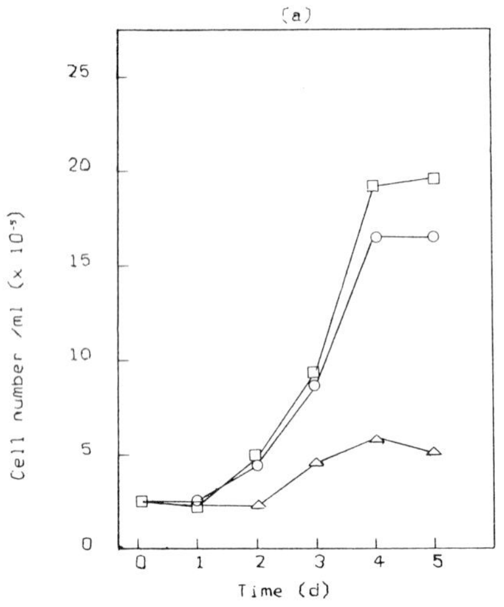

Cell counts for McCoy cells inoculated (2.5 x 10^5 cells ml^{-1}) into 100ml microcarrier cultures (5mg ml^{-1}) in GMEM supplemented with 4mM glutamine ○, 4mM glutamic acid □, and 4mM 2-oxoglutarate △. Ammonia concentrations were measured daily (closed symbols).
All cultures were supplemented with 10% bovine serum,

incubated at 37°C with a 10% CO_2 overlay. Stirring was maintained at 50 r.p.m.
n = 3.

Figure 3. McCoy cell growth (a) and ammonia production (b) in microcarrier cultures supplemented with glutamine, glutamate ., or 2-oxoglutarate.

(b)

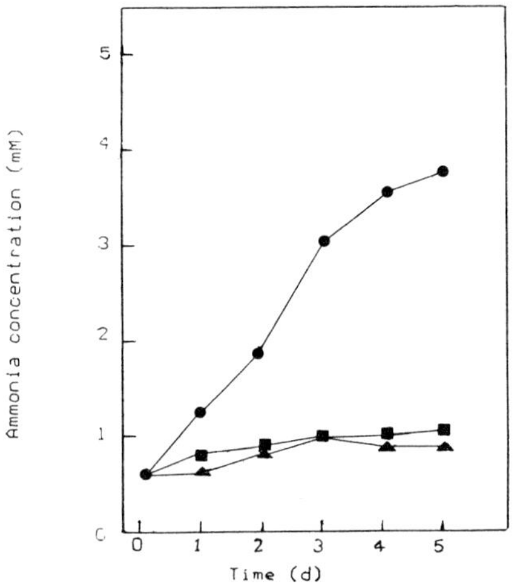

Cell counts for McCoy cells inoculated (2.5×10^5 cells ml^{-1}) into 100ml microcarrier cultures (5mg ml^{-1}) in GMEM supplemented with 4mM glutamine ○, 4mM glutamic acid □, and 4mM 2-oxoglutarate △. Ammonia concentrations were measured daily (closed symbols).
All cultures were supplemented with 10% bovine serum,

incubated at 37°C with a 10% CO_2 overlay. Stirring was maintained at 50 r.p.m.
n = 3

and pyrollidone carboxylic acid. (8,22)

(ii) Glutamate. Almost completely disappeared over a 5 day period (Figure 4) with highest specific consumption rate being in the lag and early periods of growth.

(iii) 2-Oxoglutarate. Was relatively slowly utilised over a 5 day period at the end of which it remained present in the medium at about 2mM. (Figure 4)

Selective substrate feeding.

McCoy cells grown in standard GMEM (with glutamine) on microcarriers, at 5mg ml^{-1}, produced cell yields of $1.6-1.8 \times 10^6$ cells ml^{-1} in batch culture.

In an investigation designed to determine the effects of further addition of glutamine and glutamate during microcarrier culture, these amino acids were added, on a daily basis, as concentrates during the latter part of the growth period. Each amino acid was added to increase the culture concentration by 1mM. Figure 5 shows that the effect of such selective feeding was to increase the cell yield by 13% in the glutamine suplemented cultures, and by 20% in the glutamate supplemented cells.

DISCUSSION

The first comprehensive study of nutrient requirements for cell growth was reported over thirty years ago (6) and led to the development of Eagles minimal essential medium. Eagle recognised the glutamine requirement of cells, suggesting an essential though unspecified role. Many media formulations used for animal cell culture at present are modifications of the original reported by Eagle with glutamine at 10-100 fold the concentration of any other amino acid.

During cellular metabolism glutamine can be deaminated in two steps leading to the production of 2-oxoglutarate which, in turn, may be

Figure 4. McCoy cell growth and substrate utilisation in microcarrier cultures supplemented with glutamine, glutamate , or 2-oxoglutarate.

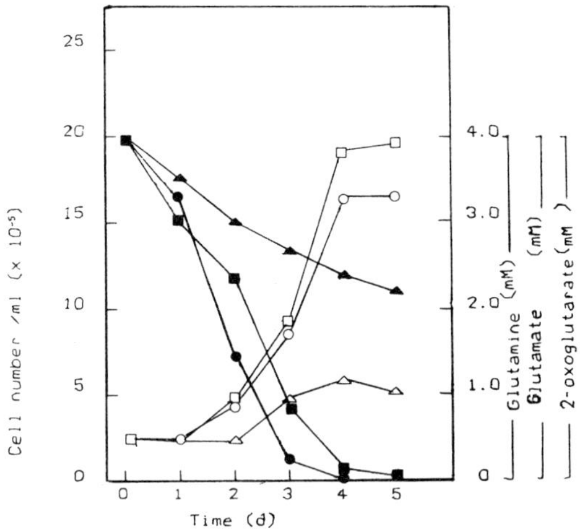

Cell counts for McCoy cells inoculated (2.5 x 10^5 cells ml^{-1}) into 100ml microcarrier cultures (5mg ml^{-1}) in GMEM supplemented with 4mM glutamine ○ , 4mM glutamic acid □ , and 4mM 2-oxoglutarate △ The substrate concentrations were measured daily (closed symbols).
All cultures were supplemented with 10% bovine serum, incubated at 37°C with a 10% CO_2 overlay. Stirring was maintained at 50 r.p.m.
n = 3

Figure 5. Growth of McCoy cells in microcarrier cultures with a regime of selective substrate feeding.

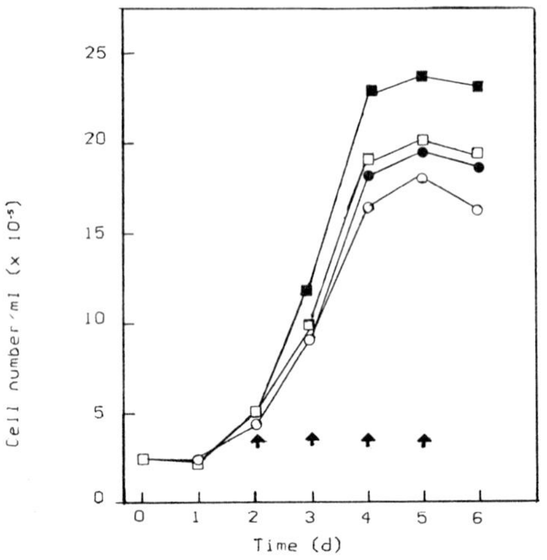

Cell counts for McCoy cells inoculated (2.5×10^5 cells ml^{-1}) into 100ml microcarrier cultures in GMEM supplemented with 4mM glutamine ○, and 4mM **Glutamate** □. At daily intervals from day 2 onwards, these amino acids were added (🡅) as concentrates to their respective cultures. Effect on cell numbers (closed symbols) of each addition (which increased culture concentration by 1mM) are shown.
All cultures were in duplicate, supplemented with 10% bovine serum, and incubated at 37°C. Stirring speed was maintained at 50 r.p.m.

completely catabolised to CO_2 by the TCA cycle. This leads to production of 2 mole of ammonia for each mole of glutamine so metabolised. Chemical decomposition of 1mole of glutamine results in the production of 1mole of pyrollidone carboxylic acid and 1mole of ammonia (8) and is affected by pH, temperature and variation in the nature and amount of serum (14,22). Measurements of ammonia concentrations in cell free glutamine supplemented GMEM used for this experimental work showed an increase of about 0.1mM day^{-1} when incubated at 37°C. Such changes are confirmed in previous reports (3,8,22).

Concentrations of ammonia in cell culture media (22) are considerably higher than those found *in vivo* (12). Thus cells may have become adapted to higher ammonia concentrations *in vitro*. It has been shown that the toxicity of ammonia varies considerably over a large number of cell lines tested (unpublished results). Here we show concentration dependent growth inhibition of both McCoy and BHK21 C13 cells but no significant inhibition of Vero cells.

Although early (6) attempts at substitution of glutamine with glutamic acid or 2-oxoglutarate were unsuccessful, later studies showed that cells would grow if high levels of glutamate were substituted for glutamine (7,8,19). Griffiths and Pirt (8) suggested replacement of glutamine by glutamate as glutamine was unstable at 37°C. Nagle and Brown (15) used glutamate as an alternative to glutamine so that the medium could be autoclaved.

In this investigation we show that the substitution of glutamine with glutamate or 2-oxoglutarate led to considerably lower ammonia accumulation without affecting cell doubling times. It was previously predicted that such a reduction in ammonia concentration may enhance cell yield (9). In microcarrier cultures of McCoy cells grown in glutamate supplemented medium the cell yield was 15% greater compared to controls grown with glutamine. A further increase, of 20%, in cell yield was obtained in these cultures by daily feeding of glutamate. The substitution of glutamine with 2-oxoglutarate resulted in lower cell yields, 6 x 10^5 cells ml^{-1}, although a reduction in ammonia concentration >50% was observed.

McCoy cells grown in glutamine supplemented medium showed a high specific rate of glucose utilisation in the lag phase a phenomenon which has been previously noted (8,17). This was accompanied by a high specific rate of lactate production. In the cells adapted to glutamate or 2-oxoglutarate significantly lower specific rates of glucose consumption and lactate production were observed. These results may indicate a more efficient utilisation of available substrate, since increases in cell numbers were similar for each of the three groups.

In the growth of McCoy cells on microcarrier cultures it is apparent that glutamine is more rapidly utilised particularly in the early stages of the culture, whereas the glutamate and 2-oxoglutarate concentrations show slower rates of utilisation. The reported reciprocal relationship between glucose and glutamine utilisation suggests that further studies may elaborate this mechanism when applied to the adapted cells (10,24)

Maximising cell yields in culture is dependent on optimising conditions for growth. Medium perfusion techniques have been reported and show considerable increases in final cell yield (3,17). However, these enhanced yields may necessitate increases in medium costs along with the additional cost of the perfusion apparatus. The possibilities of obtaining high cell yields in batch cultures by modifications of media formulations are attractive. As shown, yields may be further enhanced by selective feeding regimes during culture. Full analysis of the metabolic requirements of the cells in culture could lead to even higher productivities by this approach.

In conclusion this work has demonstrated that;

(i) Ammonia can accumulate in standard culture media to levels over 2mM.

(ii) This concentration of ammonia can be growth inhibitory for some cell lines.

(iii) A simple and rapid means of adapting cells to a glutamine free environment is possible in which glutamic acid (or 2-oxoglutarate) is used to substitute for glutamine on a mole for mole basis.

(iv) The use of such alternative substrates can result in the reduction of final ammonia concentrations by over 50%.

(v) Resultant adapted cells are capable of replication at the same rate as their glutamine grown counterparts and can result in higher final cell yields in microcarrier culture.

ACKNOWLEDGEMENTS

This work has formed part of a study for a Ph.D. thesis (T.H), and has been supported by grants made by the Science and Engineering Research Council of Great Britain. The cells used in this study were supplied by Mr. A. Bailey, and Mr. D. Coupes of Manchester Medical School.

REFERENCES

1 Ardawi, M.S.M., Newsholme, E.A. (1983). Glutamine metabolism in lymphocytes of the rat. *Biochem. J.* **212**, 835-842.

2 Bergmeyer, H.U., Bernt, E. (1963). 2-oxoglutarate. In H.U. Bergmeyer (ed), Methods in enzymatic analysis Pp. 324-327. New York: Academic Press Inc.

3 Butler, M., Imamura, T., Thomas, J., and Thilly, W.G. (1983). High yields from microcarrier cultures by medium perfusion. *J. Cell Sci.* **61**, 351-363.

4 Butler, M. Spier, R. (1984). The effects of glutamine utilisation and ammonia production on the growth of BHK cells in microcarrier cultures. *J.*

of Biotech. **1**, 187-916.

5 Cohen, S.A., Bidlingmeyer, B.A. Tarvin, T.L. (1986). PITC derivatives in amino acid analysis. *Nature* **320**, 769-770.

6 Eagle, H. (1955). Nutritional needs of mammalian cells in tissue culture. *Nature* **122**, 501-504.

7 Eagle, H. 1959). Amino acid metabolism in mammalian cell culture. *Science*, **130**, 432-437.

8 Griffiths, J.B., Pirt S.J. (1967). The uptake of amino acids by mouse cells (strain LS) during growth in batch culture and chemostat culture: the influence of cell growth rate. *Proc. Roy. Soc. B.* **168**, 421-438.

9 Iio, M., Moriyara, A., Murakami, H. (1984). Effects on cell proliferation of metabolites produced by cultured cells and their removal from culture in defined medium. *Proc. Int. Symp. on growth and differentiation of cells in defined environment.* Pp437-442.

10 Kvamme, E., Svenneby, G. (1961) The effects of glucose on glutamine utilisation by Ehrlich ascites tumor cells. *Cancer res.* **21**, 92-97.

11 Levintow, W.L., Eagle, H. (1961). Biochemistry of cultured mammalian cells. *Annu. Rev. Biochem.* **30**, 605-640

12 Lund, P., Brosnan, J.T., Eggleston, L.V. (1970). In *Essays in cell metabolism.* Eds H.L. Kornberg and J.R. Quayle. Wiley Press, New York.The regulation of ammonia meabolism in mammalian tissues. 167-188

13 Margolis, B.L., Lifschitz, M.D. (1985). Ammonia production and amino acid metabolism by rat renal papillary epithelial cells in culture. *J. Biol. Chem.* **260**, 501-507.

14 McLimans, W.F., Blumenson, L.E. Repasky,E., and Ito, M. (1981). Ammonia loading in cell culture systems. *Cell Biol Int. Rep.***5**, 653-660.

15 Nagle, S.C.Jr, Brown, B.L. (1971). An improved heat-stable chemically defined medium for growth of mammalian cells. *J. Cell Physiol.* **77**, 259-264.

16 Kovacevic,Z., Morris, H.P.(1972). The role of glutamine in the oxidative metabolism of malignant cells. *Cancer Res.* **27**, 273-279.

17 Nahapetian, A.T., Thomas, J.N. and Thilly, W.G. (1986). Optimisation of environment for high density Vero cell culture: Effect of dissolved oxygen and nutrient supply on cell growth and changes in metabolites. *J. Cell Sci.* **81**, 65-103.

18 Patterson, M.K. (1979). Measurement of growth and viability of cells in cell culuture. *Meth. Enzym.* **58**, 141-152.

19 Paul, J., Fottrell (1963). Mechanism of D-glutamyltransferase repression in mammalian cells. *Biochim Biophys Acta.* **67**, 334-335.

20 Reitzer, L.J., Wice,, B.M. and Kennell. (1979). Evidence that glutamine not sugar, is the major energy source for cultured HeLa cells. *J. Biol. Chem.* **254**, 2669-2676.

21 Stoker, M., MacPherson, I. (1964). Syrian hamster fibroblast cell line BHK21 and its derivatives. *Nature.* **216**, 1355-1357.

22 Tritsch, G.L., Moore, G.E. (1962). Spontaneous decomposition of glutamine in cell culture media. *Exp. Cell Res.* **28**, 360-364.

23 Van Wezel, A.L. (1967). Growth of cell strains and primary cells on microcarriers in homogeneous culture. *Nature* **216**, 64-65.

24 Zielke, H.R., Ozand, P.T., Tildon, J.T. Sevdalian, D.A. and Cornblath, M. (1978). Reciprocal regulation of glucose and glutamine utilisation by cultured human diploid fibroblasts. *J. Cell Physiol.* **95**, 41-48

25 Zielke, H.R., Zielke, C.L. and Ozand, P.T. (1984). Glutamine: A major energy source for cultured mammalian cells. *Fedn Proc. Fedn Am. Socs exp. Biol.* **43**, 121-125.

Discussion

J. FEDER

For the conversion of glucose to lactate in the presence of glutamine the glucose uptake rates have been decreased significantly when you went to glutamine. How did this affect the glucose to lactate conversion? Also could you

	hypothesise why you have such an increase in glucose uptake in the presence of glutamine?
HASSELL	It would need to be further investigated.
J. CLARKE	When you adapted you cells to the glutamine free environment were the effects of glutamate and oxoglutamate compared on the adapted cells or on the original cell.
HASSELL	On the original cell.
CLARKE	Did you notice any morphological changes in the adapted cells or changes in viral susceptibility.
HASSELL	We did not objectively investigate changes in morphology.
J.B.GRIFFITHS	You have two cell lines which were inhibited with ammonia and one cell line which seemed unaffected. It would seem therefore that you have a useful experimental model here for seeing what the effects of ammonia toxicity are and even overcoming them by comparing cell lines.
HASSELL	We have not done any comparative studies.
GRIFFITHS	Have you any idea why Vero cells can endure higher levels of ammonia?
HASSELL	There are a number of reasons in the literature but we cannot decide which effect applies.
GRIFFITHS	Did you increase the concentration of ammonia with the vero cell unit you had an inhibitory effect?
HASSELL	Yes, we went up to 5mm with ammonia without seeing growth inhibition. We decided to go further because we would be creating conditions which did not normally exist in cell culture

systems and which would be irrelevant for this particular study.

FEDER We have been looking at a number of cells with regard to ammonia toxicity AG1523 (diploid fibroblast) we do ot see toxicity until we are in excess of 5mM. This is above the level naturally produced in culture. We obtain 50% inhibition between 7-10mM

18 The attachment of MDCK cells to three types of microcarriers in different serum-free media

T. E. Sayer, M. Butler and A. J. MacLeod

ABSTRACT

The Madin Darby canine kidney (MDCK) epithelial cell line has been adapted to growth in a serum-free hormone-supplemented medium suitable for use in microcarrier cultures. In a novel investigation the attachment and cell growth of MDCK cells has been compared on three types of microcarrier, DEAE dextran (Cytodex 1), collagen-coated dextran (Cytodex 3) and glass-coated plastic (Ventreglas). The effect of coating the microcarriers with fibronectin was also evaluated. The basal medium (1:1 mixture of DMEM/Ham's F12) was supplemented with the following :- (i) a hormone supplement (containing insulin, transferrin, hydrocortisone, prostaglandin E_1, triiodothyronine, sodium selenite and dibutyryl cAMP), (ii) 5% human plasma protein fraction 4-1 and (iii) 10% bovine calf serum.

Measurements of cell numbers per microcarrier showed maximum attachment rates in the first 2hr. after inoculation. The rate of cell to microcarrier attachment was dependent upon the media supplement used and decreased in the order: hormone supplement > fraction 4-1 > bovine calf serum. A significantly higher attachment rate was observed with microcarriers coated with fibronectin or collagen.

The final cell yield was observed to vary with the media supplement used and decreased in the order: fraction 4-1 > hormone supplement > bovine calf serum. Improved cell yields were found with the use of fibronectin or collagen coated microcarriers. Thus, the highest cell yields were observed with MDCK cells grown on collagen or fibronectin coated microcarriers in a medium supplemented with plasma fraction 4-1.

INTRODUCTION

The use of microcarriers in cell culture allows the growth of anchorage-dependent animal cells in conditions where surface area is not limiting (2). This allows high cell yields to be obtained in unit culture systems where a large total surface area is available for cell growth. In such batch cultures final cell densities of approximately $2x10^6$ cells/ml can be routinely obtained without changes of media (5).

Serum supplemented media have been used in culture for the production of a range of commercially valuable biologicals, such as human interferon (9) and vaccines (18). The substitution of serum by hormone and other non-serum supplements is valuable for several reasons, such as reduction in batch to batch variation, decrease in cost and ease of product extraction. It also reduces contamination risks which is especially important in commercial production of animal cell products.

Few serum-free media formulations have been reported for use in microcarrier culture (6). Such formulations should contain attachment factors neccessary for cell adhesion to the microcarriers as well as growth factors to allow high final cell yields. Here we investigate the use of a hormone supplement and a plasma fraction as substitutes for serum in the culture of MDCK cells on three types of microcarriers.

Fibronectin is found on the surface of many cells and has been implicated as a specific attachment molecule (12, 21). We investigate the value of coating microcarriers with fibronectin prior to cell inoculation.

MATERIALS AND METHODS

Cells

MDCK cells (Flow Laboratories) were routinely maintained in 150cm^2 T-flasks. Before sub-culture the cells were washed with EDTA (ethylenediamine tetraacetic acid) (0.1% w/v) and removed from the surface by trypsin (0.25% w/v).

Media

MDCK cells were grown in a 1:1 mixture of Dulbecco's modified Eagle's medium (DMEM) and Ham's F12, containing penicillin (100IU/ml) and streptomycin (100ug/ml). To this basal medium one of the following supplements was added:-

(i) A hormone supplement (for T-flask cultures), containing insulin (5ug/ml), transferrin (5ug/ml), hydrocortisone (50nM), prostaglandin E_1 (25ng/ml), triiodothyronine (5pM) and sodium selenite (10pM), (8).

(ii) A hormone supplement (for microcarrier cultures), containing insulin (25ug/ml), transferrin (25ug/ml), hydrocortisone (50nM), prostaglandin E_1 (25ng/ml), triiodothyronine (5pM), sodium selenite (10pM) and dibutyryl cAMP (42ug/ml), (8).

(iii) Human plasma protein fraction 4-1 (5%), obtained from cold ethanol precipitation (Cohn fractionation) during the production of human plasma protein concentrates (16).

(iv) 10% bovine calf serum.

Adaptation of MDCK cells to growth in serum-free media

The adaptation procedures for cell growth in the hormone supplemented medium and the 5% fraction 4-1 medium were identical. All these procedures were conducted in T-flasks:-

(i) Cells grown in 10% bovine calf serum (BCS) supplemented medium were subcultured into a 5% BCS medium and grown for three passages.

(ii) At passage 4 a complete change to the serum-free medium was made at 70-80% cell confluence. Thus the cells were introduced to the serum-free medium when already attached to the substratum.

(iii) The culture was subsequently maintained in serum-free medium.

The cells used in the microcarrier experiments were at passage 5 after the first subculture into serum-free media.

Fibronectin coating of microcarriers

The microcarriers (5mg/ml) were gently shaken in serum-free medium containing fibronectin (3mg/ml) for 15min. at 22°C (13). At these concentrations, 10ug of fibronectin was available to coat $1cm^2$ of microcarrier surface.

Microcarrier cultures

The MDCK cells were grown in 100ml cultures in Bellco spinner flasks. Stirring was maintained at a constant rate of approximately 50rev/min. The microcarriers, (Cytodex 1 and Cytodex 3 from Pharmacia Chemicals, Ventreglas from Ventrex BioVentures Group), were hydrated and autoclaved at 110°C in phosphate buffered saline (without Mg^{2+} and Ca^{2+}). They were coated with fibronectin as indicated and resuspended in the appropriate medium at

5mg/ml. Cells were inoculated at $3x10^5$ cells/ml.

Growth

Daily counts were taken by the following procedure:-

(i) The microcarriers from a culture sample (1ml) were allowed to settle and the supernatant discarded.

(ii) The microcarriers were treated with 1ml of trypsin (0.25% w/v) at 37°C for at least 1hr.

(iii) The detached cells were counted with a haemocytometer.

Measurement of attachment rate

The rate at which the cells attached to the microcarriers was measured as follows:-

(i) At 2 hourly intervals after inoculation 1ml samples were observed under a light microscope (x10 magnification).

(ii) The number of cells attached to each of 100 randomly selected microcarriers was counted.

(iii) A specific cell attachment rate was calculated as the mean number of cells attaching to one microcarrier per hour.

RESULTS

Cell yield

Fig. 1. shows the results of MDCK cell growth on Cytodex 1 and Cytodex 3 in media supplemented with 10% BCS, the hormone supplement and

Fig.1. Growth curves for MDCK cells on microcarriers without fibronectin treatment.

a. Cytodex 1.

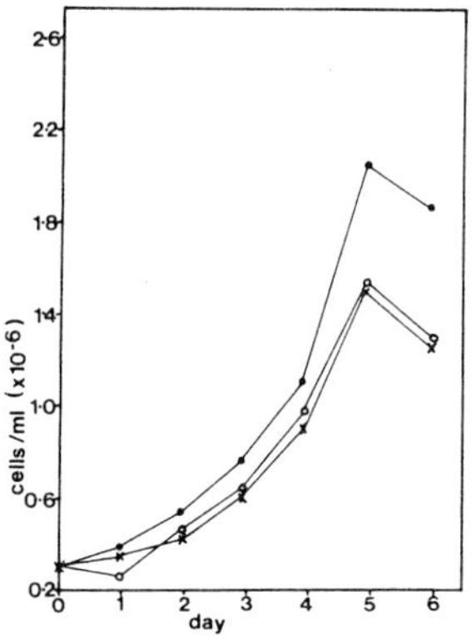

Spinner flask cultures of 100 ml, with microcarriers at 5mg/ml and MDCK cell inoculum 3×10^5 cell/ml. Basal medium:- 1:1 mixture of DMEM/Ham's F12.

Media supplemented with:-

o——o hormone supplement.

•——• 5% fraction 4-1.

x——x 10% BCS.

b. Cytodex 3.

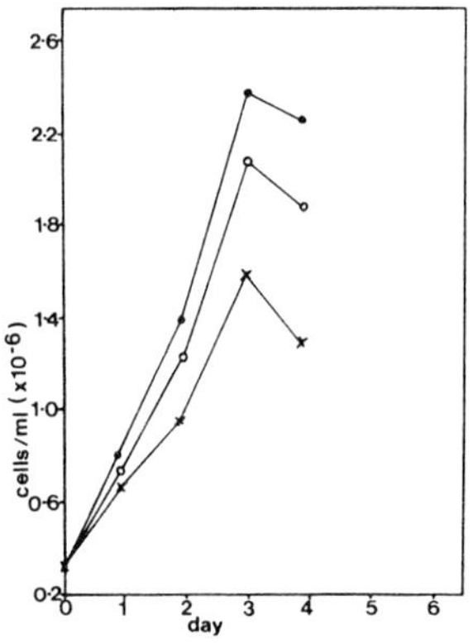

5% fraction 4-1. The cells grew more rapidly in all three types of media when attached to the Cytodex 3 with no apparent lag in growth. However, when attached to the Cytodex 1 the cells showed a lag period of approximately 24hr., and the growth rate was considerably slower. On both types of microcarriers the cells in medium supplemented with 5% fraction 4-1 grew to the highest density (2.4×10^6 cells/ml) in 3 days. The final cell density in the serum-free hormone supplemented medium was greater than that in medium supplemented with 10% BCS, but not as high as in fraction 4-1 medium. We were unable to observe surface attachment or growth of MDCK cells on untreated Ventreglas microcarriers in any of the media formulations.

Fig. 2. shows the results of MDCK cell growth on microcarriers treated with fibronectin. In these experiments the cells grew well on all three microcarriers, including the Ventreglas, and reached their maximum cell density on day 3 after inoculation. Differences in growth rate and final cell yield were due to the media and not the type of microcarriers used.

In all experiments the 5% fraction 4-1 supplemented medium produced the highest growth rate and the highest final cell yield. The 10% BCS supplemented medium gave the lowest growth rate and the lowest final cell yield.

Cell attachment

Fig. 3. shows the change in specific cell to microcarrier attachment rates. The cells attached more rapidly to the Cytodex 3 than to the Cytodex 1 in all the media used. In each case attachment was most rapid in the serum-free hormone supplemented medium and least so in the 10% BCS supplemented medium. Cell attachment in either serum-free media was most

*Fig.*2. Growth curves for MDCK cells on fibronectin treated microcarriers.

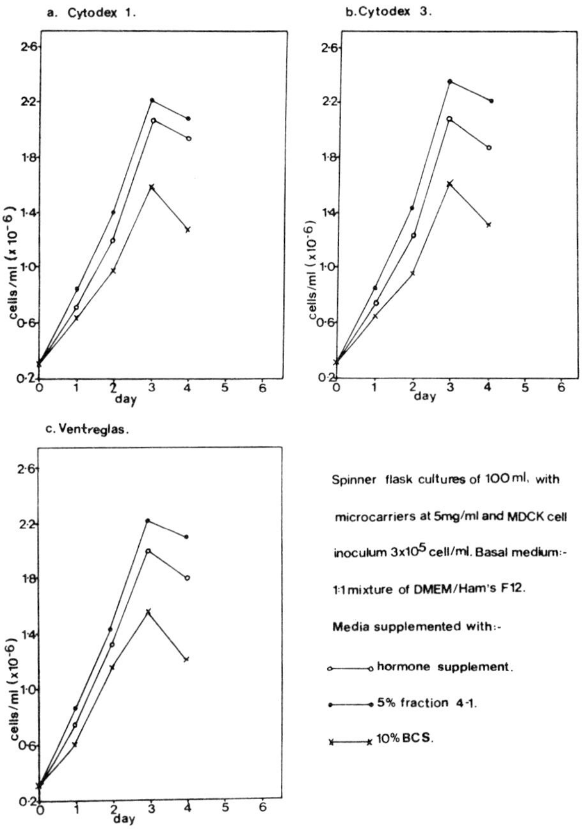

Fig. 3. Specific attachment rate on microcarriers without fibronectin treatment.

a. Cytodex 1.

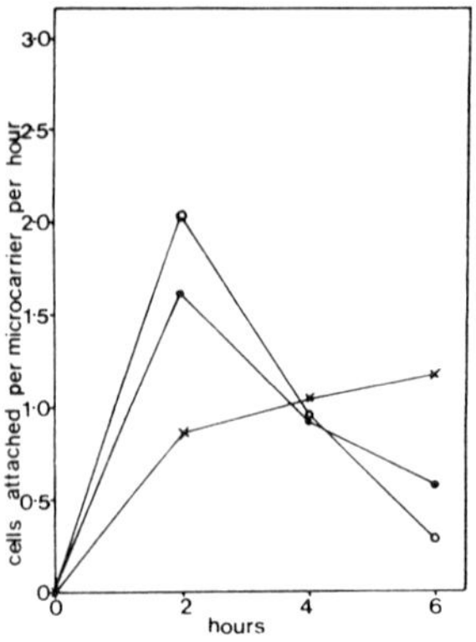

Spinner flask cultures of 100 ml, with microcarriers at 5 mg/ml and MDCK cell inoculum 3×10^5 cell/ml. Basal medium:- 1:1 mixture of DMEM/Ham's F12.

Media supplemented with:-

o——o hormone supplement.

•——• 5% fraction 4-1.

x——x 10% BCS.

b. Cytodex 3.

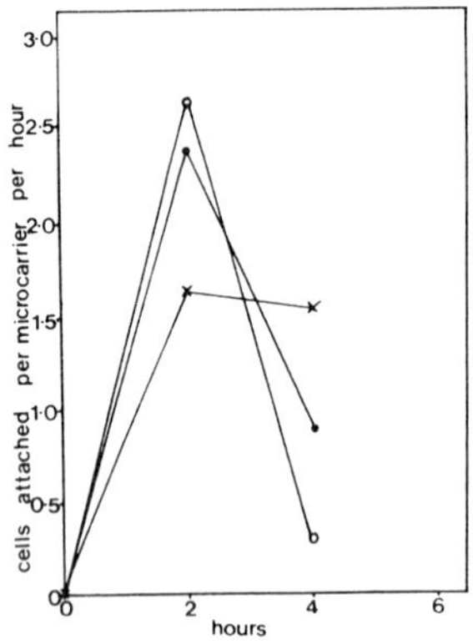

rapid in the first 2hr. after inoculation. However, in the serum supplemented medium cell attachment appeared to take longer.

With fibronectin coated microcarriers cell attachment occurred equally well to all three microcarriers (fig. 4). As with the uncoated microcarriers the rate of attachment was greatest in the serum-free hormone supplemented medium and lowest in the 10% BCS supplemented medium. In the serum-free media cell attachment was most rapid in the first 2hr. after inoculation, whereas longer attachment times were observed in serum supplemented medium.

DISCUSSION

Microcarriers provide a high surface area to volume ratio for growth of anchorage dependent cells in unit batch cultures. At least 18 different types of microcarriers of various formulations and coatings are now commercially available (4). In these experiments we have investigated the use of three microcarrier types for their ability to support the growth of MDCK cells. These microcarriers have different surfaces available for cell attachment - DEAE dextran, collagen and glass.

The highest cell attachment rates and yields were found with the collagen coated microcarriers (Cytodex 3). However, pre-treatment of the DEAE dextran and glass surface microcarriers with fibronectin increased the growth rate and resulted in cell attachment and yields comparable with those on the collagen surface. The MDCK cells did not grow on the glass surfaced microcarriers without fibronectin treatment, although reports have indicated such treatment to be uneccessary for other cell lines (19)

Fibronectin has been shown to be secreted by human fibroblasts (11) and to be important in cell spreading and division as well as attachment

Fig. 4. Specific attachment rate on fibronectin treated microcarriers.

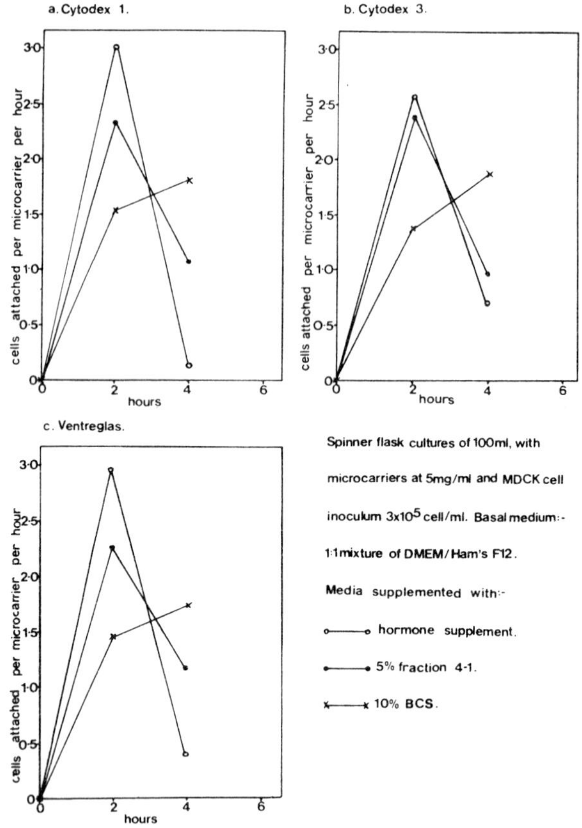

Spinner flask cultures of 100ml, with microcarriers at 5mg/ml and MDCK cell inoculum 3×10^5 cell/ml. Basal medium:- 1:1 mixture of DMEM/Ham's F12.

Media supplemented with:-

o——o hormone supplement.

•——• 5% fraction 4-1.

x——x 10% BCS.

(1, 21). B104 neuroblastoma cells have been shown to attach to a plastic substratum in the absence of fibronectin but they do not divide. However, the addition of human plasma fibronectin to the serum-free defined medium permitted cell division to occur (1). Other examples show that efficient cell attachment can occur in the absence of added fibronectin. BHK (baby hamster kidney) cells have been shown to attach faster to polystyrene culture flasks in serum-free medium not containing fibronectin than in serum containing medium (10). In such cases the cells maybe producing their own fibronectin or other attachment factor. Clearly, the value of fibronectin addition is dependent upon the cell line. In our experiments the cell attachment surface was coated with fibronectin prior to inoculation. This has previously been shown to be preferable to fibronectin addition to the culture medium (20).

There are many advantages in using serum-free media. Serum suffers from batch to batch variation in growth promoting qualities resulting in variable cell yields between cultures. It is chemically undefined, and this leads to problems in the extraction of animal cell products which are secreted into the medium. The hormone supplement used in this report for the growth of MDCK cells in microcarrier cultures is an example of a media formulation which avoids these problems.

This hormone supplement was a simple mixture of 7 ingredients which allowed adaptation of MDCK cells from a serum supplemented culture. The slow adaptation of cells to the serum-free media is a process associated with the weaning of cells from the need for serum and associated growth factors so that cells may be induced to synthesise their own essential factors. This previously developed, chemically defined formulation (8) allowed cell growth on microcarriers to a final cell yield above that obtained with the serum supplemented medium.

The plasma fraction 4-1 suuplement is not completely defined and the problems of batch to batch variation remain (14). However, it contains only about 10% as much protein as serum which eases downstream processing problems of product purification. As it is a human plasma fraction it is not always neccessary to have a completely pure product if it is for subsequent human use. This reduces the cost of production.

Fractionation of human blood plasma is performed routinely for the extraction of haemophilia factors using cold ethanol precipitation (7). Fraction 4-1 is a by-product of this process and has previously been shown to have cell growth promoting properties. A mouse-mouse hybridoma cell line, ES-6, which secretes a monoclonal antibody to human group A red blood cell antigen has been grown successfully in medium supplemented with fraction 4-1 (16). It is not completely chemically defined but is known to contain albumin, transferrin, alpha-1 antitrypsin, immunoglobulin and insulin-like growth factors and it does not contain fibronectin (17). Some of these factors are known to stimulate cell growth and are often included in serum-free formulations (3). Under the conditions used in microcarrier cultures of MDCK cells a 5% supplement of fraction 4-1 enhanced the growth and cell yield of the cultures to a greater extent than a 10% bovine calf serum supplement.

In conclusion:-

(i) MDCK cells can be adapted to a serum-free medium which can be supplemented with a plasma protein fraction or a mixture of defined hormones.

(ii) The initial attachment rate, growth rate and final cell yield were all greater in the serum-free media used than with the serum supplemented medium.

(iii) Coating the microcarriers with fibronectin or collagen had a positive effect on these growth parameters.

ACKNOWLEDGEMENTS

I acknowledge the bursary from the ESACT Organizing Committee and financial support from the Isle of Man Board of Education.

REFERENCES

1. Bottenstein,J.E. and Sato,G.H. (1980). Fibronectin and polylysine requirement for proliferation of neuroblastoma cells in defined medium. *Exp. Cell Res.* **129**, 361-366.
2. Butler,M. (1985). Growth limitations in high density microcarrier cultures. *Develop. biol. Standard.* **60**, 269-280.
3. Butler,M. (1986). 'Serum-free media' in Mammalian Cell Technology. Ed.:Thilly,W.G. (Butterworth). pp.91-108.
4. Butler,M. (1987). 'A comparative review of microcarriers available for the growth of anchorage-dependent cells' in Animal Cell Biotechnology. Vol.3. (Academic Press). **In press.**
5. Butler,M and Thilly,W.G. (1982). MDCK microcarrier cultures: seeding density effects and amino acid utilization. *In Vitro.* **18(3)**, 213-219.
6. Clark,J.M., Gebb,C. and Hirtenstein,M.D. (1982). Serum supplements and serum-free media: applicability for microcarrier culture of animal cells. *Develop. biol. Standard.* **50**, 81-91.
7. Cohn,E.J., Strong,L.E., Hughes,W.L., Mulford,D.J., Ashworth,J.N., Melin,M. and Taylor,H.L. (1946). Preparation and properties of serum and plasma proteins. IV. A system for the separation into fractions of the protein and lipoprotein components of biological tissues and fluids. *J. Am. Chem. Soc.* **68**, 459-475.
8. Crespi,C.L., Imamura,T., Leong,P., Fleischaker,R.J., Brunengraber,H. and

Thilly,W.G. (1981). Microcarrier culture: Applications in biologicals production and cell biology. *Biotech. and Bioeng.* **23**, 2673-2689.

9. Giard,D.J., Loeb,D.H., Thilly,W.G. and Wang,D.I.C. (1979). Human interferon production with diploid fibroblast cells grown on microcarriers. *Biotech. and Bioeng.* **21**, 433-442.

10. Grinnell,F. (1976). The serum dependence of baby hamster kidney cell attachment to a substratum. *Exp. Cell Res.* **97**, 265-274.

11. Grinnell,F. and Feld,M.K. (1979). Initial adhesion of human fibroblasts in serum-free medium: possible role of secreted fibronectin. *Cell.* **17**, 117-129.

12. Grinnell,F. and Feld,M.K. (1980). Spreading of human fibroblasts in serum-free medium: inhibition by dithiothreitol and the effect of cold insoluble globulin (plasma fibronectin). *J. Cellular Physiology.* **104**, 321-334.

13. Lai,C., Ankel,E.G. and Hopwood,L.E. (1984). Membrane fluidity of Chinese hamster ovary cells on plasma fibronectin-coated microcarriers. *Exp. Cell Res.* **150**, 77-83.

14. MacLeod,A.J. (1987). The recovery of proteins useful in animal cell culture from human plasma protein fractions. *Develop. biol. Standard.* **66**, 562.

15. MacLeod,A.J. (1987). The use of plasma protein fractions as medium supplements for animal cell culture. *Advances in Biochemical Engineering.* In press.

16. MacLeod,A.J. and Thomson,M.B. (1985). Use of plasma protein fractions as serum substitutes for *in vitro* cell culture. *Develop. biol. Standard.* **60**, 55-61.

17. Orly,J. and Sato,G. (1979). Fibronectin mediates cytokinesis and growth of rat follicular cells in serum-free medium. *Cell.* **17**, 295-305.

18. Spier,R.E. and Whiteside,J.P. (1976). The production of foot-and-mouth

disease virus from BHK 21 C 13 cells grown on the surface of DEAE Sephadex A50 beads. *Biotech. and Bioeng.* **18**, 659-667.

19. Varani,J., Dame,M. and Beals,T.F. (1983). Growth of three established cell lines on glass microcarriers. *Biotech. and Bioeng.* **25**, 1359-1372.

20. Yamada,K.M., Akiyama,S.K., Hasegawa,T., Hasegawa,E., Humphries,M.J., Kennedy,D.W., Nagata,K., Urushihara,H., Olden,K. and Chen,W. (1985). Recent advances in research on fibronectin and other attachment proteins. *J. of Cellular Biochemistry.* **28**, 79-97.

21. Yamada,K.M. and Olden,K. (1978). Fibronectins - adhesive glycoproteins of cell surface and blood. *Nature.* **275**, 179-184.

19 Spreading, growth and function of cells attached to a plasma clot interphase between agarose and the atmosphere

E. M. Bey

ABSTRACT

A fibrin net cast over an otherwise "cell unfriendly" surface allows cell lines like Vero, HeLa, human oesophageal carcinoma and human keratoma cells and cell strains like human foreskin fibroblasts, chick fibroblasts and secondary Vervet monkey kidney cells to attach and spread within hours and subequently grow to confluency. The fibrin net is formed when a film of re-calcified human or horse plasma is left to clot in contact with nutrient agarose or a porous polycarbonate membrane filter placed on top of the agarose. The cells then grow under polar condition exposed to the atmosphere of a CO_2 incubator and feeding from below. This system has obvious advantages when individual clones or plaques or entire cell sheets have to be manipulated.

INTRODUCTION

The nature of surfaces on which cells grow is of importance for attachment and spreading and also for

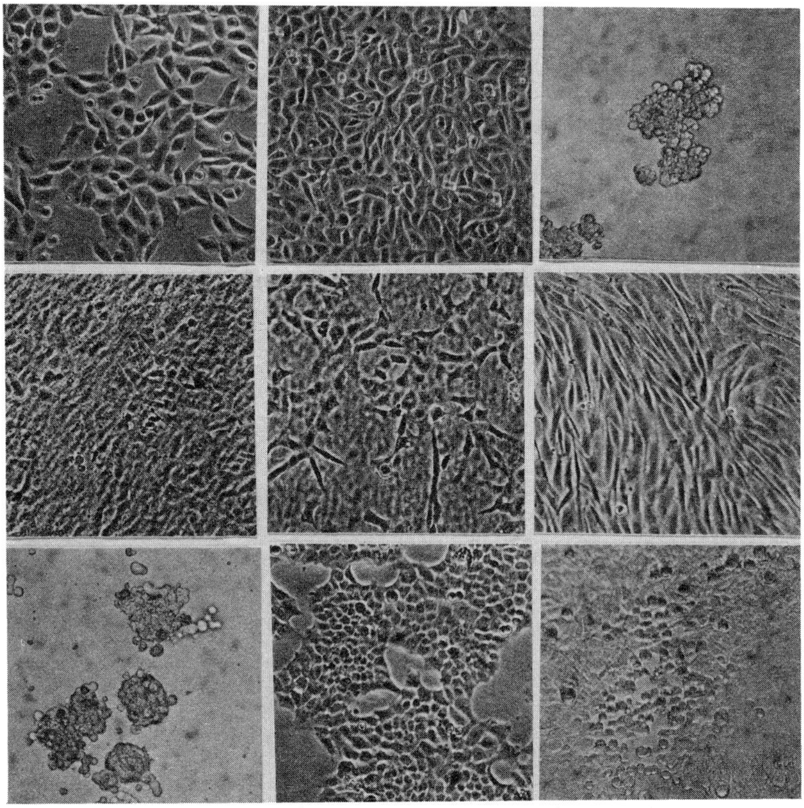

FIG. 1. Light photomicrographs of living cells maintained under conditions where they can express maximum polarity. The cells were seeded and left to attach for 3 h in medium after which time the liquid was removed and the cells left exposed to the atmosphere of the CO_2 incubator (5 % CO_2 in air) feeding from the nutrient agarose layer below. (a), (b), (f) and (h) are Vero cells, Hela cells, human foreskin fibroblasts (10th passage) and human epidermoid carcinoma cells (A431) respectively, growing under "cell friendly" conditions on a thin fibrin film biomatrix on top of agarose. Vero cells (d) and (e) were seeded on Nucleopore membranes, pore-size 30 nm which was

placed on top of the agarose and made "cell friendly" by flooding with recalcified plasma. The cells were left to grow for one day (e) and 3 days (d). A confluent sheet of Vero cells was infected under "dry" conditions with 0.2 ml of 10,000 x diluted Simian Adenovirus SA7. After 5 days, well defined plaques appeared which could be counted and individually harvested (i). Photographs (c) and (g) show Vero cells and Hela cells which were planted on top of untreated nutrient agarose (c) and Nucleopore membrane (g). The cells did not attach and spread, but remained viable in clusters for several days. x 125.

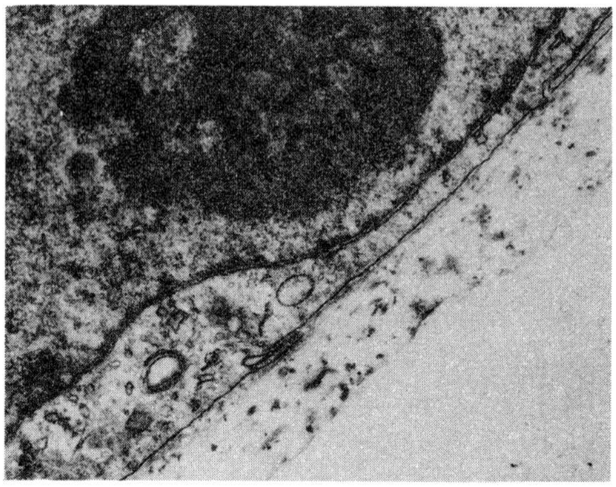

FIG. 2. Transmission electron micrograph of a Vero cell on the 5th day of culture on a thin plasma clot film cast over a Nucleopore membrane (npm), pore-size 30 nm. Note the intact layer of fibrin (fib). x 34,000

TABLE 1

ATTACHMENT AND GROWTH OVER 2 DAYS OF SEVERAL CELL STRAINS AND CELL LINES ON PLASTIC OR FIBRIN FILM COVERED SURFACES *

Cell type	Attachment after 3 h (%)		Increase of cell number after 48 h related to number of attached cells (times)	
	Type of surface			
	Plastic	Fibrin	Plastic	Fibrin
Vero	91 ± 7	93 ± 5	3.5 ± 0.4	2.6 ± 0.3
HeLa	89 ± 7	98 ± 2	2.8 ± 0.3	1.9 ± 0.4
SN	70 ± 11	80 ± 2	2.9 ± 0.5	2.3 ± 0.5
HF P12	94 ± 3	97 ± 2	6.0 ± 0.5	3.4 ± 0.6
CF P2	58 ± 12	76 ± 9	2.3 ± 0.3	1.4 ± 0.2
VK P2	76 ± 9	66 ± 12	2.3 ± 0.5	3.0 ± 0.7

* Cells were grown in 24 well Costar plates. The initial outplant was 5 x 10^4 cells / well. After count of unattached cells, the medium was replaced in cultures which grew on plastic but was nor replaced in the cultures which grew on fibrin film coverd nutrient agarose. After 48 hours the cells were either treated with trypsin/EDTA or Dispase and counted.

differentiation and function.[1-3] In vivo the growing front of cells is attached to a basal lamina or extracellular matrix and it seems likely that intimate contact between cell and non-cellular substrat rather than cell to cell contact stimulates growth.[4-8] In vitro cell monolayers usually are grown submerged under nutrient fluid or semi-solid medium and there are few systems where cells are placed above the nutrient phase, exposed to the atmosphere.[9,10] A method is described here by which the surface of nutrient agarose or of a porous polycarbonate membrane is modified by a thin film of clotted plasma which permits cells to attach, spread, grow and function without supernatant medium, under "dry" conditions. Exposed to these polar circumstances the cells are accessible to manipulations such as microsurgery, preparation for electronmicroscopy, removal of particular clones and harvesting of viral plaques.

MATERIALS AND METHODS

Plasma. As sources of plasma human or horse blood bags containing 70 ml of citrate- phosphate- dextrose-adenin buffer CPDA (citric acid H_2O 327 mg, sodium citrate 2 H_2O 2.63 g, sodium acid phosphate 2 H_2O 2.51 g, anhydrous dextrose 2.90 g and adenin 27.5 mg per 100 ml) were used. The plasma was spun at 1000 g for 15 min, platelets were removed by filtration through a non-sterilizing depth filter (Whatman No 5) and the clear plasma was then sterilized by filtration through a 0.22 um membrane (Millipore). The plasma was kept frozen in aliquote amounts.

Preparation of surface. 2% molten agarose (Sigma High Gelling Temperature Agarose Type V) was mixed 1:1 with double concentrated Earle's MEM containing 5% heat inactivated newborn bovine serum (Flow). 24 well culture trays (Costar) and 5 cm diameter plastic petri dishes were filled with 1 or 2 ml nutrient agarose respectively and allowed to set at 4°C. Icecold human or horse plasma was mixed with sterile $CaCl_2$ solution (final conc. 0.6%) to neutralize the citrate present in the CPDA buffer. Working quickly, the nutrient agarose in each culture well was covered with calcium containing plasma and the excess fluid drawn of; the trays were then incubated at 37°C until the plasma clotted (5 to 10 min). Also 2.5 cm diameter filter discs, pore size 30 nm or below (Nucleopore Co., Pleasanton, CA), were placed on top of nutrient agarose underlays and treated with recalcifying plasma in the same manner.

Cell culture. The cells were grown in Earle's MEM and 10% inactivated newborn bovine serum, trypsinized with Dulbecco's trypsin / EDTA (1:1), counted with a Coulter Counter Model ZB1 and depending on the cell type the numbers adjusted to 5×10^3 or 10^4 cells/ml. One to two ml of cell suspension was then placed on top of the modified nutrient agarose or into an empty well. After 3 h the medium was withdrawn and the unattached cells counted. The cells which had attached to the agarose button were left "dry" while fresh medium was added to the control cells on the plastic surface. After 48 h cells growing on the plastic surface were treated with 1 ml trypsin / EDTA mixture and counted. Since trypsin is inactivated by the plasma fibrin film, 1 ml 0.04% Dispase Grade II (Boehringer) in Ca^{++} free phosphate buffered saline was used for the "dry" cultures. The following cells were used: Vero cells, Hela cells, human oesophageal carcinoma cells SN[11] , human epidermoid carcinoma cells A431[12], human foreskin fibroblasts 12th and 13th passage, 12 day chick fibroblasts and Vervet monkey kidney cells 2nd passage.

"Dry" monolayers of Vero cells in 5 cm diameter petri dishes were infected with 0.2 ml Simian Adenovirus SA7 virus dilution (original suspension diluted 10^4 times). After 3 days the cultures were screened daily for the appearance of viral plaques.

Electron microscopy. The cells were fixed in 2.5 % glutaraldehyde in cacodylate buffer and stained in situ with uranyl-acetate and lead-citrate, embedded in Epon and sectioned. For the work presented here a Philips 300 transmission electron microscope was used.

Photography. Light microscope photographs were taken with a Polaroid camera mounted on an inverted Nikon microscope using Polaroid positive/negative 4 x 5 Land film type 55. Electron microscope photographs were taken with a Kodak electron image film 4463.

RESULTS

All cells attached and spread within a few hours if the surface was covered with a thin layer of fibrin while none of the cell types tested attached to plain nutrient agarose surface. The appearance and behaviour of the cells on this fibrin film surface was similar to that on "cell friendly" plastic or glass, each cell type showing characteristic morphological features (Fig.1 a,b,f,h). Within 48 h under "open" conditions and feeding from below, the cells grew almost as well on the modified surface as on the conventional plastic surface (Table 1) and formed monolayers over the entire space of the wells. The cells could be removed from the fibrin sheet with Dispase but not with trypsin, and replanted. Similarly, Nucleopore membranes - a surface to which cells generally do not attach - were made "cell friendly" by flooding with a thin layer of clotting plasma. The cells, spread on the fibrin net, drew nutrients from below through the pores and grew to confluency (Fig.1 d,e). The membrane filter together with the cells can be lifted up and either replanted or processed further.

Attached to the fibrin sheet, the cells responded normally to the lytic effects of viruses. Under "dry" condition, SA7 virus infection of Vero cells did not spread throughout the whole culture but remained localized in defined plaques. (Figure 1i).

Transmission electron microscopy showed that the

cells adhere intimately to the fibrin film and even after several days in culture do not lyse the fibrin matrix (Figure 2).

DISCUSSION

Surfaces can be modified to either inhibit or induce cell attachment, for instance coating with agar or agarose prevents attachment while certain proteins, covalently bound to agarose by CNBr, provide for surfaces to which cells attach well[13].

It was found that a thin layer of fibrin, cast over nutrient agarose or a porous membrane, not only allows many cell types to adhere but also to spread and grow while feeding from below. As can be seen from the photographs (Figure 1), the optical properties of the agarose are excellent and those of the 30 nm pore size Nucleopore membrane adequate. Membranes with smaller pore size are better optically, but cells will not feed adequately through very small pores. Leighton, in an earlier paper[14], discussed some problems of plasma clot cultures, including breakdown of the clot by cell proteases. During 7 days in culture no enzymatic destruction of the fibrin matrix was observed, only healthy growth of the cells attached to it.

It is not known whether the benficial effect of the fibrin film is only physical or whether it also provides a nutritional stimulus to the cells since it does contain a small amount of serum derived from platelet-free plasma.

Fujikawa et al.[15] have shown that during healing of corneal epithelium after superficial keratectomy in rabbits, epithelial cells migrate and grow over a "provisional" matrix of fibrin and fibronectin deposited on the bare surface after wounding. It is possible that a very thin plasma clot film is similar to a natural early wound surface which promotes cell attachment, spreading and growth.

Although no normal epithelial cells were used in this study, the method described here could perhaps enable one to grow out sheets of epidermal cells on porous membranes before grafting them directly onto wounds. Further studies along these lines are in progress.

1. Reid L.M. and Jefferson D.M. (1984). Cultured hepatocytes and other differentiated cells. Hepatology 4, 548-559.

2. Sanders E.J. (1983). Recent progress towards understanding the role of the basement membrane in development. Can. J. Biochem. Cell Biol. 61, 949-956.

3. Folkman J. and Moscona A. (1978). The role of shape in growth control. Nature 273, 345-349.

4. Gospodarowicz D.; Delgado D. and Vladavsky, J. (1980). Permissive effect of the extracellular matrix on

cell proliferation in vitro. Proc. Natl. Acad. Sci. USA 77, 4094-4098.

5. Hay E. (1981). Extracellular matrix. J. Cell Biol. 91, 205-223.

6. Van Exan R.J. and Hall B.K. (1983). Epithelial induction of osteogenesis in embryonic chick mandibular mesenchyme. Can. J. Biochem. Cell Biol. 61, 967-979.

7. Yamada K. (1983). Cell surface interactions with extracellular materials. Ann. Rev. Biochem. 52, 761-799.

8. Baker, F.L.; Spitzer,G.; Ajani, J.A.; Brock, W.A.; Lukeman, J.; Pathak,S.; Tomasovic, B.; Thielwoldt, D.; Williams, M.; Vines, C. and Tofilon, P. (1986) Drug and radiation sensitivity measurements of successful primary monolayer culturing of human tumor cells using cell-adhesive matrix and supplemented medium. Cancer Res. 46, 1263-1274.

9. Prumieras M.; Regnier M. and Woodley, D. (1983). Methods for cultivation of keratinocytes with an air-liquid interphase. J. Invest. Dermatol. 81 (Suppl. 1), 28-33.

10. Whitcutt, J.M. (1978). Disposable chamber for the co-cultivation of different cell types. S.A. J. Science 74, 301-302.

11. Bey E.; Alexander J.J.; Whitcutt J.M.; Hunt J.A. and Gear J.H.S. (1976). Carcinoma of the oesophagus in African. Establishment of a continously growing cell line from a tumor specimen. In Vitro 12, 107-114.

12. Fabricant R.M.; De Larco J.E. and Todaro B. (1977). Nerve growth factor receptors on human melanoma cells in culture. Proc. Natl. Acad. Sci. USA. 74, 565-569.

13. Carlsson J.; Gabel D.; Larsson E.; Ponten J. and Westermark B. (1979). Protein-coated agarose surfaces for attachment of cells. In Vitro 15 844-853.

14. Leighton J. A sponge matrix method for tissue culture. Formation of organized agregates of cells in vitro. J. Natl. Cancer Inst. 12, 545-561.

15. Fujikawa L.S.; Foster S.; Gipson J.K. and Colvin R.B. (1984). Basement membrane components in healing rabbit corneal epithelial wounds: immunofluorence and ultrastructural studies. J. Cell Biol. 98, 128-138.

Discussion (1)

K. JAMES We have a tremendous potential for looking at cell physiology, at growth factor receptors etc - I would make a plea to those in bulk production that they should also be looking at these things and relating them to stability of cell lines, their growth characteristics, and the amount of immunoglobulin they produce. Some MCAb are very unstable and the use of serum-free media with these cells which secrete proteases (they are cancer cells) means that an anti-protease is not present. With some immunoglobulins the presence of these proteases could be important and should always be considered. As for protein A, surely most of us are exposed to S. aureus and the dose would be so small that it is not worth bothering about.

W. TOLBERT In several talks it was recommended that production should be from dead or dying cells. Dying cells release proteases into the medium and again this is a problem in low serum media. Has anyone looked for protease damage to MCAb eg, by HPLC.

S. REUVENY I would like to ask Dr Merten if he found any differences in intracellular levels of MCAb in the three systems he described.

O. MERTEN We did not look at the internal levels. I would like to make a comment on the proteases. We once tried to make fibrin capsules. Some clones did not affect the fibrin, others disolved the fibrin.

R. ARATHOON In the 3 different cell lines you were comparing did they all produce the same class of immunoglobulin?

MERTEN Yes, all the same.

JAMES Susceptibility to proteases of mouse lgG is influenced by the isotype. One has to consider that the degradation is usually in the FC region. Binding might not be affected but complement fixing ability might be affected.

Discussion (2)

R. SPIER — When comparing bioreactors we must define what we are comparing. There was a workshop in the UK on Bioreactors a few weeks ago attended by a wide cross section of users. They all wanted to know what bioreactor to use but not in terms of specific productivity. Also there was no awareness that if you look at productivity on a per litre bioreactor per unit time you get a bioreactor performance independent of the cell. What the manufacturer of a product really wants to know is how much is it going to cost - the productivity terms are irrelevant. To achieve the best cost not the bioreactor but the whole bioprocess has to be optimised as an integrated entity ie, you cann't separate out components. For instance the saving in price using low serum might be a negative value if the process has to be intensified.

HU — I don't disagree with what you have said. In order to optimise you must know the kinetics of the cells so when you change systems you optimise to the upper limits found for the cells.

SPIER — I don't agree that it is kinetics that is important. We have seen MCAb production quoted from 10 to 1000 ug/ml - it looks as if Cell Biology is the first constraint. When that is understood the chemical engineering will need to be sorted out.

HU — I am saying we should understand the kinetics of the individual cell not population variations.

Section 3
Physiological investigations

20 Investigations of the influence of physical environment on the cultivation of animal cells

M. Wudtke and K. Schugerl

ABSTRACT

To develop better reactors for animal cell cultures, it is necessary to gain more information on cell behavior in reactors. Therefore, an insect cell line is used to study the sensitivity of the cells regarding the physical environment.
In special test systems, the cells are exposed to different physical stresses:

- permanent and variable shear forces in a rotation viscosimeter
- recurrent pressure load in roller bottles
- hydrodynamic stress in a nozzle and free jet
- gas/liquid interfacial forces in a bubble column

The influence of shear rate, duration of stress and cell age on the viability and metabolism is investigated.
Also recurrent pressure was imposed on the cells, attached on a plastic wall of roller bottles.
A defined hydrodynamic stress is obtained by expanding the cell suspension through a nozzle at different pressures.
The influence of gas/liquid interfacial forces on the cells was investigated in an aerated liquid column by varying bubble sizes and surface properties.
The sensitivity of the cells to these physical stresses is evaluated by determining the biological state of the cells by the following methods:

- estimation of cell number (viability test)
- measuring size distribution of intact cells and cell debris by laser-flow-cytometry
- investigation of cell growth

- analysis of medium composition (glucose, lactate) and intracellular enzyme activity

INTRODUCTION

The cultivation of animal cells on an industial scale presents difficulties in most cases because a sufficient nutrient and oxygen supply has to be ensured, and with this a good mixing of the cell suspension.

Problems often arise because of the sensitivity of cells to the hydrodynamic stresses created in the suspension.

The different stress loads are caused for example by the stirring mechanisms in the bioreactor as well as by the tubes, valves and pumps which belong to the complex apparatus (1-4).

In order to examine which amount of stress animal cells can stand, special test appliances have been developed which permit an investigation of the hydrodynamic stress which consists of shear forces and pressure loads(5,6).

As a kind of animal cells insect cells were used for the experiments. These cells are suitable for growing insect pathogenic viruses which are being used in the field of biological pest control (7,8).

MATERIALS AND METHODS

The cells used for the experiments consists of ovarian cells of the butterfly Spodoptera frugiperda, IPLB-Sf, kindly given by Dr. A. Röder (Hoechst AG).

The serum- and antibiotic-free culture medium was a modification of the BML-TC/7A -medium supplemented with a vitamine concentrate and 0.5% egg yolk.

Cell cultivation was carried out in monolayer flasks and spinner bottles.

The cells were exposed to 4 types of stress (Fig. 1).

In a rotation viscosimeter permanent and variable shear forces can

DIFFERENT PHYSICAL STRESS IN
CERTAIN TEST SYSTEMS:

- PERMANENT AND VARIABLE SHEAR FORCES IN A ROTATION VISCOSIMETER
- RECURRENT PRESSURE LOAD IN ROLLER BOTTLES
- HYDRODYNAMIC STRESS IN A NOZZLE AND FREE JET
- GAS/LIQUID INTERFACIAL FORCES IN A BUBBLE COLUMN

Figure 1

be created by means of using different inserts. Permanent shear stress occured in a single-gap-system as well as in a double-gap-system.
The use of non-cylindric inserts permitted the generation of shear stress varying in time and intensity.
Shear rate and shear duration were chosen as variables in order to investigate the modifications in cell viability and metabolism. A relation between the age of the cell cultures and the sensitivity of the cells was studied as well.

In plastic roller bottles cylindrical metal bars were used in order to exert a recurrent pressure load on the cells which were attached to the inner surface of the bottles.
For these investigations cylindrical metal bars fo 2 different diameters were used. After the stress tests the resulting density of viable cells was compared with that of non-stressed cells.

The next type of stress measurement undertaken was the generation of hydrodynamic stress in a nozzle and free jet.
In this test apparatus the cell suspension was forced through a nozzle into a suspension reservoir where it formed a free jet. In this flow pattern pressure changes occured as well as shear stress.
Investigations of cell sensitivity were made by varying the inlet pressure and by using cell cultures of different ages.

The gas-liquid interfacial forces affecting the cells were produced in an aerated liquid column by means of different sparger systems and by different conditions of the suspension surface.

After this short summary of the different stress measurements the first method should be explained.
Figure 2 shows the searle viscosimeter. This is a single-gap-system in which the cell suspension was subjected to shear stress.
In the gap between the stationary external and the rotating internal cylinder the cells undergo shear stress depending on the angle velocity between the cylinders.
In order to prevent death of cells caused by lack of oxygen the standard system was modified. A Teflon membrane of 0.02 mm thickness permitted the oxygen supply by diffusion into the cell suspension.
The use of different diameters for the external cylinders allowed the generation of 3 different shear rates at the same rotation speed. The gap width were 1.9 mm, 0.6 mm and 0.3 mm.
The lower part of the figure shows the concentric internal cylinder

in the central position, with the elliptic one (marked B) to the left and the splined shape to the right (marked C). These 2 non-cylindric inserts changed the stationary conditions in an increasing and decreasing way during the rotation.
The volume of the cell suspension was approx. 120 ml.
The system was kept at a temperature of 27- 28 °C which is the normal cultivation temperature for insect cells.
As the stress limit for the cells could not be reached with this type of apparatus, a different test system was selected: in this case a double-gap-system (Fig. 3).
The use of the double-gap-system resulted in a higher shear effect because the whole suspension volume was used for shear load.
A further advantage of this system consisted in the reduction of the internal volume. Only 16 ml cell suspension were required per measuring process (as opposed to 120 ml for the single-gap-system).
A thin-walled hollow cylinder rotated between the stationary cup and its equally stationary core.
The cell suspension is contained in the ring gaps which extend from the hollow cylinder towards the outside and the inside. Both gaps were 0.5 mm in width.
This commercially produced system was modified in the same way as the previous one in order to permit oxygen supply through a membrane.

In order to be able to study periodic increasing and decreasing pressure load in a separate test installation, stress was produced by introducing cylindrical metal bars into roller bottles.
First a monolayer of cells was developed on the internal surface of the roller bottles which were rotated at a speed of 0.1 revolutions per minute. After 2 days a metal bar was introduced into the roller bottles.
The diameters of the stainless steel bars were 20 mm and 30 mm which correspond to a weight of 300 g and 700 g respectively.
Assuming the average cell diameter was 18 µm, the pressure stress resulted in approx. 40 ms per revolution.
After the roller bottles with the metal bars had continued rotating for 4 days at a speed of 0.1 RPM - this corresponds to a cultivation time of 6 days - the cells were harvested from the roller bottles and the resulting final cell density was determined. This density was compared to counts obtained from unstressed cultures in two other roller bottles used for control experiments (under the same conditions, but without metal bars).

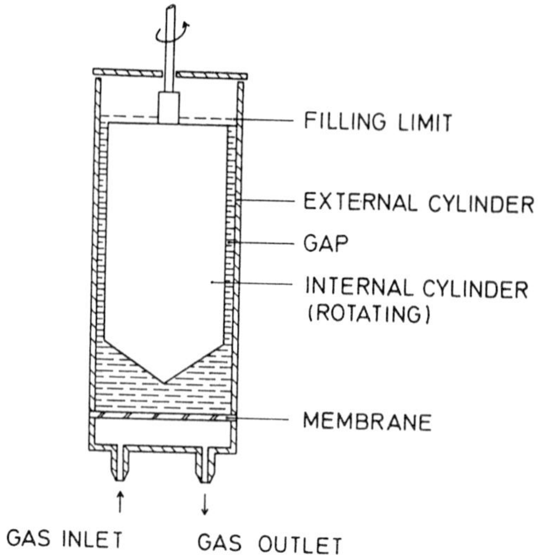

SEARLE VISCOSIMETER FOR SHEAR STRESS MEASUREMENT (SINGLE GAP SYSTEM)

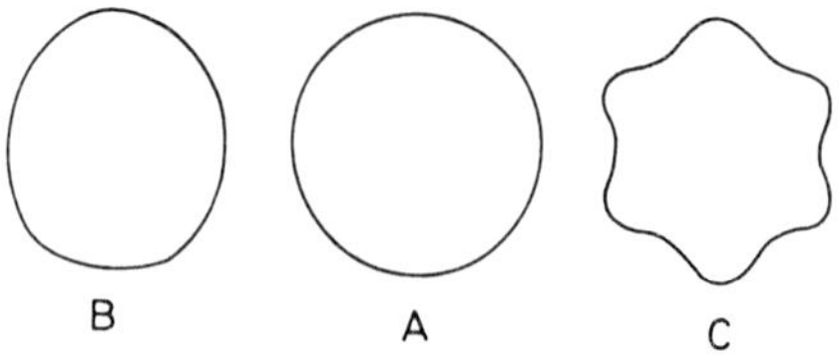

SECTIONAL DRAWING OF DIFFERENT INTERNAL CYLINDERS

Figure 2

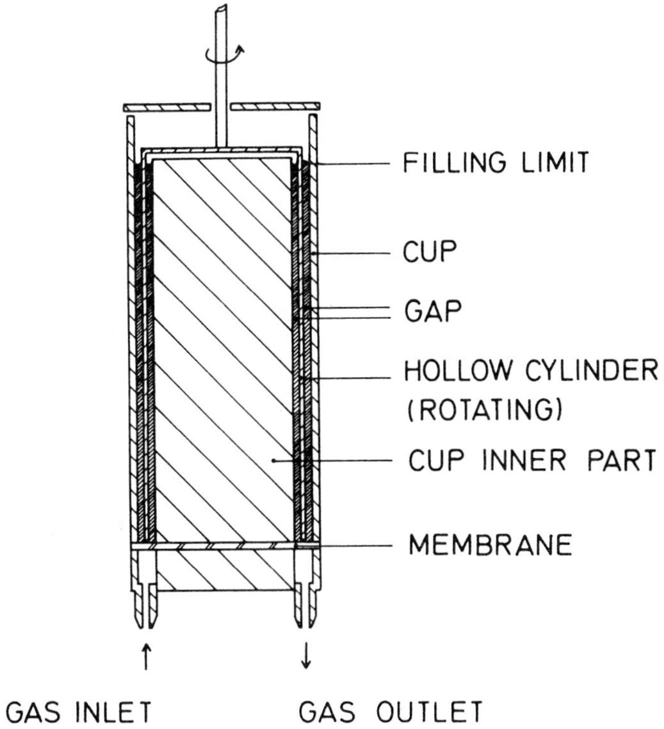

VISCOSIMETER SYSTEM
(DOUBLE GAP SYSTEM)

Figure 3

A combination of shear and pressure forces was generated in a hydrodynamic load test using a nozzle and free jet(Fig. 4).
From a spinner bottle one-third of the cell suspension was sent into a pressure vessel by increasing the pressure slightly.
High pressure air was used to adjust the pressure in the vessel.
This pressure forced the suspension via the nozzle - 0.35 mm in diameter - into the suspension left in the spinner bottle.
A free jet was formed at the orifice of the nozzle which stayed submerged in the cell suspension during the whole experiment.
In order to subject a higher number of cells to stress the free jet experiment was executed 5 times under the same conditions on the same cell suspension.
For this load tests different inlet pressure values were used:
1; 2.5; 5 and 10 bar.
After finishing the experiment cycle the cell viability was determined in the suspension taken from the sampler system. Another sample of the stressed cell culture was transferred to a spinner bottle in order to find out if the growth rate differed from the normal one.

In order to study the interaction between the gas-liquid interface and the cells experiments were carried out in a bubble column (Fig. 5).
The bubble column - working volume 0.3 litre - was equipped with a glas tube - 2 mm in diameter - for the generation of single bubbles, and with a sintered disc -pore size: 40 - 100 µm- for the generation of a bubble swarm.
In order to prevent settling of cells (especially in the case of the single bubble experiments) the cell suspension was kept moving by a magnetic stirrer.
To avoid bursting of bubbles at the interface between cell suspension and gas area, the culture was covered with a layer of liquid paraffin. The bubbles could rise in this layer and the bursting occured only at the paraffin-gas interface.
In order to ensure sufficient oxygen supply for these experiments the cell suspension was aerated with oxygen via a silicon tube
A temperature of 27 - 28 °C was maintained in the culture by means of a double jacket.
As a complement to the microscopical viability tests the following analytical methods were used for the shear stress experiment in order to permit precise informations on the conditions of cells which have been subjected to stress (Fig. 6).

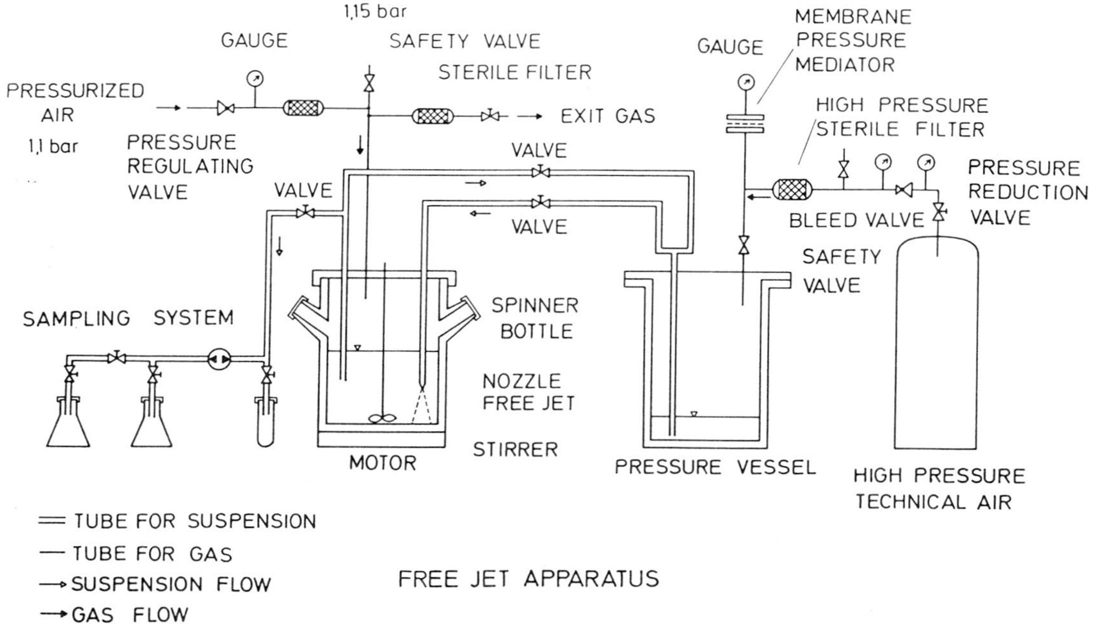
1,15 bar
GAUGE
SAFETY VALVE
STERILE FILTER
PRESSURIZED AIR
1,1 bar
EXIT GAS
PRESSURE REGULATING VALVE
VALVE
VALVE
VALVE
GAUGE
MEMBRANE PRESSURE MEDIATOR
HIGH PRESSURE STERILE FILTER
PRESSURE REDUCTION VALVE
BLEED VALVE
SAFETY VALVE
SAMPLING SYSTEM
SPINNER BOTTLE
NOZZLE FREE JET
STIRRER
MOTOR
PRESSURE VESSEL
HIGH PRESSURE TECHNICAL AIR
TUBE FOR SUSPENSION
TUBE FOR GAS
SUSPENSION FLOW
GAS FLOW
FREE JET APPARATUS

Figure 4

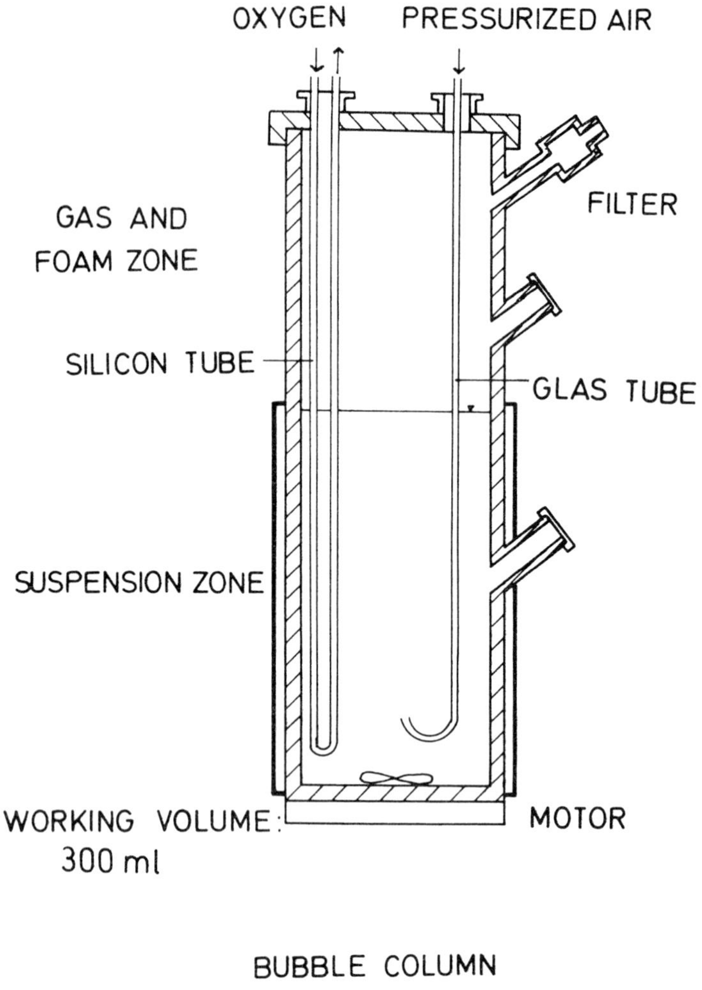
OXYGEN
PRESSURIZED AIR
FILTER
GAS AND
FOAM ZONE
SILICON TUBE
GLAS TUBE
SUSPENSION ZONE
WORKING VOLUME:
300 ml
MOTOR
BUBBLE COLUMN

Figure 5

ANALYTICAL METHODS:

-ESTIMATION OF CELL NUMBER
(VIABILITY TEST)

-MEASURING SIZE DISTRIBUTION OF INTACT CELLS AND CELL DEBRIS BY LASER FLOW CYTOMETRY

-INVESTIGATION OF CELL GROWTH

-ANALYSIS OF MEDIUM COMPOSITION (GLUCOSE,LACTATE) AND INTRACEL-LULAR ENZYME ACTIVITY

Figure 6

RESULTS AND DISCUSSION

The results of the stress measurements involving the cylindrical insert and a gap width of 0.6 mm are shown in Figure 7.
The duration of stress - ranging from 1 to 24 hours - had only a slight effect on the cell viability, which was determined in a haemocytometer using the trypan blue exclusion test.
The use of 2 shear rates - 3,100 1/s and 2,150 1/s - yielded insignificant differences in the number of viable cells. In both cases approx. 30 % of the cells were dead after 24 h.
The employment of the different external cylinders and the non-cylindric inserts produced similar results in relation to the cylindric insert. In this cases the stress duration was about 2 h.
Figure 8 shows cell cultures of different ages which were subjected to three shear rates.
A significant reduction of viable cells with increasing stress duration and shear rate can be observed. The cells aged 5 days reacted in a more uniform way than those aged 3 days because they were in the exponential phase of growth.
In some cases a part of the stressed cell suspension was cultivated again in spinner bottles. Identical growth rates and final cell counts could be seen between this cultures and control spinner bottles if the cells were exposed to the 3 shear rates for 15 min, just so after a stress duration of 1 h with the shear rates 1,800 1/s and 2,580 1/s.
The investigations of recurrent pressure load showed no difference between cell count after the stress test and the corresponding control experiments.
The diagramm (Fig. 9) shows the decrease of viable cells by increasing the inlet pressure. The experiments were carried out with cell cultures of different ages.
The cells which were forced through the nozzle at a pressure of 2.5 bar showed a lag-phase twice as long, and the final cell count only resulted in half the viable cell count in spinner bottles under standard conditions.The free jet experiment using 1 bar inlet pressure, howewer showed normal growth characteristics and the resulting cell counts were absolutly normal.
After a free jet experiment with 5 bar no cell growth in the spinner bottles could be found.
In the experiment without a paraffine-layer the cells were dead

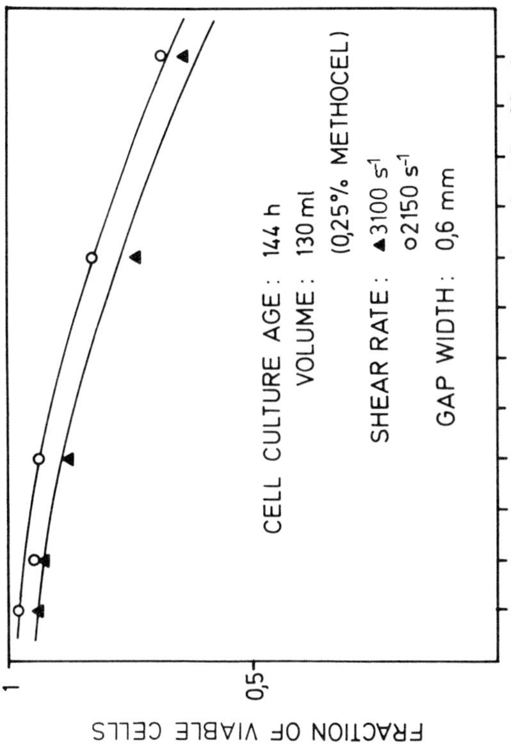

STATIONARY SHEAR STRESS: CELL VIABILITY AS A FUNCTION OF SHEAR TIME AND TWO DIFFERENT SHEAR RATES

Figure 7

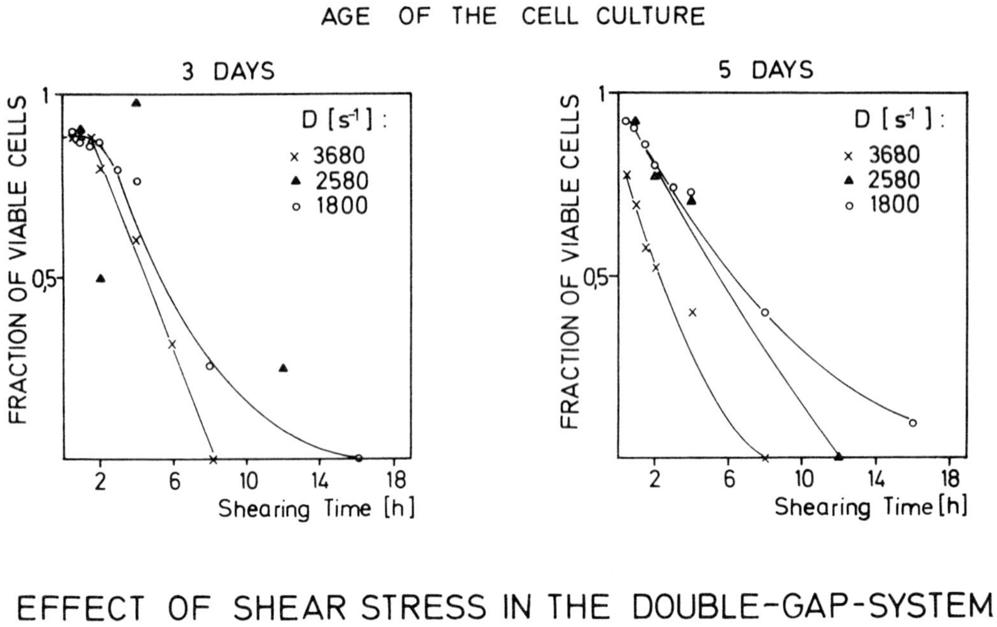

Figure 8

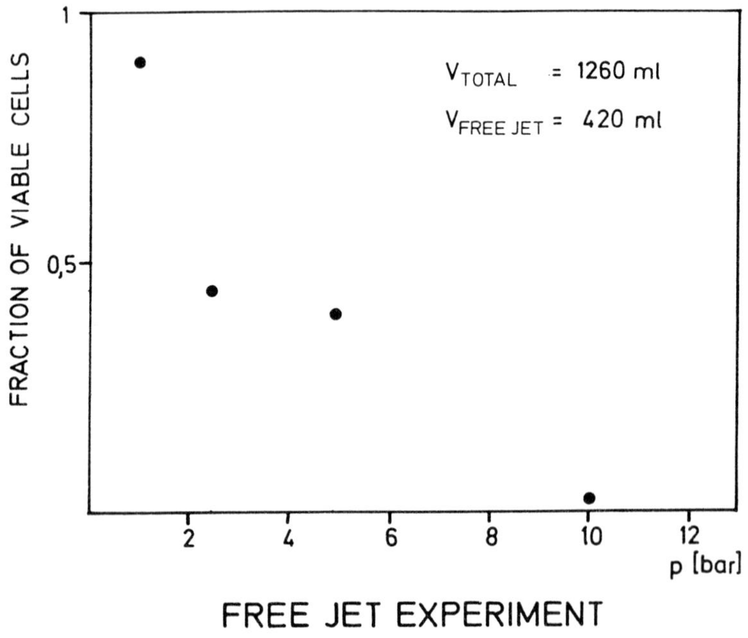

Figure 9

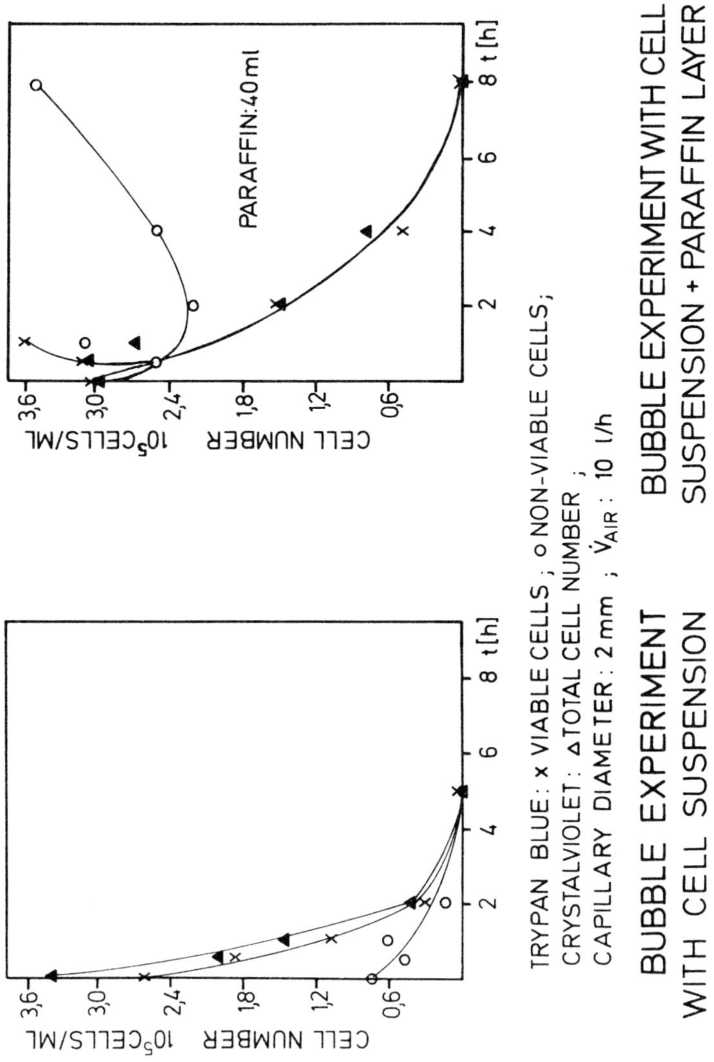
CELL NUMBER 10^5 CELLS/ML
3,6
3,0
2,4
1,8
1,2
0,6
2
4
6
8 t [h]
PARAFFIN: 40 ml
TRYPAN BLUE: x VIABLE CELLS ; o NON-VIABLE CELLS;
CRYSTALVIOLET: ▲TOTAL CELL NUMBER ;
CAPILLARY DIAMETER: 2 mm ; V̇AIR : 10 l/h
BUBBLE EXPERIMENT WITH CELL SUSPENSION
BUBBLE EXPERIMENT WITH CELL SUSPENSION + PARAFFIN LAYER

Figure 10

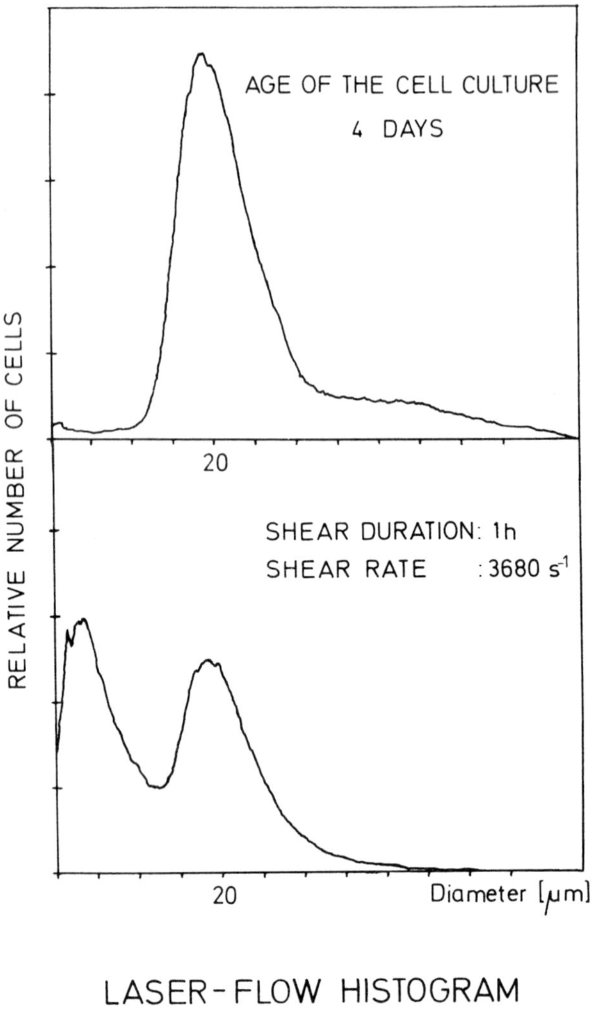
AGE OF THE CELL CULTURE
4 DAYS
RELATIVE NUMBER OF CELLS
20
SHEAR DURATION: 1h
SHEAR RATE : 3680 s-1
20
Diameter [μm]
LASER-FLOW HISTOGRAM

Figure 11

after 5 h, only cell debris was to be found then(Fig.10).
The diagramm next to it shows a decrease of the number of viable cells and an increase of the number of dead cells within 8 hours.
Hence it can be assumed that the paraffine-layer prevented the appearance of cell debris because the bubbles do not burst at the cell suspension-gas interface.
A nearly similar result yielded by using the sintered disc aerator.
The measurements of size distribution carried out with the laser-flow-cytometer correlated well with the values obtained from the viability tests. The laser-flow-cytometer, however, permitted a better evaluation of the cell debris produced by the shear stress (Fig. 11).
As mentioned the investigations of cell growth were carried out after several shear and also free jet experiments.
Determination of glucose and lactate was not suitable for the evaluation of these modifications of cell condition produced by stress.
Preliminary results of spectrophotometric enzyme measurements showed that the enzyme lactate dehydrogenase (LDH) - soluble in the cytoplasma - might be suitable as a marker-enzyme for determining the changes of the cell behavior.

CONCLUSION

The results of the investigations demonstrate that insect cells - as an example of animal cells - can withstand in some cases a larger amount of physical stress as supposed.
After a short shearing time at 3,680 1/s and 2,580 1/s respectively the cells showed normal growth characteristics.
Similar behavior could be found for cells undergo the nozzle and the free jet with 1 bar inlet pressure.
The experiments in the bubble column showed a significant decrease of cell debris when covering the suspension surface with a paraffin-layer to avoid bursting of bubbles.
In addition laser-flow-cytometry and enzyme measurements were used for analytical investigations. LDH might be used as a marker-enzyme.

REFERENCES

1 Pollard, R.; Khoshrovi,B.
Process Biochem. Mai 1978, 31-37 (1978)

2 Girard, H. C.
Cell Lulture and its Applications
Ed,: R. T. Acton, J. D. Lynn
Academic Press, New York, 111-127 (1977)

3 Tolbert, W. R.; Schoenfeld, R. A.; Lewis, C.; Feder, J.
Biotechnol. Bioeng. 24, 1671-1679, (1982)

4 Märkl, H.
Fifth Int. Ferment. Symp. Berlin 4, 79 (1976)

5 Augenstein, D. C.; Sinskey, A. J. Wang, D. I. C.
Biotechnol. Bioeng. 24, 409-418 (1971)

6 Sinskey, A. J.; Fleischaker, R. J.; Tyo, M. A.; Giard, D. J.; Wang, D. I. C.
Ann. N. Y. Acad. Sci., 47-59 (1981)

7 Miltenburger, H. G.
Develop. Biol. Stand. 46, 295-300 (1980)

8 Maramorosch, K.
Practical Tissue Culture Applications
Ed.: K. Maramorosch, H. Hirumi
Academic Press, New York, 387- 398. (1979)

Discussion

T. YAMANE What kind of paraffin did you use?

WUDTKE Paraffin oil.

SILBERKLANG You mentioned that the long term goal was to produce viruses. We have also worked with insect cells growing viruses and we find that their resistance to hydrodynamic stress in a spinner flask varies tremendously when they are virus-infected. Have you done such studies with virus-infected cells?

WUDTKE No we haven't.

SILBERKLANG We have to reduce our stirring by half when we infect the cells.

21 Shear sensitivity of three hybridoma cell lines in suspension culture

C. G. Smith, P. F. Greenfield and D. H. Randerson

SUMMARY

A couette viscometer was modified and a method developed to measure the effects of shear on mammalian cells at constant shear rates. Exhaustive testing ensured that measured effects were not the result of artefacts introduced by poor mixing, oxygen or nutrient deficiencies or cell settling.

Three hybridoma cell lines were subjected to constant shear rates within the couette viscometer for periods of up to 15 hours. The effects of shearing were measured by live cell counts, cell viability and by the release of the cytoplasmic enzyme lactate dehydrogenase into the culture medium.

The results were compared with the behaviour of equivalent cell suspensions grown in a control spinner flask. While differences were observed in the relative sensitivity to shear for the different cell lines, these differences were small. The activity of lactate dehydrogenase proved to be a useful indicator of cell damage.

The technique was shown as a useful means of examining the shear sensitivity of mammalian cell lines. It indicates that there is a critical shear rate below which negligible cell damage is likely to occur even over extended time periods. This value is of the same order as that calculated for typical culture conditions in potential bio-reactors.

INTRODUCTION

The growing demand for high value biochemicals produced by mammalian cells has lead to an increasing interest in the large-scale cultivation of such cells. Monoclonal antibodies produced by hybridomas have been in particular demand to supply the needs of many diagnostic, immunopurification and immunotherapeutic applications.

Maintaining homeostasis of environment is important for promoting cell productivity, and suspension cultures are emerging as the preferred large-scale cultivation technique. Much of the sterilization, automation and inoculation technology developed for microbial suspension cultures has been directly applicable to suspension culture bioreactors for the growth of mammalian cells. Encapsulation and hollow fibre culture techniques have, to date, failed to gain widespread acceptance since they require special startup techniques and suffer from nutrient limitation problems. These techniques compete more effectively where very large quantities of product are required.

Mammalian cells lack the tough cell wall found in bacteria, and their fragile outer membranes have not developed with a tolerance for large hydrodynamic forces. Sensitivity to shearing is a well recognised problem of suspension cultures where uniform physiological conditions are attained by effective agitation. With mammalian cell bioreactors now operating at 1000 litre capacity and with membrane based recycle systems showing that they have the potential for improving reactor productivity, equipment design requires optimisation of agitation without causing cell rupture and acceptable flowrates through the membrane systems to avoid cell damage.

Few quantitative studies of shear effects have been published. There is also a lack of understanding of the interaction of cells in suspension culture with the hydrodynamic forces resulting from the fluid motion. Midler and Finn (1) found that the survival of protozoa grown in stirred vessels could be correlated with impeller tip speed. In studying the shear sensitivity of insect cells in couette flow, Tramper et al. (2) observed a considerable drop in viability over 3 hours at shear stresses of 1.5 Nm^{-2}. However, this did not correlate well with observed cell survival in a culture vessel. Stathopoulos and Hellums (3) found that shear stresses of 0.26 Nm^{-2} did not cause loss of viability for anchorage dependent kidney cells. The interesting phenomenon of increased release of a urokinase enzyme was observed at shear rates around 0.65 Nm^{-2}, and it was postulated that this was caused either by an increase in enzyme production or in membrane permeability.

Croughan et al. (4) applied hydrodynamic theory to reevaluate previously published shear sensitivity data for microcarrier cultures. It was concluded that cell survival in agitated bioreactors could not be correlated with impeller tip speed. The size of the turbulent eddies was found to be a more appropriate basis for comparing loss of cell viability. In examining hydrodynamic effects on microcarrier cultures, Cherry and Papoutsakis (5) confirmed by theoretical analysis that turbulent eddies of similar size as microcarriers were likely to cause high shear stresses on attached cells. This paper presents an experimental investigation into the shear sensitivity of non-anchorage dependent hybridoma cells in the well defined shear environment of a couette viscometer. The results are expected to lead to a fundamental understanding of the response of mammalian cells to shear fields over extended culture times.

MATERIALS AND METHODS

A detailed description of the equipment and techniques employed in this study has been previously published (Smith et al. (6)). A well defined shear field was obtained in a coaxial cylindrical viscometer (Rheomat 115, Contraves, Zurich: MS/0 1054). Studies with culture medium established that laminar flow could be maintained up to a shear rate of 900 s^{-1}. At higher shear rates, Taylor vortexing occurred. In coaxial viscometers, the shear stress varies across the gap, however, for the geometry of the MS/0 measuring system of the Contraves viscometer this variation is minimal for fluids such as culture medium.

Three murine hybridomas were investigated in this study -

(i) B45 (MAbCo, Brisbane, Australia) producing antibodies to surface antigen of *Brucella abortus*

(ii) HSA (developed in our laboratory) producing antibodies to human serum albumin.

(iii) IgA (developed in our laboratory) producing antibodies to immunoglobulin A.

All hybridomas were the fusion products of NS-1 murine myelomas and spleen cells from immunized Balb/c female mice. The culture medium consisted of RPMI 1640 (Flow Laboratories, Sydney, Australia), supplemented with 10% foetal bovine serum (Filtron, Melbourne, Australia), 0.02mM 2-mercaptoethanol (Sigma, St. Louis, MO), 2mM L-glutamine (Gibco, Chagrin Falls, OH), 1mM sodium pyruvate (Flow), 200 I.U./ml penicillin (Gibco) and 200 μg/ml streptomycin (Gibco).

After a 36 hour conditioning period in a spinner flask (Techne, Cambridge, U.K.), cells were transferred to the Contraves viscometer. Response of cells was monitored regularly over 15 hours by cell counts as determined by trypan-blue exclusion on a haemocytometer. Release of the cytoplasmic enzyme lactate dehydrogenase (LDH) was also monitored by performing spectrophotometric determinations on cell samples. Cells remaining in the spinner flask acted as a control.

Presterilization and careful technique ensured that no contamination occurred over the 15 hour culture period. Cells inoculated into the viscometer had entered the exponential growth phase and the initial cell concentration was controlled to values between 1.5 and 3.0 x 10^5 cells/ml. The preliminary experiments reported in Smith et al. (6) determined that cells were not being subjected to limitations of oxygen or nutrients and that exponential growth was possible within the viscometer.
Hourly agitation prior to sampling ensured that significant settling of cells did not occur. Spectrophotometric studies of a dye solution revealed that evaporative loss from the viscometer at 37°C was less than 2% over 15 hours.

The three hybridoma cell lines were subjected to shear rates ranging from 210 to 870 s^{-1}. The B45 hybridoma was intentionally subjected to additional shear rates of 1240 and 1780 s^{-1} which were within the range where Taylor vortexing occurs.

Samples were taken at the start and finish of each experiment for antibody analysis by enzyme-linked immunosorbent assay (ELISA). However, as mentioned in the preliminary investigation (Smith et al. (6)), variability in the ELISA technique and low antibody production in the early exponential phase have meant that differences in antibody production have not been significant to date. Work is continuing in our laboratory to improve quantification of the ELISA. Scanning electron microscopy of sheared and control cells did not reveal any distinct differences in cell surface morphology.

RESULTS AND DISCUSSION

Regression analysis of spinner flask live cell counts resulted in a simple exponential growth model for the control. Average cell doubling time was 15 hours. Live cell survival within the Contraves viscometer was normalized for each experiment by dividing each viscometer cell count, X by the control cell count, X_e predicted by the exponential model. Figure 1 presents the cell survival, X/X_e at a range of shear rates for the three hybridoma cell lines. A trend of decreased cell survival for increasing shear rates is observed. An analysis of variance on differences in paired cell counts showed that growth at 210 s^{-1} for all three cell lines could not be statistically differentiated from the control growth. The result at 420 s^{-1} for the B45 hybridoma was also no different from control growth. The trend of decreasing live cell count was continued for the B45 hybridoma at nominal shear rates of 1240 and 1780 s^{-1} where Taylor vortexing was known to occur.

A shear rate of 210s^{-1} corresponds to a shear stress of 0.16 Nm^{-2} for culture fluid with viscosity of 0.77 cP at 37°C. This correlates well with the work of Stathopoulos and Hellums who found that viability was unchanged at a shear stress of 0.26 Nm^{-2} for anchorage dependent kidney cells.

The curvature shown on the plots of figure 1 implies that logarithmic kinetics of cell survival are inappropriate. A proposed non-logarithmic model involves inactivation from viable cells to a live but sensitive intermediate followed by cell death.

Cell viability was determined as the percentage of the cells over total cells as determined by trypan blue exclusion. No major trends in percent viability as a function of shearing time were evident. For this reason viabilities were averaged for the total 15 hour experimental period.

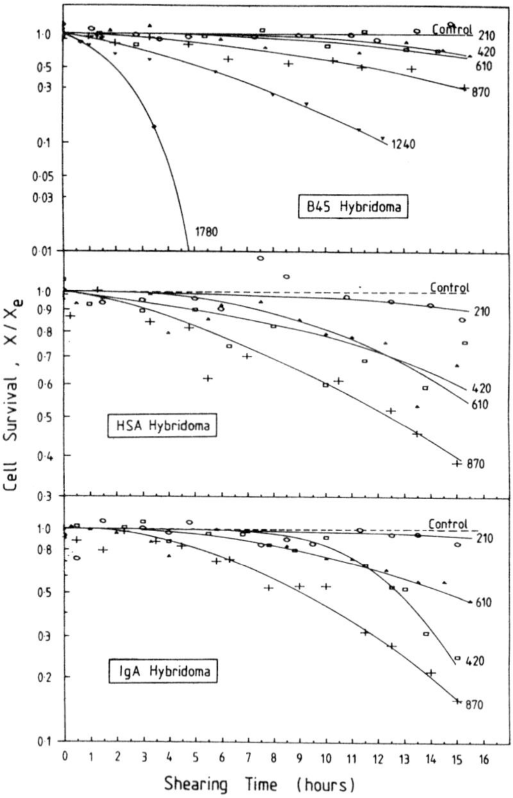

Figure 1: Cell Survival at a Range of Shear Rates for Three Hybridoma Types (shear rate in s^{-1})

Results are presented in Table 1 as viscometer viability less control viability. Although viscometer viabilities are generally less than control viabilities no particular trend is exhibited over the range of shear rates investigated. The reason for this is not clear, however, interactions between the rate of mechanical cell breakup and rates of enzymatic degradation of cell membranes may be important.

The release of the cytoplasmic enzyme LDH was monitored over the experimental period for both the viscometer and spinner flask. An initial LDH concentration of around 50 I.U./L (at 32°C) existed in culture samples prior to shearing, which was thought to originate from the foetal bovine

TABLE 1: Percent Difference in Viability between Viscometer and Control Cells (Average control viability = 86%)

Shear Rate (s^{-1})	Hybridoma Type		
	B45	HSA	IgA
210	-3	2	-1
420	-3	-8	0
610	-4	2	-1
870	-8	-2	-6
1240	-4	-	-
1780	-4	-	-

serum added to the culture medium. In the spinner flask the average increase in LDH concentration over fifteen hours was 9 I.U./L.

In order to normalize the LDH release in each experiment, the difference between LDH released in the viscometer and spinner flask was divided by the starting live cell concentration, X_o. This is shown in Figure 2 as profiles of normalized LDH release over the 15 hour culture time for the three hybridoma cell lines at a range of shear rates.

The results presented in Figure 2 confirm the expected trend of increased LDH release with increasing shear rate. The cell growth rate at 210 s^{-1} shear rate was determined to be no different from the control growth rate, however, with all hybridoma cell types the LDH release at 210 s^{-1} appears to be increasing to levels significantly greater than that of the control. The LDH assay mixture is of non-optimal osmolarity and it may be possible that weakened cell membranes are rupturing upon addition of cell samples to the assay mixture. This implies that shear damage may be occurring even at shear rates as low as 210 s^{-1}. The results of the LDH release, live cell count and viability determinations show that response to shearing is similar for all three hybridoma cell lines and this may be attributed to their common myeloma and spleen cell origins.

Preliminary investigations in a laboratory scale fermenter equipped with a Rushton turbine agitator indicate that shear damage to cells may be occurring at impeller speeds as low as 100 rpm. Shear rates in the vortices which trail turbine impellers were estimated by Van't Riet and Smith (7). For impeller Reynold's numbers of 6000 the shear rate was around 60 times the impeller rotational speed, indicating a shear rate of approximately 100 s^{-1} in the fermenter. This is of the same order as that found in the couette viscometer, however, the results of studies by Croughan et al (4) and Cherry and Papoutsakis (5) indicate that a more thorough investigation of the hydrodynamics in impeller agitated bioreactors is necessary for a full understanding of shear damage to cells.

CONCLUSIONS

The results presented in this study indicate that damage to mammalian cells

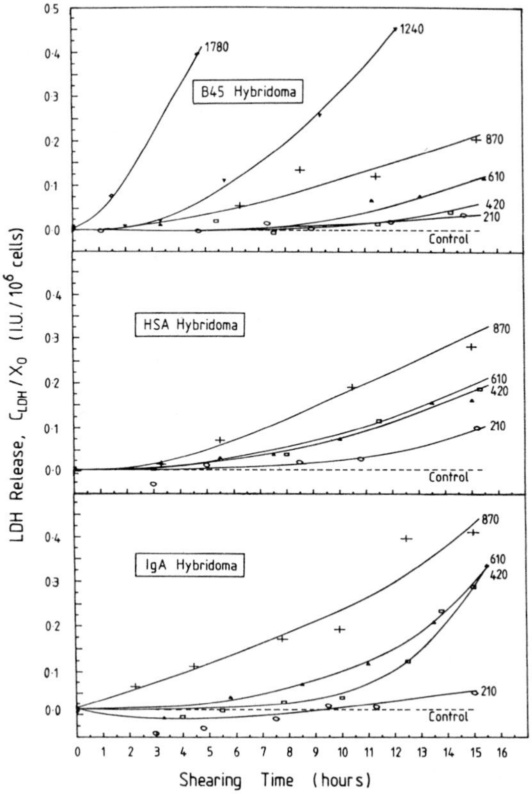

Figure 2: Lactate Dehydrogenase Release at a Range of Shear Rates for Three Hybridoma Types. (shear rate in s^{-1})

in a constant well-defined shear field may be occurring at shear rates as low as 210 s^{-1}. Cell growth at 210 s^{-1} shear rate was determined to be no different from the control growth rate and yet the LDH concentration increased significantly toward the end of the 15 hour culture period. This phenomenon may be indicative of microstructural changes to cell membranes even at this low shear rate and will be investigated further. No major differences in the response of the three hybridoma cell lines to shearing were exhibited. Preliminary shear sensitivity studies in impeller agitated fermenters indicate that damage to cells is occurring at shear rates comparable to those used in the Contraves viscometer.

REFERENCES

1. Midler, M. and Finn, R.K., "A Model System for Evaluating Shear in the Design of Stirred Fermentors", Biotech. Bioeng., Vol. 8, pp 71-84 (1966).
2. Tramper, J., Williams, J.B., Joustra, D. and Vlak, J.M., "Shear Sensitivity of Insect Cells in Suspension", Enz. Micr. Tech., Vol. 8, pp 33-36 (1986).
3. Stathopoulos, N.A. and Hellums, J.D., "Shear Stress Effects on Human Embryonic Kidney Cells In Vitro", Biotech. Bioeng., Vol. 27, pp 1021 - 1026 (1985).
4. Croughan, M.S., Hamel, J. and Wang, D.I.C., "Hydrodynamic Effects on Animal Cells Grown in Microcarrier Cultures", Biotech. Bioeng., Vol. 29, pp 130-141 (1987).
5. Cherry, R.S. and Papoutsakis, E.T., "Hydrodynamic Effects on Cells in Agitated Tissue Culture Reactors", Bioprocess Eng., Vol. 1, pp 29-41 (1986).
6. Smith, C.G., Greenfield, P.F. and Randerson, D.H., "A Technique for Determining the Shear Sensitivity of Mammalian Cells in Suspension Culture", Biotech. Techniques, Vol. 1, pp 39-44 (1987).

7. Van't Riet, K. and Smith, J.M., "The Trailing Vortex System Produced by Ruston Turbine Agitators", Chem. Eng. Sci., Vol. 30, pp 1093-1105 (1975).

Discussion

J. FEDER	Have you looked at dose effects of cells on microcarriers?
GREENFIELD	No, the difficulty with our system is that we have a gap width of 400 microns. Where we are dealing with cells of 10 microns that is fine but cells on carriers have a diameter of over 150 microns and our system is not suitable for that size.
FEDER	The shear rates to take cells off carriers is probably much less than off other substrates.
H. KATINGER	Did you measure the osmolality?
GREENFIELD	Yes, I can't remember the value but there was no problem and we are aware of its importance.
KATINGER	You several times said "What is usual for all reactors". What is usual?
GREENFIELD	At typical impeller speeds of 50-150 rpm with typical size impellers, minumum and maximum shear stresses were calculated and averaged as 0.1-0.2 newtons/m^2.
C. MORRIS	Your control viability was only 86% - why was it so low?
GREENFIELD	I do not know.

22 NMR metabolic studies of human breast cancer cells

M. Neeman, E. Rushkin, A. Kadouri and H. Degani

ABSTRACT

The use of nuclear magnetic resonance spectroscopy as a non-invasive method for characterizing the metabolism of perfused mammalian cells is demonstrated in studies of the human breast cancer T47D cell line. Three methods of cell immobilization were employed: embedding cells in agarose gel threads, growing cells on agarose polyacrolein microcarriers and on 'filters' made of non woven polyester fabric. Scanning electron microscopy showed distinct differences in the celi morphology and growth pattern while ^{31}P NMR showed differences in the cell energetics between the three methods. The rate of lactate production, measured on line by ^{13}C NMR upon adding labeled glucose (1- ^{13}C) to the perfusate, was 0.18 mM/s in cells in gel threads and 0.09 mM/s in cells on microcarriers. Modulations in the concentration of the phosphate metabolites and their interconversion rates at steady state, due to changes in medium oxygen content and in carbon source were monitored on line. The signal intensity of ATP and PCr were found to correlate with cell density. Thus NMR can be employed to study cell metabolism under various growth conditions and to investigate the response of cells to environmental changes.

INTRODUCTION

The non invasive nature of nuclear magnetic resonance (NMR) spectroscopy has been exploited extensively in recent years to monitor the metabolism of perfused cells and organs and of organs in alive animals (6). By employing multinuclei NMR (^{1}H , ^{31}P , ^{13}C , ^{15}N etc.) it is possible to determine the intracellular concentration and turnover rates of a wide range of metabolites and ions, and to monitor changes in their concentration in real time. the sensitivity of NMR is low compared to similar radioactive studies, however, it provides a unique mean to determine on a single sample, the rates of substrate utilization and label distribution and to identify the chemical composition of the intermediates and products. NMR methods can also be applied to measure reaction rates at equilibrium and at steady state conditions (1,9). For this purpose the nuclei of one of the exchanging species can be 'magnetically labeled' with radio frequency irradiation and the transfer of this 'label' due to the exchange reaction is then detected. Another distinct application refers to the determination of intracellular pH from ^{31}P NMR measurements of the chemical shift of the intracellular inorganic phosphate signal (11). Thus, NMR spectroscopy furnishes a new approach to study cell metabolism under various growth conditions and to investigate the response of cells to environmental changes.

Studies of cells by NMR require development of methods for immobilizing a high density of cells ($>5x10^7$ cells/ml) in a rapidly perfused culture vessel. We have adapted three such methods and utilized them to characterize the metabolism of human breast cancer T47D cells; 1. Embedding cells in agarose gel threads (5), 2. Growing cells on agarose polyacrolein microcarriers (10) and 3. Growing cells on 'filters' made of non woven polyester fabric (7). The mode of cell growth and the shape and surface features of the cells, were characterized by scanning electron microscopy (SEM). Here we report the use of ^{31}P and ^{13}C NMR to study the metabolism of the T47D cells (8), under aerobic and anaerobic conditions and with various carbon sources.

MATERIALS AND METHODS

Cell culture

T47D-clone 11 cells (8) were cultured routinely in monolayer using RPMI-1640 medium supplemented with fetal calf serum (10%), glutamine (2 mM), streptomycin (100 ug/ml), penicillin (100 units/ml) and mycostatin (25 ug/ml). Embedding cells in agarose gel threads was performed by the method of Foxall and Cohen (5) as described previously (12). Microcarrier culture was obtained by seeding $3x10^6$ cells in 12 ml medium containing 0.5 ml polyacrolein microsphere beads (Galisar, Israel) derivatized with poly D-lysine (10). Cells were cultured for 7 days prior to the NMR experiment. 'Filters' made of non woven polyester fabric were prepared for cell growth by the method of Kadouri (7). Cells were cultured on the 'filters' for 10 days prior to the NMR experiment. For NMR measurements the cells were prepared by one of the above methods in a conventional 10 mm NMR tube and were continuously perfused at a rate of 4 ml/min at $35^{o}C$.

Scanning electron microscopy

The cells grown on microcarriers or on 'filters' were fixed and coated with gold by a procedure described previously for fixation of cells in agarose gel threads (12). Micrographs were obtained with a Philips analytical SEM 505/515.

NMR measurements

^{31}P and ^{13}C NMR spectra were recorded at 121.5 and 75.5 MHz respectively with a Bruker CXP-300 spectrometer. The acquisition parameters are given in the figure legends. The assignments of the signals was based on previous studies (12). The fluxes from Pi and PCr to ATP and from ATP to ADP were determined at steady state by ^{31}P magnetization transfer experiments. The methods of magnetization transfer and the data analysis procedures, used in our study, were described previously in detail (12).

RESULTS AND DISCUSSION

The phosphate metabolites detected in ^{31}P NMR spectra of T47D cells are exhibited in Figure 1. Cells in gel threads (Fig 1A) and cells on microcarriers (Fig 1B) show differences in the concentration of several phosphate metabolites and in the intracellular pH. Spectra obtained from cells grown on 'filters' was similar to spectra of cells on the microcarriers. The signals due to phosphocreatine (PCr) and phosphorylcholine (PC) were lower in cells in gel threads relative to cells on microcarriers, while the signals due to glycerol-phosphorylcholine (GPC) and glycerol-phosphorylethanolamine (GPE) were higher, in the cells in gel threads relative to cells on the microcarriers. The intracellular pH, determined from the chemical shift of the intracellular Pi signal, was lower (pH=7.1) in cells embedded in gel threads relative to cells grown on microcarriers (pH=7.4). These differences are probably due to the trypsinization step and the temperature change required for the preparation of the gel threads.

The ratio between glucose utilization through glycolysis and through the Krebs cycle, under aerobic conditions, was studied by ^{13}C NMR. Glucose (5.6 mM), labeled specifically on C_1 (99% ^{13}C), was added to 50 ml of medium in a closed recirculated perfusion system. ^{13}C NMR spectra were recorded sequentially every 30 minutes for 20 h (Figure 2). The rates of glucose utilization and lactate production as well as the rate of labeling of

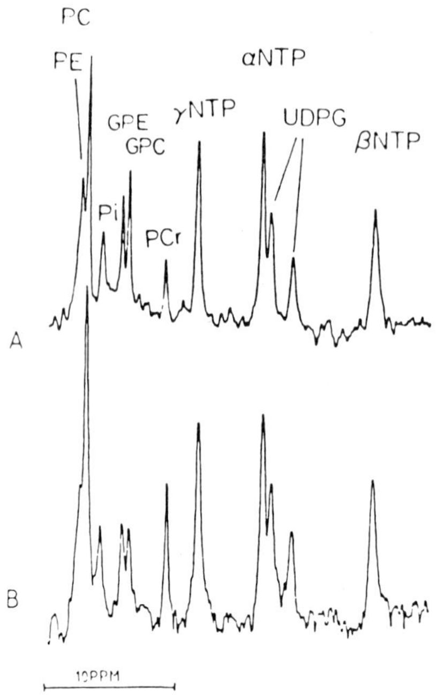

Figure 1: 121.5 MHz ^{31}P spectra of T47D cells (clone 11). Spectra were obtained using 90° pulses and 10 s repetition time. The total acquisition time was 2 h. Line broadening of 15 Hz was used in the Fourier transform. A. Cells embedded in agarose filaments. B. Cells grown for 8 days on microcarriers.

glutamate and alanine could be followed. For cells embedded in gel threads the lactate production rate was calculated to be 0.18 mM/S (3 experiments) while in cells grown on microcarriers the lactate production rate appeared to be about 50% lower (one experiment).

Scanning electron micrographs showed marked differences in the shape and surface features between the three culture methods (figure 3). The cells embedded in the gel threads were dispersed and round with a smooth cell surface (fig 3a), the cells on the microcarriers showed a multilayer growth pattern (fig 3b) while on the filters the cells formed monolayers along the fibers (fig 3c). The cells on the microcarriers and on the filters were flattened with 'hairy' cell surface.

Measurements of ^{31}P NMR spectra at various days during growth on microcarriers yielded the following results: The concentration of intracellular ATP and PCr of cells grown for 3-7 days, remained constant while the signal due to PC decreased relative to ATP with day of culture (figure 4). This suggests that the intensity of the signals due to ATP and PCr can serve to determine cell density during the logarithmic growth phase.

The modulation in the concentration of the high energy phosphate metabolites and in the intracellular pH due to aerobic-anaerobic transitions are shown in figure 5. Transfer to anaerobic conditions with glucose as the main carbon source (DMEM medium) induced a reduction in the PCr concentration (90%) and in the pH (by 0.4 units) without changing the ATP concentration. The decrease in the [PCr] and concomitant increase in [H^+] is in accord with the equilibrium of the creatine kinase reaction. Upon reoxygenation of the medium the cells recovered within 30 minutes. Subsequent deletion of glucose from the medium (DMEM) caused a rise (2 fold) in the Pi and a drop in the concentration of PCr (90%) with no change in the ATP concentration and the pH. However, in the absence of

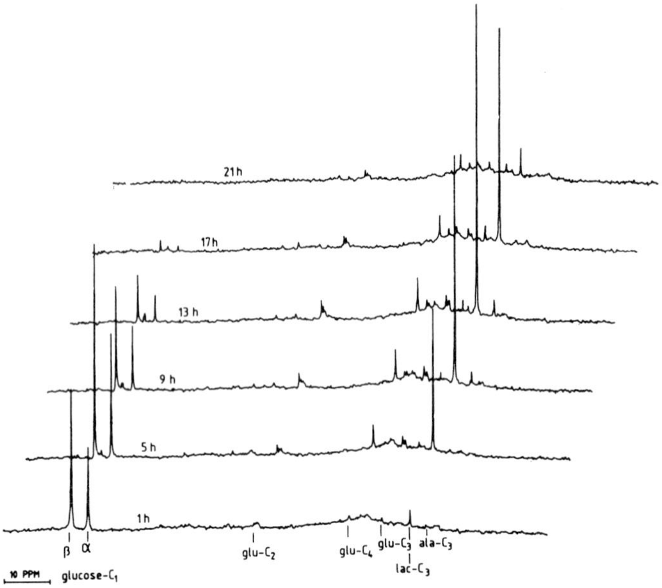

Figure 2: Glucose utilization by T47D (clone 11) cells. Cells were grown for 7 days on microcarriers. Glucose (1- ^{13}C 99%) was added to a glucose free medium (50 ml 5.6 mM). 75.5 MHz ^{13}C spectra were recorded sequentially for 21 h. Each trace was accumulated for 2 h using 45° pulses and 2 s repetition time and Fourier transformed with 10 Hz line broadening.

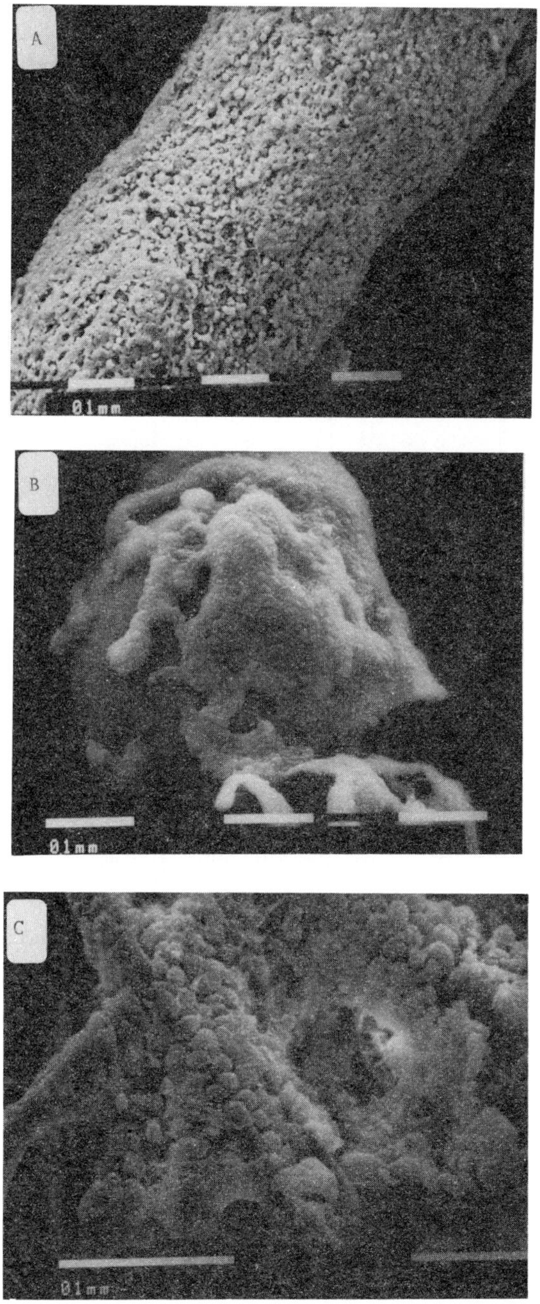

Figure 3: Scanning electron micrographs of T47D (clone 11) cells. A. Cells embedded in agarose filaments. B. Cells grown for 5 days on microcarriers. C. Cells grown for 12 days on filters.

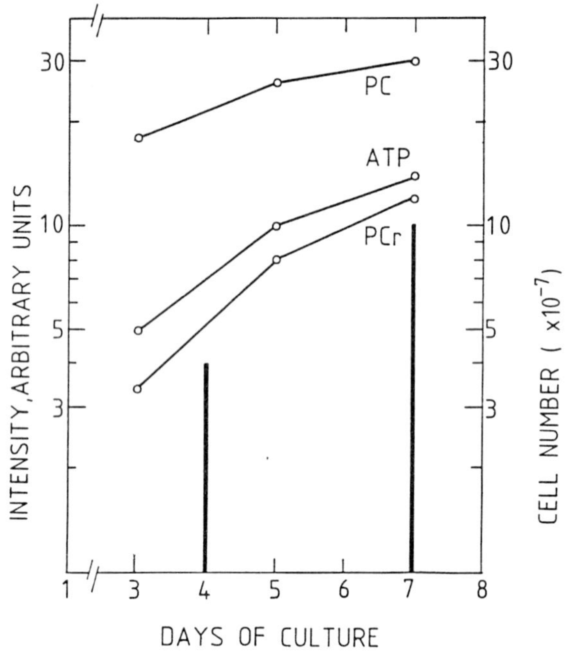

Figure 4: Growth of T47D cells on microcarriers. Open circles represent the levels of phosphate metabolites as obtained from ^{31}P spectra of cells cultured on the microcarriers for the indicated number of days. 3 h spectra were recorded by applying 90° pulses and a 10 s repetition time. The bars are the cell counts after trypsinization.

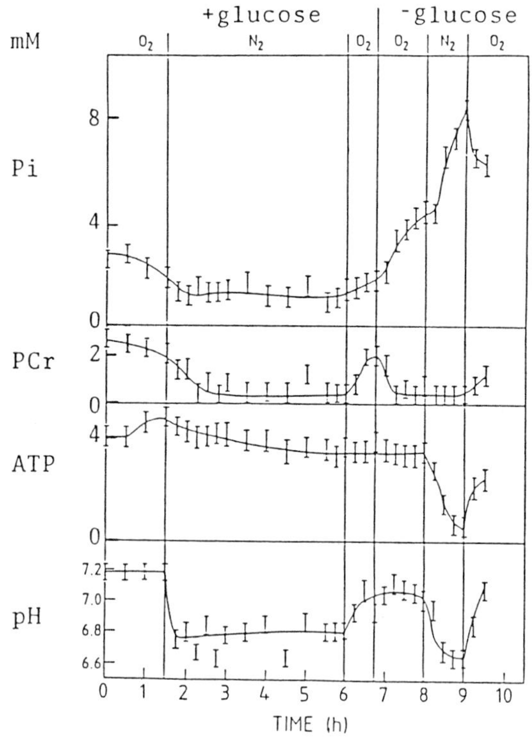

Figure 5: The effects of carbon source and oxygen content on the ^{31}P spectrum of T47D (clone 11) cells. Cells were embedded in agarose filaments. 15 min ^{31}P spectra were recorded using 60° pulses and 1 s repetition time.

glucose, anaerobic conditions led to a marked decrease in the ATP concentration (90%) and an increase in the concentration of Pi (2 fold increase relative to aerobic conditions). Upon reoxygenation of the medium,the concentrations of the intracellular phosphates were restored to their initial values.

Under aerobic conditions the transfer from glucose containing medium to media containing other carbon sources which enter directly into the Krebs cycle, such as glutamine, alanine and lactate, was accompanied not only by changes in the steady state concentrations of the high energy phosphates and metabolites derived from phospholipids, but also by an increase in the fluxes from Pi to ATP and from ATP to ADP. For example, substituting the carbon source from glucose to alanine in cells embedded in gel threads caused the changes in the phosphates concentration and fluxes as summarized in table 1. Similar changes were observed also with cells grown on microcarriers. It is clear that the changes in the energetics result from modulations in the main pathway for energy production. Glucose can be utilized through aerobic glycolysis and through the Krebs cycle while alanine can be utilized to form ATP only through the Krebs cycle. Although the concentration of ATP decreased by only 12% upon transfer from glucose to alanine, the ratio ATP/Pi and the phosphorylation potential were markedly reduced (by a factor of 4) and the flux from Pi to ATP increased by a factor of 3. In addition to the changes in the energetics an unpredicted decrease in the concentration of GPC and GPE was observed. Modulations in the concentration of these metabolites were observed following differentiation of Friend erythroleukemia cells (3), interferon treatment of Friend leukemia tumors in mice (13). The content of GPC and GPE varies in different muscle types (2) and between adriamycin resistant and sensitive human breast cancer (MCF7) cells (4). The overall regulation and the mechanisms responsible for the changes observed during transfer between

TABLES

Table 1: Concentration and dynamics of the phosphate metabolites in T47D cells (clone 11), embedded in agarose filaments and perfused with medium containing glucose or alanine as the carbon source. Data was obtained from ^{31}P spectra (3 experiments) recorded with 90° pulses and a 10 s repetition time.

	concentration (mM)			
	ATP	PCr	Pi	GPC+GPE
glucose	4.0	1.5	1.4	4.8
alanine	3.5	0.1	7.4	0.2

	flux (mM/s)		
	Pi->γATP	βATP->βADP	PCr->γATP
glucose	0.2	1.3	0.2
alanine	0.5	4.7	-

carbon sources are now under investigation.

Conclusions

1. NMR can monitor on line and noninvasively the metabolism of cells in a perifusion system.
2. Modulations in the energetics under different conditions; both in the metabolite concentrations and in the flux rates can be determined by ^{31}P NMR.
3. Quantitative measurements of the utilization of substrates through the various metabolic pathway can be measured by ^{13}C NMR using labeled metabolites.

Similar NMR studies can be extended to other systems and other nuclei (^{1}H , ^{23}Na , ^{39}K , ^{15}N etc.).

REFERENCES

1. Alger, J.R. and Shulman, R.G. (1984). Q. Rev. Biophys. 17, 83-124.
2. Burt, C.T., Kleps, R.A., Kapin, M., Danon, M.J. and Perurena, O. (1982) Life Sciences 30, 39-44.
3. Carpinelli, G., Podo, F., Di Vito, M., Proietti, E., Gessani, S. and Belardelli, F.V. (1984) FEBS Lett. 176, 88-92.
4. Cohen, J.S., Lyon, R.C., Chen, C., Faustino, P.J., Batist, G., Shoemaker, M., Rubalcaba, E. and Cowan, K.H. (1986) Cancer Res. 46, 4087-4090
5. Foxall, D.L. and Cohen, J.S. (1983). J. Magn. Reson. 52, 346-349.
6. Gadian, D.G. (1982) in: Nuclear Magnetic Resonance and its Application to Living Systems. Oxford, Clarenton.
7. Kadouri, A., Bohak, Z., Maroudas, N.G. and Sussman, M.V. (1987) Abst. Europian Society for Animal Cell Technology 8th meeting. pp 33.

8. Keydar, I., Chen, L., Karby, S., Weiss, F.R., Delarea, J. and Radu, M. (1979). Europ. J. Cancer 15, 659-669.

9. Koretsky, A.P. and Weiner, M.W. (1984) in: ^{31}P NMR Magnetization transfer Measurements of Phosphorous Exchange Reactions in vivo. (James, T.L. and Margulis, A.R. eds.). Biomedical Reasearch and Education Foundation, San Francisco, CA.

10. Margel, S. and Offarim, M. (1983) Anal. Biochem., 128,342-350.

11. Moon, R.B. and Richards, J.H. (1973) J. Biol. Chem. 248, 7276-7278.

12. Neeman, M., Rushkin, E., Kaye, A.M. and Degani, H. Biochim. Biophys. Acta, in press.

13. Proietti, E., Carpinelli, G., Di Vito, M., Belardelli, F., Gresser, I. and Podo, F. (1986) Cancer Res. 46, 2849-2857.

23 Optimization of glass-sphere immobilized bed cultures

D. Looby and J. B. Griffiths

ABSTRACT

Immobilised glass sphere cultures have a considerable scale-up potential plus certain advantages that make them an alternative to microcarrier culture. Optimisation studies were carried out in order to increase the poor productivity of these cultures. The discovery that a bed of 7mm diameter spheres yielded significantly more cells than 3mm spheres despite the large reduction in available surface area, led to this system being used as a model for the development of culture matrices. Cell yields per cm^2 increased with sphere diameter from 2mm to the maximum tested (8mm). The highest total culture yield was obtained with 5mm spheres, a compromise between surface area per unit volume and sphere diameter. A variable factor in beds of different sphere size is the useful channel size through the bed. 5mm spheres produce an average channel size of $1mm^2$ and it is proposed, therefore, that this is a minimal requirement for large volume beds. It is of interest to note that other cell systems based on a matrix, and which support high densities per unit of surface area, have channel sizes of $1mm^2$. The reasons for this are reduced surface shear permitting higher perfusion rates, and avoidance of channelling within the bed.

INTRODUCTION

Glass sphere immobilised bed systems are a useful alternative to microcarriers for large scale production of anchorage dependent cells. The advantages are (1) glass is a cheap and re-utilisable substrate, which is acceptable for the production of biologicals for human use, (2) since media mixing and aeration take place in a reservoir vessel, the cells are subjected to low shear, which facilitates multilayering and long term production from suitable cell lines, and finally, (3) media can be changed easily with minimum stress on the cells. This is useful for changing from cell growth media to production media, multiple harvesting, or infection of cells with virus. Glass sphere immobilised bed systems have been scaled up to production levels. Foot and Mouth Disease virus has been produced in batch culture in a 100 litre system made up of a 70 litre reservoir vessel and a 30 litre culture vessel packed with 50 Kg of 3mm beads (1). Also, long term production of tissue plasminogen activator has been achieved in a glass sphere culture system with continuous media feed and harvest of product rich media (2). However, we have found the technique based on 3mm beads to be below the expected productivity based on other static systems. The aim of these studies was to increase the efficiency of the process.

MATERIALS AND METHODS

Medium

The growth medium was Eagles minimal essential medium supplemented with 10% foetal bovine serum, 2.0 mM glutamine, 15 mM hepes buffer, 14 mM sodium bicarbonate, gentamycin (25Iu ml^{-1}) and glucose (2g L^{-1}).

Cell Line

The cell line used was Vero

Inoculum

The cell seed was taken from Roller bottles in late log phase of growth. The cells were harvested with 0.25% Trypsin 0.025% E.D.T.A. and inoculated into the glass bead columns at 2.0 $x10^4$ cells cm^{-2}.

Growth Monitoring

Cell growth was monitored indirectly by glucose analysis (Beckman Ltd.). Cell counts were carried out at the end of the culture period by removing aliquots of beads and counting by the Nuclei staining method or by trypsinization, followed by counting by the trypan blue dye exclusion method. The onset of stationary phase was indicated by the glucose consumption rate (G.C.R.) becoming constant

Description and Operation of Culture System

The glass sphere culture system (Fig. 1) is a recirculating perfusion system, consisting of a water jacketed immobilised bed reactor for cell growth and a reservoir vessel for media conditioning, mixing, aeration etc. The vessels are compatible with LH 500 series fermenter head plates and control equipment (L H Fermentation Ltd). Media Perfusion can be upflow or downflow by peristaltic pump (Watson Marlow Ltd). The system can be operated as a batch, fed-batch or semi-continuous system (ie, multiple harvest with media change). The physical characteristics of the packed bed are given in Table I.

Operation

Media was introduced to the system and allowed to circulate for a few hours to equilibrate with respect to temperature, pH, dissolved oxygen etc. The cell seed, which was resuspended in sufficient media to cover the beads was inoculated into the bottom of the column. The column was drained and refilled twice and left without perfusion for 8 to 12 hours to allow the cells to attach - this was followed by upflow perfusion at 2 linear cm/min increasing to 5 linear cm/min with cell growth.

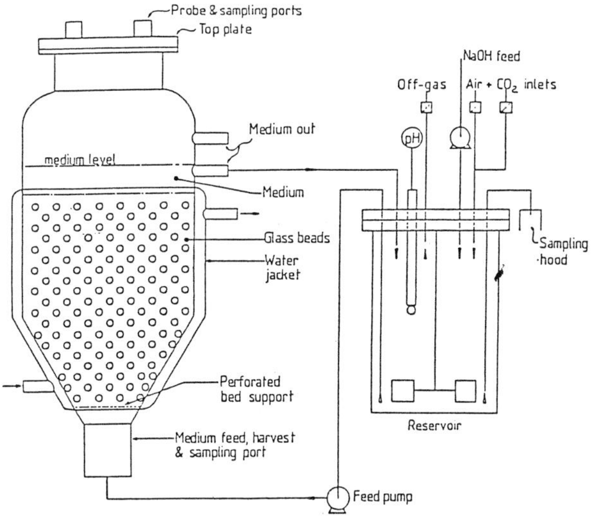

Fig. 1. Schematic of glass sphere immobilised bed culture system.

3 mm

5 Kg

0·66	0·55	0·54
0·74	0·64	0·73
0·96	0·88	1·10
1·05	1·30	0·98

Average Cells/cm^2	$0{\cdot}84 \times 10^5$
Cells/Kg	$4{\cdot}4 \times 10^8$
Cells/ml	$1{\cdot}7 \times 10^6$

Cells/cm^2 $\times 10^5$

Fig. 2. The distribution of Vero cells in a 5 Kg reactor packed with 3mm spheres. The cell seed was introduced into the bottom of the bead bed.

RESULTS

Preliminary results indicated that the glass bead culture system was considerably less efficient than either Roller bottle or Microcarrier culture systems (Table 2).

Investigations into the distribution of cells in the traditional 3mm bead beds, showed that the poor yields were due to excessive concentration gradients of cells. Particularly, axial gradients decreasing from the point of inoculation to the top of the bed (Fig. 2). This has been noticed by other workers (3).

The concentration gradients were overcome by (a) inoculating the beds through a perforated tube placed centrally through the whole vertical height, (b) precoating the beads with the attachment promoting glycoprotein fibronectin, and (c) packing the columns with larger diameter beads (7mm), (Fig.3).

The highest cell density per unit area was achieved in the 7mm bead bed. Therefore, a detailed study was carried out to determine the effects of sphere diameter on cell yield. The results showed that cell density per unit area increased with diameter, but that this only compensated for the total reduction in surface area up to 5mm diameter (Table 3). The 5mm diameter bead represented the best compromise between yield per unit area and decreasing total surface area.

Further studies into the use of 5mm spheres in 5 and 10 Kg beds revealed that there were no significant gradients (Fig. 4), and no decrease in yield with increasing bed size (Table 4).

Experiments to determine the scale up limits of the system revealed that homogeneous scale up was achievable up to at least 40 Kg (available surface area 12.8 m^2) with no decrease in yield efficiency (Fig. 5) . The cell yield achieved (Total 3.75 $x10^{10}$) is comparable to that from 220 Roller bottles or 54 grams of cytodex 3 microcarriers (Table 5).

Table I. Characteristics of packed bed of 5mm spheres.

Total Surface area cm^2/Kg	4570			
Available Surface area (70% cm^2/Kg	3200			
Total Volume ml/Kg	625			
Void Fraction	0.4			
Void Volume ml/Kg	250			
Channel size (mm^2)	1.0			
		Height	x	Diameter
Dimensions of packed beds (cm)	1.5 Kg	19	x	8
	5.0 Kg	18	x	15
	10.0 Kg	20	x	20

Table II. Comparison of cell yield from different culture systems.

System	Cell Yield ($x10^5$ $cells/cm^2$)
*Glass bead 1.5 Kg	0.91
*Glass bead 5.0 Kg	0.83
Roller bottle	2.0
#Microcarrier	1.52

*3 mm borosilicate glass spheres
#Cytodex 3 microcarriers

TABLE III. Growth of cells on spheres of different diameters

BEAD DIAMETER (mm)	CHANNEL SIZE (mm^2)	COUNT $x10^5/cm^2$	COUNT $x10^8/Kg$	COUNT $x10^6/ml$
2	0.16	0.86	6.7	2.7
3	0.32	0.78	4.0	1.6
4	0.68	1.25	5.0	2.0
5	1.0	2.5	8.0	3.2
6	1.4	2.2	5.0	2.0
7	1.8	2.4	5.0	2.0
8	2.2	2.3	4.3	1.7

Table IV. Comparison of cell yield from 5mm spheres in a range of bed sizes.

	CELL YIELD		
BED Size (Kg)	$x10^5/cm^2$	$x10^8/Kg$	$x10^6/ml$
1.5	2.5	8.0	3.2
5.0	2.9	9.3	3.7
10.0	3.1	9.9	4.0

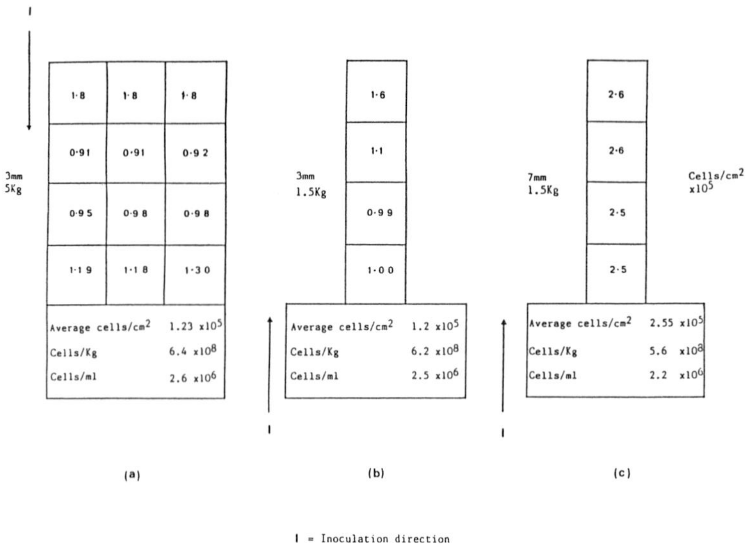

Fig. 3. The distribution of Vero cells in glass bead reactors (a) 3mm beads inoculation through a centrally placed perforated tube, (b) 3mm beads precoated with fibronectin and, (c) 7mm beads.

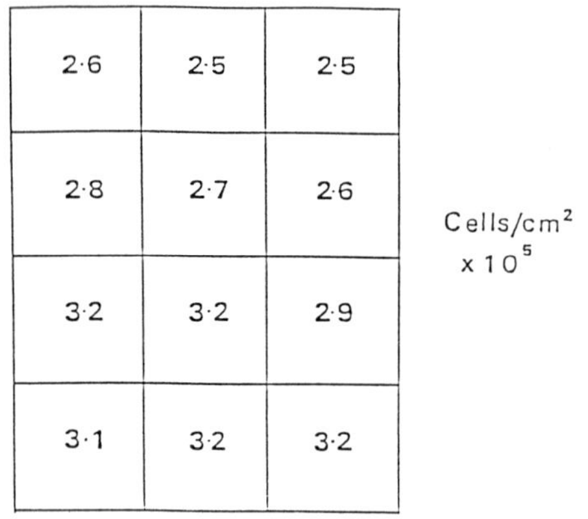

Fig. 4. Distribution of cells in a 5Kg reactor packed with 5mm spheres. The inoculum was introduced into the bottom of the bead bed.

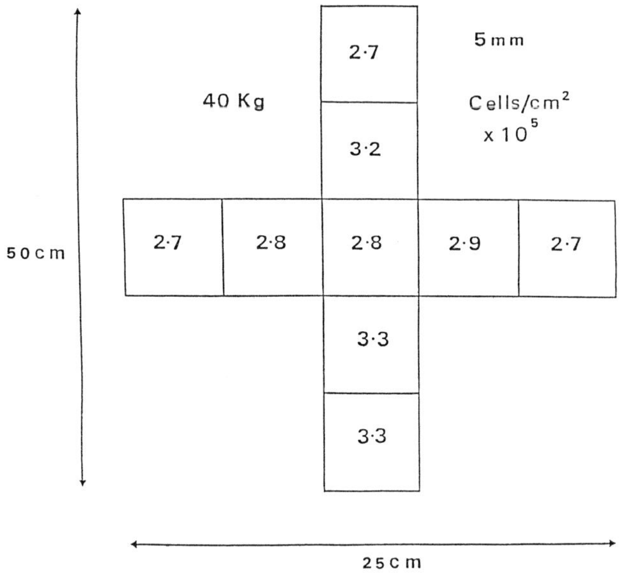

CELL YIELD

Average cells/cm^2	2.9 $\times 10^5$
Cells/Kg	9.3 $\times 10^8$
Cells/ml+	3.7 $\times 10^6$

Fig. 5. Scale up of cell growth from 5mm spheres. Cells were inoculated into the bottom of the bead bed.

Table V. Scale of system required to produce 3.75 x 10^{10} Vero cells

System	Scale	Total Media Vol. (L)	Cells $\times 10^5$/cm2
Glass bead	40 Kg (5mm spheres)	40	2.93
Microcarrier[4] (cytodex 3)	54 g	27	1.5
Microcarrier[5] (cytodex 1)	90 g	45	0.7
Roller bottles	220	55	2.0

DISCUSSION

Considerable improvement in cell yield was achieved over the traditional bottom inoculated 3mm bead beds. The problem with these beds was excessive concentration gradients of cells, probably caused by filtration effects of the bead bed on the cell inoculum, and also settling out of the cells during the attachment stage of growth ie, before media circulation was initiated.

Certain modifications to the system, which independently overcome these gradients were (1) central inoculation, through a perforated tube, along the vertical length of the bed, (2) using beds of larger diameter spheres (7mm) which gave a more open bed configuration (channel size 1.8mm^2), and (3) precoating the 3mm spheres with fibronectin. Treatments (1) and (2) brought about even distribution of the cell inoculum, and (3) promoted rapid attachment of the cells thus minimizing the percolation of cells down through the bead bed in the attachment stage of growth.

Gradient free beds, with a significant increase in overall cell yield, were achieved by all three modifications, however the highest yield was achieved by using larger bead sizes. A systematic study of a range of bead sizes revealed that increasing sizes, despite having a lower surface area, brought about higher yields. The conclusion was that a channel size of at least 1mm^2 was required (provided by 5mm beads) in large volume cultures (Table 3).

Channel sizes in excess of 1mm^2 offered no further increase in cell yield per unit surface area and therefore represent a decrease in cell yield per unit volume. Channel sizes less than 1mm^2 resulted in a significant decrease in yield per unit surface area and per unit volume, probably due to channelling effects. Indeed, homogeneous scale up to 40 Kg culture systems is achievable with 5mm bead beds, without any decrease in yield or appearance of cell gradients. It is concluded, therefore, that channel size is a critical factor in the efficient scale up of bead beds. The reason for this is probably the reduced shear and turbulance, due to the unimpeded flow through the more open bed matrix, allowing cells to multilayer and form aggregates, whilst maintaining a gradient free bed. Also, channelling/short circuiting of media flow as seen in the smaller bead beds was prevented.

Cell growth in glass bead beds compares favourably with microcarrier and Roller bottle production systems. However, improvements are envisaged, which could increase its potential as a production system especially as the maximum practical scale up limit of the system has not yet been reached. Also the surface area to volume ratio could be greatly increased by replacing the solid spheres with sintered glass spheres, thus allowing cell growth to take place within the bead matrix as well as on the surface.

ACKNOWLEDGEMENTS

This work was supported by the Biotechnology Group, Department of Trade and Industry.

REFERENCES

(1) Whiteside, J.P., Whiting, B.R. and Spier, R.E., (1979). Development of methodology for the production of Foot and Mouth Disease virus from BHK 21 C13 Monolayer cells grown in a 100L (20M^2) Glass sphere propagator. Develop. Biol. Standard. 42, 113-119.

(2) Bioresponse, inc. Hayward, CA 94545 USA.

(3) Whiteside, J.P. and Spier, R.E., (1985). Factors affecting the productivity of Glass sphere propagators. Develop. Biol. Standard. 60, 305-311.

(4) Griffiths, J.B., Thornton, B. and McEntee, I.D. (1982). The development and use of microcarrier and glass sphere culture techniques for the production of herpes simplex virus. Develop. Biol. Standard. 50, 103-110.

(5) Billig, D. (1987). Scaling up microcarrier cultures of mammalian cells for production purposes. In "Process possibilities for plant and animal cell cultures" (Webb, C. and Mavituna, F. ed.) pp 277-284, Ellis Horwood Limited.

24 Examination of a simple model for the diffusion of oxygen into dense masses of animal cells

A. D. Murdin, R. Wilson, N. F. Kirkby and R. E. Spier

ABSTRACT

The equations describing the diffusion of oxygen into respiring tissue are well known. A simplified solution is presented for each of the three geometries encountered in animal cell culture, namely diffusion into a spherical particle, from a cylindrical capillary or into a flat slab. The limits and assumptions inherent in the solutions are discussed in the light of animal cell culture conditions and the solution for diffusion into a flat surface is compared with experimental measurements of the metabolism of hybridomas immobilised in a slab of agarose.

INTRODUCTION

The oxygenation of animal cells in culture is of major importance to their growth and production. Monitoring and controlling the supply of oxygen to cells in suspension is relatively simple and has been dealt with previously (4). Monitoring the distribution of oxygen within masses of immobilised cells is more difficult. Relevant studies have been made of fungal pellets (1,2) and of tumours (5) but little data is available for immobilised mammalian cells. The problem with all such work is measuring oxygen concentration or cell metabolism in situ. A simple method for assessing the metabolism of immobilised cells has recently been described (6). This study compares results obtained using this method with a theoretical model for the diffusion of oxygen into masses of immobilised cells.

THEORETICAL MODEL

A greatly simplified treatment of the problem of oxygen diffusion within a mass of immobilised cells can be made by allowing the following assumptions:-

1) Cellular respiration rate is independent of oxygen concentration.

   This is unlikely to be true, though it may be that the change in respiration rate is small over an appreciable range of oxygen concentration. The likely effect of this assumption is to make the model a poor fit to reality at low oxygen concentrations.

2) Oxygen moves within the cell mass by diffusion only.

   This is true if the extracellular fluid in the cell mass is stagnant. For instance, if the cells are immobilised in agarose or some other form of polymeric matrix, or are at concentrations approaching those found in tissues (10^9/ml) there is little opportunity for mass flow. The assumption is certainly not true in, for example, those designs of hollow fibre cartridge bioreactors which generate significant pressure gradients within the cartridge.

3) The immobilised cell mass is homogeneous.

4) There is no significant diffusion barrier between the cell mass and the surrounding medium.

There may be a stagnant boundary layer at the medium/cell mass interface, which could form a barrier. The use of a membrane to separate cells from medium may present a further barrier. The significance of either of these would have to be determined for each system examined.

The analysis can be further simplified by considering only the steady-state situation (i.e. that situation in which there is no net change of oxygen concentration in the system). This is most likely to hold true in continuous culture, which is the desired state for immobilised cell systems.

A more detailed treatment of diffusion in relevant systems can be found in references 1, 2 and 5.

TERMS

The following terms are used:-

D = diffusion coefficient (cm^2/s)

c = concentration ($\mu g/ml$)

c_m = concentration in surrounding medium ($\mu g/ml$)

R = respiration rate ($\mu g/ml/s$)

r = distance (cm)

r_o = distance at which $c = 0$ (cm)

DIFFUSION INTO A FLAT SLAB

Consider a thin layer within the slab of unit area, thickness δr and at a distance r from the interface with the medium.

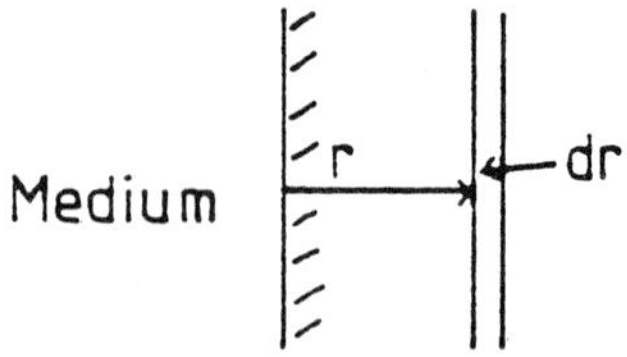

At steady state the mass balance equation for this layer is :

OXYGEN IN = OXYGEN OUT + OXYGEN USED

$$-D\left[\frac{dc}{dr}\right]_r = -D\left[\frac{dc}{dr}\right]_{r+\delta r} + R.\delta r \qquad (1)$$

Which may be rearranged to give

$$R = D\left[\frac{\left[\frac{dc}{dr}\right]_{r+\delta r} - \left[\frac{dc}{dr}\right]_r}{\delta r}\right] \qquad (2)$$

i.e.

$$R = D\left[\frac{d^2c}{dr^2}\right] \qquad (3)$$

The general solution to equation (3) is

$$c = ar^2 + br + e \qquad (4)$$

where a, b and e are constants.

Therefore

$$\frac{dc}{dr} = 2ar + b \qquad (5)$$

$$\frac{d^2c}{dr^2} = 2a \qquad (6)$$

Solving for a, b and e

$$a = \frac{R}{2D} \qquad \text{from (3)}$$

Using the boundary conditions

$$c = 0\ , \ \frac{dc}{dr} = 0 \quad \text{at} \quad r = r_o$$

$$b = -\frac{R\, r_o}{D} \qquad \text{from (5)}$$

and $$e = \frac{R\, r_o^2}{2D} \qquad \text{from (4)}$$

Thus equation (4) becomes

$$c = \frac{R\, r^2}{2D} - \frac{R\, r r_o}{D} + \frac{R\, r_o^2}{2D}$$

i.e.

$$c = \frac{R}{2D}\left[r_o - r\right]^2 \tag{7}$$

From the boundary condition

$$c = c_m \quad \text{at} \quad r = 0$$

$$r_o = \sqrt{\frac{2Dc_m}{R}} \qquad \text{from (7)}$$

Thus

$$c = \frac{R}{2D}\left[\sqrt{\frac{2Dc_m}{R}} - r\right]^2 \tag{8}$$

DIFFUSION FROM A CYLINDRICAL CAPILLARY

Consider a thin shell around the capillary of unit length, radius r and thickness δr

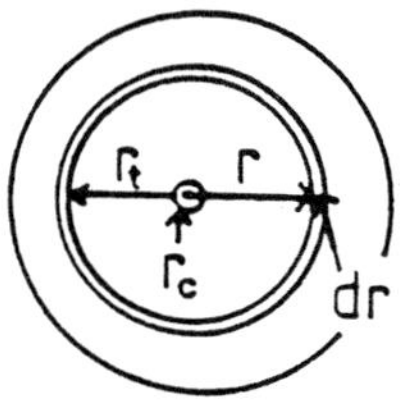

At steady-state the mass balance equation for this shell is

OXYGEN IN = OXYGEN OUT + OXYGEN USED

$$-D.2\pi r\left[\frac{dc}{dr}\right]_r = -D.2\pi(r+\delta r)\left[\frac{dc}{dr}\right]_{r+\delta r} + R.2\pi r\ \delta r \tag{9}$$

which may be rearranged to give

$$R = D\left[\frac{\left[\frac{dc}{dr}\right]_{r+\delta r} - \left[\frac{dc}{dr}\right]_r}{\delta r}\right] + \frac{D}{r}\left[\frac{dc}{dr}\right]_{r+\delta r} \qquad (10)$$

i.e.

$$R = D\left[\frac{d^2c}{dr^2} + \frac{1}{r}\frac{dc}{dr}\right] \qquad (11)$$

The general solution to equation (11) is

$$c = ar^2 + b \log r + e \qquad (12)$$

where a, b and e are constants

Therefore

$$\frac{dc}{dr} = 2ar + b/r \qquad (13)$$

$$\frac{d^2c}{dr^2} = 2a - b/r^2 \qquad (14)$$

Solving for a, b and e

$$a = \frac{R}{4D} \quad \text{from (11)}$$

Using the boundary conditions

$c = c_m$ at $r = r_c$ (radius of the capillary)

$c = 0$, and $\frac{dc}{dr} = 0$ at $r = r_o$

then $$b = -\frac{R\,r_o^2}{2D} \quad \text{from (13)}$$

and $$e = c_m - \frac{R\,r_c^2}{4D} + \frac{R\,r_o^2}{2D}\cdot\log r_c \quad \text{from (12)}$$

Thus equation (12) becomes

$$c = c_m + \frac{R}{4D}\left[r^2 - r_c^2 - 2r_o^2 \log \frac{r}{r_c}\right] \qquad (15)$$

A value for r_o may be obtained by substituting the boundary conditions $c = 0$ at $r = r_o$ into equation (15) and using Newton's method to obtain a numerical approximation to r_o.

DIFFUSION INTO A SPHERE

Consider a thin shell within the sphere of radius r and thickness δr.

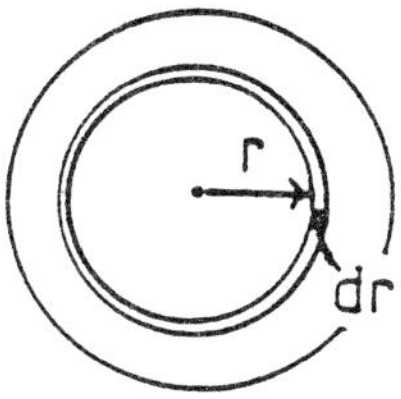

At steady state the mass balance equation for this shell is

OXYGEN IN = OXYGEN OUT + OXYGEN USED

$$D.4\pi(r+\delta r)^2\left[\frac{dc}{dr}\right]_{r+\delta r} = D.4\pi r^2\left[\frac{dc}{dr}\right]_r + R.4\pi r^2\delta r \qquad (16)$$

Neglecting terms in δr^2, this may be rearranged to give

$$R = D\left[\frac{\left[\frac{dc}{dr}\right]_{r+\delta r} - \left[\frac{dc}{dr}\right]_r}{\delta r}\right] + \frac{2D}{r}\left[\frac{dc}{dr}\right]_{r+\delta r} \qquad (17)$$

i.e.

$$R = D\left[\frac{d^2c}{dr^2} + \frac{2}{r}.\frac{dc}{dr}\right] \qquad (18)$$

The general solution to equation (18) is

$$c = ar^2 + b + e/r \qquad (19)$$

where a, b and e are constants.

Therefore

$$\frac{dc}{dr} = 2ar - e/r^2 \qquad (20)$$

$$\frac{d^2c}{dr^2} = 2a + 2e/r^3 \qquad (21)$$

Solving for a, b and e

$$a = \frac{R}{6D} \qquad \text{from (18)}$$

For a sphere which is aerobic throughout its volume the boundary conditions are :

$$c = c_m \quad \text{at} \quad r = r_s \quad \text{(radius of the sphere)}$$

$$\frac{dc}{dr} = 0 \quad \text{at} \quad r = 0$$

thus $e = 0$

and $b = c_m - \frac{R\, r_s^2}{6D}$ from (19)

Thus equation (19) becomes

$$c = c_m - \frac{R}{6D}\left[r_s^2 - r^2\right] \qquad (22)$$

For a sphere which contains an anaerobic volume at its centre the boundary conditions change to

$$c = c_m \quad \text{at} \quad r = r_s$$

$$c = 0 \quad \text{and} \quad \frac{dc}{dr} = 0 \quad \text{at} \quad r = r_s - r_o$$

b and e can then be obtained by substitution as above, and r_o can be obtained by the method described for the cylinder.

MEASUREMENT OF THE METABOLISM OF IMMOBILISED CELLS

The metabolism of immobilised cells was assessed by determining their ability to reduce 3-(4,5-dimethylthiazol-2-yl) -2,5 diphenyltetrazolium bromide (MTT). MTT is soluble in water, forming a pale yellow solution. The reduced form is blue and insoluble in water. Reduced MTT is deposited as blue crystals within actively metabolising cells (9,6).

Cells immobilised in 1% agarose at 5 x 10^6/ml were incubated under RPMI 1640 plus 1% foetal calf serum in an atmosphere of 95% air, 5% CO_2. After 42 hours 500 µg/ml of MTT was added to the supernatant medium, and the incubation was continued overnight. The agarose slab was then recovered and sectioned vertically (6). Oxidised, yellow, MTT had diffused throughout the slab. Blue, reduced, MTT had been deposited close to the surface of the slab. The sections were photographed and the photographs analysed to determine the brightness as a function of depth below the surface. Brightness was inversely related to the amount of reduced MTT present. The brightness levels were used as an index of cellular metabolism (3).

RESULTS AND DISCUSSION

Figure 1 shows a comparison of the measured metabolic activity of immobilised hybridomas with the calculated steady-state oxygen concentration. Despite the numerous assumptions made in deriving the mathematical model there is a good fit between the two sets of data in this instance.

The use of the cytometabolism diffusion test of Wilson and Spier (6) in conjunction with photographic image analysis provides a simple but powerful means for assessing the metabolism of immobilised cells. The combination of this technique with the mathematical model (which may be adapted to molecules other than oxygen) provides the basis of a method for studying the effect of various factors on the metabolism of immobilised cells in-situ and for predicting the performance of cells immobilised in a defined situation.

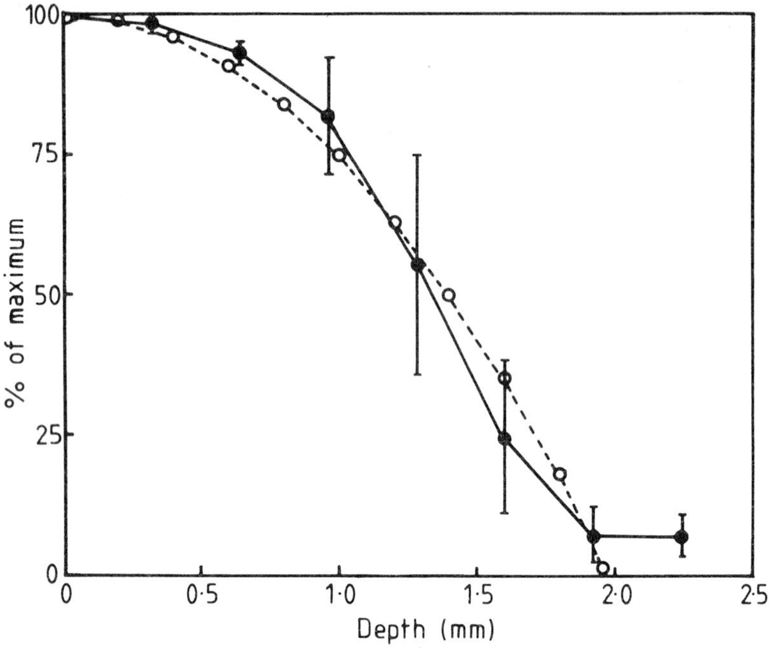

Figure 1. Comparison of the metabolic activity of cells of a hybridoma line immobilised in an agarose slab (●——●) with the calculated steady state oxygen concentrations within the slab (○---○). Both sets of data are expressed as percentages of the values at the surface of the slab (depth = 0 mm).

Metabolic activity was assessed by measuring MTT reduction as described in the text. The results shown are the means of four determinations. The error bars show standard deviations.

Steady-state oxygen concentrations were calculated using equation (8), assuming that c_m = 7·22 μg/ml (i.e. saturated with respect to the gas phase), R = $9{\cdot}7 \times 10^{-3}$ μg/ml/s (i.e. cellular respiration rate = 7 μg oxygen/10^6 cells/hour) and that D = $2{\cdot}65 \times 10^{-5}$ cm^2/s (i.e. the value in water at 37°C).

The measured metabolic activity is described by the relationship

$$M = \left[1 - \frac{r^2}{r_o^2}\right] \times 100\%$$

where M is metabolic activity expressed as a percentage of maximum. The metabolic activity predicted from this relationship is shown (●---●).

REFERENCES

1) B. Metz and N.W.F. Kossen (1977). The growth of moulds in the form of pellets - a literature review. Biotechnology and Bioengineering 19, 781-799.

2) M. Moo-Young and H.W. Blanch (1981). Design of biochemical reactors - mass transfer criteria for simple and complex systems. Advances in Biochemical Engineering 19, 1-69.

3) T. Mosmann (1983). Rapid colorimetric assay for cellular growth and survival: application to proliferation and cytotoxicity assays. Journal of Immunological Methods 65, 55-63.

4) R.E. Spier and B. Griffiths (1984). An examination of the data and concepts germane to the oxygenation of cultured animal cells. Developments in Biological Standardization 55, 81-92 (Karger, Basel).

5) I.F. Tannock (1968). The relation between cell proliferation and the vascular system in a transplanted mouse mammary tumour. British Journal of Cancer 22, 258-273.

6) R. Wilson and R.E. Spier (1987). A cytometabolism diffusion test (CMDT) for the determination of animal cell bioreactor parameters. Submitted for publication to Enzyme and Microbial Technology.

25 Kinetics of prourokinase production by mammalian cells in culture

W. S. Hu, T. Y. Tao, M. A. Bohn and G. Y. Ji

ABSTRACT

The kinetics of prourokinase production by mammalian cells was examined in different cultivation systems. The prourokinase produced was secreted in the medium and accumulated in a relatively high concentration. The production of prourokinase could be sustained for a period of six weeks or gelatine coated microcarriers. When the volumetric productivity was compared, no significant differences were observed among the three systems examined.

INTRODUCTION

In the last few years there has been an increasing interest in the design of bioreactors or in the selection of bioreactors for large scale cultivation of animal cells. However the information on the direct comparison of cell growth and product formation in different types of reactors has been scarce. In this study we compared the growth of TCL 598 cells and the production of prourokinase in three different types of reactors, namely microcarrier culture, roller bottles, and ceramic systems. In the study of reactor performance, one should try to distinguish between the intrinsic kinetics and the kinetic behavior caused by the reactor configurations. The intrinsic kinetics are the response of cells to their environmental conditions. The kinetic behavior of cells may be different in different types of reactors due to differences in the residence time distribution and the mass transfer which in turn cause the variation in the chemical environment of cells. If we can control both environmental factors to be the same in different types of reactors, in principle the kinetic behavior of cells in all those types of reactors should be identical. Here we report the comparison of TCL 598 kinetic behavior of cells in three different types of reactors. The kinetic behavior of those cells appears to be very similar in all the three systems.

MATERIALS AND METHODS

Cells

The TCL 598 cell line was obtained from Collaborative Research Inc. (Lexington, MA). Stock cultures were maintained in T-flasks and roller bottles using Dulbecco's Modified Eagle (DME) medium supplemented with 5% fetal bovine serum (FBS).

Microcarrier culture

Microcarrier cultures were seeded from roller bottle stock cultures. The 100 ml cultures were carried out in 250 ml spinner vessels and the 500 ml cultures were carried out in one liter round bottom vessels. The procedure of cell cultivation in these vessels has been described previously (Hu et al, 1985). Both Cytodex 1 and Cytodex 3 microcarriers were purchased from Pharmacia Inc. (Sweden).

Cultivation in a ceramic system

The ceramic system used in the experiment was 16 KC and KC Biologicals, Inc. The configuration and operation of this ceramic system was descrived previously (Lyderson et al., 1984). The surface area available for cultivation was 16 m^2 according to the manufacturer. A cell suspension of approximately 10^9 TCL cells was obtained from roller bottoles by trypsinization and used as the inoculum. A 20 liter medium in the reservoir was attached to the system and was recirculated continuously.

Prourokinase assay

The urokinase activitiy was determined by its amidolytic effect on the substrate pyro-Glu-Gly-Arg-pNA (Kohno et al., 1984). A mixture of 90 μl of ice-cold buffer A (50 mmole/l Tris, 38 mmol/l NaCl, pH 8.8) or plasmin (2.5 units/ml) was incubated with 710 μl of buffer A in test tubes at 37°C for one min. Then, 100 μl of the standard urokinase at different dilutions or the sample to be assayed was added to the reaction mixture and incubated at 37 °C for 10 min. Subsequently, 100 μl of the substrate (1.5 mg/ml) was added and the mixutre was incubated at 37°C for exactly 5 min. The reation was stopped by adding 100 μl of

50% acetic acid. The rate at which p-nitroaniline (pNA) was released was measured photometrically at 405 nm. The standard curve was obtained by plotting the A_{405} versus the concentration (units/ml) of the serially diluted standard solutions. The concentration of prourokinase (zymogen) in the sample is the activity with plasmin activation (the concentration of urokinase plus prourokinase) minus that without plasmin activation (the concentration of urokinase). The concentrations of proruokinase are all expressed in relative units/ml.

RESULTS

The growth of TCL 598 cells on two types of microcarriers, namely Cytodex 1 and Cytodex 3, was first compared. Cells grew equally well on both types of microcarriers after attachment. In the first stage cells grew rapidly until a monolayer was covered and then the growth slowed down. No difference in cell growth in the two types of microcarriers was observed (Figure 1). However, after 270 hours of cultivation the cells on Cytodex 1 began to fall off microcarriers and a decrease on cell concentration was observed. On the other hand, cells on the Cytodex 3 microcarriers continued to grow and eventually reached a density of approximately 5-6 x 10^6 cells/ml. In a typical experiment carried out with Cytodex 3, the cultivation period can be as long as seven weeks. For products which are constitutively secreted into the cell culture medium, it is essential that cells be maintained on microcarriers for an extended time to prolong the production period. Therefore Cytodex 3 was chosen as the type of microcarrier for our subsequent studies.

The kinetics of cell growth and product formation on Cytodex 3 microcarriers in the medium DME supplemented with 5% fetal calf serum is shown in Figure 2. Cells grew rapidly to approximately 1.8 x 10^6 cells/ml to form the monolayer and then the growth rate decreased, eventually reaching a cell concentration of approximately 5 x 10^6/ml. The glucose concentration decreased at a faster rate immediately after medium change. As the glucose concentration decreases to approximately 1 g/l, the rate of consumption decreased as shown by the decreased slope in the glucose concentration profile. This phenomenon is observed repeatedly every time immediately before medium exchange. The prourokinase concentration increases during the cultivation. When the glucose consumption rate decreased, the rate of prourokinase production also decreased. In our subsequent experiment the medium was replenished when the glucose concentration decreased to below 1.0 g/l. Shown in Figure 3 are the cumulative prourokinase production, cumulative glucose consumption and the cumulative lactate production. The volumetric rate of prourokinase production appears to be relatively constant over a 600-hr period. The overall conversion of glucose to lactate is approximately 30-40% on a weight basis.

The effect of serum concentration on the production of prourokinase was then compared (Figure 4). A one liter microcarrier culture with 5 g/l of Cytodex 3 was established using 5% fetal calf serum. The culture was allowed to reach 2.5 x 10^6 cells/ml before it was divided into four cultures, each with a different serum concentration. No difference in the growth kinetics was observed, eventually all four cultures reached a concentration of approximately 6 x 10^6 cells/ml. The amount of glucose consumed is affected by the serum concentration. At a higher serum concentration, a slightly higher glucose consumption rate was observed (Figure 5). However when the production of prourokinase was compared, a very significant difference among different serum concentrations was observed (Figure 6).

The kinetics of prourokinase production was then examined in a ceramic module (Figure 7). In this case, the cell concentration in the system could not be measured easily and only the prourokinase production, the lactate concentration and the glucose concentration were measured. The medium was recirculated through the ceramic module and was harvested and replenished periodically. The glucose concentration decreased and the lactate concentration and prourokinase concentration increases in between medium replenishments. The prourokinase concentration increases approximately linearly during each medium exchange period. At times occasional decrease of the rate of prourokinase production was observed when the glucose concentration decreases below 1.5 g/l. In both microcarrier culture and ceramic system there appeared to be a correlation between a glucose concentration below 0.5 g/l and a lower prourokinase productivity. Although this correlation may not represent a cause and effect relationship, it was used as an operating criterion both microcarrier

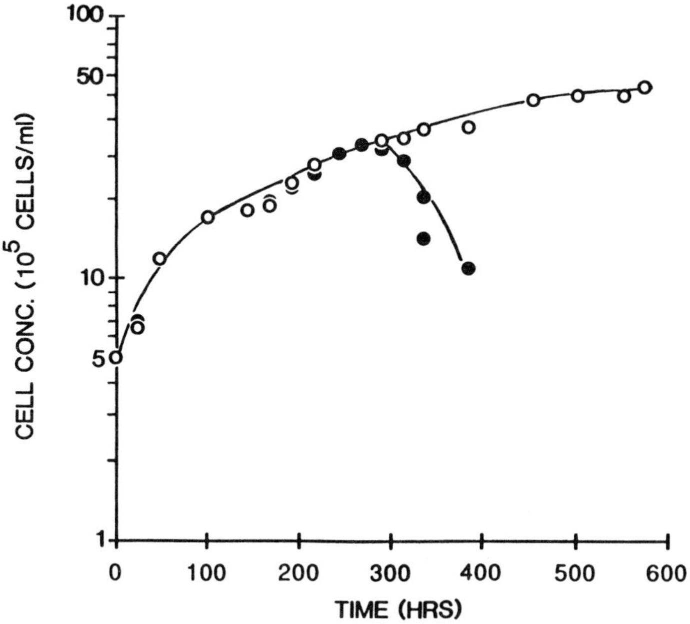

Figure 1. Comparison of growth kinetics of TCL 598-77 cells on Cytodex 1 and Cytodex 3 microcarriers. (•) : Cytodex 1; (o) : Cytodex 3. The growth medium was DME without phenol red supplemented with 5% FBS. The microcarrier concentration was 5 g/l in both cases.

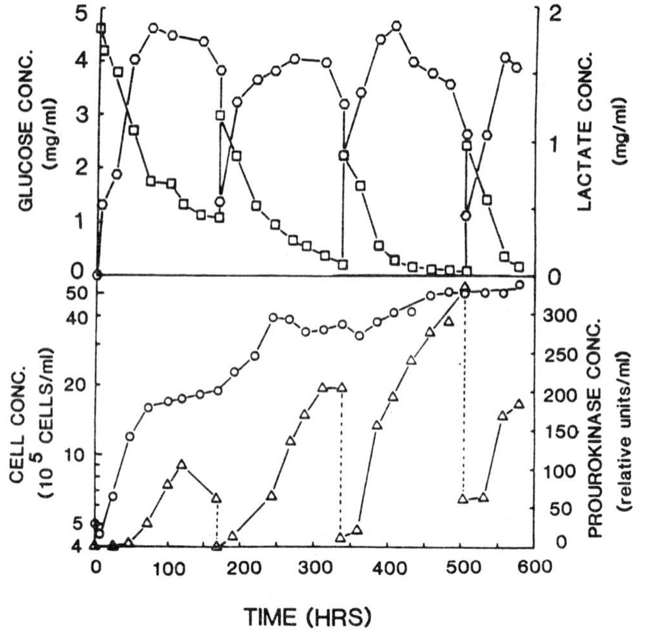

Figure 2. Kinetics of prourokinase production by TCL 598-77 cells grown on Cytodex 3 microcarriers. A microcarrier concentration of 5.0 g/l was used. (o) : cell concentration ; (Δ) : prourokinase concentration; (□) : glucose concentration; (⬡) : lactate concentration.

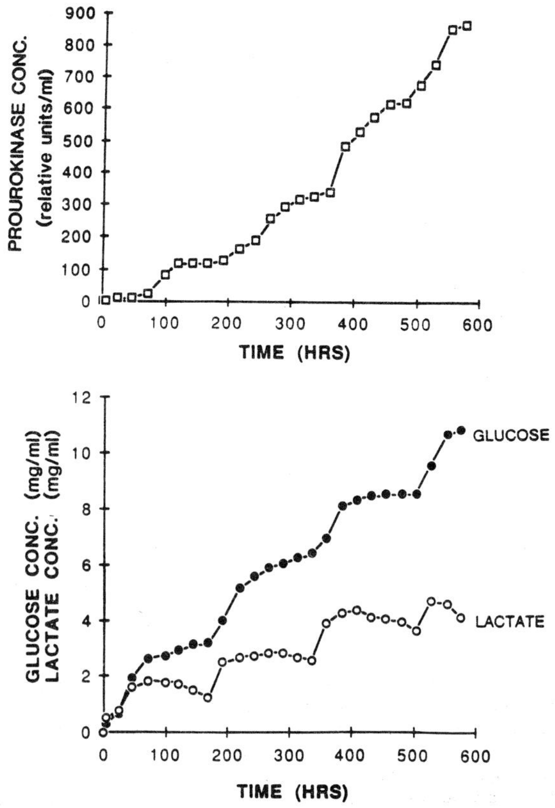

Figure 3. Cumulative prourokinase production, glucose consumption and lactate production of TCL 598-77 cells grown on Cytodex 3 microcarriers. A microcarrier concentration of 5.0 g/l was used. (□) : prourokinase production; (•) : glucose consumption; (o) : lactate production.

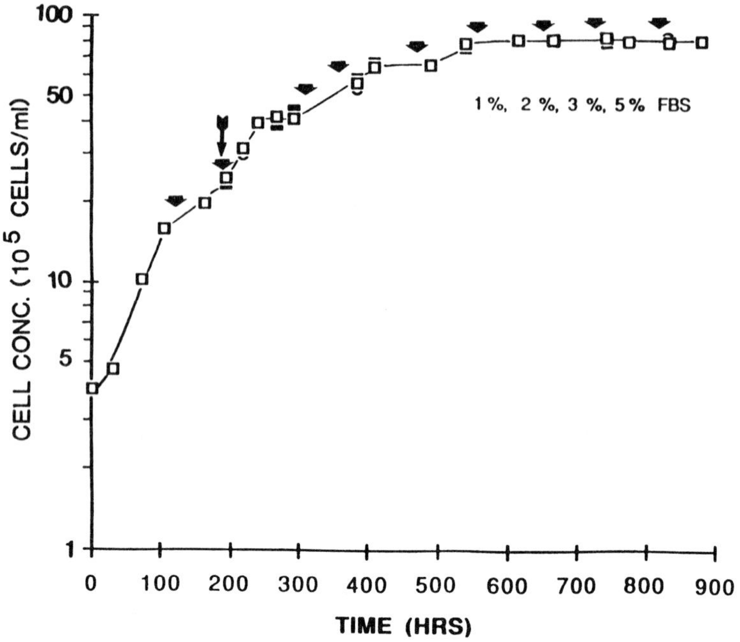

Figure 4. The effect of serum concentration on growth of TCL 598-77 cells on Cytodex 3 microcarriers. A microcarrier concentration of 5.0 g/l was used. The growth medium was DME without phenol red supplemented with 5% FBS at the beginning of a one-liter culture. The culture was divided into four 250 ml cultures at the eighth day (as indicated by the large arrow) with different serum concentrations. Medium was replenished periodically (as indicated by the small arrows) to maintain a glucose concentration above 0.5 mg/ml. The medium for all four cultures were exchanged at the same time. (□) : 5% FBS; (□) : 3% FBS; (o) : 2% FBS; (•) : 1% FBS.

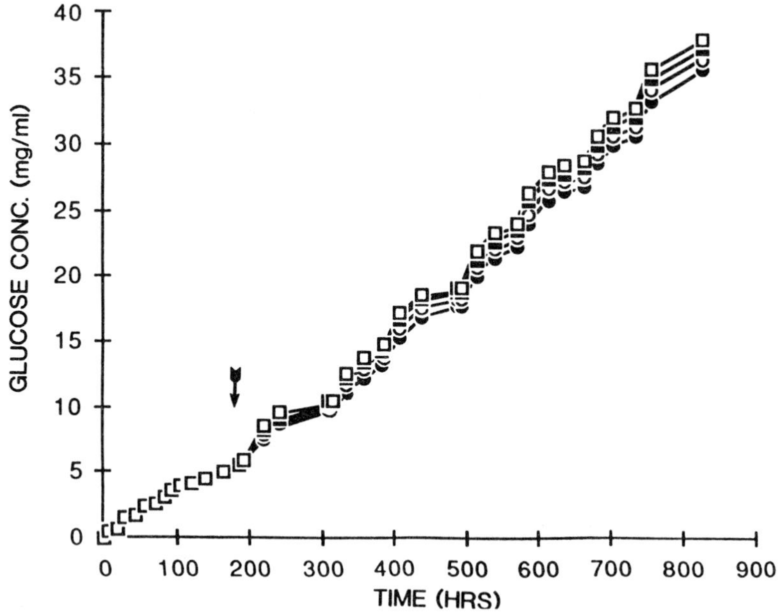

Figure 5. The effect of serum concentration on cumulative glucose consumption of TCL 598-77 cells grown on Cytodex 3 microcarriers. (□) : 5% FBS; (□) : 3% FBS; (o) : 2% FBS; (•) : 1% FBS.

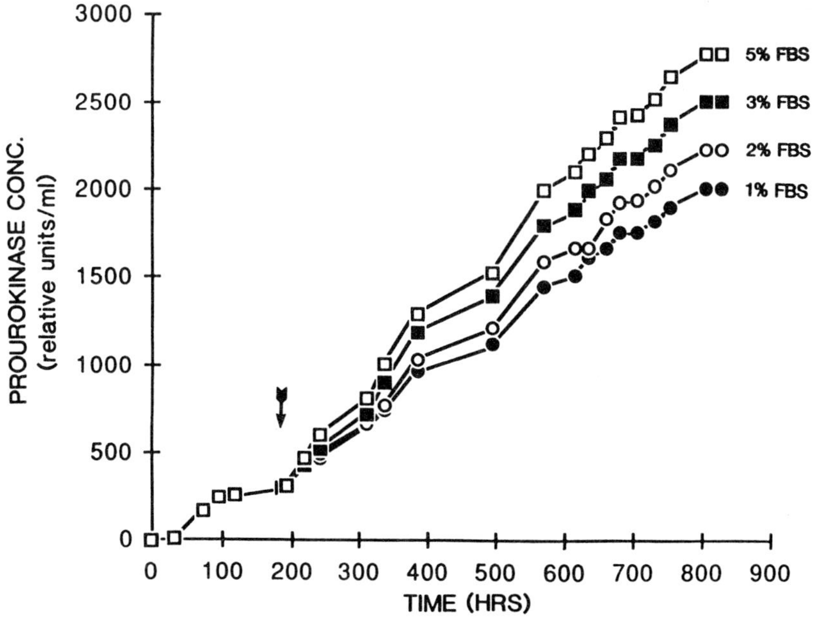

Figure 6. The effect of serum concentration on cumulative prourokinase production of TCL 598-77 cells grown on Cytodex 3 microcarriers. (□) : 5% FBS; (□) : 3% FBS; (o) : 2% FBS; (•) : 1% FBS.

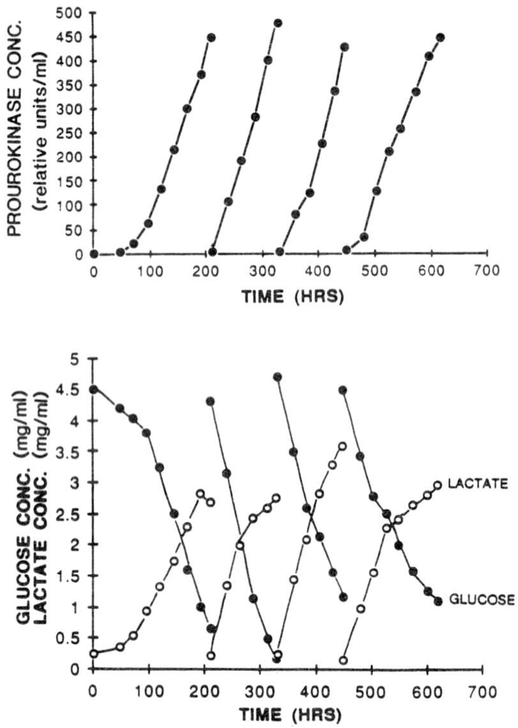

Figure 7. The kinetics of glucose consumption, lactate, and prourokinase production by TCL 598-77 cells cultivated in a ceramic system. The growth medium was DME supplemented with 1% FBS. (•): glucose; (o): lactate; (•): prourokinase.

culture and in ceramic system. The medium was replenished when the glucose concentration decreased to between 0.5-1 g/l.

The cumulative prourokinase and lactate production as well as cumulative glucose consumption are shown on Figure 8. In all three cases the cumulative concentration increased initially at a exponential rate for a very short period and then the increase appears to be at a constant rate over the cultivation period. The conversion of glucose to lactate in the ceramic module is approximately 60-70% which is higher than that observed in the microcarrier culture system.

Shown on Figure 9 are the results from the roller bottles. In this case the cell concentration also cannot be measured easily and only glucose, prourokinase, and lactate concentrations are reported. When the glucose consumption and lactate production were compared 60-70% of conversion of glucose to lactate was observed.

In the three cases examined above, namely microcarrier culture, ceramic system and roller bottles, the cell concentration per unit of volume of medium are different in all three cases. The durations in between medium exchanges are also different in the three cases. Therefore a rational approach has to be taken to compare the three systems. The amount of prourokinase accumulated during each medium exchange period can be related to the specific productivity of the cells and the cell concentration as shown in Equation 1.

$$V(P-P_e) = \int_{t_{i-1}}^{t_i} q_p X dt \tag{1}$$

In Equation 1 P is the concentration of prourokinase accumulated in the medium at the time of medium replenishment, P_e is the residual prourokinase concentration after the previous medium replenishment. In our case, P_e is negligible. X is the total number of cells in the system, V is the amount of medium changed each time of the medium exchange. After i times of medium exchange the total amount of prourokinase produced can be expressed as Equation 2.

$$\sum_{i=1}^{i} V_i P_i = \sum_{i=1}^{i} \int_{t_{i-1}}^{t_i} q_p X dt = \int_{0}^{t_i} q_p X dt \tag{2}$$

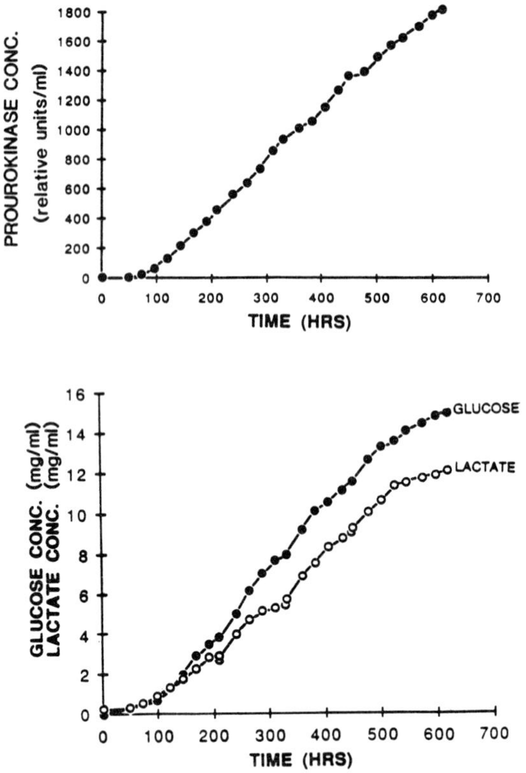

Figure 8. Cumulative prourokinase production, glucose consumption and lactate production in a ceramic system. (•): glucose; (o): lactate; (•): prourokinase.

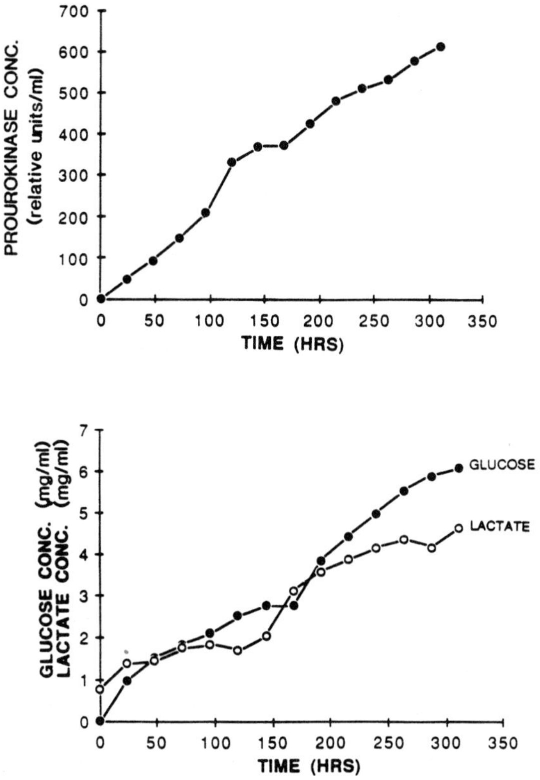

Figure 9. Cumulative glucose, lactate and prourokinase concentrations of TCL 598-77 cultures in roller bottles. (•): glucose; (o): lactate; (•): prourokinase.

Since the amount of medium replenished each time is constant ($V_i = V$ = constant), Equation 2 becomes

$$\sum_{i=1}^{i} P_i = \frac{\int_0^{t_i} q_p X dt}{V} \qquad (3)$$

From Equation 3 it can be seen that a better comparison of the three systems is to compare the cumulative prourokinase concentration, $\sum P_i$ in the medium (Figure 10). As can be seen, those three kinetic curves from roller bottle, microcarriers, and ceramic systems, are very similar to each other. The difference between ceramic module and microcarrier culture is in the range of the standard deviation of prourokinase assay. In all three cases the increase in prourokinase concentration appears to be linear over an extended period. However, during that period the cell concentration was not constant but increasing. Therefore, the specific rate in the three cases was most likely to decrease during cultivation. Since the cell concentration is available only in microcarrier culture systems, only the specific productivity from the microcarrier culture system was calculated (Table 1). The specific productivity of prourokinase at all four different serum concentrations decreased gradually to below 50% of the initial value after 600 hours.

CONCLUSION

We have examined the prourokinase production in three different cultivation systems using the same cell line. We have demonstrated that if the intrinsic kinetics of growth and production are maintained in different reactor systems, the performance of the same cell type in these reactors are the same. However in such a system the specific productivity of prourokinase decreases as a function of time.

REFERENCES

Hu, W-S., Meier, J. and Wang, D.I.C. (1985). Biotechnol. Bioeng. 27, 585.

Lydersen, B.K., Pugh, G.C., Paris, M.S., Sharma, B.S., and Noll, L.A. (1985). Biotechnol, 3, 63.

Kohno, T., Hopper, P., Lillquist, J.S., Suddith, R.L., Greenlee, R., and Moire, D.T. (1984). Biotechnology, 2, 688.

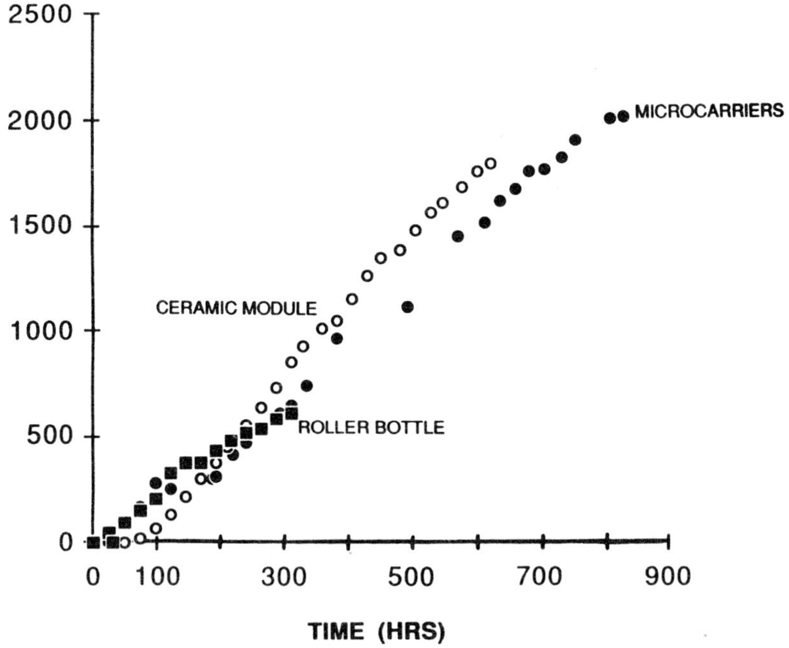

Figure 10. Comparison of the cumulative prourokinase production in the three cultivation systems. The growth medium was DME supplemented with 1% FBS. (•) : 5 g/l Cytodex 3 microcarriers; (o) : ceramic module K16; (□) : roller bottle.

TABLE 1. SPECIFIC PROUROKINASE PRODUCTIVITY (10^{-6} UNITS/CELL - DAY)

	CULTIVATION PERIOD		
% SERUM	192-387 HOURS	387-570 HOURS	570-827 HOURS
1%	20	9	6
2%	22	11	7
3%	27	12	8
5%	30	14	9

Discussion

R. SPIER	When you did your pro-urokinase assay with different quantities of serum did you compensate for your serum?
HU	Yes, but in fact the background was very small.
W. TOLBERT	What concentration of microcarrier did you use?
HU	$5_g/1$.
TOLBERT	Did you look at the activation of pro - to urokinase in the different systems?
HU	Very little difference - always between 90 to 95%.
O. MERTEN	It appeared that production was growth associated.
HU	No, I wouldn't agree with that.
MERTEN	When I used this cell line in T-flasks it was very growth associated - the higher the growth rate the higher the production per cell.
HU	My data is not sufficient to conclude that.

26 Specific kinetic patterns of production of monoclonal antibodies in batch cultures and consequences on fermentation processes

O-W. Merten, G. E. Pälfi, G. Klement and F. Steindl

Summary:

Studying the kinetic of production of monoclonal antibodies in batch cultures three different specific production patterns (expressed as pg IgG/cell x hour) were found: I. the highest cell specific IgG-production during lag and beginning of log phase; II. like I., but in addition a second cell specific peak production at the end of the exponential and the onset of the stationary and death phase; III. quite constant cell specific production during the growth. The IgG-secretion stopps when the stationary phase is reached.

The consequences of these production kinetics on two different process modes (continuous, repeated batch culture) of low cell density fermentations (up to $3x10^6$ cells per ml) are shown for all groups. Cells of group III should be cultivated in a continuous mode, cells of group II in a repeated batch mode (for example). Cell lines of group I should be cultivated in a repeated batch mode or in a continuous process mode, in dependence on their production and growth kinetics.

Introduction:

The development of the monoclonal antibody technology by Köhler and Milstein (1) provided, for the first time, the possibility of producing monoclonal antibodies recognizing specific antigens of choice. The exquisite and nearly invariable specificity of the monoclonal antibodies combined with the potential of producing them in unlimited quantities made them very interesting for many aeras of biological, biochemical, and medical research. With the begin of their commercialization high amounts of more

or less highly purified monoclonal antibodies are required, whose production is done by the use of different fermentation systems (for review, see: Merten (2)). In any case, the fermentation process has to be designed in such a way that the biological potential of the cell is exploited in an optimal way. Besides the optimization of the technique and the medium formulation it is necessary to know how the production and growth kinetics of a certain cell line look like. This knowledge provides the information necessary for the design of the process mode (e.g.: continuous or discontinuous culture, batch, fed batch mode, etc.) using low cell density (up to $3x10^6$ cells/ml) or high cell density (more than $3x10^6$ cells/ml) production systems. For investigating this field, three different cell lines, showing one of the three different batch production kinetics, were chosen for the realization of fermentation processes, run in a repeated batch and/or continuous process mode. The results, obtained from these fermentations, were compared with each other and with published data.
The probable consequences for high cell density production processes are discussed in brief because of the lack of sufficient information.

Materials and Methods:

Hybridoma cell lines: Three different mouse-mouse hybridoma cell lines, producing monoclonal antibodies (IgG) were used: 371, 1663, and U0208 (Katinger, pers. commun.). The cells were cultivated in DMEM (GIBCO: 041-1965), supplemented with 5% fetal calf serum (FCS)(GIBCO) in Roux bottles (Nunc 1-53732) at 36.5°C in a 5%CO_2-atmosphere. The cell line 1663 was cultivated in DMEM supplemented with 10%FCS. Every third or fourth day the cultures were split in a ratio of 1/3 to 1/5. The cell lines were free of mycoplasma, which was tested using the method of Chen (3).
Fermentor cultivation: The cell lines 371 and 1663 were cultivated in 4.7 l Chemap airlift fermentors, the cell line U0208 in a 10 l Chemap-CF airlift fermentor. The medium was DMEM-5%FCS for the cell lines 371 and U0208. Cell line 1663 was cultivated in DMEM-10%FCS. Generally, the fermentors were run using the following conditions: pH 6.9-7.1; temperature 36.5°C, pO_2 about 50-60% air-saturation for the cell lines 371 and U0208 and 30% air-saturation for the cell line 1663. The pH regulation was done via the injection of CO_2 into the gas flow which was about 0.6

fermentorvolumes per hour. The fermentors were run in a continuous process mode with a dayly continuous medium change of 30-64% or in a repeated batch mode with a discrete medium change of 10-70% every first to fourth day, with respect to the condition of the culture.

Assays: For the determination of the cell density, viability, and IgG-concentration dayly samples of 5ml of the cell suspensions were taken. These determinations and the following calculations (viability, growth rate, and specific production rate) were done, according to Merten (4). In the case of the continuous fermentations, the dilution rate was taken into account for the calculation of the growth rate and the specific production rate.

Results:

Batch production and growth kinetics of hybridomas: Three different batch production kinetics of hybridoma cell lines were found up to now (4), which are summarized in fig.1. These kinetic patterns can be distinguished by the following characteristics: Cell lines, showing production pattern I (fig.1A), show the highest cell specific Ig-production during lag and beginning of the log phase. The cell specific production is lowering down and does not reincrease during the stationary growth and death phase of the batch culture. Cell line 371 is showing this production pattern (4).

Cell lines, which follow the production pattern II (fig.1B), show pattern I at the beginning of the batch culture, but in addition they have a second cell specific peak production at the end of the exponential growth and the onset of the stationary growth and death phase. The cell line U0208 shows the production pattern II (Merten, unpublished results).

The cell specific production pattern III (fig.1C) is characterized by a quite constant cell specific production during the growth. The Ig-secretion stopps when the stationary growth phase is reached, no considerable reincrease of the production is detectable during the death phase. Cell line 1663 shows this production pattern (4). The possible influences on these production kinetics were already discussed by Merten (4).

Airlift-fermentations: The influences of these batch production kinetic patterns on fermentation processes and the process modes were tested by the use of the three above mentioned cell lines,

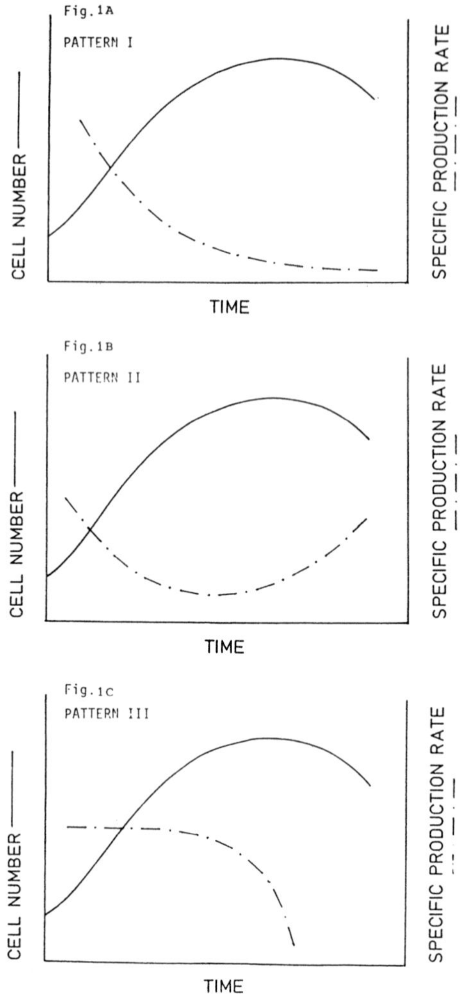

1. Different production kinetics of hybridomas in batch culture systems: A: Pattern I; B: Pattern II; C: Pattern III. In all cases, the cell growth and the cell specific production rate are shown.

all showing one of the production kinetic patterns. In all cases two different process modes were used, the repeated batch mode and the continuous culture. In the case of cell lines 371 and 1663, this was done subsequently during the same fermentation process, in the case of U0208, two different fermentations were done.

Cell line 371 was cultivated in a fermentor for about 900h (fig.2). This process was started with three repeated batch cycles during the first 260h, which were followed by a continuous cultivation process up to 870h. The top diagramme in fig.2 which presents the viability and the viable cell count per ml, shows apparently that the cell density increased during the continuous cultivation. The maximal cell densities were about $7.5x10^5$ and $1.45x10^6$ cells per ml for the repeated batch and the continuous culture, respectively. The viability ranged between 70 and 90% normally, between 130 and 250h it was lower because of a too late change of the medium. The product accumulation (mg IgG/l) and the specific productivity (pg IgG/cell x hour) are shown in the middle diagramme (fig.2).It is evident that this cell line showed a really better IgG-accumulation during the repeated batch (average 35.5mg/l) than during the continuous culture (average 21.4 mg/l). This is also valid for the cell specific productivity. The average was 0.658 and 0.279 pg IgG/cell x hour for repeated batch and continuous culture, respectively. The growth rate of the cells (bottom diagramme of fig.2) was influenced by the different process modes. According to the characteristics of the batch production pattern I (4) the maximum specific productivity was always some hours in advance compared with the maximal growth rate at the onset of the repeated batch cycles.

The repeated batch culture of U0208 (batch production pattern II) is shown in fig.3, the continuous culture in fig.4. The repeated batch culture lasted for 700h, whereby 5 cycles were done. Because of the irregular medium change the cell density and the viability ranged from $1x10^5$ to $2x10^6$ cells/ml and from 30 to 90%, respectively (top diagramme of fig.3). The IgG-accumulation ranged between 30 and 600 mg/l, whereby the production was increased during the stationary growth and death phase of the culture (compare top and central diagramme of fig.3). The cell specific productivity ranged between 0.2 and 9.4 with one exception of 63.7 pg IgG/cell x hour. This very high value and the peak of 9.4 at 684h were accompanying the death phase of the culture, which is

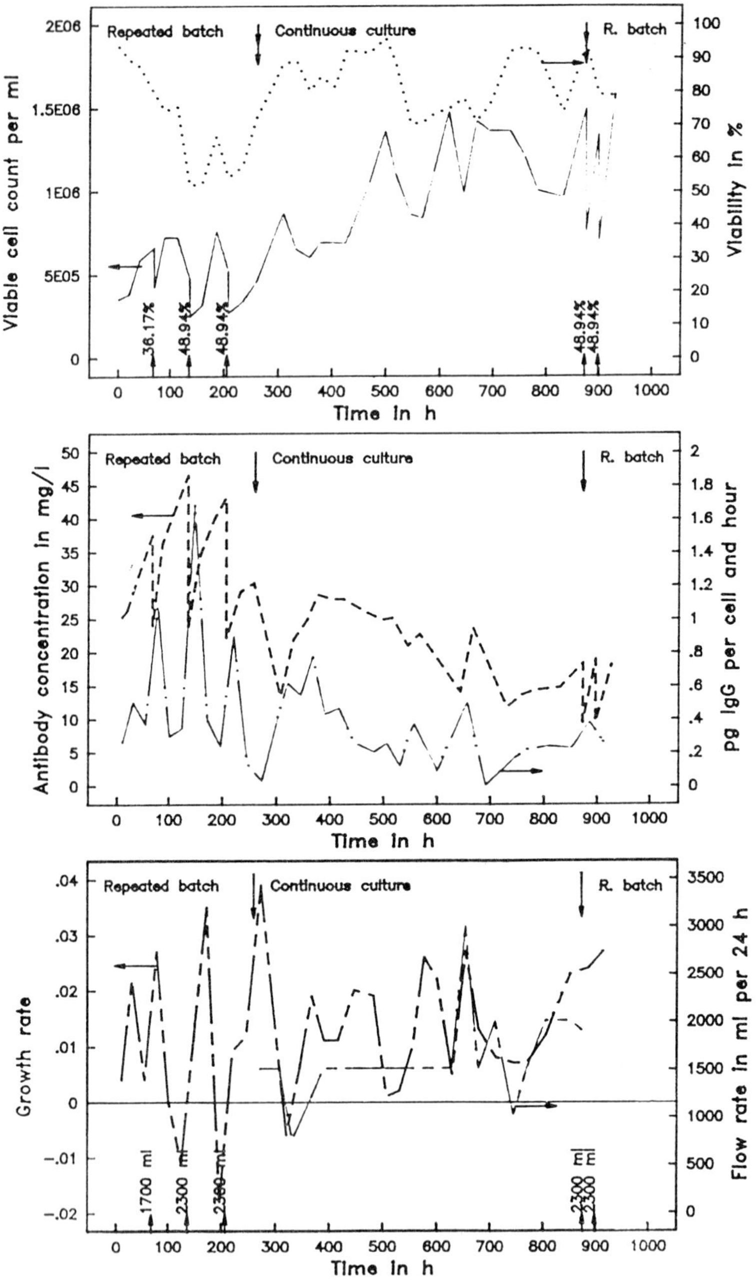

2. Cultivation of cell line 371 in a 4.7 l airlift fermentor.
Top diagramme: The viable cell count per ml and the viability (%) of the cells are shown. The percentage of the medium change during the repeated batch cycles is indicated.
Middle diagramme: The antibody concentration in mg/l and the

cell specific productivity (pg IgG/cell x hour) are shown. Bottom diagramme: The growth rate and the flow rate in ml per 24h (only during the continuous cultivation) are shown. The volume of the changed supernatant during the repeated batch cycles is indicated.

Note: In all diagrammes, the change from repeated batch to continuous culture and back is indicated.

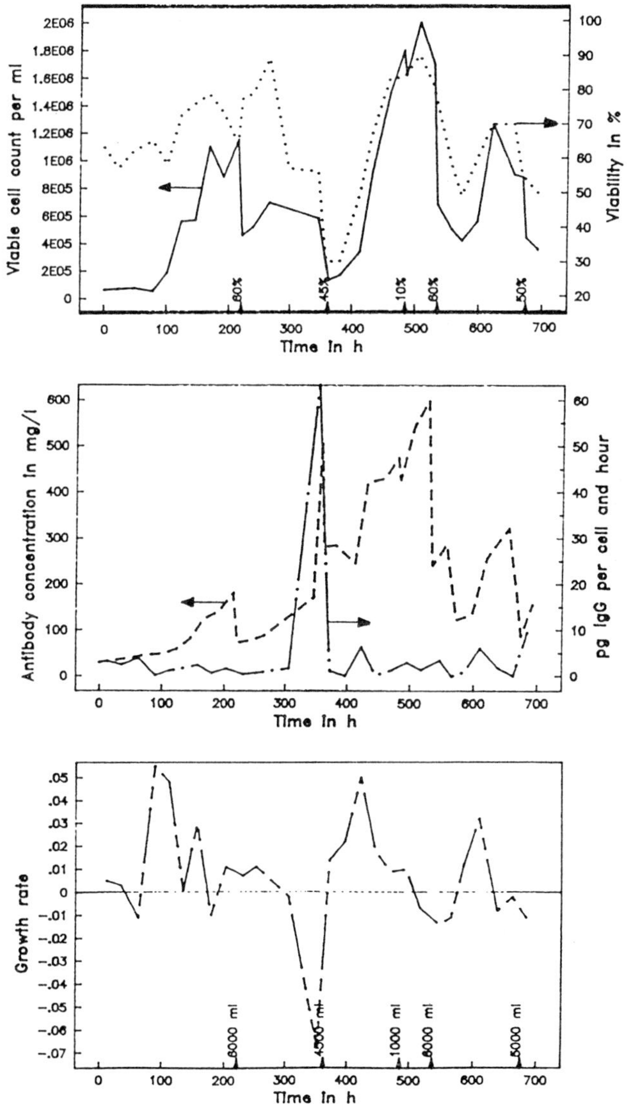

3. Cultivation of cell line U0208 in a 10 l airlift fermentor in repeated batch mode. (For description, see fig.2)

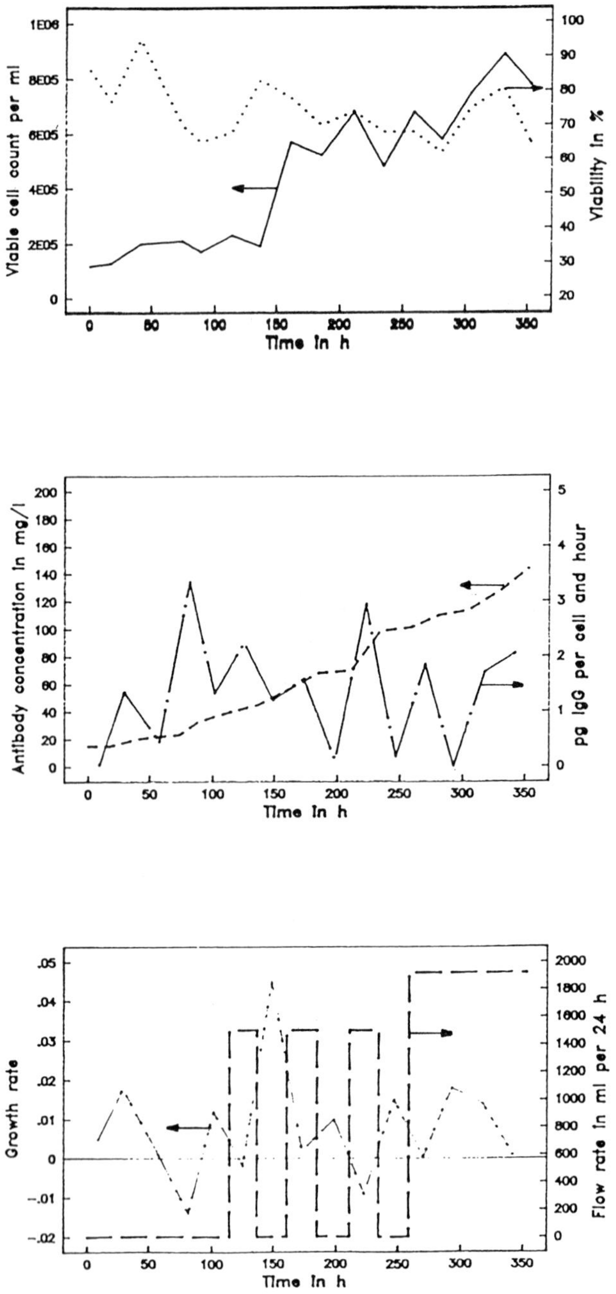

4. Continuous cultivation of cell line U0208 in a 10 l airlift fermentor.(For description, see fig.2)

indicated by the lowering down of the cell viability (top diagramme of fig.3). The specific production peaks of 6.3 at 423h and 6.1 at 610h were accompanying the onset of a new repeated batch cycle after a long dying phase. These four cell specific peak productions are in agreement with the characteristics of the batch production pattern II (4).

A second fermentation of U0208 was a continuous culture of 350h (fig.4). After a long lag phase the cells started to grow, cell densities of up to 0.9×10^6 cells/ml were achieved. The viability ranged between 60 and 80% (both shown in the top diagramme of fig.4). The IgG-accumulation increased during the culture up to 140mg IgG/l, the specific productivity ranged between 0.001 and 3.35 pg IgG/cell x hour. The comparison of the specific productivity and the growth rate (middle and bottom diagramme of fig.4) showed clearly that both curves were inverse to each other (except for the first peak of the growth rate and the specific productivity), which is in accordance with the production pattern II characteristics (4).

The comparison of both process systems for this cell line shows the advantage of the repeated batch mode for U0208: It is evident that the cell density was superior in the repeated batch system than in the continuous culture. The average IgG-accumulation was 208.5 and 67.3 mg IgG/l and the average specific productivity was 4.409 and 1.364 pg IgG/cell x hour for the repeated batch and the continuous culture, respectively.

Cell line 1663, showing batch production pattern III, was cultivated in an airlift fermentor (fig.5) in a repeated batch mode for about 580h (8 repeated batch cycles) and subsequently in a continuous process mode for about 1000h. It is evident that the cell densities were more or less equal in both different process modes, the viability of the cells was slightly higher during the continuous culture in comparison with that of the repeated batch system (top diagramme of fig.5). Both, the IgG-accumulation and the cell specific productivity, were elevated during the continuous process (middle diagramme of fig.5). The average IgG-concentration was 71.3 mg IgG/l for the repeated batch and 153.8 mg IgG/l for the continuous culture. The average specific production rates were 0.866 and 1.566 pg IgG/cell x hour for the repeated batch and the continuous process mode, respectively.

During the continuous culture the product concentration was decresing after 1350h because of the increase of the dilution rate.

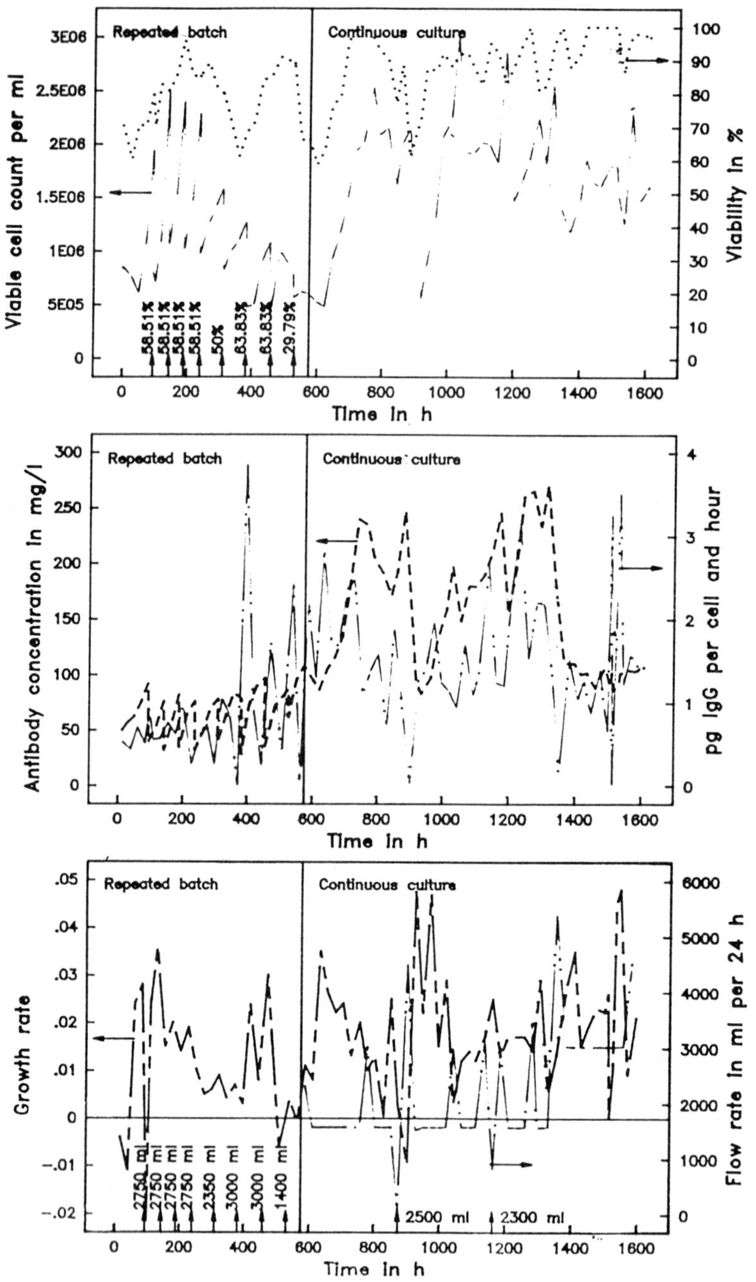

5. Cultivation of cell line 1663 in a 4.7 l airlift fermentor. (For description, see fig.2)

Note: In all diagrammes, the change from repeated batch to continuous culture is indicated.

The growth rate was not affected by the different process modes (bottom diagramme of fig.5). During the continuous culture two additional medium changes took place (870h and 1160h) which became necessary because of the failure of the medium pump.
The summarized results of the fermentations and batch cultures (4; Merten, unpublished results), done with the three abovementioned cell lines, are presented in table I.

Discussion:

The three different batch production kinetics were described and summarized by Merten (4): Pattern I and III were only found relatively rarely (4-7: pattern I; 4,7,8, Henno, pers. commun.: pattern III), whereas most of the published hybridoma cell lines showed production kinetic pattern II (4,9-17). This may be one of the reasons why only few attention was paid for the design of a low cell density fermentation process (a homogeneous process without cell retention), and why only few considerations were done about the optimal process mode of a hybridoma fermentation up to now. However, the results (table I) proove that the variation of the process mode can significantly change the output of this fermentation. The implications of the three different batch production patterns on the design of a fermentation process are apparent when comparing the fermentation results (fig.2-5) with the batch growth curve and cell productivity patterns, shown in fig.1. Generally, the cells have to be kept in a physiological state where they show their highest product secretion. This means for cell lines, showing production pattern I (cell line 371) that they have to be kept in a physiological state which is equivalent to that of the beginning of the log phase of a growing batch culture. These conditions are provided by batch and repeated batch systems (see fermentation of 371 in fig.2 and table I), and to some extent by a continuous culture, which is depending on the steepness of the specific production curve during the log-phase, however.
Cell lines, showing production pattern II (like U0208), have to be kept in a physiological state which is equivalent to that of the beginning of a batch culture or more practically which corresponds to that of the stationary growth and death phase of a batch culture. The second case is preferable because this condition can be provided by a lot of different process modes, easily:

Table I. Comparison of batch (Rouxbottle-culture), repeated batch, and continuous culture (both: cultures in airliftfermentors) without cell retention:

Process mode	cell line (Pr.Pat[1])	Max. cell-density/ml	Max. mg IgG/l	average mg IgG/l	Specific production pg IgG/cell x hour[2]		Reactor-vol.(l)	Time (h)	Perfusion ml/l,d
batch	371 (I)	$2x10^6$	50.3	38.3	0.066-1.212	(0.307)	0.05	125	-
	U0208 (II)	$0.4x10^6$	43.7	30.8	0-2.593	(1.851)	0.05	115	-
	1663 (III)	$1.24x10^6$	74.9	49.8	1.249-1.905	(1.614)[3]	0.05	138	-
repeated batch	371	$0.75x10^6$	46.5	35.5	0.260-1.724	(0.658)	4.7	288	135
	U0208	$1.8x10^6$	599	208.5	0-63.68	(4.409)	10	936	about 60, variable
	1663	$2.5x10^6$	108	71.3	0-2.426	(0.866)	4.7	600	192, but variable
continuous	371	$0.8-1.5x10^6$	30.4	21.4	0-0.774	(0.279)	4.7	576	331
culture	U0208	$0.6-0.9x10^6$	143.4	67.3	0-3.354	(1.364)	10	384	70, but variable
	1663	$2-3x10^6$	271.4	153.8	0-3.525	(1.566)	4.7	1000	403

Notes: 1) Pr.Pat.: Production pattern of the batch cultures is indicated in parenthesis behind the name of the cell lines used.

2) Values in parenthesis indicate the average values

3) Only the values during lag-, log-, and end of the log- phase; including the values of the stationary and death phase, this average value is about 1,000

batch, fed batch, repeated batch, or systems with 2-stages, whereby the first one provides the maximal cell growth and the second one the optimal/maximal production during the death phase (2-stage repeated batch like done by Reuveny et al. (12) or 2-stage continuous culture). In any case, the growth and the cell specific production have to be separated from each other, to get an optimal cell specific productivity. This is prooved by the fact that the cell line U0208 showed always a higher cell specific production when the cell growth was retarded (in the repeated batch and continuous culture)(fig.3, 4). The comparison of the repeated batch and the continuous culture (fig.3, 4, table I) shows that a non continuous culture system is preferable because of the better productivity and the higher achieveable product concentrations. This is in agreement with others, like Birch et al. (10,18), Reuveny et al. (12), and Boraston et al. (15), who used non continuous cultivation systems for the production of monoclonal antibodies with throughout good results, because of the probable fact that only cell lines, showing batch production pattern I or mainly II, were used (Reuveny et al. (12)'s cell line followed production pattern II).
Summarizing: discontinuous process modes, like batch, fed batch, repeated batch, etc., are always a good choice for the production of monoclonal antibodies, using cell lines showing production pattern II and to some extent pattern I. Because of the fact that most of the published hybridoma cell lines show production pattern II, it was not necessary to pay much attention to the production kinetics (already abovementioned). Low cell density cultivation systems were used and scaled up to thousand liters to produce monoclonal antibodies in a batch mode (Birch et al. 10, 18). Others, like Reuveny et al. (12) compared different process modes for a cell line, showing production pattern II, by ignoring the classical continuous culture. The process modes: batch, fed batch, repeated batch with/without a second stage were supposed as to be superior which is correct in the case of cell lines, showing production pattern II (fig.1).
Another fact prevented and prevents the use of a continuous culture partly. The production of monoclonal antibodies for human therapy has to be controlled rigidly. The possibility of a change of the cell population during the continuous fermentation process requires a lot of additional control of the cells itself, at the onset, the ongoing, and the end of the process, which in-

creases the production costs. Therefore, a limited production process, like a batch process, is preferable to some extent. Cells, showing production pattern III (1663, fig.5), have the best production during their growth phase. The best cultivation mode is a continuous culture because it provides a continuous growth and therefore a good continuous production (table 1).
A minority of cell lines show production pattern III. Only Fazekas de St. Groth (8) published the application of a continuous culture for the production of monoclonal antibodies and received good results. However all cell lines used showed production pattern III, and the continuous culture was preferable, therefore. Unfortunately, no comparison between different process modes (continuous versus discontinuous) was done.
The probable future process will be a high cell density cultivation system (for review, see: Merten 2). Hereby, the production kinetics may be changed. Altshuler et al. (9), Reuveny et al. (12), Emery (19), and van Wezel et al. (20) found a 2 to 10 fold increased cell specific production in high density systems (agitated perfusion, hollow fiber), in comparison with low density systems. Assuming, that the differences between high and low density cell cultures are not too great, the abovementioned implications of the batch production kinetics might be extended for high cell density systems, too. In any case, the cells have to be hold in a physiological state which is optimal for the overall production. The cell lines used by Altshuler et al. (9), Reuveny et al. (12), Emery (19), and van Wezel et al. (20) showed batch production pattern II, and promising results were obtained from high cell density cultures in all cases. Feder and Tolbert (21) used successfully high density culture systems, too; unfortunately, the batch production patterns of their cell lines are not known.
However, up to now, it is too early to give a more detailed view for high cell density production processes because the cell physiology might differ in these systems from that in low cell density systems, which was already mentioned.
Final remarks: The growth and production kinetics of cell lines are the most important parameters to be considered when monoclonal antibodies have to be produced in an industrial scale. Like for microbial fermentations, the most optimal process has to be designed for economical reasons. Hereby, the production kinetics in combination with the growth kinetics of a batch culture give already implications for the preferable process mode of an indu-

strial fermentation process.

References:

(1) Köhler, G., Milstein, C. (1975). Continuous cultures of fused cells secreting antibody of predefined specificity. Nature 256, 495-497.

(2) Merten, O.-W. (1986). The use of different types of fermentor for the production of monoclonal antibodies. Presented at:"2nd European Symposium on Protein Purification Technologies", 29.9.-2.10.86, Nancy, F.

(3) Chen, T.R. (1977). In situ detection of mycoplasma contamination in cell cultures by fluorescent Hoechst 33258 stain. Exp.Cell Res. 104, 255-262.

(4) Merten, O.-W. (1987). Batch production and growth kinetics of hybridomas. in press.

(5) Williams, JA. (1984). Effects of medium concentration on antibody production. J.Tissue Culture Methods 8, 115-118.

(6) Lavery, M., Kearns, M.J., Price, D.G., Emery, A.N., Jefferis, R., Nienow, A.W. (1985). Physical conditions during batch culture of hybridomas in laboratory scale stirred tank reactors. Develop. biol. Standard. 60, 199-206.

(7) Merten, O.-W., Reiter, S., Himmler, G., Scheirer, W., Katinger, H. (1985). Production kinetics of monoclonal antibodies. Develop. biol. Standard. 60, 219-227.

(8) Fazekas de St. Groth, S. (1983). Automated production of monoclonal antibodies in a cytostat. J.Immunol.Method. 57, 121-136.

(9) Altshuler, G.L., Dziewulski, D.M., Sowek, J.A., Belfort, G. (1986). Continuous hybridoma growth and monoclonal antibody production in hollow fiber reactors-separators. Biotechnol. Bioeng. 28, 646-658.

(10) Birch, J.R., Thompson, P.W., Boraston, R., Oliver, S., Lambert, K. (1986). The large scale production of monoclonal antibodies in airlift fermenters. Presented at:"UMIST symposium on process possibilities for plant and animal cell cultures", 25.3.-26.3.86, Manchester, GB.

(11) Emery, A.N., Lavery, M., Williams, B., Handa, A. (1986). Large-scale hybridoma culture. Presented at:"UMIST-symposium on process possibilities for plant and animal cell cultures", 25.3.-26.3.86, Manchester, GB.

(12) Reuveny, S., Velez, D., Miller, L., Macmillan, J.D. (1986).

Comparison of propagation methods for their effect on monoclonal antibody yield in fermentors. J.Immunol.Method. 86, 61-69.

(13) Velez, D., Reuveny, S., Miller, L., Macmillan, J.D. (1986). Kinetics of monoclonal antibody production in low serum growth medium. J.Immunol.Method. 86, 45-52.

(14) Reuveny, S., Velez, D., Riske, F., Macmillan, J.D., Miller, L. (1985). Production of monoclonal antibodies in culture. Develop.biol.Standard. 60, 185-197.

(15) Boraston, R., Thompson, P.W., Garland, S., Birch, J.R. (1984). Growth and oxygen requirements of antibody producing mouse hybridoma cells in suspension cultures. Develop.biol.Standard. 55, 103-111.

(16) Reuveny, S., Velez, D., Macmillan, J.D., Miller, L. (1986). Factors affecting cell growth and monoclonal antibody production in stirred reactors. J.Immunol.Method. 86, 53-59.

(17) Merten, O.-W., Reiter, S., Katinger, H. (1985). Stabilizing effect of reduced cultivation temperature on human x mouse hybridomas. Develop.biol.Standard. 60, 509-512.

(18) Birch, J.R., Boraston, R., Wood, L. (1985). Bulk production of monoclonal antibodies in fermenters. Trends Biotechnol. 3, 162-166.

(19) Emery, A.N. (1986). Growth of hybridomas and secretion of monoclonal antibodies in vitro. Presented at the Society of Chemical Industry-Symposium on Large-scale production of monoclonal antibodies, 9.12.86, London, GB.

(20) van Wezel, A.L., van der Velden-de Groot, C.A.M., de Haan, H.H., van den Heuvel, N., Schasfort, R. (1985). Large scale animal cell cultivation for production of cellular biologicals. Develop.biol.Standard. 60, 229-236.

(21) Feder, J., Tolbert, W.R. (1985). Mass culture of mammalian cells in perfusion systems. Am.Biotechnol.Lab. 3, 24-36.

27 A model system to study gene expression and scale up of microcarrier cultures of recombinant DNA manipulated cells

B. Lundgren and S. Nilsson

ABSTRACT

In our model system clones of Hygromycin B resistant C127 cells, constitutively expressing a secretable form of Staphylococcus nuclease, were cultured on Cytodex 3. Serial subcultivations of nearly confluent cell cultures were performed to simulate a scale up of the production culture volume. Following the final subcultivation step, cells were maintained on the microcarriers at stationary phase for production studies both in the presence and absence of a serum supplement. No changes in cell doubling times were observed during the subcultivation period. Constant final cell densities were obtained and maintained througout the production period. The gene expression and the stability of the clones were monitored by assaying the amount of secreted S. nuclease. The rate of gene expression remained at a constant level throughout the production phase in the presence of 5% serum but was slightly decreased in the absence of a serum supplement.

INTRODUCTION

Recombinant DNA techniques have made it possible to produce large amounts of desired proteins expressed from cloned genes in an appropriate host cell. At first it was assumed that prokaryotic hosts like E.coli would be the best choice for production. However, this has not turned out to be so. Several problems have emerged e.g.accumulation of expressed proteins in a denatured, insoluble form which increases the complexity of purification. Furthermore, there is a lack of posttranslational modifications such as glycosylation, which is sometimes necessary for biological activity, and secretion cannot generally be performed in bacteria. The alternative hosts for production of valuable proteins are then yeast or mammalian cells. Both yeast and mammalian cells are capable of secretion and glycosylation of

proteins. However, the patterns of glycosylation and macromolecular assembly in yeast are sometimes different from those which occur in mammalian cells (1,2)
We have set up a model system to study the expression of a recombinant gene, kappa-nuclease (knuc), from stably transfected mouse C127 cells cultured on microcarriers in lab-scale spinner cultures. The microcarrier cultures were subcultivated three consecutive times to simulate a scale-up of production culture volumes of up to 160 litres. The following parameters were studied during the time of the experiment: 1) the growth properties during exponential phase of the recombinant cell line C127:pPyL-knuc compared with unmodified C127 cells, 2) serum requirement at cell confluence without loss of cells, 3) stability of the gene expression between subcultivations and at the stationary phase with or without serum.

MATERIALS AND METHODS

Cells

C127 cells (5) and the recombinant cell line C127:pPyL-knuc were propagated in Dulbecco's modified Eagle medium (DMEM, Gibco) supplemented with 5% fetal calf serum (FCS, Gibco).

Plasmids

The reporter gene (knuc) is a fusion between the coding sequences of the rat immunoglobulin kappa light chain signal peptide and the Staphylococcus nuclease A.
In the pPyL-knuc recombinant, transcription of the chimeric gene is controlled by the Polyoma virus enhancer and late promoter (Fig.1). A more detailed description of the construction of the chimeric reporter gene (knuc) and the Polyoma virus based expression vector, pPyL-knuc, is to be published elsewhere.
The plasmid vector pKSV-Hyg contains the Hygromycin B resistance gene (4) expressed from the SV40 early promoter.

Transfection and isolation of stabile transfected cells

Semiconfluent cultures of C127 cells, in petri dishes, were co-transfected with 20 ug of pPyL-knuc and 2ug of pKSV-Hyg by the calcium-phosphate coprecipitation method (7). Transfection efficiency was enhanced by a 3 minutes 15% glycerol shock as described (3). 24 hours post transfection the cells were trypsinised and seeded on a 10 cm plastic petri dish in 8 ml DMEM + 10% FCS. The cells were then allowed to grow for another 24 hours prior to the addition of Hygromycin B (Calbiochem) at a final concentration of 0.5 mM. The selective medium was changed every 3:rd day. After 14 days of growth in selective medium the Hygromycin B resistant clones were trypsinised and pooled.

Staphylococcus nuclease assay

The amount of S. nuclease secreted from the cells was determined from 100 ul portions of culture supernatant. The volume of the samples were adjusted to 250 ul by addition of distilled water. The reactions were started by adding 250 ul of 2 x Nuclease buffer (100 mM sodium borate, pH 8.8, 10 mM NaCl, 5 mM $CaCl_2$ and 2.4 mg/ml salmon sperm DNA (Sigma Type III)) and incubated at 37°C. The reactions were terminated after 30 minutes by the addition of 250 ul of 7% perchloric acid. The samples were thoroughly mixed and placed on ice for 10 minutes before centrifugation for 10 minutes. The supernatants were transferred to new tubes and diluted with 750 ul of distilled water. The amount of acid soluble oligo- and mono-nucleotides were determined spectrophotometrically at 260 nm. A calibration curve was run in parallel with commercially available Staphylococcus nuclease (Pharmacia).
Samples of medium were taken once every time before subcultivation during scale up and every fifth day before medium change during production phase.

Microcarrier cultures

The cells were grown on microcarriers (Cytodex^R 3, Pharmacia), at a concentration of 4 g/l, in 75 ml spinner cultures (Techne). Prior to cultivation the microcarriers were suspended in phosphate buffered saline (PBS) pH 7.4 (Ca^{++} and Mg^{++} free), and autoclaved at 120°C at 1 bar for 30 minutes. The microcarriers were then washed twice with DMEM (1:2) and transferred to the culture vessels with DMEM containing 5% FCS and 6 mM glutamine (Flow). After a 30 minutes preincubation, at 37°C, 7.5% CO_2, cells were added at a density of $6x10^4$ cells/ml. Two parallel cultures per cell line were set up. Cell numbers were determined by counting extruded nuclei (6).

Subcultivation of microcarrier cultures

When cultures approached confluence, at a cell density of $8\text{-}10x10^5$ cells/ml, the two parallel spinner cultures were pooled and the microcarriers were allowed to sediment. The medium was aspirated and each volume of microcarriers were washed with four volumes of 100 mM phosphate buffer, pH 8.0, containing 1 g/l of glucose, 3 g/l of NaCl and 0.02% EDTA. The washing solution was carefully removed and one volume of trypsin (360 U/ml, Pharmacia) in PBS, pH 8.0, was added per volume of microcarriers and incubated at 37°C for 10 minutes. The trypsin activity was inhibited by adding one volume of culture medium to one volume of microcarrier-cell suspension. The amount of viable cells was determined and then seeded at about a 1:15 dilution on microcarriers.

RESULTS AND DISCUSSION

The cell doubling time of the four exponential cultures of C127:pPyL-knuc was compared with the control cultures of normal C127 cells. As shown in Fig.2 there was no difference in the doubling time of C127:pPyL-knuc cells and the C127 cells, indicating that the integrated copies of vector DNA did not affect the growth rate of the cells. Fig.2 also shows that the growth rate as well as the cell densities, reached after each exponential growth of the C127:pPyL-knuc cells, remained constant over the three consecutive steps of subcultivation.
The fact that we could start each spinner culture with a cell number of $6x10^4$ cells/ml made it possible to use a split ratio of close to 1:15 at each subcultivation. This is an advantage since it makes it possible to quickly come up to larger culture volumes.
At confluence, following the final subcultivation step after 560 hours, the cultures were replenished with fresh medium with (5% FCS) or without serum. The cultures were maintained at stationary phase over a time period of 515 hours for production purposes. 95% of the medium was exchanged every fifth day (the remaining 5% consists of the bead volume). Those cultures maintained in 5% FCS continued to increase in cell number whereas the serum free cultures remained at a constant cell density throughout the production phase (Fig.2).
As shown in Fig.3, the production rate of S. nuclease from the C127:pPyL-knuc cells was basically constant during the scale up procedure. At stationary phase the production remained at a constant level in the presence of 5% FCS whereas without serum it showed a slight decrease at later stages (Fig.4). When looking at molecules nuclease produced per cell and day, the difference between serum containing and serum free stationary cultures is less pronounced (Fig.5,Table 1).
The data presented in this paper show that the integrated vector, in the C127:pPyL-knuc cells, had no effect on the growth properties of the cells during at least three steps of subcultivation. Furthermore, the data show it is possible to keep the cells in stationary phase for production purposes, with or without serum, for prolonged periods of time. Although the level of gene expression

VECTOR CONSTRUCTION

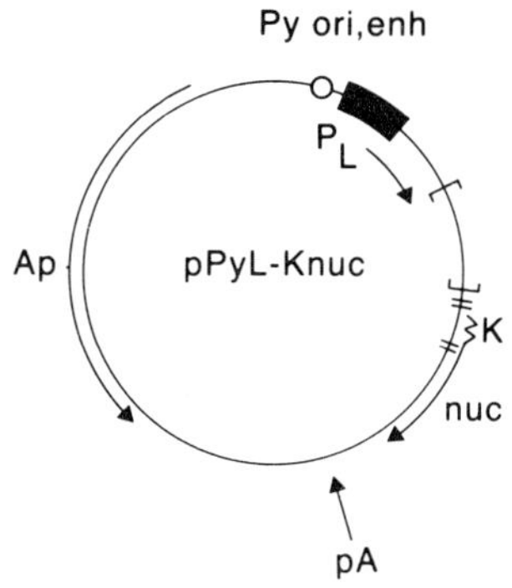

Fig. 1

A schematic picture of the vector construction used. The vector contains the polyoma virus ori and enhancer region including the late promotor, the SV40 VP1 intron and the SV40 late poly-A site. The bacterial sequnces are from pML.

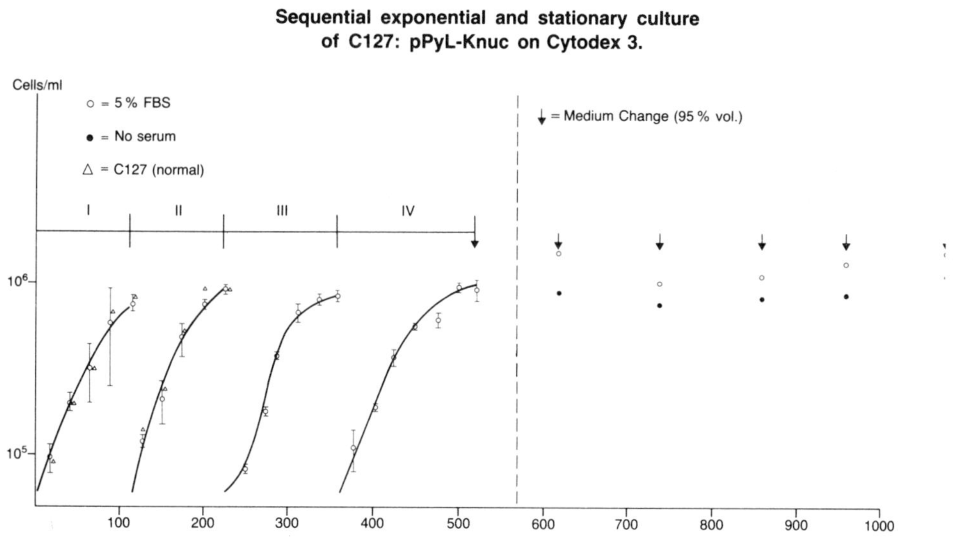

Fig. 2

Sequential exponential and stationary culture of C127:pPyL-knuc on Cytodex 3.
The transfected cells were cultured in four culture steps before they were kept at stationary phase. At the end of each culture, samples were taken to determine the amount of nuclease produced. Samples were also taken during the

stationary phase just before medium changes (arrows). Normal C127 were used as controls during the first two culture steps.

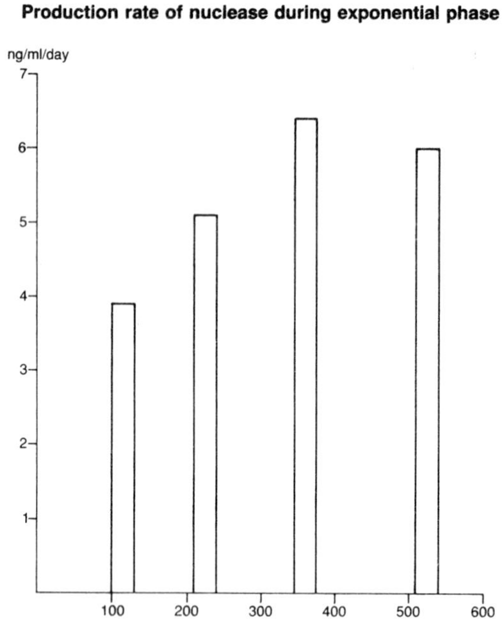

Fig. 3

Production rate of nuclease during exponential phase. The bars represents the final concentration of nuclease produced during the four culture steps (ng/ml/day).

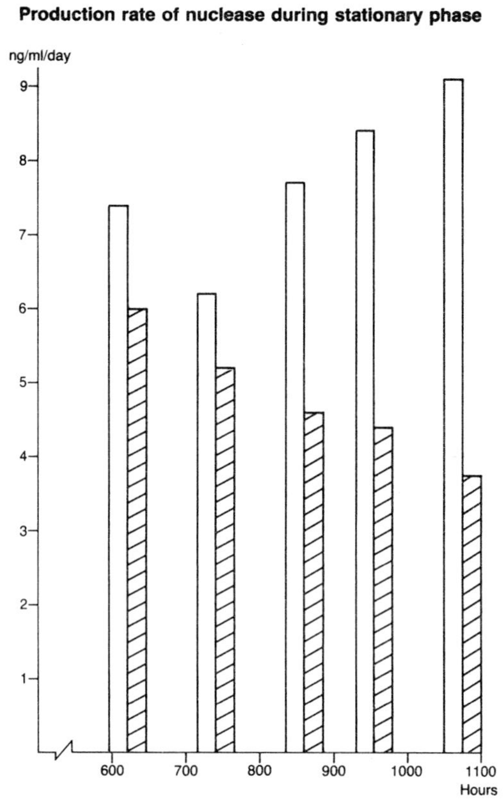

Fig. 4

Production rate of nuclease during stationary phase.
The five doublets of bars represents the final concentration of nuclease produced between each of the five medium changes □ cultures containing serum, ▨ serum free cultures (ng/ml/day).

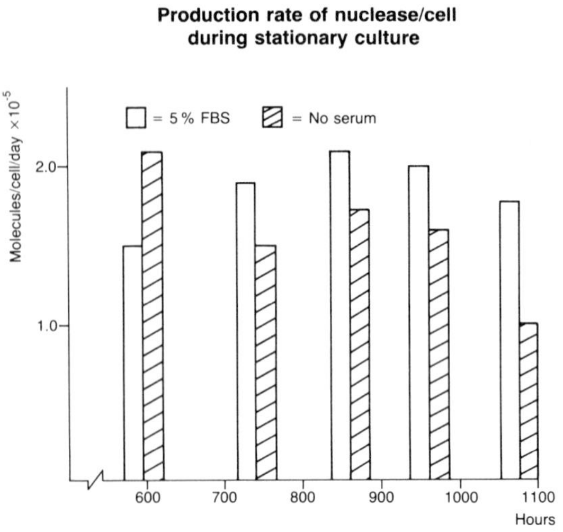

Fig. 5

Production rate of nuclease per cell during stationary phase. The five doublets of bars represet the final concentration of nuclease produced per cell between each of the five medium changes (molecules/cell/day, the Mw of the nuclease is 19 800 d).

Productivity

	Exponential culture (5 % FBS)	Stationary culture (5 % FBS)	Stationary culture (serum free)
Final product concentration			
ng/ml	32.3±7.3 (8)	35.7±6.5 (5)	21.9±3.8 (5)
Volume productivity			
ng/ml/day	5.4±1.1 (8)	7.8±1.1 (5)	4.8±3.8 (5)
Cell productivity			
ng/10^6/day	n.d.	6.1±0.8 (5)	5.2±1.3 (5)
Cell productivity			
mol/cell/day x 10^{-5}	n.d.	1.9±0.2 (5)	1.6±0.4 (5)

Table 1

A summary of the production data.
The numbers in brackets show the number of values used to calculate the mean value. n.d. = not determined.

was lower than other reported systems (1) we believe that the system we have used offers possibilities to further study gene expression and scale up of microcarrier cultures of recombinant-DNA manipulated cells.

REFERENCES

1. Bebbington, C. and Hentschel, C. (1985). The expression of recombinant DNA products in mammalian cells. Trends in biotechnology **3**, 314-317.

2. Emtage, J.S. (1985). DNA makes protein makes money. Nature **317**, 185-186.

3. Frost, E. and Williams, J. (1978). Mapping temperature-sensitive and host-range mutations of adenovirus type 5 by marker rescue. Virology **91**, 39-50.

4. Gritz, L. and Davies, J. (1983). Plasmid-encoded hygromycin B resistance: the sequence of hygromyocin B phosphotransferase gene and its expression in Escherichia coli and Saccaromyces cerevisiae. Gene **25**, 179-188.

5. Lowy, D.R., Rands, E. and Scolnick, E.M. (1978). Helper-independent transformation by unintegrated Harvey Sarcoma virus DNA. Journal of virology **26**, 291-298.

6. Sandford, K.K., Earle, W.R., Evans, V.J., Waltz, H. and Shannon, J. (1951). The measurement of proliferation in tissue cultures by enumeration of cell nuclei. Journal of the National Cancer Institute **11**, 773-795.

7. Weber, F. de Villiers, J. and Schaffner, W. (1984). An SV40 "enhancer trap" incorporates exogenous enhancers or generates enhancers from its own sequences. Cell **36**, 983-992.

Discussion

C. BEBBINGTON Do you know what the copy number of the vector is in there cells?

LUNDGREN No, we haven't got that far.

General Discussion

W. TOLBERT — What do people feel is the optimum oxygen concentration of cells in a bioreactor?

W. SCHEIRER — At cell densities above 10 million oxygen limitation is the biggest problem - even with bubbling pure oxygen, or bubble free aeration etc. we are always at the limit. We thus have to look for the lowest point rather than the optimum. This we find is 10% saturation - below 10% we get changes in metabolism resulting in more acid production, often less product. Optimum of around 50% saturation is probably best.

A. MILLER — With human epidermal keratinocytes low oxygen levels causes a high PE but low growth rate. High oxygen partial pressure results in rapid growth. We thus initiate our cultures at a low oxygen tension then change to a high tension.

D. KILBURN — Taking you back in history when I looked at oxygen sensitivity on rubella virus replication it had very little effect above 10% and below 150% saturation so oxygen was not important.

B. MAIORELLA — Apart from the effect of oxygen tension on the cells we should look at its effect on the medium and its components. Using a sensitive H_2O_2 assay we have found H_2O_2 is continuously and spontaneously produced on culture medium with or without light. Hydroxyl radicals attack cell lipids or oxidise the reduced medium components. I would therefore tend to the lowest tension possible to reduce oxidation of the medium.

W. TOLBERT — That's a good point as we do see oxygen toxicity. Dr Parsons mentioned containment in his talk - are there any points on containment of mammalian cells

B. GRIFFITHS — We are building a 150 l P3 contained fermenter at the moment. The fermenter has triple seals, steam tracing, steam barriers against addition vessels, grouping of probes and sampling ports behind a glove compartment. The problem is one of valves - there is no failsafe valve available. The steam condensate lines are a problem and either have to be grouped in a containment box or led into a kill-tank. So the reactor is no problem but to achieve a completely contained process needs a lot of plumbing in specialised facilities.

A. MILLER — What is the price of the fluorescent based biomass system that Dr Parsons spoke about?

D. PARSONS — The price is in the order of £12000.

W. SCHEIRER — If you use this probe with phenol red in your medium this probe is useless - as the excitation is in the same range as phenol red absorbtion. Also, fluorescence of riboflavin is in the same range as NADH. It may be possible to use this probe in very dense cultures where the background interference would become negligble.

O. MERTEN — There was a publication on using the NADH probe for suspension cells and they found a correlation between the signal from the fluourescent probe and the cell number but only during the lag and log phase. In the stationary phase there was no correlation so there could be trouble with this probe in continuous processes.

S. REUVENY — Several researchers have shown that hybridoma cells under the same agitation conditions show different degrees of sensitivity to shear force. Why should cells differ in their sensitivity to shear force?

R. SPIER — Everybody in their laboratory do things differently.

REUVENY — We have 3 lines against the same antigen all brought up in identical media etc, but they show different shear sensitivities.

SPIER — The cells must be different - if you looked at other physiological parameters eg, glucose utilisation rate, you would see other differences and considering the way in which hyridomas are made it should not be suprising. We should turn things around and say things are different, so what are the differences that manifests itself as sensitivity to shear.

P. GREENFIELD — The differences I showed on my slide were not that great, there were differences but they were small compared to other cell lines.

S. REUVENY — We have found big differences, even some that could not be grown in a spinner flask at all.

A. MILLER — When you obtain clones you have different chromosome contents - you select for antibody secretion only, not for all the other cell characteristics. So you would expect a big difference in cells from the same cloning experiment.

A. BEN-MICHAEL — The fluidity of the membrane and cytoskeletal differences are all variable factors in shear sensitivity.

Section 4
Bioreactor developments

28 Membrane-stirrer-reactor for bubble free aeration and perfusion

H. Büntemeyer, B. G. D. Bödeker and J. Lehmann

Introduction

The advantages of perfusion systems for the cultivation of mammalian and human cells in vitro are well known. With a perfusion system higher cell densities than in batch or fed-batch systems can be obtained. Therefore, higher product concentrations and productivities are possible /2/.

Up to now there are several different perfusion systems commercially available. One group of these systems is constructed for homogeneously stirred cultures with continuous cell retaining (New Brunswick, Bilthoven). These systems almost use fast rotating spin filters mounted on the stirrer shaft. The spin filters are made of microporous stainless steel or synthetic material with a pore size of 5 μm or greater /2, 3, 4/. Another construction is a satellite filter vessel for the reprocessing of the cells /5/. In all cases aeration is done by sparging air and/or oxygen into the culture /2, 3, 5/ or by using membrane aeration with a silicone tube /6/. The possibility of monitoring cell growth and behavior is one advantage of these systems. One great disadvantage is the clogging of the spin filter systems after about 1 - 2 weeks /2/. The other group of perfusion systems demonstrates the inhomogeneous type of high density cell culture like hollow fibre reactors /7/, dense cell reactors or reactors with immobilized cells /8/. In these systems the medium is circulated fast through the cell containing matrix. The advantages of these systems are very high cell densities up to 5·10 cells/ml and high product concentrations. The disadvantages are uncontrolled cells and high perfusion rates.

In this paper a new homogeneously stirred reactor with bubble free aeration and perfusion system is described. The core of this system is a stirrer on which two hollow fibre membranes are mounted. The first, a hydrophobic membrane, is used for bubble free aeration described earlier /1/. It is connected to a gas supply system creating the necessary gas mixture depending on pH and DO. The second, a hydrophilic membrane, is used for supplying the culture with nutrients and for removing the spent medium and waste products. This membrane is connected on one side via a feeding pump to the medium reservoir and on the other side via a harvest pump to a harvest container. Feeding and harvesting pump are running alternatively. The advantage of this construction is the possibility to avoid clogging by back flushing the filtration system with fresh medium. The pore size of this membrane is 0.3 μm. Therefore, only very small particles can clog the pores. With the whole system several cell lines have been successfully cultivated. The system has been scaled up without any problems to a volume of 20 litre until now.

Material and Methods

Cell propagation system:

The cells are cultivated in a homogeneously stirred fermentor. The core of the system is a membrane stirrer. On a glass support in case of a 1 litre fermentor or on a teflon support in a 20 litre fermentor two separate hollow fibre membranes out of polypropylene (Accurel®, ENKA, Wuppertal) are mounted. The membranes have a specification shown in Table 1. The hydrophobic membrane used for bubble free aeration of the culture is connected via a sterile filter to a gas supply unit. This unit creates a gas mixture of carbon dioxide, oxygen, nitrogen and air depending on pH and DO of the culture. At the outlet of the aeration membrane a sterile trap is connected. With this aeration system it is possible to supply the culture at almost any time of the fermentation period with the optimal gas mixture. At the beginning of a fermentation process, when cell densities are low, the culture can be supplied with CO_2, N_2 and air to perform the optimal culture conditions. Later on, when cell densities are high, the culture can be additionally supplied with oxygen. In addition, the aeration membrane can be used to remove carbondioxide produced by the cells. Specific oxygen transfer rates in dependence of gaflow and stirrer speed obtained in a 23 litre cell culture fermentor with 2.3 m hydrophobic membrane per litre culture are shown in Fig. 1. The second

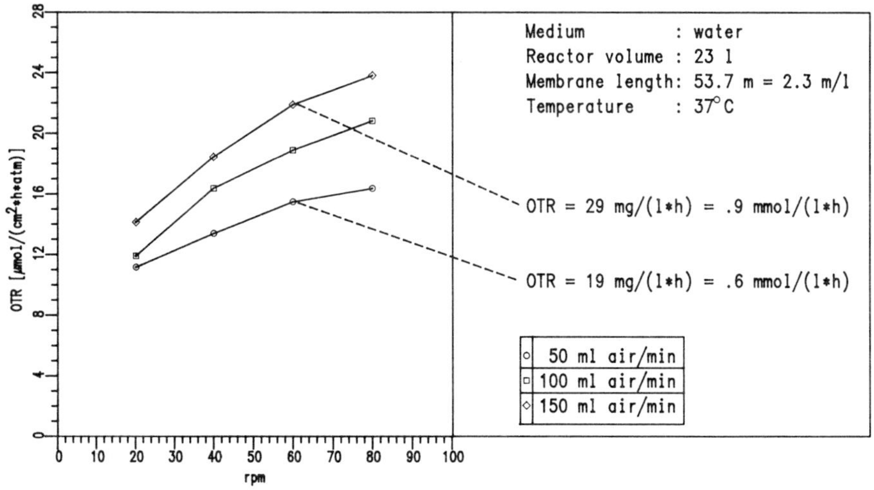

Fig.1. Oxygen transfer rates at a membrane stirrer for a 23 litre fermentor

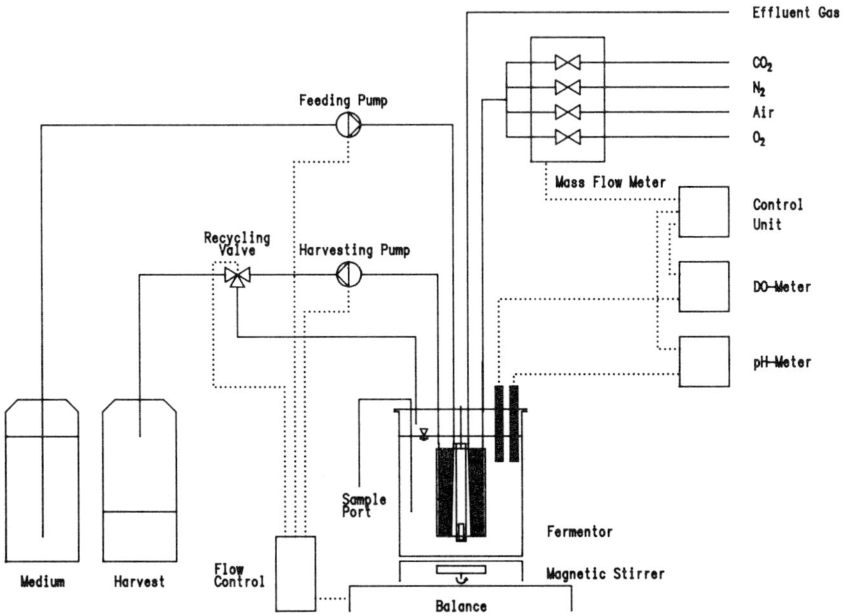

Fig.2. Scheme of the fermentor system with bubble free aeration and perfusion

membrane mounted on the support consists of the same material as shown in Table 1. Before mounting it is hydrophilized with ethanol for 1 hour and washed with pure water to remove the solvent. With this membrane it is possible to carry out continuous microfiltration. The whole system is schematically shown in Figure 2.

Perfusion mode:

The perfusion system is constructed as shown in Fig. 3. The pumps are connected at each side of the hollow fibre membrane. The liquid level of the fermentor is controlled by a level or weight control.

The perfusion process starts with filling the fermentor using the feeding pump. At that time the harvest pump stops and closes the outlet of the hydrophilic membrane. Therefore, the medium pumped by the feeding pump into the fermentor has to pass the membrane. When the maximum level is reached, the feeding pump is stopped and the harvest pump is started. In this case the feeding pump closes the inlet of the membrane. The medium removed from the fermentor by the harvest pump has to pass the membrane. When the minimum level is reached, the harvest pump is stopped and the feeding pump is started to fill the fermentor again. This backflushing procedure makes it possible to prevent an early clogging of the membrane. Each pumping cycle, feeding and harvesting, lasts about 15 to 20 minutes. The perfusion rate is controlled by setting the pump rates of the two pumps. With the recycling valve in the harvesting stream it is possible to recycle fresh medium which is in the membrane tube at the end of each feeding cycle.

Fermentation strategy:

The cells are seeded out in the fermentor at viable cell densities of about 2 to $4 \cdot 10^5$ cells/ml. In a batch mode the cells were cultivated to viable cell densities of about $1 \cdot 10^6$ cells/ml. At that state the perfusion mode is started with a small perfusion rate. Depending on cell density the perfusion rate has to be changed to higher values. For process guide values easily measurable components of the feeding medium or of the waste medium are possible like glucose, special amino acids, lactate or else. Which guide value is chosen depends on each cell line.

Cells and culture media:

With this cultivation system several different cell lines were

Table 1.
Membrane Data of the Accurel® hollow fibre (ENKA, Wuppertal)

d_e = 2.6 mm, d_i = 1.8 mm

Pore volume :	75.0 %
Medium pore size :	0.3 µm
Burst pressure :	6.5 bar
Implosion pressure :	1.5 bar
Specific surface, outside:	81.7 cm^2/ m
Specific surface, inside :	56.5 cm^2/ m
Specific volume, outside :	5.3 ml / m
Specific volume, inside :	2.5 ml / m

cultivated. One group of cells was antibody producing in suspension cultured cell lines such as mouse-mouse-hybridomas, rat-mouse-hybridomas and EBV-transformed human B-lymphocytes. The other group of cells was in suspension cultivated adherent cells such as Hela, BJAB and genetically manipulated BHK 21 cells for different purposes. The media used for these cultivations based on commercially available compositions like DMEM (Dulbellco's modified Eagle Medium), IMEM (Iscove's modified Eagle Medium), HAM F12 and RPMI 1640. They were often used serum free with some additional supplements or in the other cases with fetal calf serum up to 10 % (v/v).

Cell count:
Viable and dead cells were estimated microscopically using trypan blue exclusion.

Results and Discussion

Depending on cell line and cultivation purpose viable cell densities up to $2 \cdot 10^7$ cells/ml with perfusion rates of $D = 1.2\text{-}1.5\ d^{-1}$ could be obtained.

Fig. 3 shows a time course of a cultivation of EBV-transformed B-lymphocytes over a period of about 2 months in a 1.2 litre scale. For aeration a 2.4 m hydrophobic membrane was mounted on the stirrer support in addition to a 2 m hydrophilic membrane corresponding to a specific length of about 2 m hydrophobic membrane and 1.7 m hydrophilic membrane per litre culture, respectively. At the beginning the cells were cultivated in batch-mode to a cell density of about $1 \cdot 10^6$ cells/ml in a period of 5 days. The perfusion was started with an initial rate of $D = 0.45\ d^{-1}$ for one week. The cells reached viable densities of about $3 \cdot 10^6$ cells/ml. In the next 7 days the perfusion rate was set to $D = 0.9\ d^{-1}$. In this case a viable cell density of about $7 \cdot 10^6$ cells/ml could be obtained. In order to measure cell specific utilisation rates of glucose and amino acids and production rates of waste products the perfusion rate was set down to $D = 0.45\ d^{-1}$ for a period of 2 weeks. In the last 3 weeks the perfusion rate was set to several values up to $D = 1.3\ d^{-1}$. It could be shown that the viable cell density is strongly correlated with the perfusion rate. However, at the last month of the fermentation period the concentrations of glucose and some amino acids in the fermetor were nearly zero. Over the complete fermentation period only one double membrane stirrer unit was used. In order to prevent an early clogging of the hydrophilic membrane, maximum flow rates of about 0.7 litre medium per meter membrane and day were performed in addition to the back flushing procedure.

In Fig. 4 a time course of a 3 week cultivation of EBV-transformed B-lymphocytes in a 1.2 litre scale is shown. The membrane length was the same as in Fig. 3. After a one day batch phase the perfusion was set with increasing cell density up to a maximum value of $D = 1.3\ d^{-1}$. At 3 times (at day 9, 12 and 14) cells were harvested for other purposes. Except the first 2 days the cells grew with constant growth rate. After this fermentation period of 3 weeks the filter unit was still at full flow efficiency.

Fig. 5 shows a time course of a 16 days cultivation period in a 23 litre cell culture fermentor with the same cell line used in the figures above. In this case 2.3 m hydrophobic and 0.9 m hydrophilic membrane per litre culture were built on to the support, respectively. The fermentation strategy was the same as described above. The cells were cultivated in batch mode to a viable cell density of about $1 \cdot 10^6$ cells/ml (day 4). At that time the perfusion is increased with increasing cell density. In this case a viable cell density of about $1 \cdot 10^7$ cells/ml could be obtained with a perfusion rate of $D = 0.95\ d^{-1}$ corresponding to 20 litre medium per day.

<u>Summary</u>

The results shown above demonstrate that the cultivation system described here is a valuable tool for mammalian cell culture. Bubble-free aeration with microporous moving membranes is an excellent possibility for culture aeration.
With cell free medium exchange through a hydrophilic membrane high viable cell densities of about $2 \cdot 10^7$ cells/ml can be obtained.
The back flushing procedure is necessary to prolong membrane stability to more than 2 months. A clogged membrane can be cleaned by treating with pure water and biochemical or soft chemical reagents like pepsine/HCl, H_3PO_4, $NaHCO_3$. To our experience the membrane

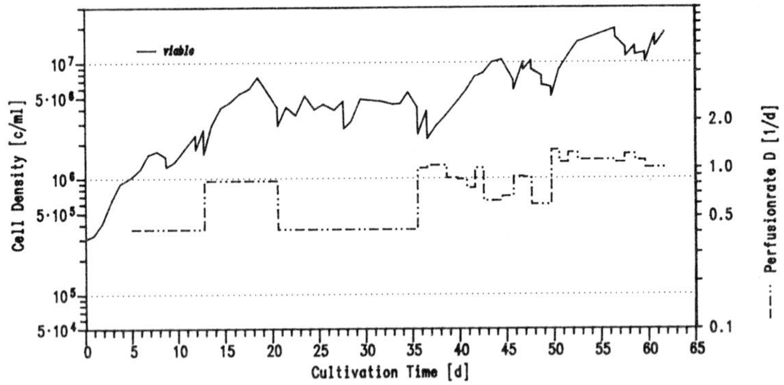

Fig.3. Continuous cultivation of EBV - transformed B-Lymphocytes in a 1.2 litre fermentor with 2.4 m hydrophobic membrane and 2 m hydrophilic membrane

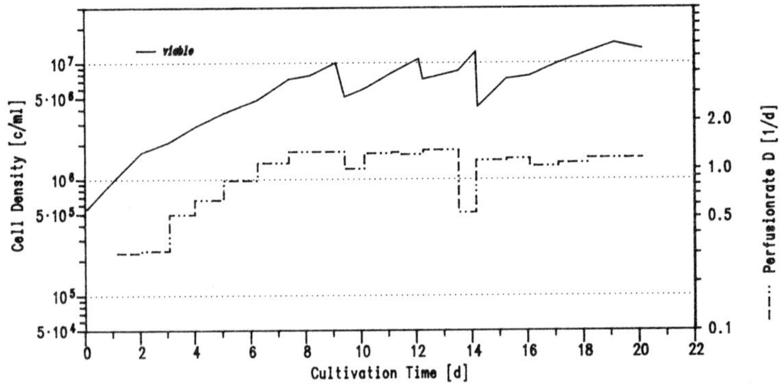

Fig.4. Continuous cultivation of EBV-transformed B - Lymphocytes in a 1.2 litre fermentor (same conditions as in Fig.3.)

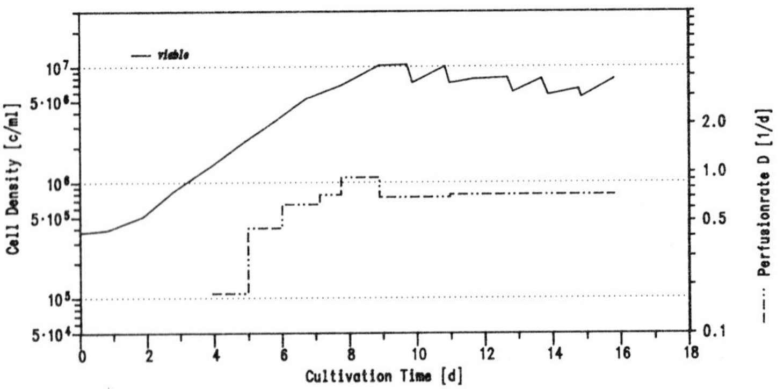

Fig.5. Continuous cultivation of EBV-transformed B - Lymphocytes in a 23 litre fermentor with 2.3 m hydrophobic membrane and 0.9 m hydrophilic membrane per litre, respectively.

should not filtrate more than one litre culture broth per meter membrane and day to guarantee a stable behavior for weeks.

References

/1/ Lehmann, J.; Piehl, G.W.; Schulz, R.
Develop. biol. Standard. 66, 227 (1987).

/2/ Reuveny, S.; Velez, D.; Miller, L.; Macmillan, J.D.
J. Immunol. Methods 86, 61 (1986).

/3/ Van Wezel, A.L.; van der Velden-de Groot, C.A.M.; de Haan, H.H.; van den Heuvel, N.; Schasfoort, R.
Develop. biol. Standard. 60, 229 (1985).

/4/ Himmelfarb, P.; Thayer, P.S.; Martin, H.E.
Science 164, 555 (1969).

/5/ Tolbert, W.R.; Feder, J. in: Annual Reports on Fermentation Processes, Vol. 6, eds. G.T. Tsao, M.C. Flicker and R.K. Finn (Academic Press, New York), p 35.

/6/ Kuhlmann, W. Develop. biol. Standard. 66, 263 (1987).

/7/ Schönherr, O.T.; van Gelder, P.T.J.A., van hees, P.J.; van Os, A.M.J.M.; Roelofs, H.W.M.
Develop. biol. Standard. 66, 211 (1987).

/8/ Littlefield, S.G.; Gilligan, K.J.; Jarvis, A.P.
Hybridoma 3, # 1, 75 (1984).

Discussion

S. REUVENY — Can you specify which amino acid was limiting?

BUNTEMEYER — Serine, tryptophan and methionine.

H. KATINGER — Can I confirm that you used an EBV transformed lymphocyte?

BUNTEMEYER — Yes that is correct

M. KAYLANPUR — How do you sterlise your membranes, and how many times can you re-use them?

BUNTEMEYER — By steam in the fermenter and they are used 5-10 times.

29 Immobilization of hybridomas in packed-bed bioreactors

A. D. Murdin, J. S. Thorpe and R. E. Spier

ABSTRACT

Packed bed bioreactors have been used for the culture of adherent monolayers of animal cells for a number of years. They offer a simple and convenient system for the culture of adherent cells but have not been used to date for the culture of hybridomas. This paper describes the use of packed beds of polyester foam particles and other materials for the culture of a hybridoma cell line. Cells trapped within the porous particles produced levels of antibody comparable with levels in static flask or airlift cultures.

INTRODUCTION

Packed bed bioreactors for the cultivation of anchorage dependant animal cells have been in use for over twenty five years (2). The commonest support material has been a matrix of 3mm glass beads (1,4,8). Reactors using a packed bed of such beads have been successfully run at the 100 litre scale for the growth of BHK21 C13 monolayer cells and the subsequent production of foot-and-mouth disease virus (aphthovirus) vaccine antigen (5). The growth and productivity of anchorage dependant cells in such systems has been widely studied (1,5,6).

To date, however, the use of packed bed bioreactors for the cultivation of non-adherent mammalian cells, especially hybridomas, has attracted little attention. With the use of an appropriate support matrix the system would appear to have many advantages. It is mechanically simple, robust and well studied. Since the cells would be passively trapped within an inert matrix the process should be gentle towards the cells yet have all the advantages of other immobilization systems - the ability to perfuse the cells with fresh medium, provision of a conditioned and protected environment for the cells, increased process intensity, simplified downstream processing and so on.

This paper describes preliminary studies of packed bed reactors using a variety of support matrices for the culture of hybridomas and the production of monoclonal antibodies.

MATERIALS AND METHODS

Bioreactor

The general configuration of the reactor is shown in Figure 1, and follows the design of Spier and Whiteside (4). A vessel holding a bed of the support matrix was connected to a medium reservoir which acted as an airlift. The reservoir

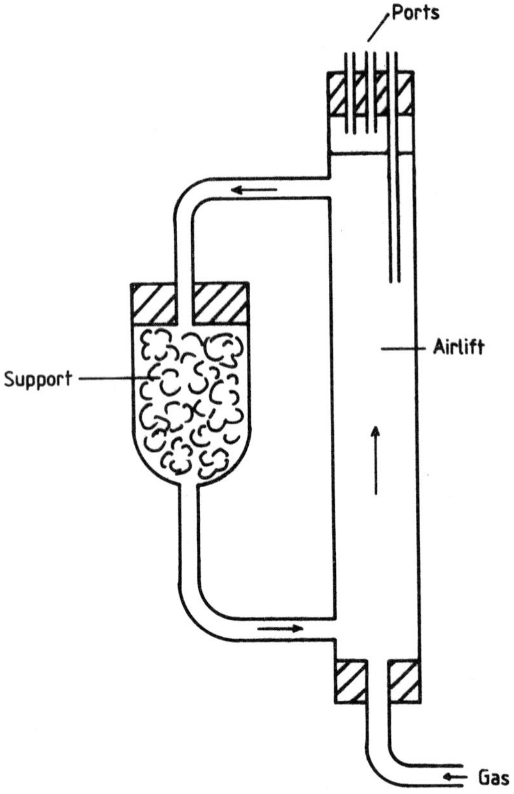

Figure 1. The packed-bed bioreactor design of Spier and Whiteside (4).

was sparged with 5% CO_2 in air. Ports for sampling, antifoam addition and gas outflow were let into the medium reservoir.

Cells

An NS-1 derived hybridoma cell line, 321, was the kind gift of Andrew Wright, of ICI plc. The cells secrete IgG specific for the herbicide paraquat (8). The cells were cultured in RPMI 1640 (Flow Laboratories Ltd., Rickmansworth, UK) plus 2mM glutamine and 0.02% w/v sodium hydrogen carbonate, with 5% adult bovine serum (Advanced Protein Products Ltd., Brierley Hill, UK).

Support materials

Six bed materials were examined. These were stainless steel wool (Radleys, Sawbridgeworth, UK), nylon wool, nylon fibre (Travenol Laboratories, Thetford, UK), sintered glass discs (Pyrex, J.Bibby Science Products Ltd, UK) and polyether or polyester foam (Flexiplas Ltd., Wisbech, UK). The foam was cut into rough 3-5mm cubes before use. All materials were thoroughly washed in 7X detergent (Flow Laboratories Ltd.) and rinsed in distilled, deionised water prior to use.

Assays

Where possible, cell numbers and viability were assessed by trypan blue dye exclusion.

Glucose was assayed enzymically, utlizing glucose oxidase to form peroxide from glucose, and 4-aminoantipyrine plus phenol to act as oxygen acceptors and form a coloured product.

Lactate and lactate dehydrogenase (LDH) were measured enzymically using the reaction

$$\text{lactate} \underset{\text{LDH}}{\overset{\text{NAD} \rightarrow \text{NADH}}{\rightleftharpoons}} \text{pyruvate}$$

Ammonia was measured using an EIL ammonia electrode.
Glutamine was measured enzymically, using glutaminase to form ammonia from glutamine.
Antibody was measured using a sandwich ELISA specific for murine IgG.

RESULTS AND DISCUSSION

Entrapment of cells in support matrices.

The ability of the various support matrices to trap cells was assessed (Figure 2). Cells were introduced into the medium reservoir of the reactor and counts of cells remaining in suspension were taken over the next few hours.
Polyester foam trapped cells most efficiently. Stainless steel wool and sintered glass were also useful supports, trapping cells more slowly but having a high capacity. After twenty four hours few cells could be observed in suspension in reactors in which these materials formed the packed bed (data not shown). The nylon fibre had a high affinity for cells but a limited total capacity, becoming saturated with cells only 30 minutes after the start of the experiment. In reactors utilising nylon fibre packed beds the cells were

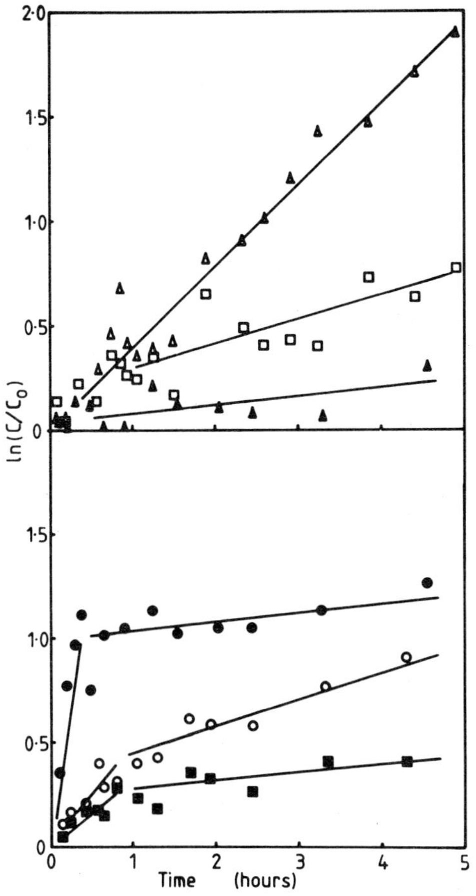

Figure 2. The entrapment of cells by support matrices formed from polyester foam (Δ), sintered glass (○), nylon fibre (●), nylon wool (▲), polyether foam (■) or stainless-steel sponge (□).

effectively growing in suspension. Polyether foam and nylon wool were very poor support materials, trapping few cells. The polyether foam was toxic to this cell line. Cell death could be observed in circulating cells within ten hours of inoculation, and no antibody was produced (Figure 3).

Antibody productivity.

The amount of antibody produced by cells in various culture systems was assessed (Figure 3). There was no difference in overall antibody production by cells cultured in 175cm^2 static flasks, 1500ml airlift reactors or 150ml packed bed reactors, with the exception of the reactor in which the packed bed was constructed from polyether foam. As noted above, this material was toxic to this cell line. These results indicate that a support material may be chosen to meet other design criteria (e.g. mechanical strength, reuseability) without adversely affecting antibody productivity.

Growth of cells.

The release of LDH was used to estimate the viability of cells trapped in a packed bed (3). Figure 4 shows the relationship between cell growth, viability and the release of LDH into the culture medium in a static flask culture. The rise in LDH activity corresponds precisely to the decline in cell viability.

Figure 5 shows the relationship between LDH activity and

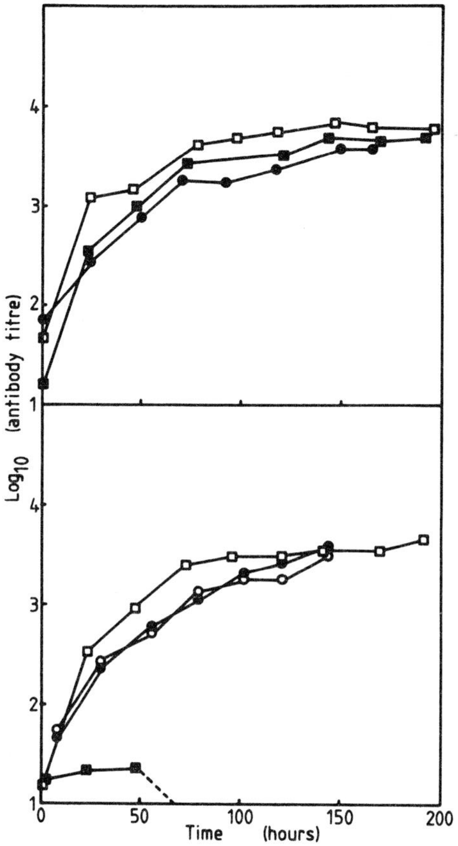

Figure 3. Comparison of IgG production by 321 hybridomas in various bioreactors. Panel A shows IgG production in a packed bed of polyester foam (□), in a 150ml static flask (■) and in a 1500ml airlift reactor (●). Panel B shows IgG production in packed beds of stainless steel sponge (□), polyether foam (■), sintered glass (○) and nylon fibre (●).

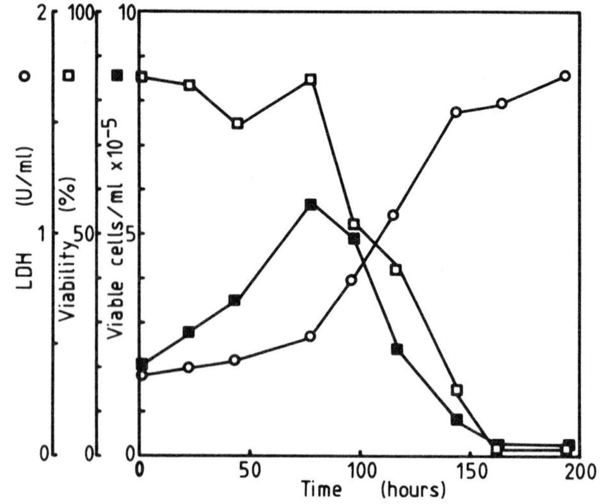

Figure 4. Production of lactate dehydrogenase (LDH) by 321 hybridomas in a 150ml static flask culture, and its relationship to cell growth and viability.

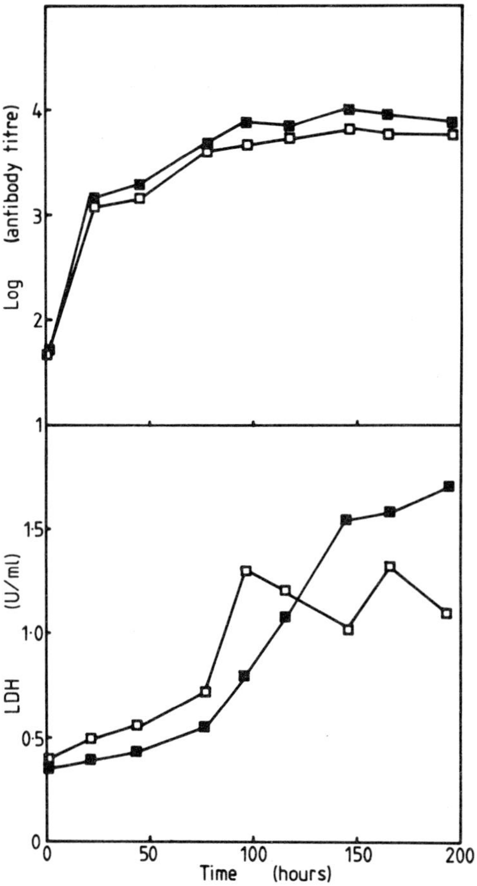

Figure 5. Comparison of the production of IgG and LDH by 321 hybridomas in a 150ml static flask culture (■) and in a polyester foam packed-bed bioreactor (□).

antibody production by the hybridomas in a static flask and a polyester foam packed bed reactor. In both cases LDH activity rises after about four days in culture, indicating the death of a previously viable population of cells. Antibody productivity is greatest during the first two to three days in culture, preceding the rise in LDH activity. This indicates that antibody is secreted by viable cells rather than released by dying cells as the cell membrane loses its integrity. This conclusion is supported by the absence of significant levels of antibody in the culture medium from the polyether foam packed bed reactor (Figure 3). Cells from a three day old culture inoculated into this reactor died within 24 hours but did not release antibody. Cells from similar cultures inoculated into other reactors remained viable and antibody titres in the culture medium rose rapidly. Thus the initial rise in antibody titre observed is due to the post-inoculation synthesis and release of antibody by viable cells and is not due to the release by dying cells of antibody formed pre-inoculation.

Cellular metabolism in the packed bed reactor.

The consumption of glutamine and glucose and the production of ammonia and lactate by cells in polyester foam packed bed reactors and in static flasks were assessed.

There was no difference in the consumption of glutamine or in the production of ammonia between the two systems. Glutamine concentrations declined steadily until all glutamine had been consumed. This occurred after four to five days. Ammonia concentrations rose steadily to 1.7mM by four days and remained constant thereafter.

In contrast, the two systems exhibited marked differences in their consumption of glucose and production of lactate. In both systems the consumption of glucose was most rapid in the first two to three days and much reduced after this. Approximately one third of the available glucose was used in the packed bed reactor, one half in the static flask. Final lactate concentrations were higher in the static flask than in the packed bed reactor. Similar differences have been

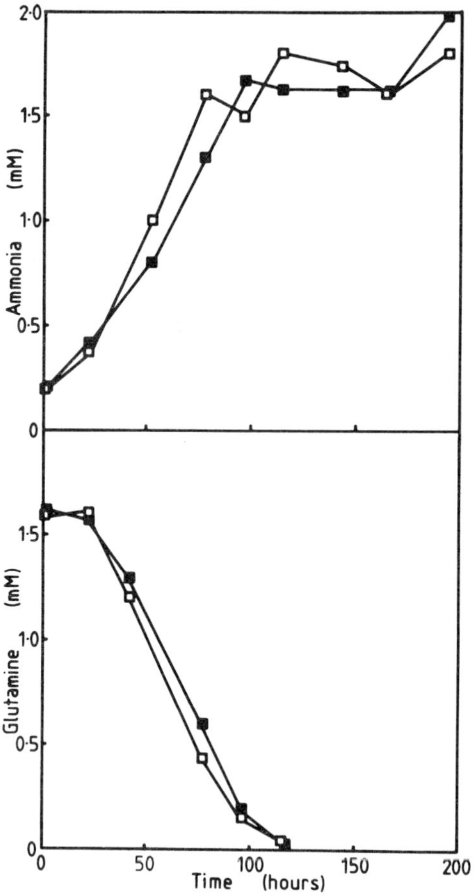

Figure 6. Comparison of the consumption of glutamine and the production of ammonia by 321 hybridomas in a 150ml static flask culture (■) and in a polyester foam packed-bed bioreactor (□).

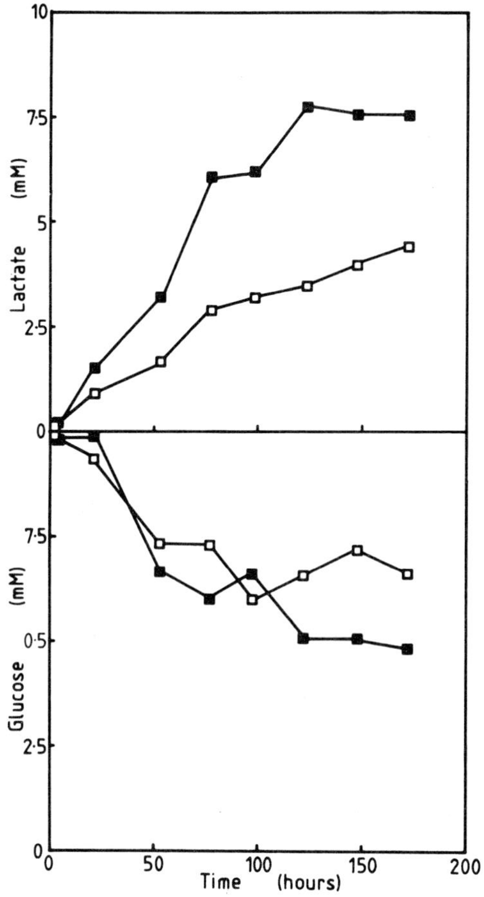

Figure 7. Comparison of the consumption of glucose and the production of lactate by 321 hybridomas in a 150ml static flask culture (■) and in a polyester foam packed-bed bioreactor (□).

observed in a number of experiments and are probably due to the inhibition of glycolysis by the Pasteur effect in the better oxygenated packed bed system. This may prove to be a useful way of controlling the accumulation of a potentially toxic metabolite in reactors of this kind.

CONCLUSIONS

Packed bed bioreactors of the design described are able to support the productive growth of immobilised cells of a hybridoma cell line, using a variety of materials to form the packed bed. Cell growth and productivity in such systems are comparable to the growth and productivity observed in conventional static and suspension systems.

Similar packed-bed reactors have been scaled up to 100 litres capacity for the production of adherent BHK cells (5) so there is the very definite possibility that scale-up to this level can be achieved for hybridomas. The reactor design can easily be modified to allow perfusion with fresh medium, with the potential this has for long-term culture. Experiments to study factors affecting the growth, metabolism and productivity of hybridomas in long-term cultures are planned.

ACKNOWLEDGEMENTS

The authors wish to thank Dr. S.C.Musgrave and Dr. N. Kirkby for their helpful advice and criticism.

REFERENCES

1) C.Burbidge (1979) The mass culture of human diploid fibroblasts in packed beds of glass beads, Develop. biol. Standard. 46, 169-172

2) T.A.McCoy, W.Whittle and E.Conway (1962) A glass helix perfusion chamber for massive growth of cells in vitro, Proc. Soc. Exp. Biol. (NY) 109, 235-237

3) R.E.Spier (1977) Determination of the time to harvest foot-and-mouth disease virus cultures by measurements of the supernatant concentration of lactic dehydrogenase, Biotech. and Bioeng. 19, 929-932

4) R.E.Spier and J.P.Whiteside (1976) The production of foot-and-mouth disease virus from BHK 21 C13 cells grown on the surface of glass spheres, Biotech. and Bioeng. 18, 649-657

5) J.P.Whiteside and R.E.Spier (1981) The scale up from 0.1 to 100 litres of a unit process system for the production of four strains of FMDV from BHK monolayer cells, Biotech. and Bioeng. (1981) 23, 551-565

6) J.P.Whiteside and R.E.Spier (1984) Factors affecting the productivity of glass sphere propagators, Develop. biol. Standard. 60, 305-311

7) A.F.Wright, T.P.Green and L.L.Smith (1987) The development of mouse monoclonal antibodies that bind and neutralize paraquat, Develop. biol. Standard. 66, 495-501

8) W.Wohler, H.W.Rudiger and E.Passarge (1972) Large scale culturing of normal diploid cells on glass beads using a novel type of culture vessel, Exp.Cell.Res. 74, 571-577

Discussion

M. de KORT — I noticed you used polyester foam for your bed packing. Isn't this rather too soft and becomes compressed during use?

MURDIN — We did not see significant compression in use over 10 days batch run. We pack the bed so that it is slightly compressed but don't get further compression.

30 Production of biologicals by animal cells immobilized in polyurethane foam matrix

A. Lazar, S. Reuveny, A. Muzrahi, M. Avtalion, J. P. Whiteside and R. E. Spier

ABSTRACT

A closed loop perfusion system for cultivation of animal cells in polyurethane (PU) foam matrix is described. The container for cell growth is packed with PU sponges in the form of 0.5-1 cm^3 cubes. The porous matrix leaves a path for nutrients to flow through the packed bed. Medium is continuously circulated by an airlift pump mounted in the cell-free chamber.

The system can be used for both anchorage dependent and independent cells, where the former grow attached and the latter immobilized in the foams. The foam matrix reactor has been tested with a variety of cell lines producing different natural products. These include HeLa-derived growth factor (HDGF) from a human cervical carcinoma, carcinoembryonic antigen (CEA) from a human colon carcinoma, B-cell stimulating factor (BSF) from a human hybridoma and monoclonal antibodies from mouse hybridoma cell lines.

INTRODUCTION

Immobilized cell system have several distinct advantages which make them attractive targets for development of commercial scale production systems (11,14). First the cells are grown in small volumes of culture. This enables them to achieve high cell densities and concentrated products which facilitates downstream processing. Second, the immobilized cells can be perfused with culture medium in a vessel separate from the growth chamber where coarse environmental adjustments, such as high rates of sparging can be made before returning the medium to the culture vessel (1,2), and finally, culture medium can be changed easily and rapidly without the difficulty of separating the cells from the medium allowing continuous harvesting of secreted product over a prolonged periods.

Several systems for mammalian cell immobilization have been described. These include: hollow-fiber reactors (4), ceramic matrix (7) microencapsulation in agarose (8) and alginate (6) and packed bed reactors (PBR, 13).

We present here, a system for cell immobilization by entrapment in PU foam matrix. PU foams have so far been used, mainly to cultivate plant cells in culture to increase the levels of production of secondary metabolites (10,11). A few reports also indicate the use of PU sponges in in-vivo studies of allograph-infiltrating cells (9,12). The system is based on the packed bed glass sphere propagator described by Spier and Whiteside (13,16). The system can be used for both anchorage dependent and independent cells, is simple to operate and can maintain cells in culture for long time periods.

MATERIALS AND METHODS

Cell cultures

The following cell lines have been used in this study: MRC-5, normal human diploid cells, BHK, baby hamster kidney cells. HeLa, a human carcinoma cell line. HuCCl, a human colon carcinoma cell line. TH5, a human T hybridoma and G10G11 a mouse B hybridoma. The cells were grown in RPMI 1640 medium (Gibco, U.S.A), supplemented with fetal, bovine serum (FBS) concentrations ranging from 15% for G10G11 cells, 10% for MRC-5, BHK and HeLa cells and 2% for TH5 cells. HuCCl cells were grown in medium containing 10% bovine serum dialyzate supplemented with 0.25% primatone-RL (Humko, Sheffield) and 0.01% Pluronic F-68 (BASF, Wyandotte)

Foam matrix reactor: Cells were immobilized by entrapment in polyester-PU foam matrix (Alpha Laboratories, U.K.) and cultured in a special glass reactor. A schematic description of the reactor is presented in Figure 1. The system consists of a cell growth chamber (300 ml total volume) in which PU cubes (0.5-1 cm^3) are packed. The medium is circulated by an airlift pump, mounted in the cell-free mixing chamber (700 ml total volume). Prior to cell inoculation, the PU foam sponges were rinsed three times in distilled water, boiled in distilled water for 30 min., immersed in acetone for 30 min., immersed in 95% ethyl alcohol for 30 min., followed by a 30 min. immersion in boiling distilled water to remove traces of alcohol. Finally, the sponges were autoclaved in phosphate-buffered saline. The reactor was inoculated with the cells at a concentration of 5-7x10^5 cells/ml. The culture medium was circulated by the airlift pump at a flow rate of 20-30 ml/min.

The perfusion rate was 50-60 ml/hr. This rate was sufficient to maintain a steady state of glucose, lactate and pH over a long time period. Samples were removed from the reactor through a sampling port mounted at the top of the cell growth chamber.

Analytical methods: Glucose levels in cultures were measured by Dextrostixs (Aimes, U.S.A), and by the enzymatic colorimetric determination method (Sigma # 510 method). Lactic acid concentrations were measured by the lactate dehydrogenase method (Sigma # 826 method). CEA was assayed by the enzyme-linked immunosorbent assay (ELISA) using CEA Abbott kit and technique. BSF was assayed as described by Gozes et al. (3). HDGF assay was described in our previous publication (5). Anti-CEA IgG was determined by the radial immunodiffusion assay (15).

RESULTS

PU culture sustems for anchorage dependent cells

The potential of growing anchorage-dependent cells on PU foam matrix was studied with MRC-5, BHK and HeLa cells. Since there is no direct way of determining cell numbers in the foam matrix, we monitored the cell growth by analyzing the glucose consumption of the cells. Figure 2, presents the kinetics of cell growth in PU foam

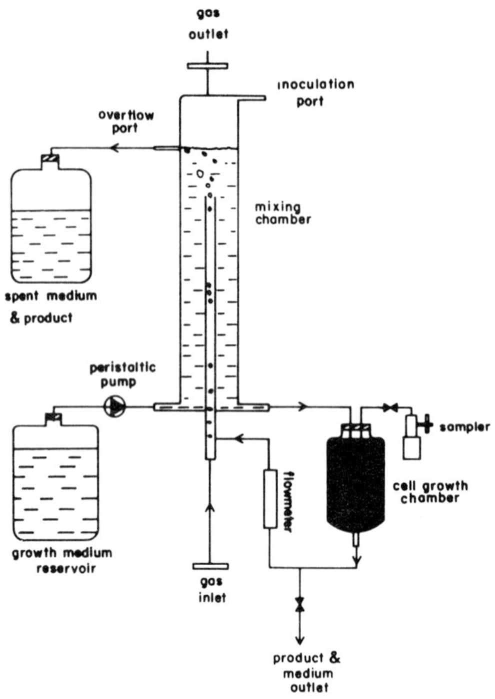

Fig. 1. Diagram of PU packed bed reaction.

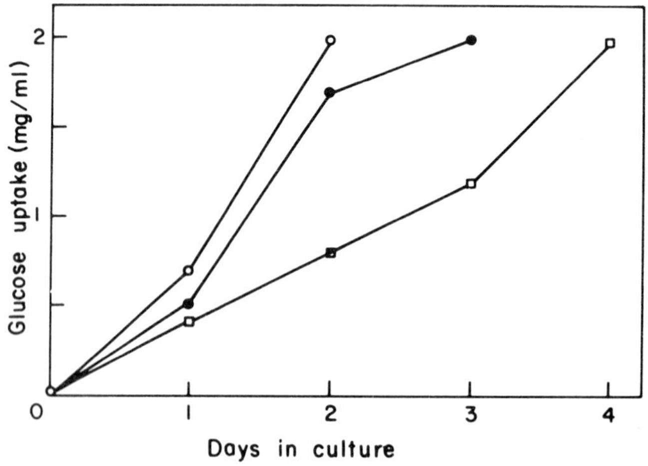

Fig. 2 Kinetics of growth of various anchorage dependent cells in PU foams.

○—○ BHK cells; ●—● HeLa cells; □—□ MRC-5 cells.

matrix in petri dishes. The cells attach, spread and grow in the foam at similar kinetics obtained by conventional culturing on T flasks. An important feature of PU-PBR is the capacity to maintain populations of cells after the growth phase in a non-proliferating state for extended time periods, expressing a specific cell product. We maintained HuCCl cells in a serum-free medium for several weeks (Fig. 3). After 20 days in culture, a fairly constant level of CEA was continuously produced. The perfusion rate was sufficient to maintain steady state levels of glucose and lactate. HuCCl cells grow in the PU foam as large aggregates penetrating into the sponge matrix (Fig. 4).

Similar results were obtained with HeLa cells producing HDGF (Fig. 5). The cells were initially grown in medium containing 10% FBS. During reactor operation, the serum concentration was reduced to zero. The conditioned medium was harvested free of contaminating serum proteins which simplifies subsequent purification processes.

PU culture systems for anchorage independent cells:

Unlike anchorage dependent cells that grow attached to the foam matrix and are thus physically separated from the perfusion medium, anchorage independent cells, such as leukemic cells and hybridomas need to be immobilized for perfusion. The PU-PBR has the advantage of immobilization by physical entrapment without the involvement of chemical interactions. This system avoids the use of membranes, cages and settling bottles to separate the cells from the medium.

The entrapment of cells in the PU matrix is very fast, and by 2 hrs, over 95% of the cells were immobilized (Fig. 6).

The production of BSF by the human T hybridoma TH5 in PU-PBR is presented in Fig. 7. This cell product is secreted at high yields during the cell growth phase. Perfusion was initiated at the 4th. day at a rate of one culture volume per day. By that time, unfavourable growth conditions, such as low glucose, low pH and high lactate started to develop in the culture. The perfusion level employed, did

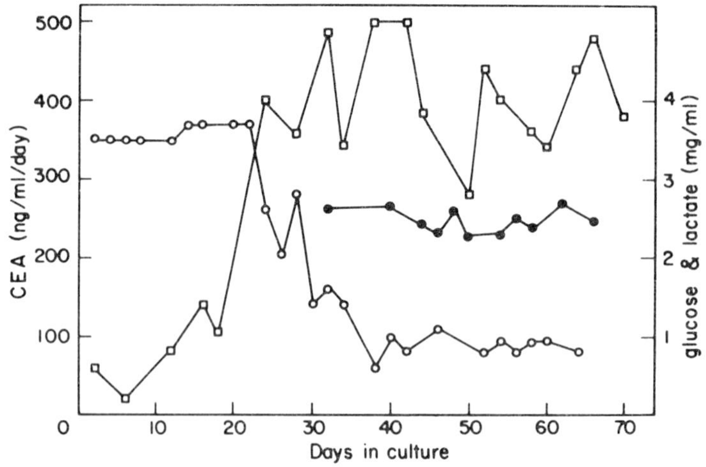

<u>Fig. 3.</u> Production of CEA by HuCCl cells in a PU packed bed reactor. ○—○ Glucose; ●—● Lactate; □—□ CEA.

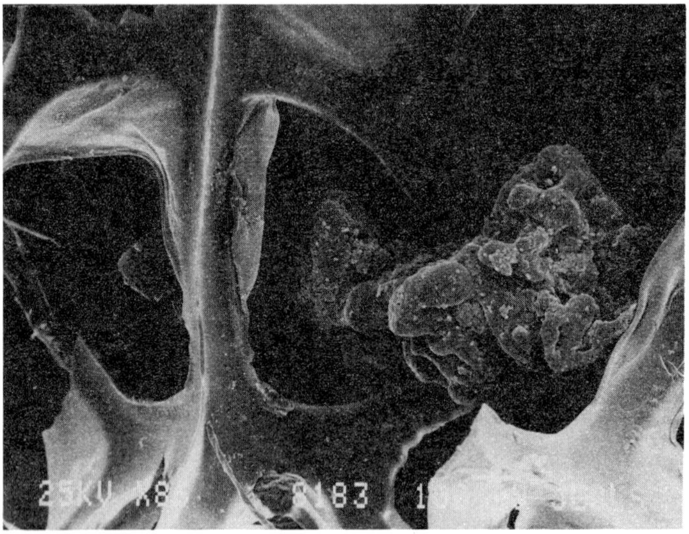

<u>Fig. 4.</u> Scanning electron microscopy of HuCCl cells grown on PU foam matrix.

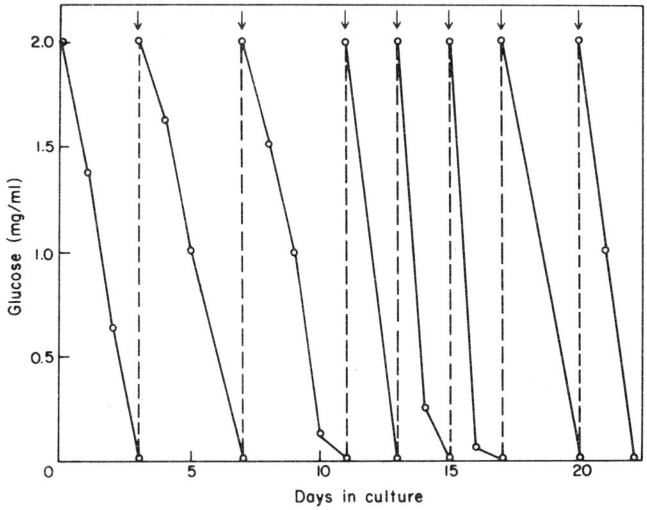

Fig. 5. Growth of HeLa cells in a PU packed bed reactor. Dashed line indicates complete medium change.

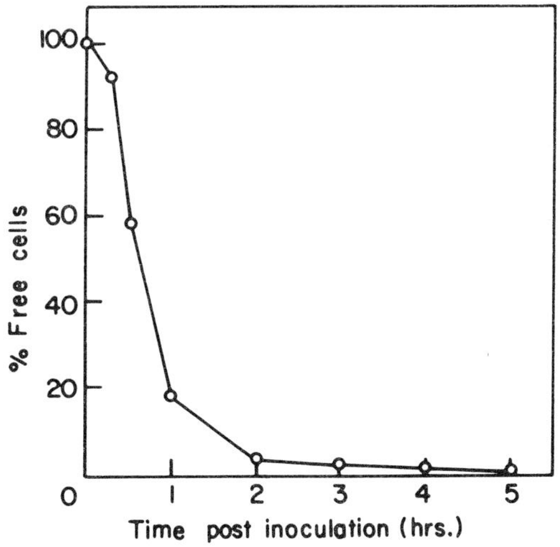

Fig. 6. Kinetics of cell entrapment in PU foam matrix. TH-5 cells were inoculated at a concentration of 5 x 10^5 cells/ml. Free cells were counted by microscopic observation in a haemocytometer.

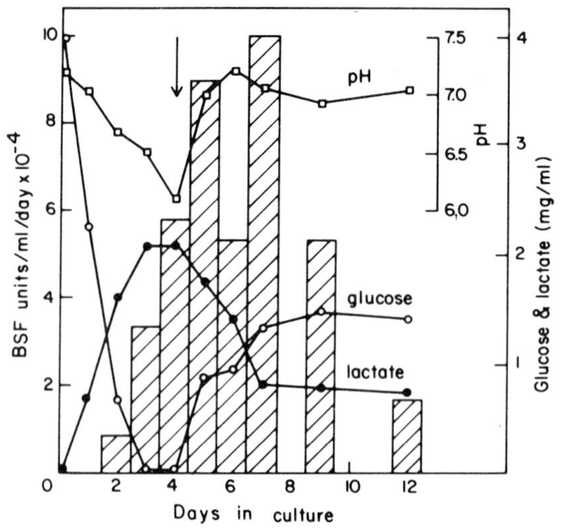

<u>Fig. 7.</u> Growth of TH-5 cells and BSF production in a PU packed bed reactor. Perfusion operation started at day 4. Columns present BSF production.

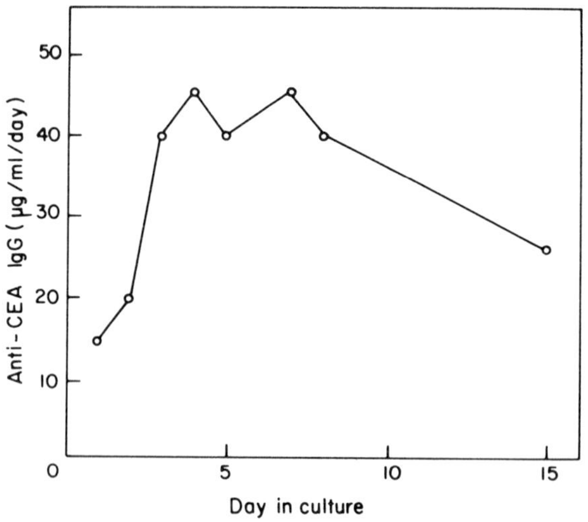

<u>Fig.</u> 8. Production of anti-CEA IgG by G10G11 mouse hybridoma cells.

maintain good cell viability as measured by glucose, lactate and pH, BSF production increased with time during the growth phase of the cells and then gradually decreased as the cells entered into a non-proliferative phase.

A similar perfusion system was used to produce monoclonal antibodies by a mouse hybridomna cell line (Fig. 8). As the TH5 cells, these cells also produce their specific product during the continuous growth phase. The productive rate of the monoclonal antibodies slowly decreased with time as the cells aged.

DISCUSSION

The perfusion culture system described above presents a new approach to the production of biomolecules by non-proliferating cells. The system enables to maintain such productive cells in a biochemically healthy and active state for extended time periods. However, the system is so far less suitable for production of biologicals from cells in a continuous growth phase, probably due to unfavourable growth conditions that may develop in the culture when cells reached high concentrations. Such growth-limiting conditions may be generated in the system in several ways. A. Production of gradients of glucose, lactate, pH and oxygen across the packed bed by an unequal distribution of nutrients and removal of metabolitic waste within the reactor. B. Generation of microenvironment pockets by highly active cells that may produce large amounts of lactic acid and ammonia; and C. Generation of anoxic pockets at certain areas in the packed bed.

These growth limitations may be solved by introduction of process control systems that will control gas flow and perfusion rate.

ACKNOWLEDGEMENTS

The authors wish to express their gratitude for the skillful technical assistance of Mrs. L. Silberstein and Mr. G. Pollack. We also wish to thank Prof. A. Shahar for the scanning electron microscopy.

REFERENCES

1. Feder, J. (1987). The development of large-scale animal cell perfusion culture systems. In: "Advances in Biotechnological Processes. (Ed. A. Mizrahi) (In press).

2. Feder, J. and Tolbert, W.R. (1985). Mass culture of mammalian cells in perfusion systems. Int. Biotech. Lab. June 40-53.

3. Gozes, Y., Epstein, N., Helmann, M., Bino, T., Rosenberg, H. and Kobiler, D. (1987). Secretion of BCGF by human hybridoma and B cell lines. Develop. biol. Stand. (In press).

4. Knazek, R.A., Gullino, P.M., Kohler, P.O. and Dedrick, R.L. (1972). Cell culture on artificial capillaries: An approach to tissue growth in vitro. Science 178, 65-67.

5. Lazar, A., Wilson, R. and Spier, R.E. (1987). Growth promoting materials derived from HeLa cell culture supernatants. Enzyme and Microb. Ferment. (In press).

6. Lim, F. and Moss, R.D. (1981). Microencapsulation of living cells and tissues. J. Pharm. Sci. 70, 351-354.

7. Marcipar, A., Henno, P., Lentowojt, E., Roseto, A. and Brown, G. (1983). Ann. N.Y. Acad. Sci. 413: 416-420.

8. Nilsson, K., Scheirer, W., Merten, O.W., Ostberg, L., Liehl, E., Katinger, H.W.D. and Moshbach, K. (1983). Entrapment of animal

cells for production of monoclonal antibodies and other biomolecules. Nature 302, 629-630.

9. Roberts, P.J. and Hayry, P. (1976). Sponge matrix allografts. Transplantation 21, 437-445.

10. Rosevear, A. and Lambe, C.A. (1983). Immobilized plant and animal cells. In: "Topics in enzyme and fermentation biotechnology. (Ellis Horwood Ltd.), 7, 13-16.

11. Rosevear, A. and Lambe, C.E. (1985). Immobilized plant cells. Adv. Biochem. Eng. Biotechnol. 31, 37-58.

12. Sirinek, L.P., Zinn, N.E., Perguson, R.M. and Orosz, C.G. (1986). In vivo mechanism of alloreactivity. Transplantation 41, 349-356.

13. Spier, R.E. and Whiteside, J.R. (1976). The production of foot and mouth disease virus from BHK 21 C 13 cells grown on the surface of glass spheres. Biotechnol. Bioeng. 18, 649-657.

14. Van Brunt, J. (1985). Immobilized mammalian cells. The gentle way to productivity. Biotechnology 4, 505-510.

15. Velez, D., Reuveny, S., Miller, L. and MacMillan, J.D. (1986). Kinetics of monoclonal antibody production in low serum growth medium. J. Immunol. Methods.86, 45-52.

16. Whiteside, J.P. and Spier, R.E. (1985). Factors affecting the productivity of glass sphere propagators. Develop. biol. Stand. 60, 305-311.

Discussion

W-S. HU — How big is your bed and reservoir?

LAZAR — 300 ml in the culture reactor and 700 ml in the reservoir.

HU You therefore get about 10^7 cells per ml - is that correct?

LAZAR Yes, that is approximately right.

U. STEINER What is the pumping rate.

LAZAR 20-30 ml/minute.

F. HOFFMAN Could the higher rate of lactate production be due to lack of oxygen?

REUVENY There is channelling inside the polyurethane foam and so some regions are probably anaerobic.

H. MURAKAMI In one of your slides you showed that although cell numbers did not vary; CEA production did.

REUVENY The fluctuation was in the system in which we continuously harvested the cells by letting the aggregate settle out. So the fluctuations were due to the harvesting technique.

MURAKAMI Did you try to cultivate the cells in low serum media?

REUVENY No.

31 Comparison of oxygenation methods in a 39L stirred bioreactor

D. Looby and J. B. Griffiths

The 39 litre (30L working volume) spin filter reactor has been designed for high density microcarrier perfusion culture (Fig. 1) (1), capable of supporting in excess of 10^7 cells/ml, at microcarrier concentrations of 10 to 15 grams (Cytodex 3) per litre. Oxygen solubility in cell cultures is low (7.6 mg O_2 L^{-1}) and despite the low oxygen demand of animal cells (2 to 6 $\mu g/10^6$ cells/hour) (2), oxygen supply to high density cultures is a limiting factor. A microcarrier culture of 1.0 - 1.5 x 10^7 cells/ml would require 60 to 90 mg O_2/L/hour.

The factors, which set the upper limits of agitation and sparging rates are (a) excessive foaming trapping microcarriers and blocking filters and (b) increased shear stress on the cells causing them to be stripped from the microcarriers.

The results (Table 1) show that surface aeration is clearly not sufficient to supply high density cultures (even with the 2:1 aspect ratio), and although direct sparging is more efficient it is unacceptable. Spin filter sparging, however, has considerable potential despite the decreased efficiency (1.5 fold compared to direct sparging). Shear damage to cells on microcarriers, foaming and high localised oxygen concentrations (which can be toxic to cells) are prevented because the bubbles do not come into direct contact with the cells, but are vented from the lower compartment to the head space, therefore high rates of sparging with air or oxygen can be tolerated. High oxygen transfer rates can be achieved in the reservoir vessel since sparging and agitation rates are limited only by foaming. Perfusion of the

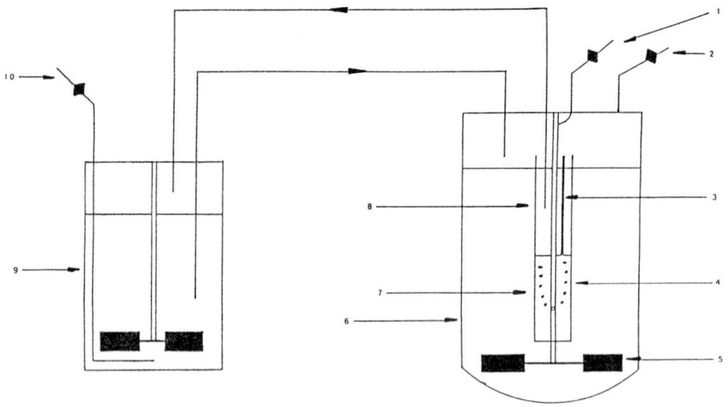

FIG. 1: Diagram of the 39 litre spin filter perfusion system (LH Fermentation) for high density microcarrier culture.

(1) air/oxygen sparge (2) top air/oxygen, (3) bubble vent to head space, (4) spin filter (60 µm mesh), (5) impeller, (6) culture vessel, (7) lower compartment (sparging), (8) upper compartment, (9) reservoir vessel, (10) air/oxygen sparge. Oxygen transfer rates were obtained by the dynamic gassing out method in water at 37C.

Table 1. Oxygenation of the 39 l spin filter reactor with 10 l perfusion vessel

Aeration Systems	Stirrer Speed (RPM)	Gas	Flow Rate (ml/min)	OTR mg/L/h	Number of cells supplied x 10^6 ml^{-1}*
Surface Aeration	40	Air	300	0.5	0.083
Direct sparging	40	Air	300	4.6	0.76
Spin Filter Sparging	40	Air	300	3.0	0.5
Perfusion (1 vol/hour)	40	Air	300	12.6	2.1
Spin Filter sparging + Perfusion (1 vol/h)	40	Air	300	15.9	2.65
Spin filter sparging	80	Oxygen	1000	51.0	8.5
Spin filter + perfusion (1 vol/h)	80	Oxygen	1000	92.0	15.3

*Based on an oxygen requirement of 6 µg/10^6 cells/hour (2) culture vessel with oxygen rich media from the reservoir (1 vol/hour) leads to a significant increase in the OTR and this was even higher when used in conjunction with spin filter sparging. However, the oxygen supplied (15.9 mg/l/h) was only enough for 2.65×10^6 cells /ml. Increasing the agitation rate to 80 rpm, and spin filter sparging with 1 litre pure oxygen per minute, allows sufficient oxygen to be supplied for up to 1.5×10^7 cells /ml.

The results from this study gives guidelines to the aeration strategies necessary to supply high density microcarrier cultures and in particular show the importance of medium perfusion as an aeration method. The system is capable of supporting a perfusion rate of 2 volumes/hour and higher sparging rates into the spin filter so its total capability has not yet been reached.

ACKNOWLEDGEMENTS

This work was supported by the Biotechnology Group, Department of Trade and Industry.

REFERENCES

1. Griffiths, J.B., Cameron, D.R. and Looby, D. (1987). A comparison of unit process systems for anchorage dependent cells. Develop. Biol. Standard., 66, 331-338.

2. Spier, R.E. and Griffiths, J.B. (1983). An examination of the data and concepts germane to the oxygenation of cultured animal cells. Develop. Biol. Standard. 55, 81-92.

Discussion

M. SILBERKLANG — Is the output per cell the same?

GRIFFITHS — The output of the cells we have looked at, which is mainly enzymes, is constant between 3 and 10 million. A problem we are studying is that the yield per cell at any of those densities is lower than we get in a roller bottle.

32 Development of large-scale mammalian cell perfusion culture systems

J. Feder

The production of biologically important molecules by animal cells *in vitro* has presented a challenge for the development of efficient large-scale culture systems. The focus must extend beyond achieving quantities of cells and products and address the downstream processing problems to be encountered. A number of different reactor systems have been employed ranging from the conventional batch or semi-continuous low cell density reactors to continuous steady state systems. The former tend to be simpler, large size, stirred reactors based upon traditional microbial fermentation experience. Continuous fermentors that attempt to maintain a homeostatic culture environment are generally more complex. However, under steady state conditions high cell density cultures are achieved with increased cell/product yields per unit volume of medium used and significant reduction of the serum concentration on the medium. This results in smaller reactors by at least a factor of ten, increased product concentration and in many cases as significant increase in the specific activity of the product in the crude product stream.

Tables 1 and 2 illustrate this in a simplified manner.

Cell Biology and genetic engineering address the issues of developing high expressing cells and the optimization of medium and growth conditions to maximize that which has been engineered into the cell. This represents the most important intrinsic determinant of the production potential of a given cell. The cell culture system must optimize culture conditions and operation to maximize this potential.

TABLE 1. IDEAL PRODUCT STREAM

- Maximum Product Concentration
- Maximum Specific Activity
- Maximum Product Stability
- Contaminants Defined

TABLE 2. TO ACHIEVE OPTIMUM PRODUCT STREAM

- Maximize Cell Expression (Cell, Medium)
- Maximize Cell Density
- Minimize Serum Requirements
- Defined Medium
- Maintain High Cell Viability

This includes maximization of the cell density, simplification of the medium components particularly serum or other growth factors and the maintenance of a culture at a high viability in a productive physiological state.

Table 3 compares the relative specific activity of a protein produced in a conventional batch reactor with that obtained in a perfusion reactor. This, of course, assumes no product feedback inhibition, a phenomenon occasionably encountered that must be addressed.

The difference in the product concentration, a factor of 5 (compared to a 10 fold difference in cell density) takes into account the dilution due to the perfusion.

All of these considerations have motivated the development of non-conventional, steady state type animal cell culture reactor systems.

A number of approaches have been employed to maintain a homeostatic culture environment for suspension grown cells including fed batch (1-3), Monod type chemostat (4-8), cytostat (9), and perfusion systems (10-18).

Fed batch culture is characterized by a continuous feed of medium to a culture of cells in a closed system at a rate which both maintains a near constant nutrient level and dilute metabolic product levels. Efficient operation requires methods for monitoring and controlling the concentration of the limiting component. Throughout the growth process the culture volume increases until the reactor is filled and the last phase of culture is that of a closed batch system. Nevertheless, over the greater portion of the cell growth phase a near steady state culture environment is maintained and higher cell densities and cell yields per unit volume of medium used are obtained than with conventional closed batch reactors. Reuveny and coworkers (3) reported obtaining a 5-8 fold increase in viable cell density of hybridoma cells in a fed batch reactor over that of an unfed batch system. About twice the monoclonal antibody yield was obtained. Although fed batch cultures can be operated in a semi-continuous or solera mode they are essentially non-continuous cultures.

In contrast, chemostat culture provides an opportunity for maintaining steady state culture conditions and cell density in a continuous mode of operation. In chemostat culture, the nutrient medium contains saturating concentrations of all of the growth components with the exception of one, the concentration of which determines the cell density and the feed rate of which determines the growth rate. Systems have been developed for chemostat culture of animal cells. However, unlike bacterial systems where the growth can be controlled by a single limiting medium component, the control of animal cell density and growth is more complicated. Metabolic products can effect the physiological state of the cell. Animal cells consume nutrients such as glucose at reasonable rates even when maintained in a non-proliferating state. Nevertheless, various mammalian cells have been grown in continuous chemostat culture, often under glucose limitations. The steady state in chemostat operation is maintained by a continuous dilution of culture so that the specific growth rate, μ, is equal to the dilution rate, D, defined as the number of volume charges per unit time. Monod (4) showed that the specific growth rate was a function of the concentration of the limiting growth component such that

$$\mu = \mu_{max} \left[\frac{S}{S + Ks} \right]$$

where μ_{max} is the maximum specific growth rate, S is the steady state concentration of the limiting component and K_s describes a dissociation constant for the cell substrate association. Thus, for a Monod type chemostat culture the specific growth rate at equilibrium is less than the maximum.

It has been shown by Herbert, Elsworth and Telling (19) that for a given substrate concentration and dilution rate there are unique steady state substrate concentrations, S_{out} and cell densities given by the following equations.

$$S_{out} = K_s \left[\frac{D}{\mu_{max} - D} \right]$$

$$\text{cell concentration} = Y\,(S_{in} - S_{out}) = Y \left[S_{in} - K_s \left(\frac{D}{\mu_{max} - D} \right) \right]$$

where Y is a yield constant defined as the amount of cells obtained per unit of substrate consumed. They also showed that the maximum output of a continuous culture is given by the following expression.

$$\text{Output}_{maximum} = \mu_{max} \cdot Y \cdot S_{in} \left[\sqrt{\frac{K_s + S_{in}}{S_{in}}} - \sqrt{\frac{K_s}{S_{in}}} \right]^2$$

When S_{in} is much greater than K_s a maximum output is obtained which is a function of the substrate concentration and the μ_{max}. This type of culture system provides an efficient means for continuous production of cells and derived products in a homeostatic environment. As an example, Tovey (7, 20) reported establishing stable steady state cultures of mouse L1210 cells in a chemostat at dilution rates ranging from 0.2 day^{-1} to 2.0 day^{-1} that were characterized by constant cell density, constant cell volume, constant lactate concentration and constant rates of incorporation of labelled thymidine, uridine and amino acids. An eight fold greater production rate of cells per unit reactor volume was obtained in the chemostat over batch operation with a 2.7 greater cell yield per liter of medium used.

The cytostat type of continuous culture also maintains a steady state by continuous dilution (9). However, in contrast to the chemostat all growth components are in excess concentration and, hence, maximum specific growth rates are maintained. Under these conditions, a range of steady state cell densities can be established at the same dilution rate. However, as the cell density is increased the steady state substrate concentration will be decreased to a limiting concentration at which the condition of $S \gg K_s$ will not be satisfied and the specific growth rate will decrease. The limiting cell densities can be estimated as will be shown below.

Perfusion introduces another degree of freedom for increasing the steady state cell concentrations and, hence, the maximum cell or product output of a continuous culture system. In perfusion the cells are physically retained in the reactor such that a continuous nutrient medium feed and spent medium removal is achieved without dilution of the cell density. This process can be coupled with a cytostat type of operation to provide a perfused continuous culture. Like the cytostat, all substrates are provided in the feed medium at saturating concentrations. The steady state substrate concentrations, cell density and uptake rate functions can be calculated from the mass balance equations.

Under steady state conditions:

$$\text{Substrate}_{in} = \text{Substrate}_{consumed} + \text{Substrate}_{out}$$

assuming staturating substrate concentration and maximum uptake rates.

$$S_{in} \times (D + P) \times V = k_s \times \text{cell number} + S_{out} \times (D + P) \times V$$

$$k_s = \frac{(S_{in} - S_{out})\ (D + P)}{(cell/V)}$$

Both the perfusion rate, P, and the harvest dilution rate, D, are expressed as the number of volume changes per unit time. The specific consumption constant, k_s, is defined in terms of the amount of substrate consumed per unit time per cell. From the values of D, P, S_{in}, k_s and the estimated minimum steady state substrate level, S_{out}, which still satisfies the condition $S_{out} >> K_s$ ($S \sim 10K_s$) one can estimate the maximum cell density from the expression

$$(cells/L) = \left(\frac{S_{in} - S_{out}}{k_s}\right)(D + P)$$

Thus, perfusion increases the limiting cell density that can be maintained at maximum growth in a cytostat by the term

$$\frac{(S_{in} - S_{out})\ P}{k_s}$$

The maximum output for the perfusion cytostat would be equal to the product of the maximum cell density and (P + D). For a cytostat culture this is the limit of the Herbert et al (19) equation.

This discussion, of course, applies to suspension grown cells in a continuous harvest mode of operation. Anchorage dependent cells that grow attached to a substrata material provide a ready configuration for perfusion. A number of such systems have been reported including perfused tissue flasks, roller bottles, hollow fibers and microcarriers. Generally, perfusion of such cultures resulted in increased rates of proliferation, significant increase in cell yield and generally multilayered cultures. Studies carried out in our laboratory with perfused hollow fiber reactors with different configuration also yielded high cell densities (21). The basic design of these reactors incorporated delivery of oxygen and carbon dioxide through the lumen of the fiber and medium perfusion on the shell side of the fibers. A flat bed hollow fiber reactor system was designed which addressed some of the limitations of the cartridge type configuration and provided an opportunity for scale-up.

Figure 1 shows a schematic diagram of a flat-bed hollow fiber cell culture reactor. A 30 cm by 30 cm fiber bed consisting of about 3000 fibers (outside diameter 340 μm) would provide about one square meter of surface area for cell attachment and growth. A number of cells have been grown in these reactors including SV3T3, WI-38, HeLa, rhesus monkey kidney (MK-2), baby hamster kidney (BHK), Vero and primary fetal kidney. As an example, SV3T3 cells were inoculated at a level of about 10^4 cells/cm^2 of fiber surface area and at harvest a density of 9×10^5 cells/cm^2 was achieved. Although no attempts were made to optimize the perfusion rates, good cell growth was maintained and the cells expressed significant quantities of macrophage migration factor, MIF, and a plasminogen activator.

An important feature of the perfused hollow fiber reactor was the capacity to maintain a population of cells after growth in a non-proliferating state for extended periods of time expressing a specific protein. Unlike the shear stress to which cells on microcarriers are continuously exposed the fiber-attached cells are in a relatively undisturbed homeostatic environment under gentle perfusion. An an example, primary human fetal kidney cells were maintained in a seum-free medium for a number of weeks during which time a fairly constant level of urokinase (about 500 units/ml) was continuously produced (21).

Thus, even anchorage dependent cells, the proliferation of which is substrata area dependent, achieve significantly higher cell densities and viabilities under perfusion.

Unlike cells that attach to a substrate matrix and are thus

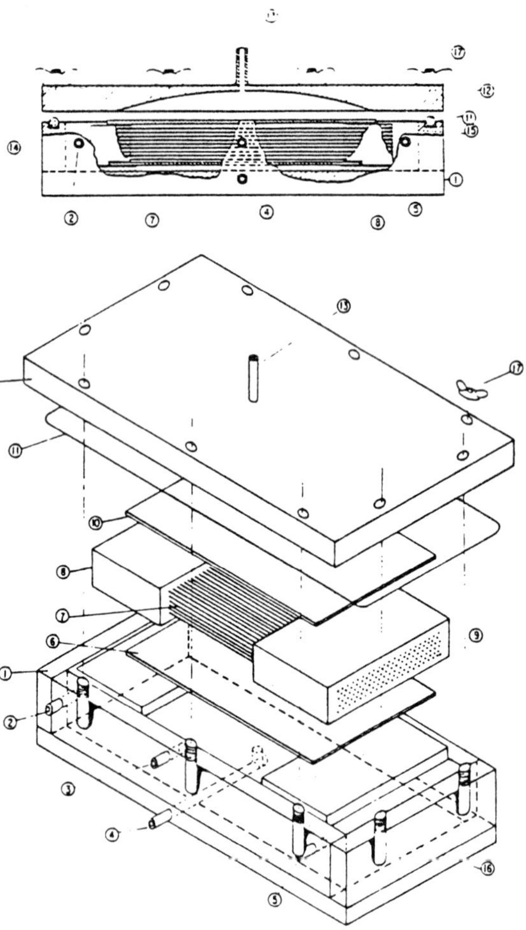

Figure 1 Schematic diagram of a flat-bed hollow-fiber cell culture reactor.
(1) Lower reactor housing; (2) air/CO_2 inlet; (3) cell inoculation port; (4) fresh medium inlet; (5) air/CO_2 outlet; (6) lower stainles-steel microporous filter; (7) hollow-fiber bed; (8) polypropylene hollow-fiber potting; (9) open ends of hollow fibers; (10) upper stainless-steel microporous filter; (11) gasket; (12) upper reactor housing; (13) spent medium outlet; (14) prereactor chamber; (15) postreactor chamber; (16) bolt; (17) winged nut. Reprinted from Ku et al., (18) with permission of the publisher.

physically separate from the perfusion medium, cells that grow in suspension present a particular challenge for perfusion. Mechanical filtration devices that allow cell free medium withdrawal without clogging by cells and cellular debris must be devised. Himmelfarb, Thayer and Martin (12) introduced a spin filter device for perfusing animal cell suspension cultures. The filter was rotated by a magnetic stirrer at about 300 rev/min and the shear forces at the boundary between the rapidly rotating filter surface and more slowing moving liquid minimized clogging of the filter during operations.

The spin filter concept was the basis for the large-scale perfused suspension culture systems developed in our laboratory (14-16, 18). In our case, a stationary filter surrounded by a rotating concentric agitator located in a separate satellite vessel was developed.

A diagram of the early perfusion cytostat reactors is shown in figure 2.

As expended medium was withdrawn through the filter in the satellite vessel a capacitive level system controlled the fresh medium pump. The pH was monitored by an autoclavable INGOLD® electrode and controlled by varying the CO_2 concentration of the overlay gas and by addition of sodium bicarbonate when necessary. Reservoirs for fresh and expended medium were connected to the reactor through peristaltic pumps. Agitation of the cell culture suspension was provided by the use of large area slowly rotating flexible sheets or "sails." These minimized the shear effects on the cells, a particularly valuable feature for microcarrier suspension cultures to be discussed later. Aeration was supplied by a 5-10% CO_2 in air mixture overlay of the medium surface and a very low sparge of 100% oxygen as required. Temperature control was maintained by locating the reactor system in a 37°C warm room. Reactor sizes from a few liter research models to 100 liter production models

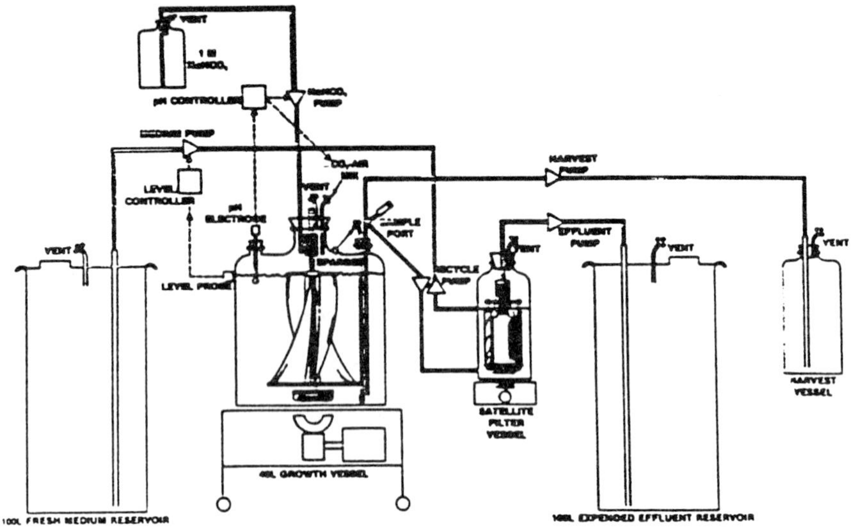

Figure 2 Schematic diagram of the perfusion cytostat system. Cell-free expended medium was removed from the interior of the porcelain filter at a rate set on the effluent pump to provide the required perfusion of fresh medium. Cell suspension was recycled to and from the satellite vessel by a double-headed pump at a rate sufficient to maintain similar component concentrations in the two vessels. While operated as a cytostat, a harvest pump removed cell suspension directly to a harvest vessel contained in an adjacent 4°C cold room. The resulting reduction of the liquid level in the growth vessel was compensated by an exterior capacitive level probe which actuated the medium pump to supply fresh nutrient medium. An autoclavable pH electrode penetrating the growth vessel allowed control of the pH by varying the CO_2 in the mixture of overlay gas and by the addition of sodium bicarbonate when necessary. A sterile air-shielded sample port and a means for low rate oxygen sparge through a sintered glass dispersion tube were also provided. Except for the harvest vessel, the entire system was contained with a 37°C warm room. Reprinted Tolbert and Feder (23) with permission of the publisher.

have been developed and employed for production of various cells and expressed products. The perfusion suspension system could be operated both in batch and continuous cytostat-type modes. The cultures were characterized by high cell density, significant increase in cell yield per unit volume medium and often reduction in required serum concentration in the medium. As an example, it was observed early that rat Walker carcinosarcoma cells grown in continuous perfusion culture reached cell densities as high as 30.8×10^6 cells/ml with viabilities near 100% compared to maximum cell densities of about 1.5×10^6 cells/ml in batch culture. The exponential growth rates were about the same (~ 0.042 hr^{-1}). In addition, the yield of cells per liter of medium used was about 3 fold that obtained without perfusion (Table 4). No attempt was made to optimize the perfusion rate in these studies. However, at rates below 5 mls/hour/10^9 cells the growth rate decreased significantly. The glucose consumption rates for the rat Walker cells was about 50-80 micrograms/hrs/10^6 cells. Since the DMEM medium used in these studies contained 4.5 mg/ml of glucose, a feed rate of 18.5 mls/hr/10^9 cells would be required to maintain a steady state glucose level.

Another observation under perfusion growth was the reduction in required serum concentration in the medium. Figure 3 shows the growth of the human hepatoma cell line, SK-Hep-1, in continuous perfused cytostat culture with the cell density maintained at 10 mls/liter, packed cell volume. The cells were initially grown in 5% bovine calf serum supplemented DMEM but during the cytostat operation the serum concentration was reduced to 1% with minimal effects on cell growth and viability.

Perfused cytostat culture has been used to produce both cells and products in a long term continuous operation. These include monoclonal antibodies, growth factors, tissue plasminogen activator (tPA) and cell

TABLE 3. INTEGRATION - Cell Cultured and Downstream Processing

EXAMPLE

CONVENTIONAL REACTOR   cell density, 10^9 cells/liter

10% FBS = 3.9 gm protein/liter

Product = 10 mg/liter (cell productivity, 10 μg/10^6 cells/day)

$$\underline{\text{S.A.}} = \frac{10 \text{ mg/liter}}{3900 \text{ mg/liter}} = \underline{0.00256}$$

PERFUSION REACTOR   cell density, 10^{10} cells/liter

1% FBS = 0.39 gm protein/liter

Product = 50 mg/liter (cell productivity, 10 μg/10^6 cells/day)

$$\underline{\text{S.A.}} = \frac{50 \text{ mg/liter}}{390 \text{ mg/liter}} = \underline{0.128}$$

$$\underline{\text{RELATIVE S.A.}} = \frac{0.128}{.00256} = \underline{50}$$

TABLE 4. COMPARISON OF PRODUCTION OF RAT WALKER CARCINSARCOMA CELL IN CONVENTIONAL AND PERFUSION SUSPENSION CULTURE

	Total Medium[a]	Total cells [a]	Yield
	liters	x 10^{12}	x 10^4 cells/liter
Conventional Suspension Culture [b]	14,150	13,422	0.95
Perfusion culture system [c]	612	1,705	2.80

a. This represents the sum of medium and cells from a number of experimental runs.

b. 100 liter vibromixer suspension reactor.

c. 4 and 40 liter perfusion suspension culture reactors.

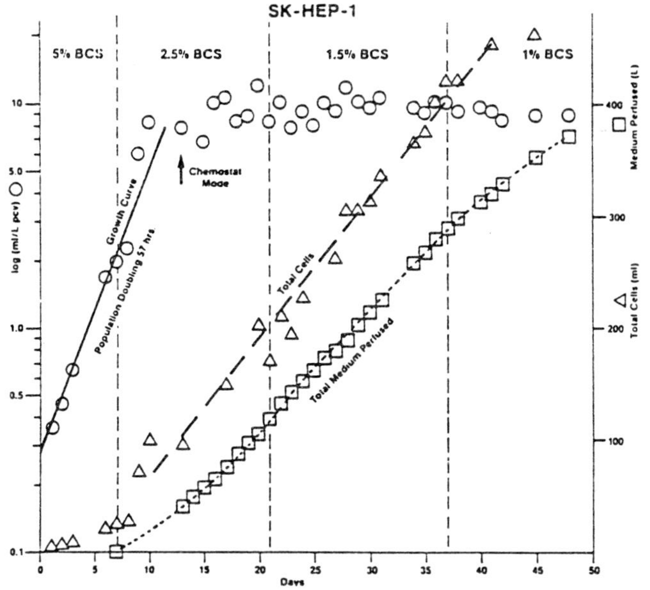

Figure 3 Growth of human hepatoma cells in the 12 liter perfusion cytostat system.
Initial growth was in 5% bovine calf serum supplemental DMEM. Serum concentration was reduced as indicated. Cytostat operation was initiated on day 10 to maintain a cell density of 10 ml/L packed cell volume. Reprinted from Feder and Tolbert (26) with permission of the publishers.

derived molecules such as membrane proteins. For example, the 9.2 27 hybridoma which produces a monoclonal antibody against an antigen derived from the M14 human melonoma was cultured in a 40 liter perfusion cytostat for 18 days producing 40 gram of antibody (600 liters medium). The steady state cell density was 8.4 mls/liter, packed cell volume and the serum concentration was reduced during the run to below one percent.

To this point, we have addressed the maintenance of a steady state environment with respect to nutrients and accumulated metabolic products. However, providing adequate oxygen particulary at high cell densities is a major problem. The early perfusion reactors provided a gas sweep over the culture surface of a few liters per minute and a slow sparge of pure oxygen. To prevent foaming problems the sparge was limited to a few milliliters per minute. For oxygen consumption rates of about 0.045 to 0.060 cm^3 (STP)/min/10^9 cells for a number of cells, oxygen quickly becomes limiting when supplied in this manner. A silicone rubber tubing (O.D. 2 mm; I.D., 1.0 mm) "lung" permeator was developed to address this problem.

This was incorporated into the design of a new 14 liter computer controlled research model perfusion reactor. The initial evaluation studies wre carried out with a lung surface area of about 100 cm^2 per liter of culture volume. During cell growth the dissolved oxygen was maintained at about 0.11 atmospheres under computer control. The inlet and outlet gases from the reactor were analyzed by gas chromatography together with the dissolved oxygen and headspace gas. This yielded oxygen consumption rates and provided control data for regulating the oxygen content of the feed gas mixture to maintain the established dissolved oxygen levels in the medium. The per cent oxygen in the lung feed gas mixture ranged from an initial 10-15% to a maximum of about 75%. Thus, the maximum partial pressure of oxygen in the lung (maintained at atmospheric pressure) was about 0.69 atmospheres. Using an oxygen permeability constant for silicone rubber of 2.17×10^{-4} cm^3 (STP)/(min x P (atmos) x cm), the maximum oxygen diffusion flux would be about 2.5×10^{-3} cm^{-3} (STP) per minute per cm^2 of surface area (wall thickness, 0.05 cm). Thus, for a total of 100 cm^2 of lung surface per liter of culture a maximum oxygen supply rate of 0.25 cm^3 (STP)/minutes could be maintained. Oxygen consumption rates for Bowes melanoma cells in perfused cytostat culture of 0.045 to 0.048 cm^3 (STP)/(min x 10^9 cells) have been measured in our laboratory. Ehrlich and coworkers (22) reported consumption rates of 0.047 cm^3/ (min x 10^9 cells) for BHK and L929 cells. Consequently, a maximum cell density of about 5.4 x 10^9 cells/liter could be supported by this lung permeator configuration with these types of cells. By doubling both the lung surface area and the Lung gas pressure the oxygen transport capacity could be increased about 4 fold and a cell density approaching 3×10^{10} cells/liter could be supported without compromising the oxygen supply or the metabolic state of the cells.

Table 5 describes the growth of Bowes melanoma cells, various metabolic parameters and the production of tPA in the 14-liter computer controlled perfusion cytostat reactor. Runs 1 and 2 compared the effects of decreasing the perfusion rates. The steady state cell density was maintained at about 2.7×10^6 cells/ml. However, since these cells have a large cell volume, 3.75 $cm^3/10^9$ cells, the actual cell volume was about 10 mls/L or near 1.0% of the reactor volume. At the lower perfusion rate the steady state glucose concentration dropped to 1.65 mg/ml from 2.68 mg/ml at the higher perfusion rate and the lactate concentrations almost doubled from 1.07 mg/ml to 1.92 mg/ml. However, this did not effect the specific rates of glucose uptake, oxygen consumption or tPA production. The uptake rate function were calculated from the mass balance relationships discussed earlier. Assuming that at a steady state glucose concentration of

TABLE 5. **Perfusion Cytostat Production of Bowes Melanoma Cells and Tissue Plasminogen Activator (tPA).**

	Run 1	Run 2		Run 1	Run 2
pH	7.14	7.00	Glucose uptake ($mg/hr/10^9$ cells)	25.3	24.8
Dissolved O_2 (Atm)	0.11	0.11	Lactate Production ($mg/hr/10^9$ cells)	18.4	19.1
Cell density ml/l (PCV)	9.7	10.2	$[glucose]_{in}$ mg/mℓ	4.15	4.15
cell/L x 10^{-9}	2.6	2.7	$[glucose]_{out}$ mg/mℓ	2.68	1.65
Total cells $\times 10^{-9}$	39.0	40.8			
Doubling Time (hrs)	51.2	51.2	$[lactate]_{out}$ mg/mℓ	1.07	1.92
$D=\mu$ (hr^{-1})	0.0135	0.0135	Oxygen Uptake cm^3(STP)/hr/10^9 cell	2.7	2.9
D + P $(hr)^{-1}$	0.0448	0.0268	tPA Production ($mg/day/10^9$ cells)	0.28	0.26
P (hr^{-1})	0.0313	0.0133			

TABLE 6. Production of Human Foreskin Fibroblast Cells in Perfusion Microcarrier Reactors (23).

	Surface Area m^2	Cell Density cells/ml	Cell Density cells/cm^2	Total Cells	Yield cell/liter medium[b]
Roller Bottle	0.069	3.0×10^5	4.3×10^4	3.0 10^7	3.0×10^8
4 Liter Reactor	21.6[a]	1.0×10^7	1.8×10^5	4.0×10^{10}	1.4×10^9
44 Liter Reactor	188.0[a]	8.0×10^6	1.8×10^5	3.4×10^{11}	1.3×10^9

[a] Bio-Carriers

[b] DMEM, 10% Fetal Calf Serum

Table 7 **Comparison of the Growth of a Variety of Cells in Perfusion Microcarrier Reactors**

Cell	Microcarrier	Area m^2	Culture Volume liters	Cell Density cells/ml	Cell Density cells/cm^2	Perfusion Rate* mls/hr/10^9 cells	Total Cells	Cell Yield cells/liter	Reference
MDCK	Superbeads	2.021	0.5	9.0×10^6	2.2×10^5	30	4.5×10^9	5.2×10^8	27
VERO	Superbeads	1.078	0.1	1.0×10^7	9.2×10^4	41	1×10^9	1.7×10^8	28
		1.078	0.1	3.0×10^7	2.8×10^5	22	3.0×10^9	2.3×10^8	28
SV-HSF-153	Borosilicate glass tube segments	2.8	1.5	1.1×10^6	6.1×10^4	100	1.7×10^9	4.8×10^7	25
HFF	Bio-Carriers	188.0	44.0	8.0×10^6	1.8×10^5	5	3.4×10^{11}	1.3×10^9	15

*estimated perfusion ratio (average cell no.)

0.15 mg/ml the conditions of S>>Ks are still holding, maximum cell densities of 7.08 x 10^9 cells/liter and 4.3x10^9 cells/liter could be maintained at the higher and lower perfusion rates respectively. This, of course, assumes that glucose depletion is the limiting process. Optimization of perfusion culture systems requires detailed rate information of a number of metabolic function including both specific nutrient depletion as well as accumulation of metabolic products that might inhibit growth and expression. Such studies are in progress in our laboratory. Empirical approaches involving rates of medium turnover, however, might be operationally quite useful until more definitive metabolic data are available.

Microcarrier systems are one of the most effective methods for large scale culture of anchorage-dependent cells (23, 24). There have been various reports on perfused microcarrier culture (15, 25-28). The microcarrier culture system developed in our laboratory employed a modification of the perfusion culture reactor described earlier. This involved the addition of a settling chamber in the recycle line between the cell reactor and the satellite filter vessel (15, 26) (Figure 4). As cell microcarrier suspension is pumped to the satellite filters vessel it flows through the settling chamber in which the relatively dense slurry of cells and microcarriers were allowed to settle outside the reactor vessels agitated volume. Essentially cell-microcarrier free medium continues in the recycle circuit to the satellite filter vessel. This operation could be maintained by directly removing a fraction of the spent medium without need for the filter. During the settling procedure beads and cells at high density were brought into close contact resulting in a bridging of cells from one carrier to another and the formation of aggregates of microcarriers. Under perfusion conditions the cell density increased significantly beyond that projected from the available bead surface area.

Figure 5 shows photomicrographs of the growth of human foreskin

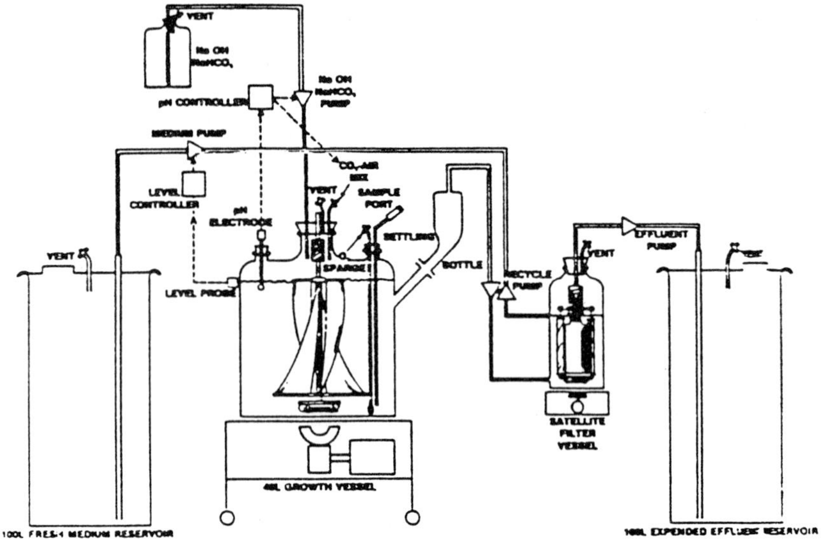

Figure 4 Schematic diagram of the perfusion microcarrier reactor system.
It differs from that shown in Fig. 2 by the addition of a settling bottle in the recycle circuit between growth and satellite filter vessels. This sequestering of the cells and microcarriers outside of the agitated volume of the growth vessel enhances formation of cell-bead aggregates and allow essentially cell-carrier free medium to be pumped to the filter. Reprinted from Tolbert and Feder (23) with permission of the publishers.

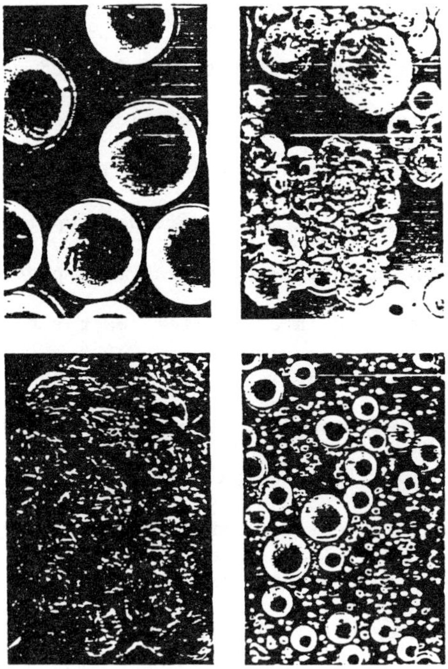

Figure 5 Photomicrographs of human diploid forescin fibroblasts on Bio-carriers.
Upper left, one day after inoculation of the 4 liter reaction, 240x; upper right, after 15 days in culture, 80x; lower left, cell aggregate at higher magnification; lower right, dissociated cells and microcarriers after brief trypsin treatment, 80x. Reprinted from Tolbert and Feder (23) with permission of the publisher.

fibroblast cells on the Bio-Carrier microcarriers. Cell densities approaching 10^7 cells/ml and greater than 1.8×10^5 cells/cm^2 of surface area were achieved. Cell dissociation from the beads was obtained readily by a 10 minute treatment at 37°C with Hank's balanced salt solution containing 0.5 mg/ml of porcine trypsin and EDTA (0.2 mg/ml) allowing for expansion and scale-up. Using this procedure, a total of 3.4×10^{11} cells were produced in a 44 liter reactor containing 400 grams of Bio-Carriers with a total surface area of 188 square meters. A 4-fold increase in cell yield was achieved over that obtained in an unperfused roller bottle production (Table 6). A variety of different non-transformed human fibroblast cells as well as transformed and genetically engineered animal cells and secreted products have been produced efficiently in these perfused microcarrier reactors. Growth vessels up to 100 liter have been developed equipped with lung permeator oxygen delivery and computer control.

A number of studies have shown that perfusion increased cell density of microcarrier cultures (15, 17, 25, 28). However, there has been no attempt to optimize the perfusion rates with respect to cell or product yield. This is illustrated in Table 7.

The perfusion rates expressed as milliliters of medium perfused per hour per 10^9 cells are based upon estimated average cell numbers. In almost all cases under perfusion conditions both higher cell densities per unit volume of medium as well as higher cell densities per unit surface area were obtained. However, less impressive cell yields per unit volume of medium used were reported probably reflecting less than optimum perfusion rates. There is a need for basic metabolic data on nutrient uptake rates and generation of metabolic products that inhibit cell growth and expression. This type of information provides the basis for designs of medium for perfusion reactors and rational perfusion protocols.

The perfusion culture systems described above support a popula-tion of cells in a continuous growth phase during which the expression of specific products often occurs. For example, the production of monoclonal antibodies by hybridoma cells is achieved in a perfusion cytostat during the continuous growth phase of the process. Many biologically important molecules *in vivo*, however, are produced by non-proliferating cells. This also has been observed *in vitro* with a variety of cells, particularly those types of molecules that are not involved in regulating growth such as autocrines. In practice, a population of cells is grown in a serum supplemented nutrient medium, the medium removed and a serum-free medium added for the production phase of the process. This conditioned medium is free of contaminating serum proteins and consequently simplifies the purification processes that must follow. The production of biomolecues by non-proliferating cells and the maintenance or such productive cells in a biochemically healthy and active state has been the focus of research efforts in our laboratory in recent years. Since the cell growth phase is partitioned from the maintenance and expression phase of the process, the medium component requirement for the latter are often quite simplified. A computor controlled perfusion maintenance reactor system has been developed which can maintain both anchorage dependent and anchorage independent cells at high density for extended periods of time (29).

Figure 6 presents a schematic cross sectional diagram of a perfusion maintenance reactor. Cells are maintained in a semi-rigid matrix within a reactor vessel chamber. Fresh medium is supplied through relatively low porosity tubes dispersed throughout the reactor chamber and expended medium is withdrawn through relatively high porosity tubes positioned to minimize the flow path and maximize the medium perfusion. A silicone rubber tube lung permeator is provided to supply oxygen. Anchorage dependent cells are grown on microcarriers to high density and

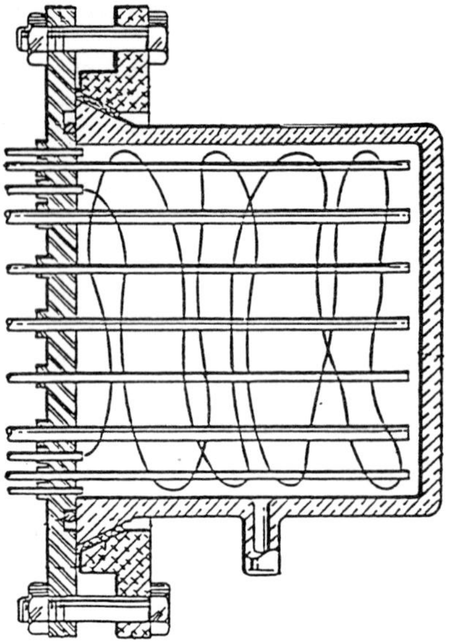

Figure 6 Schematic cross sectional diagram of perfusion maintenance reactor. This system maintains high density cultures of non-proliferating cells for continuous production of secreted molecules (25,67).

inoculated as such into the reactor. This provides a ready matrix for the cell packing. Suspension grown cells are mixed with microcarriers or other material to provide a perfusion matrix. Cell densities up to 10^8 cells/ml can be achieved in this reactor. A 16.5 liter perfused maintenance reactor has been developed with a total capacity in excess of 10^{12} cells. At these high cell densities depletion of medium nutrients and accumulation of metabolic products rapidly occur. Nevertheless, successful maintenance of growth arrested cells in a homeostatic environment has been achieved for extended periods of time. Among the various cells studied were included anchorage independent hybridomas and anchorage dependent fibroblasts producing monoclonal antibodies and tissue plasminogen activator respectively.

Although impressive results have been obtained in early studies using serum free or low serum media, little is known about the maintenance requirements of cells in vitro. Generally, metabolic rates are significantly less for growth-arrested cells as compared to growing cells. However, little is known whether there are specific maintenance factors that are required for continued cell viability and expression that are distinct from growth factors. This is an important area for research that can impact the utility of such an approach for the production of important molecules.

To summarize, the introduction of perfusion provides significant advantages for large-scale culture of both anchorage dependent and suspension grown cells. High cell densities with good viability are achieved, cell yields per unit volume of medium are maximized and generally lower concentrations of medium components including serum are required under steady state conditions The increase in cell density particularly when coupled with continuous cytostat operation and the reduction in serum provides a concentrated crude product stream having a significantly greater specific activity. This can simplify the downstream purification process and, thus, impact the overall production efficiency and yield. Also, high cell density reactors can be considerably smaller than those employed for conventional culture. Optimization of such reactors, however, requires much basic metabolic information. Culture media have been developed for batch systems. High cell density culture in a steady state environment provides a special challenge for optimization of medium. Perfusion depends upon some mechanical device to retain cells while removing spent medium and refeeding with fresh medium. It is important to re-examine methods to achieve this. Aeration is a generic problem for all types of animal cell reactors. However, at high cell densities the oxygen requirements are accordingly increased further magnifying the oxygen delivery problem.

The maintenance of a steady state culture environment by means of perfusion minimizes to some extent the advantages gained by achieving high cell densities. As the cell density increases, the perfusion rates increase. Figure 7 illustrates this by showing the effects of cell density on both total protein and protein concentrations and specific activity. A 100-fold increase in cell density, 10^9 cells/liter to 10^{11} cells/liter, results in a 100 fold increase in the total protein produced. However, due to the dilution by the perfusion the effective increase in product concentration and specific activity is less than 8-fold. Further optimization must address additional approaches to establishing homeostatic culture conditions.

Finally, the potential advantages of perfusion maintenance reactors should motivate research directed at defining the maintenance requirements of non-proliferating, productive animal cells.

REFERENCES

1. Pirt SJ: The theory of fed batch culture with reference to the

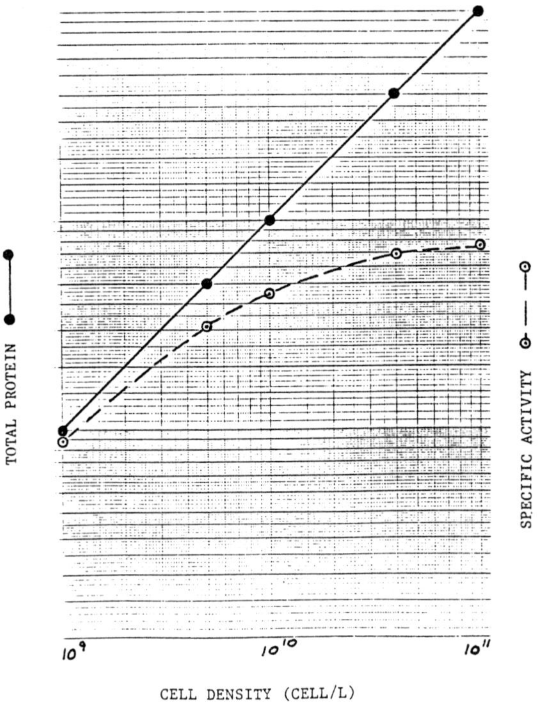

Figure 7 Effect of cell density on productivity

pencillin fermentation. J. Appl Chem Biotechnol 24:415, 1974.

2. Glacken MW, Fleischaker RJ, Sinskey AJ: Large-Scale production of mammalian cells and their products: Engineering principles and barriers to scale-up. Ann NY Acad Sci 413:355, 1983.

3. Reuveny S, Velez D, Miller L, Macmillan JD: Comparison of cell propagation methods for their effect on monoclonal antibody yield in fermentors. J Immunol Methods 86:61, 1986.

4. Monod J: La technique de culture continue theorie et applications. Ann Inst Pasteur 79:390, 1950.

5. Cohen EP, Eagle H: A simplified chemostat for the growth of mammalian cells: Characteristics of cell growth in continuous culture. J Exp Med 113:467, 1961.

6. Moser H, Vecchio G: The production of stable steady states in mouse ascites mast cell cultures maintained in a chemostat. Experientia 23:120, 1967.

7. Tovey MG, Brouty-Boye D: The cultivation of animal cells in a chemostat. J Appl Chem Biotechnol 26:345, 1976.

8. Tovey MG: The cultivation of animal cells in continuous-flow culture. In Spier RE Griffiths JB (eds): "Animal Cell Biotechnology," London: Academic Press, 1985, Vol. 1, p. 195.

9. de St. Groth SF: Automated production of monoclonal antibodies in a cytostat. J Immunol Meth 57:121, 1983.

10. Kruse PF Jr, Keen LN, Whittle WL: Some distinctive characteristics of high density perfusion cultures of diverse cell types. In Vitro 6:75, 1970.

11. McCoy TA, Whittle W, Conway E: A glass helix perfusion chamber for massive growth of cells in vitro. Proc Soc Exp Biol Med 109:235, 1962.

12. Himmelfarb P, Thayer PS, Martin HE: Spin filter culture: The propagation of mammalian cells in suspension. Science 164:555, 1969.

13. Kruse PF Jr: Use of perfusion systems for growth of cell and tissue cultures. In Rothblatt GH, Cristofalo VJ (eds):" Growth, Nutrition and Metabolism of Cells in Culture," New York: Academic Press, 1972, Vol. II, p. 11.

14. Tolbert WR, Feder J, Kimes RC: Large-scale rotating filter perfusion system for high-density growth of mammalian suspension cultures. In Vitro 17:885, 1981.

15. Tolbert WR, Feder J: Large-scale cell culture technology. In Tsao GT (ed): "Annual Reports on Fermentation Processes," New York: Academic Press, 1983, Vol 78, p 35.

16. Feder J, Tolbert WR: The large-scale cultivation of mammalian cells. Scientific American 248:36, 1983.

17. Feder J, Tolbert WR: The development of perfusion culture systems for anchorage dependent animal cells. New York Acad Sci: in press.

18. Feder J: The development of large-scale animal cell perfusion culture systems. In Mizrachi A: "Advances in Biotechnological Processes," New York: Alan R Liss, Inc. in press.

19. Herbert D, Elsworth R, Telling RC: The continuous culture of bacteria; a theoretical and experimental study. J Gen Microbiol 14:601, 1956.

20. Tovey MG: The cultivation of animal cells in the chemostat: Application to the study of tumor cell multiplication. Ad Cancer Res 33:1, 1980.

21. Ku K, Kuo MJ, Delente J, Wildi BS, Feder J: Development of a hollow-fiber system for large-scale culture of mammalian cells. Biotechnol Bioeng 23:79, 1981.

22. Ehrlich KC, Stewart E, Klein E: Artificial capillary perfusion cell culture: metabolic studies. In Vitro 14:443, 1978.

23. van Wezel AL: Growth of cell-strains and primary cells on microcarriers in homogeneous culture. Nature (London) 216:64, 1967.

24. van Wezel AL, Van der Velden-de Grost CAM: Large scale cultivation of animal cells in microcarrier culture. Process Biochem 13:16, 1978.

25. Harms E, Wendenburg J: Large scale perfusion culture of cells growing on surfaces with automatic gas and medium control. Cytobiologie 18:67, 1978.

26. Tolbert WR, Hitt MM, Feder J, Kimes RC: Method of growing-anchorage dependent cells. US Patent 4, 335, 215, 1982.

27. Butler M, Imamura T, Thomas J, Thilly WG: High yields from microcarrier cultures by medium perfusion. J Cell Sci 61:351, 1983.

28. Nahapetian AT: Growth and maintenance of anchorage-dependent mammalian cells in perfused systems and metabolism of nutrients. In Thilly WG (ed): "Mammalian Cell Technology," Boston: Butterworths, 1986, p 151.

29. Tolbert WR, Feder J, Lewis C Jr: Static cell culture maintenance system. US Patent 4,537,860:1985.

Discussion

M. BUTLER — How many parameters did you look at in your Bowes melanoma culture to ensure that you had an efficient perfusion system?

FEDER We looked at product, glucose and lactate, oxygen and ammonia. I have an emperical value obtained from repeated work of the point at which spent medium no longer supports growth. Cells/unit volume/time. In the past we have looked at amino acids and glutamine ie, a very complex package to determine the threshold after which medium is of no use. It is remarkable how consistent oxygen utilisation is - it also is good for the early detection of contamination.

L. FORMAN How did you know your cells stopped proliferating in the hollow fibre reactor?

FEDER How do you measure cell proliferation in a hollow fibre, especially ours which are stainless steel so you can't see anything? We chose rate of lactate production as our parameter and translated it to cells.

33 Microcarrier DORMACELL does not detectably alter viral or plasmid DNA upon direct exposure

I. Keller, U. Brüggemann, H. Mansi-Wothke, D. Schwengers and W. Doerfler

Large-scale cultivation of anchorage-dependent cells using microcarriers has become a widely applied technique. Several raw materials have been used to produce microcarriers since the introduction of this technique by van Wezel (1) in 1967, however, only microcarriers based on dextran and substituted with diethylaminoethyl (DEAE) groups have succeeded to a larger extent. DORMACELL, produced by Pfeifer & Langen, is a microcarrier of this type.

It has to be expected from microcarriers that they do not affect the cells that are cultivated. In order to investigate this requirement for the microcarrier DORMACELL, well proven for example for the production of ß - interferon (2), we have at first tested linear DEAE-dextran (having the same type of substitution like DORMACELL) with the Ames test (3):

Salmonella test with S-9 mix

The product was tested in Petri plates with several specially selected mutants of Salmonella Typhimurium. The test was performed with liver microsomal metabolizing enzymes from Aroclor induced rats to determine whether DEAE-dextran can be metabolized to one or more mutagenically active intermediates. Positive controls were done with 2-aminoanthracene and benzo-(A)-pyrene.

In the bacterial mutagenicity test and under the experimental conditions, DEAE-dextran caused neither base-pair substitution, nor frameshift mutation (see Table 1).

Table 1

SALMONELLA TEST WITH S-9 MIX

NUMBER OF REVERTANT COLONIES PER PLATE WITH GROUP MEANS

Dose/Plate	TA 98		TA 100		TA 1535		TA 1537		TA 1538	
0 µg control										
	33		149		20		19		19	
	45		162		19		17		17	
	27		127		32		14		11	
	48		146		17		11		14	
	36		141		18		12		29	
	34		136		22		30		19	
	35		116		24		23		24	
	36		130		16		21		16	
	25	35	139	138	18	21	28	19	23	19
1,58 µg	46		140		19		20		19	
	35		99		12		24		19	
	30	37	135	125	29	20	25	23	21	20
5 µg	37		144		23		15		13	
	36		129		13		13		14	
	27	33	142	138	24	20	14	14	26	18
15,8 µg	32		129		15		18		18	
	33		150		24		24		18	
	24	30	135	138	16	18	20	21	10	15
50 µg	27		135		14		13		12	
	33		129		16		21		18	
	30	30	135	133	25	18	26	20	26	19
158 µg	33		130		23		17		18	
	38		141		16		17		24	
	30	34	139	137	16	18	13	16	17	20
500 µg	22		128		23		17		14	
	29		140		27		12		22	
	33	28	139	136	20	23	18	16	22	19
1580 µg	38		123		23		22		17	
	31		126		20		19		14	
	40	36	141	130	22	22	21		14	15
5000 µg	38		141		22		22		12	
	34		153		18		16		17	
	36	36	142	145	18	19	16	18	21	17
2-AA 0,5 µg	725		1129		---		---		513	
	820		1264		---		---		650	
	730	758	1085	1159	---	---	---	---	603	589
2-AA 1,0 µg	---		---		462		198		---	
	---		---		522		221		---	
	---	---	---	---	473	486	226	215	---	---
BaP 5 µg	168		516		---		68		109	
	166		516		---		47		95	
	163	166	525	519	---	---	96	70	88	97

--- Not determinded

However, since a negative result of the Ames test does not guarantee the absence of mutagenicity we have developed the following test procedure and applied it at first again on DEAE-dextran:

Measurement of infectiosity of pBR322 plasmid DNA by means of transfection tests.

For the transfection experiments, pBR322 plasmid DNA, either treated with DEAE-dextran or left untreated and then purified, was used in suitable dilutions (1 mM EDTA). HB 101/λ bacteria were employed as receptor strain. The Ca^{2+} transfection method was applied (4,5,6). Transfected bacteria were plated out on tetracycline containing (20 μg/ml) agar plates and cultured at 37°C.

Even in high concentrations, DEAE-dextran treatment does not change the infectiosity of pBR322 DNA (see Table 2).

All these experiments were carried out in homogenous solution with the linear and soluble DEAE-dextran having the same type of substitution like DORMACELL.
Further experiments were carried out to answer the question:

"Does treatment of DNA with DORMACELL cause changes in the restriction patterns of a large DNA molecule?"

The genome of human adenovirus type 2 (Ad2) was used as the test system to examine possible effects of DORMACELL on DNA. Ad2 DNA is a linear, double-stranded molecule of 35,937 nucleotide pairs in length.

An amount of 20 μg of Ad2 DNA was incubated for 16 to 18 h at 37°C in 300 μl of 10 mM Tris-HCl, pH 7.5, 1 mM EDTA as a control. At the same time, 40 μg of Ad2 DNA were incubated under the same conditions in 600 μl of 1 mM EDTA containing 6.7 mg DORMACELL carrier per ml. At the end of the incubation period, the solutions were adjusted to 1 M NaCl, 0.5 % sodium dodecyl sulfate (SDS) and treated at 37°C for 30 min. The microcarrier was removed by low speed centrifugation, and the DNA was reextracted by conventional phenol treatment followed by ethanol precipitation. The DNA was then resuspended in 100 μl 1 mM EDTA.

Figure 1

DNA of Human Adenovirus Type 2 : Treatment with Dormacell Carrier

Analysis: Cleavage with restriction endonucleases

Examples:

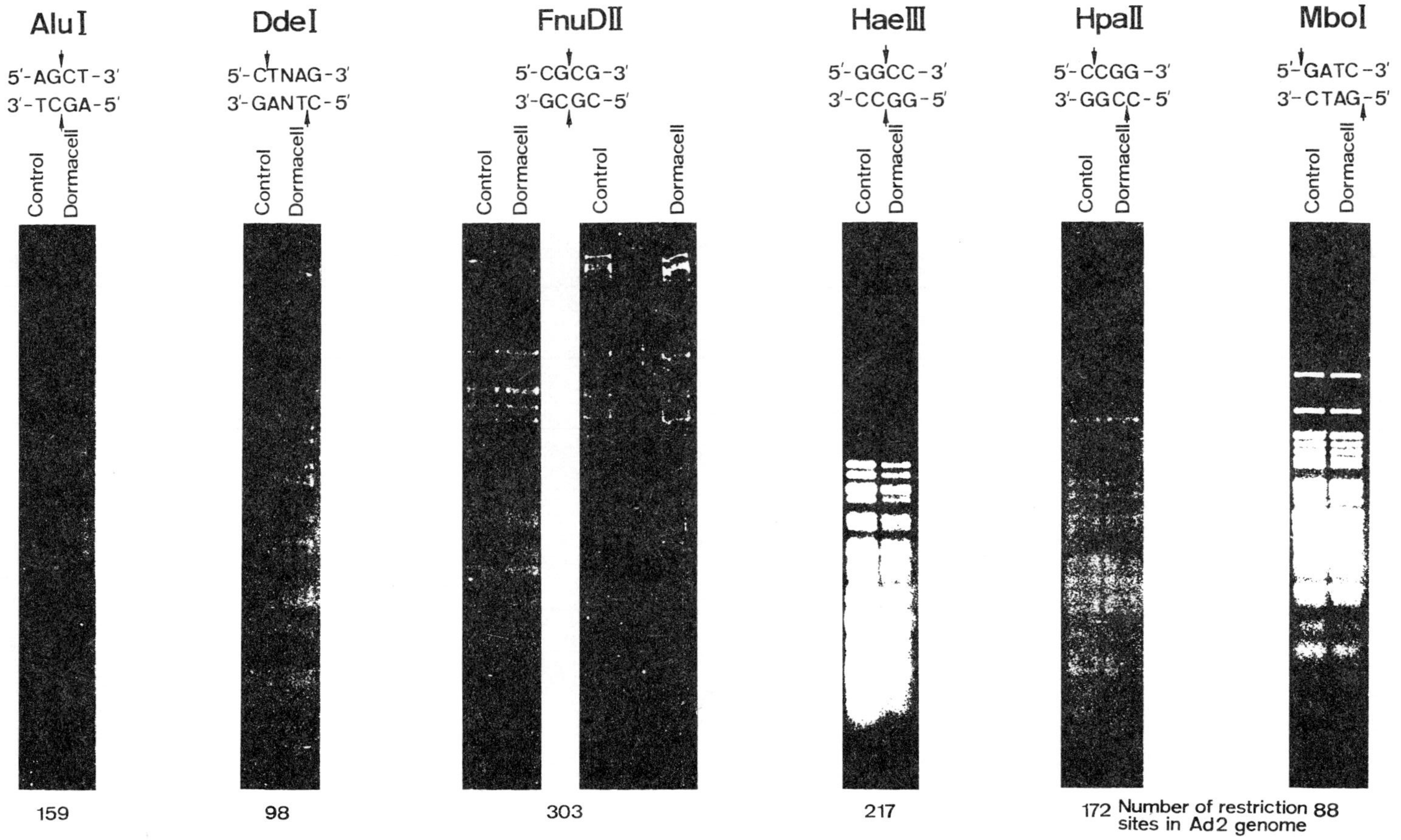

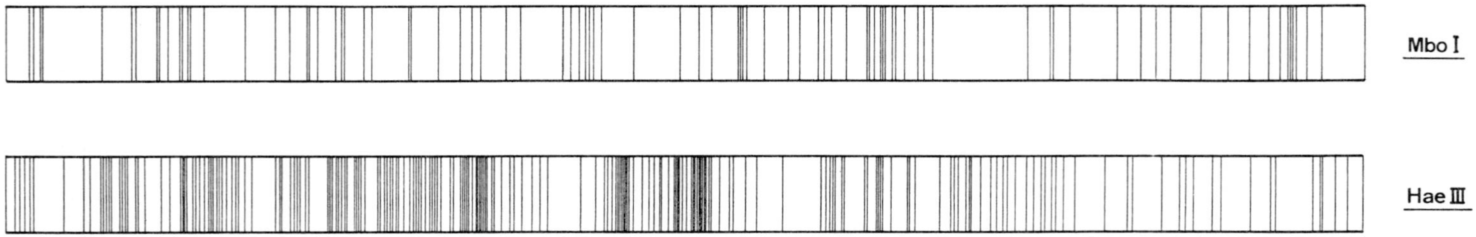

Figure 2 Restriction patterns for endonucleates.

Table 2: Infectiosity[a] of pBR322 DNA incubated with DEAE-dextran or with control buffer

Experiment No.	pBR322 DNA[b] (degree of purity)	Conc. of DEAE-dextran	Number of colonies counted[c]	Relative infectiosity
1	1 x CsCl	0	292	1.0
		100 μg/ml	281	0.96
		10 mg/ml	241	0.83
		100 mg/ml	248	0.85
2	2 x CsCl	0	444	1.0
		100 mg/ml	471	1.06
		0	386	1.0
		100 mg/ml	352	0.91
3	1 x CsCl	0	135	1.0
		100 mg/ml	205	1.52

[a] Standard transfection methods were used.

[b] In the DNA samples , purified 1 x or 2 x , the [^{3}H] activity was counted in aliquots. On the basis of the DNA quantities thus determined, equal quantities of DNA treated with DEAE-dextran or of control DNA were diluted 5 - 100x in 1 mM EDTA. Before dilution, CsCl was removed by means of dialysis of the individual preparations against 1 mM EDTA.

[c] Duplicates in all experiments.

The DORMACELL-treated and the untreated control DNA were cut with several restriction endonucleases, and the resulting DNA fragments were separated by electrophoresis on 2 % agarose or on 10 % polyacrylamid gels. DNA fragments were visualized by ethidium bromide staining and UV photography.

The described method cannot determine minor changes that could have occurred at random in individual DNA molecules.

Out of the total of 35,937 nucleotides of Ad2 DNA 1037 nucleotides were tested by this randomized assay.

The scheme shows the restriction patterns for endonucleases as indicated. The maps for Mbo I and Hae III designate all restriction sites for these enzymes on the Ad2 DNA molecule.

Conclusions

Treatment of the 35,937 nucleotide pair DNA molecule of human adenovirus type 2 with DORMACELL for 16 to 18 h at 37°C did not lead to detectable changes in restriction patterns when a number of frequently cutting restriction endonucleases were used. More than 1000 nucleotides were challenged in this randomized assay.

Comparable results had been obtained with plasmid DNA pBR322 which was exposed to DEAE-dextran and in which 1069 nucleotides out of 4363 had been challenged by restriction endonuclease analyses.

(1) van Wezel,A.L. Nature 216 (1967),64-65

(2) Ames,B.N.,McCan,J.,Yamasaki,E. Mutation Res.31 (1975),347-364

(3) Schulz,R.,Krafft,H.,Lehmann,J. Biotech.Letters 8 (1986),557-560

(4) Cohen,S.N.,Chang,A.C.I.,Hsu,L. Proc.Natl.Acad.Sci.USA 69 (1972), 2110-2114

(5) Doerfler,W.,Stabel,S.,Ibelgaufts,H.,Sutter,D.,Neumann,R., Groneberg,J.,Scheidtmann,K.H.,Deuring,R.,Winterhoff,U. Cold Spring Harbor Symp.Quant.Biol.44 (1979), 551-564

(6) Mandel,M.,Higa,A. J.Mol.Biol.53 (1970), 159-162

34 Macroporous microcarriers

K. Nilsson, S. Birnbaum, F. Buzaky and K. Mosbach

Abstract

A new type of microcarrier has been prepared. The microcarrier is characterized by a large number of interconnected cavities thus making its interior available for cell growth. In addition to the obvious increase of surface area, a number of other advantages were obtained. These include, higher plating effiency, higher number of cell doublings, mechanichal protection and a decrease of the required inocula.

Introduction

The first report on the use of small beads(microcarriers) for the culture of anchorage-dependent animal cells was in 1967 by van Wezel[1]. This microcarrier, a commercial ion exchanger, was later found to be toxic when used in high concentrations. The reason for this toxicity was not eliminated until 1979 through the introduction of low-charged dextran microcarriers by Levine and coworkers[2]. This opened the field for a large number of charge optimized microcarriers. Microcarriers which promote growth have thus been prepared with a positive charge(dextran, cellulose, polyacrylamide, trisacryl etc) as well as with a negative charge(polystyrene, glass, etc).

A different approach has been to use "natural" materials. The most promising material seems to be gelatin. Gelatin has thus been used either for surface coating of dextran beads(Cytodex 3, Pharmacia) or for the preparation of a homogeneous bead[3].

These microcarriers share the common feature that only their surface is used for cell growth. This type of cell immobilization, adsorption, has a number of drawbacks. As the cells are located at the surface they are subjected to mechanical stress. The loading capacity is low which results in a low increase of biomass. The microcarriers prepared so far all have a very smooth surface which makes cell attachment inefficient.

We have prepared a totally new type of microcarrier. The microcarrier, based on gelatin, has a large number of interconnected cavities available for cell growth. The method of immobilization has thus shifted from surface adsorption to a more or less pronounced semi-entrapment.

Materials and Methods

Preparation*

Toluene, containing Span 85(6% w/v) is added continuously to a 100 ml aqueous solution of gelatin(8% w/v, 60°C) containing Tween 80(6% w/v) under stirring . The added toluene forms droplets in the gelatin solution until saturated. Further addition of toluene results in a phase inversion thus producing droplets of gelatin in toluene. As the emulsion is stable, the gelatin droplets will contain droplets of toluene. The structure is conserved by cooling below 20°C which results in solidification of gelatin. Ethanol is added and the beads are collected on a glass filter after which the organic solvents are removed by washings with ethanol and then acetone. subsequently, the beads are dried at room temperature overnight. After sieving, the fraction between 125-180 μm is further processed. The beads are reswelled in phosphate buffer(0.1 M, pH 7.0) and crosslinked with glutardialdehyde(0.8% w/v) for 30 minutes at 15°C. After removal of excess glutardialdehyde they are autoclaved at 121°C for 20 minutes. This treatment reduces the volume to about 50%. After washing with water and acetone the macroporous microcarriers are dried overnight at

60°C. Solid beads of gelatin were prepared as described earlier[3].

Cells and Culture Media

BHK-21 and VERO cells were maintained in 75 cm^2 plastic flasks with DMEM(10% FCS, 4 mM glutamine, 50 ug/ml gentamicin) at 37°C and 95%/5% air/CO_2.

Spinner Culture

Beads, swelled in PBS, were sterilized by autoclaving and washed with PBS and media. All experiments were performed in 50 ml spinner bottles(Bellco) at 60 rpm under the same conditions as described above.

Cell Enumeration

For determination of cell growth, duplicate samples of 0.5 ml were taken from a well mixed suspension of beads. After sedimentation of the beads, 0.3 ml supernatant was removed and 0.8 ml Dispase(Boehringer, grade II, 5 mg/ml in PBS) was added. The mixture was incubated at 37°C until the beads were dissolved(about 30 minutes) after which the cells were counted microscopically with a hematocytometer.

Results

The described preparation results in the formation of beads which a large number of cavities as seen in figure 1. The size of these cavities, which can be controlled through the mixing conditions during the dispersion of toluene in the gelatin solution, makes the interior of the beads available for cell growth.

The results with respect to cell yield were found to be superior as compared to the yield obtained on solid beads for the two tested cell lines, BHK 21 and Vero. As might be expected, it was also found that

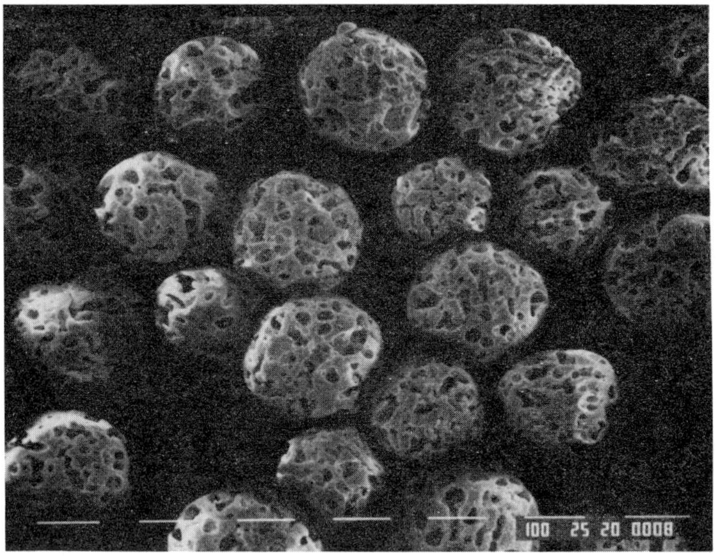

Figure 1.

SEM-picture of macroporous beads. Bar size 100 μm.

cell attachment was higher due to the rough surface of the macroporous beads, see figures 2 and 3.

The SEM-picture, seen in figure 4, shows a proportional comparison between the cells and the cavities, in this case BHK 21 cells after two days of culture. After seven days of culture, the outer surface is completely covered as seen with Vero cells in figure 5.

In order to investigate the possibility of inocula reduction, a constant amount of beads was mixed with different amounts of Vero cells. As seen in figure 6, the final cell yield was not affected even though the inocula was reduced from 280.000 cells/ml to 28.000 cells/ml.

Discussion

No attempts have been made to obtain extremely high cell concentrations. With the simple test system used, 50 ml spinners without oxygenation, perfusion or pH-control, we instead reduced the amount of microcarriers from that normally applied 10%(v/v) to 2%. Even with this drastic reduction in microcarrier concentration, the available surface area was not the limiting factor for cell growth. In order to fully exploit the potential of this novel carrier, it is obvious that optimized control and reactor design is imperative.

This type of carrier will substantially reduce the problems associated with scale up of animal cell culture due mainly to two reasons. First, the cells are protected, although to different extents, from mechanical stress thus facilitating rector design. Secondly, the inocula requirement is reduced as a result of higher plating efficiency and increased surface area per bead.

The higher plating efficiency, due to the rough surface, reduces the amount of cells necessary for initiating the culture. To ensure an even distribution of cells on the beads in the final stages of

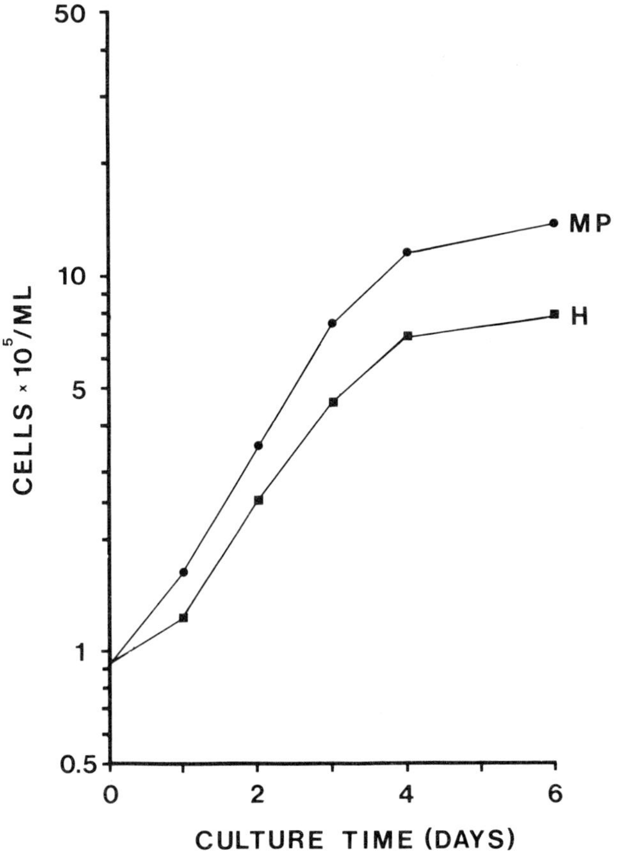

Figure 2.

Growth of BHK 21 cells on macroporous(MP) or homogeneous(H) gelatin microcarriers. Microcarriers(1 ml) and cells(95.000/ml) were mixed in a volume of 50 ml. Beginning on the second day 40 ml of medium was exchanged daily.

Reproduced with permission from ref. 4.

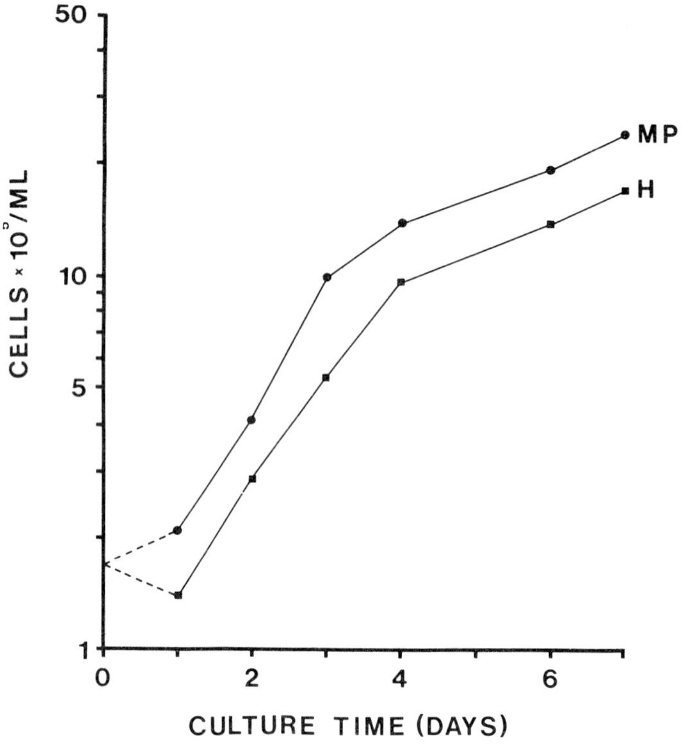

Figure 3.

Growth of Vero cells on macroporous(MP) or homogeneous(H) microcarriers.

Microcarriers(1 ml) and cells(160.000/ml) were mixed in a volume of 20 ml. After one day of culture the medium volume was adjusted to 50 ml. Beginning on the second day 40 ml of medium was exchanged daily. Reproduced with permission from ref. 4.

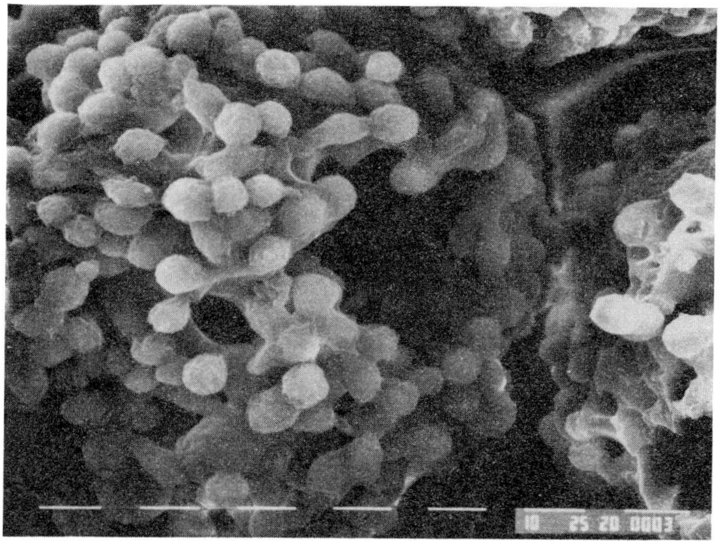

Figure 4.

SEM-picture of BHK cells after two days of culture. Bar size 10 μm.

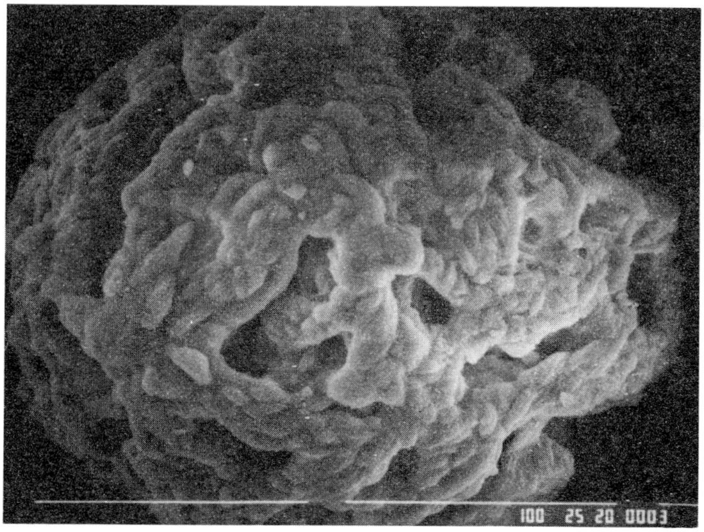

Figure 5.

SEM-picture of Vero cells after seven days of culture. Bar size 100 μm.

culture, it is necessary to inoculate with a minimum number of cells per bead for conventional microcarriers, usually 5-10, as the cells usually cannot migrate from one bead to another. In the case of macroporous microcarriers, it was possible to reduce this number to 2-3 cells/bead.

For non-porous microcarriers, the maximum total surface area for a given volume of beads can only be obtained if the bead size is as small as physically possible but still supporting cell attachment and growth. But, as the bead size gets smaller, the amount of inocula necessary for a given volume of beads increases since the number of beads is larger. Furthermore, the number of cell doublings that can be obtained is reduced as the bead size decreases since the surface area per bead is less, see figure 7. We have solved this contradiction by preparing and applying macroporous beads where the interior of the bead is available for cell attachment and growth thus significantly increasing the surface area per bead. We can then use larger as well as fewer beads and thereby reduce the inocula requirement and at the same time still obtain a high cell yield which is not limited by the surface area. The number of transfer steps necessary to reach the final production stage is thus reduced.

Recent results, not shown, have shown that the mechanical stability of the beads withstands more than 1000 hours continuous culture in a perfused 1 litres reactor.

This method for preparation of macroporous gelatin beads is also applicable for the preparation of macroporous beads from other materials like dextran, polyacrylamide or agarose[9].

References

1. van Wezel, A. L. (1967). Growth of cell strains and primary cells on microcarriers in homogeneous culture. Nature 216,64-65.

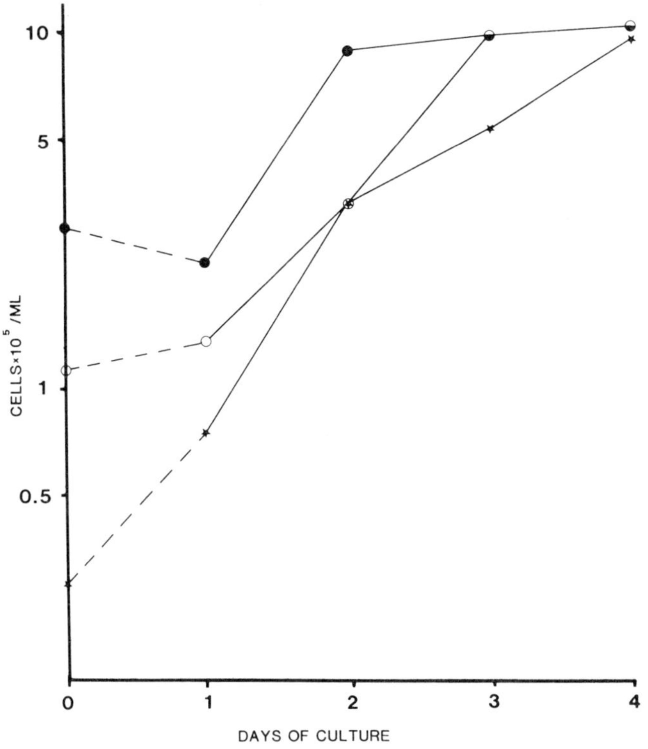

Figure 6.

Growth of Vero cells at different amounts of inocula.

Microcarriers(2 ml) and Vero cells were mixed at different cell concentrations in 25 ml of medium. At day one medium volume was adjusted to 50 ml.

●——● 280.000 cells/ml

○——○ 112.000 cells/ml

★——★ 28.000 cells/ml

SMALLER ←	BEAD SIZE	→ LARGER
LARGER ←	SURFACE AREA / VOLUME	→ SMALLER
SMALLER ←	NO OF DOUBLINGS / BEAD	→ HIGHER
HIGHER ←	INOCULUM SIZE	→ SMALLER

Figure 7.

Inborn contradiction in the preparation of conventional microcarriers.

2. Levine, D. W., Wang, D. I. C. and Thilly, W. G. (1979). Optimization of growth surface parameters in microcarrier cell culture. Biotechnology and Bioengineering XXI, 821-845.

3. Nilsson, K. and Mosbach, K. (1980). Preparation of immobilized animal cells. FEBS Letters 118, 145-150.

4. Nilsson, K., Buzaky, F. and Mosbach, K. (1986). Growth of anchorage-dependent cells on macroporous microcarriers. Biotechnology 4, 989-990.

5. Nilsson, K. and Mosbach, K. (1985). Macroporous particles. Swedish Patent application no 8504764-5.

Discussion

M. BUTLER — One limitation of anchorage-dependent cell growth is curvature. Could this be a problem with your microcarriers?

NILSSON — I think this problem exists at bead diameters of 40 microns, at least for epithelial cells. Probably need a larger size for fibroblasts.

BUTLER — Can you tell me the maximum number of cells you have observed per bead?

NILSSON — No, I haven't used controlled conditions, or perfusion, which is needed to maximise cells.

A. KADOURI — Do you get clumps of coagulated cells when you digest the gelatin to re-seed the cells?

NILSSON — Some cell lines, eg, MDCK, do clump. Others eg, Vero give nice single cell suspensions.

35 Growth of animal cells on microbeads III: Semi continuous in situ estimation of numbers

D. Thebline, J. Harfield, O. Hanotte, D. Dubois and A. O. A. Miller

Abstract

Electronic sizing of microbeads covered with anchorage-dependent cells can advantageously be used to in situ estimate the cell density (average number of cells per microbead). The technique however is discontinuous and requires sampling at regular intervals.

As a first step towards the construction of a fully automated probe, a specially-designed glass probe made to fit into a COULTER Fermenter Sampler, has been linked to a SETRIC Set 2C cell fermenter.

Using this installation, not only can the growth kinetics of various established cell lines be easily, rapidly and accurately be determined but also the very moment when a culture reaches confluency and a second layer forms on top of the first.

Introduction

Probes constructed to evaluate inside a reactor the number of microorganisms in suspension, must meet several requirements(3) such as fast response and high sensitivity. They must also be chemically inert, biocompatible, able to perform in colored or opaque media and ought to be autoclavable in situ.

Measuring under the same experimental conditions the number of cells adhering on microbeads (MB) presents additional difficulties.

Electronic particle sizing of Cytodex 3 MB[5] laden with McCoy cells yields a parameter, the PCN (Peak Channel Number) whose fluctuations are a mixed function reflecting variations in the number of cells and in the distribution of volumes of individual cells[1].

This type of experiments has been generalized by including the results obtained with other anchorage-dependent cells from several established mammalian cell lines.

During these experiments, the need has arisen to streamline the procedure. This work also reports the semi continuous in situ estimation of PCN.

Materials and methods

The four cell lines used are HeLa 229 (human), HTC (Hepatoma Tissue Culture cell-rat), 3T3 NIH and McCoy, the two last ones established from mouse. Growth of McCoy-, HTC- and HeLa 229 cells have been described earlier[4,5]. Mouse 3T3 NIH fibroblasts are grown in monolayer in DMEM (Gibco ref 074-02100) supplemented with glutamine, non essential amino acids (Gibco ref 114) and inactivated (30 min at 56°) horse- and newborn calf serum both at 5 % (v/v).

Evaluation of the mean number of cells per microbead (cell density ρ) and estimation of the PCN have been described earlier[2,5]. For semi continuous estimation of PCN, the MB suspension is withdrawn pneumatically from the SETRIC 2C reactor through a fine plastic tubing (2 mm inside diameter), into a specially miniaturized version of the standard 560 micron probe made to fit into a COULTER Fermenter Sampler, the orifice of the glass probe being positioned along the direction of the inflowing medium and no longer on the side of the tube. As MB flow through the sizing-measuring orifice, they are hydrodynamically focussed and diluted by a stream of MB-free growth medium aspirated from the reactor through a 100 micron stainless steel filter immersed deep in the culture. The ratio of incoming fluids (with and without MB) can be modified at will. All movements of liquids to and from the SETRIC 2C reactor are achieved by two diaphragm pumps.

Results

Using McCoy cells propagated on Cytodex 3 MB, earlier results[1] have shown the PCN of laden microcarriers to vary linearly with ρ. As shown in fig. 1 and as long as the cultures are in their exponential phase of growth, this linearity can also be observed with cells of human (HeLa 229), mouse (3T3) and rat (HTC) origin.

Depending upon the size of individual cells, these lines which all originate at a PCN of 31 a value which corresponds to the size of naked MB, display different slopes.

In fig. 1, increased scatter of the experimental results can be observed with old cultures when the cells leaving the exponential phase of growth, progressively enter the stationary phase and the culture from sparse becomes confluent.
Microscopic examination of the fate of such cultures reveals very different situations indeed to occur.

HeLa 229 cells grow until they form a mechanically very stable monolayer of tightly interacting cells. From time to time a cell detaches, is shed in the surrounding medium where it dies. The stable monolayer thus formed is in a state of dynamic equilibrium : nor the PCN nor the cell density undergo further changes.

The situation is rather different with 3T3 NIH cells which stop multiplying upon reaching confluence. At that stage, they tend to agglutinate, making MB to agregate even when the agitation is increased up to 60 rpm.

Upon reaching confluence, HTC cells succeed in piling to a certain extent. However, these local aggregates which protrude from the MB become very sensitive to local shear forces. Fragments of monolayer are being teared away leading to the formation of bare regions on the surface of the microbead and to an increased proportion

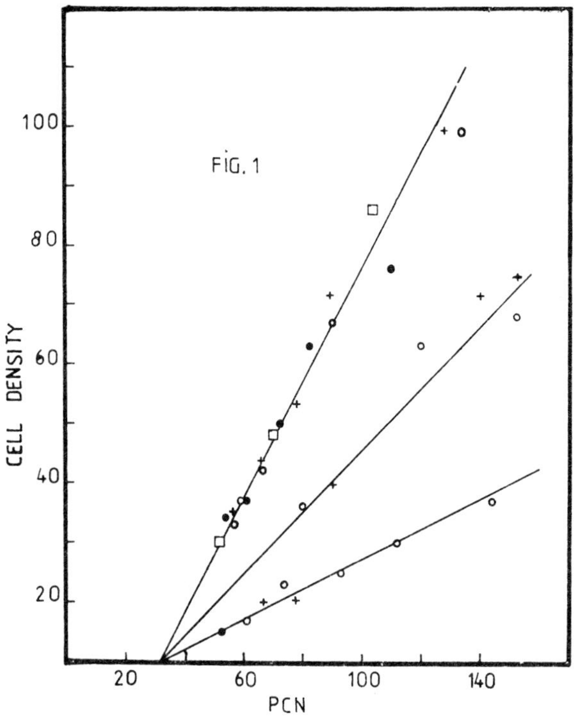

Figure 1. ρ/PCN double plot of the growth kinetics of HTC, HeLa 229 and 3T3 NIH cells

Five-hundred ml Techne culture vessels filled with 200 ml of complete growth medium containing 0,5 g MB and 6 10^7 McCoy cells are kept continuously agitated at 35 rpm during the first 8-10 hours then at 50 rpm with (HeLa 229, 3T3 NIH) or without (HTC) continuous gassing at 37°. The PCN and cell densities in the cultures are determined at various times. The various symbols represent independant cultures.

A : HTC ; B : 3T3 NIH ; C : HeLa 229.

of cells growing in suspension.

McCoy cells offer the most interesting case since they can form additonal layers on top of the first, some 10 h being necessary between completion of the first- and begin of the formation of the second layer[1].

Drawing curves such as those displayed in fig. 1 is a very cumbersome process which, for the repeated samplings it necessitates, ultimately put the very existence of the culture in jeopardy.

Alternative means have been sought to streamline the whole procedure. Using the modified version of the COULTER Fermenter Sampler linked to a SETRIC 2C reactor, samples can be withdrawn, analyzed for PCN and returned to the culture without actually opening the reactor.

Fig. 2 shows the results obtained with cultures of Cydodex 3 MB laden with HeLa 229 or McCoy cells.

As mentioned above, HeLa 229 cultures, upon reaching confluence form a sturdy monolayer whose cells shed in the surrounding medium are exactly compensated by newcomers, a situation reflected by the plateau reached by PCN at later times.

For McCoy cells repeated experiments of which only two have been given (fig. 2) show the PCN to reproducibly undergo two sudden bursts situated 10 h apart. The first of them coincides in time with the baso-apical reorganisation underwent by the cells as they reach confluence, the second reflects the formation of the second layer on top of the first.

Discussion

The results show that exponentially growing anchorage-dependent cells whether derived from several animal species or from different tissues of the same species (murine 3T3 NIH fibroblasts and

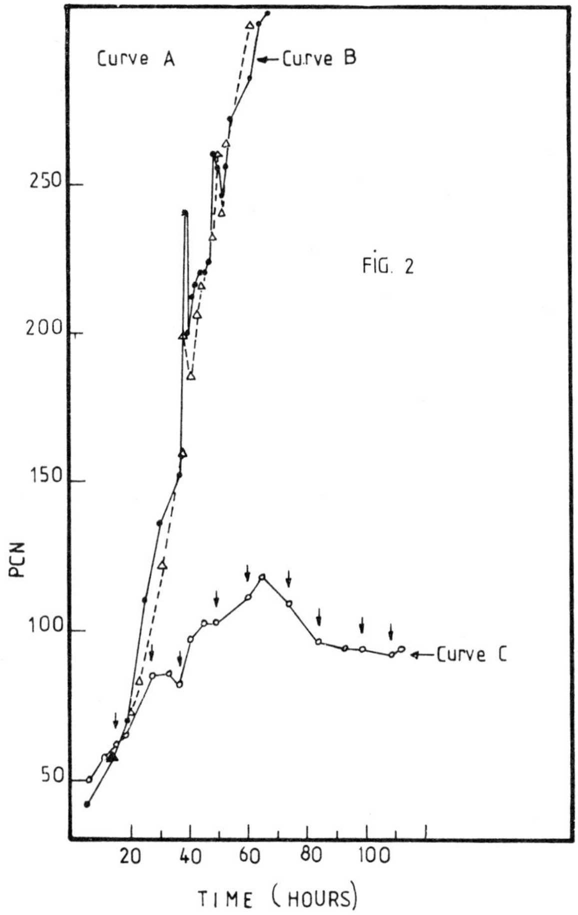

Figure 2. <u>Semi continuous in situ evaluation of the time-dependent variations of PCN</u>

One-liter two hundred ml MB cultures of HTC and HeLa 229 inoculated as described under the legend to fig. 1 into a SETRIC 2C reactor are then processed as described in Materials and Methods.

A,B : HTC ; C : HeLa 229.

For cultures A and B, the arrows indicating medium changes have been omitted.

McCoy synovial cells), always display a linear relationship between cell density and PCN.
Given standardized experimental conditions such as size of the MB and the chemical composition (type) of growth medium used, it is sufficient to determine for every cell line in current use the slope of its density versus PCN line so as to construct a library of growth kinetics which can be stored on diskette. Provided that the actual semi continuous probe is developed into a fully on-line sensor, then, control of the biomass in microbead based bioreactors becomes a trivial matter.

Although several applications can be anticipated from developments along these lines, the sudden increase of voltage pulses observed with McCoy cells, which most probably reflect the transient increased impedance accompanying the basoapical organization underwent by these cells, came to us as a complete surprise. This observation makes it possible for the first time to detect subtle changes in cell volume which until now, for lack of appropriate measuring instrumentation, escaped our attention.

Sensitivity of the method is so high that it has drawbacks of its own. For instance, measuring the PCN every time the spent medium is replaced with fresh growth medium, induces small but consistent oscillations of PCN. Obviously, it becomes mandatory when using this kind of probe, to perfuse the culture.

With this caveat in mind, it is reasonable to assume that electronic sizing based on either volume or/and impedance changes, will find useful applications not only in applied (biotechnology) - but also in basic research.

Acknowledgments

We gratefully acknowledge financial support from SETRIC Génie Industriel and COULTER Electronics and the gift of Cytodex 3 by E. Rijckebosch (Pharmacia Belga).

The excellent technical assistance of M. G. Deneubourg has been greatly appreciated.

References

1. Hanotte, O., Dejaiffe, C., Dubois, D., Vanderpoorten, P., Menozzi, F.D. and Miller, A.O.A. (1987) Growth of animal cells on microbeads. II. Estimation of biomass in situ and observation of cytopathic effects after infection of McCoy cells with Chlamydia trachomatis. Biology of the Cell (accepted for publication).

2. Harfield, J.G., Knight, P., Podgorney, H. and Wharton, T. (1985) Improved multichannel size distribution analysis. Presented at : Particle Size analysis Conference held at Bradford University 16-19 September.

3. Merten, O.W., Palfi, G.E. and Steiner, J. (1986) On line determination of biochemical, physiological parameters in the fermentation of animal cells in continuous or discontinuous mode. Adv. Biotechn. Proc. 6,

4. Miller-Faurès, A., Michel, N., Aguilera, A., Blave, A. and Miller, A.O.A. (1981) Laser flow cytofluorometric analysis of HTC cells synchronized with hydroxyurea, Nocodazole and Aphidicolin. Cell Tissue Kinet 14, 501-514.

5. Miller, S.J.O., Henrotte, M. and Miller, A.O.A. (1986) Growth of animal cells on microbeads. I. In situ estimation of numbers. Biotechn. Bioengn. 28, 1466-1473.

Discussion

U. STEINER From your histograms it looked as if the beads doubled in volume from no cells per bead to a confluent bead. What is the reason for this as obviously the cell volume is low compared to the bead volume?

MILLER — You have to note that the linear relationship is valid as long as the cells grow exponentially. With the slightest modification in growth conditions the cell can stop multiplying but still enlarge. The probe can monitor increase in volume as a result of cytopathic effect which results from intracellular multiplication of a pathogen - in this case Chlamydia. You have to be sure that the cells are in an exponential phase of growth. Also small microbeads capture more cells than the larger microbeads, and some beads remain completely cell-free. We program the computer for the size distribution of naked microbeads and subtract the size distribution of beads covered with cells. Did I answer your question?

STEINER — Not quite.

MILLER — It is clearly an increased impedance which explains the two blips shown on the slide.

36 Evaluation of the continuous perfusion culture system for the production of monoclonal antibodies

T. van der Velden-de Groot, W. Witterland, E. C. Beavery and T. L. van Wezel

ABSTRACT

The continuous perfusion system is routinely applied in our laboratory for the cultivation of hybridoma cells for the production of monoclonal antibodies. We compared the cell growth and antibody production of 13 different hybridoma cells. All cells gave acceptable yields (between 50 and 230 μg of antibody protein per ml). In addition the effect of a combination of serum substitutes and of Primatone RL as alternative for fetal calf serum was studied. Primatone RL was able to replace fetal calf serum to a large extent in a continuous perfusion culture. The mechanical damage of cells in the continuous perfusion system was, however, a drawback of this culture system. Further improvement may be obtained by the addition of compounds as Pluronic F-68.

INTRODUCTION

For the cultivation of animal cells for the production of vaccines and other biologicals as monoclonal antibodies Van Wezel and co-workers (1) have chosen for the concept of homogeneous culture systems. These systems have - besides the possibility of optimal control of environmental conditions - the advantage that scaling up problems are limited. Based upon this concept Van Wezel and his group developed a number of cell cultivation systems (batch, continuous chemostat, recirculation and continuous perfusion) for the growth of anchorage dependent as well as suspension cells. In this paper, we summarize our data on the production of monoclonal antibodies (McAbs) by a number of hybridomas in a continuous perfusion system (Fig.1) with culture volumes of 3, 15 and 40 L and compared the data with those obtained in stationary monolayer culture. The hybridomas produced antibodies belonging to the IgG1, IgG2a, IgG2b and IgM subclasses.

The growth medium consisted of Iscove's DMEM supplemented with either fetal calf serum (FCS), serum substitutes, a combination of serum substitutes and FCS or a combination of Primatone RL and FCS.

MATERIALS AND METHODS

Growth media.

Iscove's DMEM (Gibco) was supplemented with FCS (Bocknek) or with a combination of serum substitutes as bovine serum albumin, soybean lipids (Boehringer Mannheim), transferrin, insulin, selenium, ethanolamin (Sigma) or with Primatone RL (Sheffield).

Cultivation equipment.

For the continuous perfusion system the bioreactor was equipped with a stainless steel spinfilter with pore sizes of 10 or 75 μm mounted on the stirrer shaft. For the control of the cultures either the Bilthoven Unit NOVO-Paljas (Contact-Flow) or the computerized Fermentrol 5100 (Contact-Flow) was used. Both control units have been developed in The Netherlands.

Culture conditions.

Inoculation density: about 2 x 10^5 cells/ml.

Conditions in the homogeneous culture: pH, 7,2; pT, 36°C and DO level, 25% air saturation.

Perfusion rate: 0.5 - 1.0 culture volume per day. The perfusion was started when the culture reached a density of 1-2 x 10^6 cells/ml.

Samples were taken daily for cell count (trypan blue exclusion), for the control of glucose content (Ames) and for the assay of McAb by ELISA.

Assay of McAb.

An ELISA sandwich assay was applied for the determination of produced amount of McAb. The sandwich was composed of:

i. coat -> sheep anti-mouse IgG (Ig fraction of antiserum obtained after caprylic treatment).

ii. sample or reference (subclass specific immunoglobulin preparations).

iii. conjugate -> goat anti-mouse IgG (Fc specific) conjugated with horse radish peroxidase. The conjugate was purchased from Cappel.

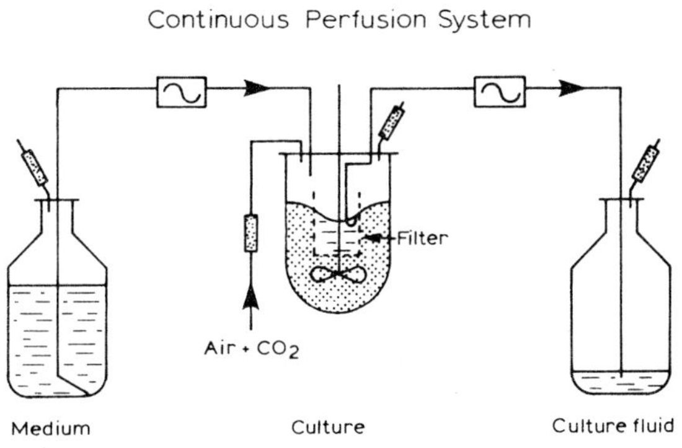

FIGURE 1

Continuous Perfusion System

RESULTS

In the past two years we have cultivated 13 different hybridomas in a continuous perfusion system for the production of McAbs. We studied the influence of several alternatives for FCS. The culture volumes ranged from 3 to 40L depending upon the desired amount of antibodies.
Before the start of the homogeneous culture, growth curves were made in stationary monolayer systems in which maximum cell densities and specific expression (μg of McAb/10^6 cells) were determined. Most cells reached densities of 2-3 x 10^6 cells/ml within five days. In Table 1 data of three representative cultures are presented. As could be expected the specific expression by the hybridomas showed great differences. The data also demonstrated a lower specific expression in the homogeneous systems probably due to mechanical damage (shear forces) in the bioreactor.
Both from economical and practical point of view attention was paid to a reduction of the needed amount of FCS. We started to omit FCS completely and used a combination of serum substitutes as an alternative. This serumfree medium gave in stationary cultures results comparable to the medium containing 10% FCS. The yields in homogeneous culture systems, however, were disappointing, even with the addition of 2.5% FCS (Table 2). Experiments with reduced serum concentrations (7.5 and 2.5% in stead of 10%) and addition of Primatone RL, as described by Reuveny (2) gave at least comparable and for most cell lines even better cell yields than those in medium with 10% FCS. In general, the specific expression in the media containing Primatone RL (Table 2) was comparable to the expression in FCS containing media. Some cells showed even a 1.5 times increased production of antibodies in media with Primatone RL (data not shown).
Table 3 shows production levels of the hybridomas producing antibodies in the subclasses IgG1 and IgG2a. Also this table shows clear differences in specific expression between the cells, irrespective of the class of the produced antibodies.

CONCLUSIONS

From the comparison of the data obtained in stationary and homogeneous culture systems it can be concluded that attention has to be paid to diminish the effect of shear forces in order to increase cell density and level of expression at least to the levels found in stationary monolayer cultures. The addition of Pluronic F-68 or comparable compounds may protect the cells from mechanical damage.
With the media presented, increase in production is only possible by increasing cell density. Experiments were performed to find optimal

TABLE 1

Comparison of expression for three hybridomas in stationary monolayer system and homogeneous culture system.

McAb subclass (code hybridoma)		specific expression monolayer	suspension
IgG1	(2)	31.4	17.1
IgG2a	(7)	22.0	16.4
IgG2b	(12)	8.4	8.4

Composition medium: Iscove's DMEM + 10% FCS.

TABLE 2

Comparison of cell densities and McAb production levels of four hybridomas producing different antibody subclasses in media with different compositions.

McAb Subclasse (code of hybridoma)	Culture volume 1	Medium[a]	Average cell density (x 10^6) vital / dead	% dead	Average (McAb) μg/ml	μg/10^6 cells
IgG1 (2)	3	1	5.8 / 3.1	34.8	167	19
	15	2	2.5 / 0.6	18.8	47.7	16.7
	15	3	9.5 / 3.2	25.5	174.2	14.7
	40	3	6.2 / 0.4	6.5	132.0	20.0
IgG2a (7)	40	1	4.2 / 0.5	9.7	93	20
	40	2	6.1 / 5.2	45.8	155	13.7
	40	3	7.4 / 2.1	22	145	15.4
IgG2b (12)	40	1	4.0 / 1.6	28.6	39.1	8.0
	40	3	7.2 / 1.2	14.3	75.3	9.0
	40	3	6.3 / 2.6	29.2	57.8	6.7
	40	3	6.1 / 1.2	16.4	72.4	9.9
IgM (13)	3	1	3.9 /<0.1	1	100	25
	40	4	4.0 / 0.7	14.9	103.7	21.8

[a]Medium : 1. Iscove's DMEM + 10% FCS.
2. Iscove's DMEM + 2.5% FCS + serum substitutes.
3. Iscove's DMEM + 2.5% FCS + 0.25% Primatone RL.
4. Iscove's DMEM + 7.5% FCS + 0.25% Primatone RL.

medium-composition or -additions not only to increase the cell density to about 2.0 x 10^7 cells/ml, but also the specific expression. In small scale experiments levels of about 500 μg/ml were obtained.
Although the reduction in FCS concentration is substantial, our attention is still focussed on complete omitting FCS in the medium and use only low molecular weight substitutes to simplify the down stream processing of the McAbs present in the harvest.

REFERENCES

1. A.L.van Wezel, C.A.M.van der Velden-de Groot, H.H.de Haan, N.van den Heuvel and R.Schasfoort.
Develop.Biol.Stand. 60: 229-236 (1985).

2. S.Reuveny, D.Velez, J.D.MacMillan and L.Miller.
Develop.Biol.Stand. 60: 185-197 (1985).

TABLE 3

Production levels of a number of IgG1 and IgG2a producing hybridoma's in continuous perfusion systems.

McAb subclasse	Cell code	McAb mg/liter
IgG1	1	142
	2	197
	3	148
	4	101
	5	163
	6	157
IgG2a	7	191
	8	126
	9	138
	10	138
	11	330

37 Recent advances in the control of monitoring of bioreactors

D. F. Parsons

SYNOPSIS

Improvements made in the last few years fall into three main categories:- vessel designs, new sensors and control methods. A plethora of bioreactor designs have been suggested but few, except a range of versatile stirred vessels, prove capable of successful commercial scale up. Closer control of cell environments using the latest generation of sensors have inevitably led to better yields and early studies of NADH fluorescence probes show that encouraging results can be obtained for the on-line estimation of biomass as well as providing useful information on cell metabolism. Other steam sterilisable probes under development include those for glucose and ammonia measurement.

Perhaps the major advance in recent years has been the use of computers for data logging and control of bioreactors. It has been possible to monitor an increasing range of process parameters which, together with off-line analytical data and derived variables can be used for control in either direct digital (DDC) or supervisory set point control (SSPC) modes. The continued development of innovative computer software utilitising signals from a new generation of biosensors will provide even greater rewards in the future.

INTRODUCTION

A wide range of bioreactor designs have been suggested over the years, each claiming advantages for the user having an appropriate application. This paper will concentrate on the implications of vessel design, control and monitoring equipment for bioreactors specifically intended for the culturing of animal cells. The overall success or otherwise of the culturing process will also depend on the careful choice of media and the purification technique adopted but these subjects will not be discussed here.

Importance of Animal Cells

The gathering pace of the commercial utilisation of animal cells as a route to such valuable products as viral vaccines, monoclonal antibodies, growth factors, enzymes, hormones etc. not to mention the large scale production of animal cells themselves for further research purposes, has put much more pressure on bioreactor manufacturers to produce designs that can match the requirements and be capable of effective scale-up. Since both anchorage dependent and independent cells are involved it is necessary for bioreactors to be capable of propagating cells on micro-carriers as well as in conventional suspended culture.

The Cell Culture Bioreactor

A range of vessels from one litre to 600 litre working volumes has been developed and found to be suitable for the culturing of a wide range of animal cells. Significant design changes from the basic bacterial fermenter were found to be required and these include an aspect ratio (working height to vessel internal diameter) of 1:1 and a hemispherical base. A marine type propellor with three helical blades has been specially designed which, together with the hemispherical profile of the vessel base, produces homogeneous cell suspensions at very low agitation speeds.

The preferred method for temperature control in animal cell culture vessels is the circulation of thermostatted water around a jacket which covers as much as possible of the working area. The major advantage of this system over electrical immersion or band heaters is the complete absence of hot spots which kill cells in their immediate vacinity and produce non-uniform growth in the gently agitated conditions. Another advantage of jacketed vessels becomes evident during power cuts, where the mass of water in the jacket maintains the culture temperature for a much longer period than does a rapidly cooling electrical heating element.

Maintenance of sterility and effective containment are important aspects in bioreactor design especially for processes involving genetically engineered cells. Stainless steel vessels with highly polished interiors and in-situ steam sterilisation facilities, including independent steam supplies to sampling and harvest valves, have considerable disadvantages over glass autoclaveable units and precautions such as double '0' ring seals on all ports together with steam jacketed air inlet and exit filters are also desirable. Full containment can be further guaranteed by the utilisation of remotely operated valves for both process and service lines making it possible to isolate the vessel from its control instrumentation.

Much greater cell densities can be achieved in vessels fitted with perfusion accessories which permit the concentration of cells by allowing the constant addition of fresh media whilst removing 'spent' media and waste products via a stainless steel mesh filter of a pore size selected to pass media but retain cells.

Control Parameters

Temperature and control of the minimum agitation required to maintain cell homogeneity have already been mentioned but other parameters may have to be controlled in order to optimise the process.

If pH control is necessary this will involve the use of a pH electrode designed for use in high protein media i.e. either a double junction type or one containing a special electrolyte. The outputs of the pH controller are normally interfaced to a solenoid valve which adds $C0_2$ as acid titrant and a peristaltic pump which adds basic media.

Control of oxygen in animal cell culture can be rather complex with variation of agitation speed clearly not an option.

If total gas flow into the culture needs to be kept constant, so as to keep the agitation constant, then control should be by enrichment of oxygen into a flow of air which itself is controlled to a pre-set level. This control scheme should include an oxygen electrode that does not have its teflon membrane exposed to the culture media because protein coating will affect the reading due to variation in oxygen permiability across the membrane. Polarographic electrodes are available which have 'sandwich' membranes consisting of a silicone layer on top of the teflon and strengthened by a stainless steel mesh. With this type of electrode low levels of oxygen can be measured and coating of the silicone has little or no effect on the reading. An oxygen controller operates on a proportional gas valve which maintains the required oxygen level whilst a mass flowmeter feeds its signal to a mass flow controller which adjusts another proportional gas valve to keep the overall gas flow constant.

Control of pCO_2 levels in the culture can be controlled using a carbon dioxide electrode, similar in construction to the oxygen electrode described above, connected to a controller, which operates essentially as an accurate pH meter with an output to a solenoid valve mounted in line with a CO_2 cylinder.

Attempts have been made to estimate the number of viable cells in bioreactors and to use this value in a continuous control loop to control the feed of nutrient. The measurement of inlet and exhaust gas O_2 and CO_2 can lead to an estimation of respiratory quotient and hence an indication of biomass changes.

Redox measurement in the culture media, assumed to be largely due to cell related oxidation/reduction reactions such as NAD - NADH have been used to control rate limiting substrates like oxygen and nutrient with some degree of success but a universally applicable on-line method of biomass estimation is still being sought. Early results from steam-sterilisable fluorescence probes, which are tuned to the wavelength of the NADH fluorescence, are very encouraging although more work has been done on yeasts and bacteria than on mammalian cells.

NADH is present in all living cells and plays an important part in metabolism. Where metabolic reactions consume or produce NADH at a constant rate it is possible to directly relate NADH concentration to living biomass. Furthermore, since substrate and oxygen depletion both markedly and rapidly affect NADH concentration it is possible to use the measurement of NADH fluorescence as part of a control loop to control critical substrate feeds or oxygen levels. The chances of optimising such a control scheme are enhanced by the use of a suitably configured computer system.

Computer Control

The advantages of computers in cell culture fall into the three categories:-

(a) data collection and manipulation

(b) sophisticated control and
(c) alarm generation

In order to collect all relevent process information it is often necessary to monitor six to eight parameters such as temperature, agitation speed, air flow, pH, pO_2, pCO_2, redox, exhaust gas analysis as well as off-line analytical data such as dry weights and glucose values. A computer system intended for integration into a multiple bioreactor plant should be capable of collecting all of this data at user selectable intervals and logging it onto a suitable long term storage device e.g. a hard disc. Data manipulation should include the ability to report the logged values in alphanumeric form as well as graphically with options for hard copies. Graphical displays can be particularly useful if multiple graphs can be plotted on common axes especially if combinations of current and historical data can be produced.

The modern computer system not only offers the choice of direct digital control (DDC), where amplified signals may be accepted by the computer interface and digital signals used for control of heaters, pumps and valves, or supervisory set-point control (SSPC) where conventional analogue controllers are retained but watched over by the computer, but further extends control capabilities into the areas of cascaded loops and control by derived variables.

Cascade control is where one measured parameter may be used to control another for example using pH measurements to adjust pO_2 levels.

A derived variable results from feeding a measured value or values into an algorythm. Respiratory quotient is a common example of a derived variable and comes from the measurement of inlet and outlet O_2 and CO_2 levels (or the measurement of outlet levels and assumptions of gas composition in the inlet air). It is possible to use derived variables in control schemes with the example of respiratory quotient being used to control a rate limiting nutrient so as to maintain a pre-determined growth pattern.

Computers permit parameters to be controlled according to varying set-points i.e. a set-point profile which can take the form of a linear ramp, stepped ramp or a smooth curve.

When the operator mixes his control options of cascaded loops, derived variables and profiles, together with the potentially extremely useful inputs from new sensors like the NADH fluorescence probe, a very sophisticated control strategy becomes available for the first time.

The ability of computers to oversee a process and generate alarms when necessary is extremely useful in animal cell culture particularly when carried out in large scale vessels where lost product and media can be very expensive. Foreground/background operation is essential to guarantee immediate notification of alarm conditions irrespective of the task being carried out on the VDU. Alarms should be clearly indicated on a reserved screen display as well as being logged onto disc and output onto a printer. Further benefits can be obtained if the computer has been programmed to generate a message for each alarm condition so that suitable corrective action can be initiated with the minimum delay. The ability to anticipate alarm situations is a particularly useful feature of a computer programmes trend watching routine and may be put to good use in the saving of valuable product when unforeseen failures occur.

Conclusions

Further advances will be made in bioreactor design with containment considerations and perfusion techniques no doubt having high priorities but, in the culturing stages at least, the main scope for improved performance is in the further development of computer control using software able to accept and manipulate signals coming from new sensors which will more closely monitor the state of the cell, a condition that could well be called - high performance physiological control.

38 Automated process control for culturing animal cells

J. Rosenquist and C. Wallberg

ABSTRACT

A process control system, consisting of the controller Alert® 50 and the plant management system ALCOM, has been installed to monitor and control simultaneously several fermenters designed for animal cell cultivation. Culture parameters are regulated via feedback and feedforward loops. Process displays illustrating control functions and continuosly updated trendcurves are available from every fermenter. All run data saved in files during the process can be retrieved during or after the cultivation. Cell growth and production data can be related to the culture parameters in diagrams, which makes it easier to optimize the process to obtain for maximal cell growth and/or maximal production.

Two cultivations, controlled by the installed process control system, with run data for the production of monoclonal antibodies from hybridoma cells in suspension are presented. Both cultivations were run in a pilot scale reactor in a volume of 100 litres; one was performed as a batch culture and the other was performed as a continuous culture.

The regulation of the culture parameters pH, temperature and dissolved oxygen worked well and the use of this process control system has considerably facilitated optimizing the performance of our animal cell cultivations.

INTRODUCTION

Animal cell culture has become an important process for the production of both natural and recombinant proteins for the pharmaceutical industry. Increasing emphasis must be placed upon process control and automation for the development and optimization of animal cell cultivation.

The recently developed controller Alert 50 together with the plant management system ALCOM is a user friendly process control system offering high flexibility and has been installed to monitor and control simultaneously seven fermenters designed for animal cell cultivation.

Two cultures with run data for the production of monoclonal antibodies from hybridoma cells in suspension are presented: one batch cultivation performed over a period of 140 hours and one continuous cultivation performed over 600 hours. The continuous culture had a constant flow of medium into the reactor and the cells present in the medium outflow were not recycled back into the reactor. Both cultivations were run in a pilot scale reactor in a volume of 100 litres and the culture parameters were regulated with the new process control system. Measured values for dissolved oxygen (% air saturation), pH and temperature are presented together with cell growth and antibody production data.

MATERIALS AND METHODS

Process control system

A process control system consisting of the controller Alert 50 and a personal computer DEC380 loaded with the plant management system ALCOM has been installed to control one pilot scale fermenter (150 l working volume) and six laboratory scale fermenters.

The controller was loaded with the configuration for the control loops, for controlling the processes independently of the plant management system ALCOM, as shown in figure 1. The ALCOM system could then be used for:

* showing dynamic process displays and continuously updated trendcurves for each running process
* calculations with collated data
* changing parameters and setpoints in running processes, either manually or according to conditions set in user algorithms.

Both the batch and the continuous cultures were run in the pilot scale fermenter. Both feedback and feedforward control have been used for controlling the culture parameters dissolved oxygen (DO = % air saturation), pH and temperature. Sequence control has been used for sterilization and emergency cooling.

The DO feedback control loop uses surface aeration with oxygen and nitrogen to control the oxygen level in the culture. The pH feedback control loop uses CO_2 and 1.0 M sodium hydroxide to control pH.

All control loops contain PID (Proportional Integral Derivative) controllers. Figure 2 shows the feedback control loop for the DO-regulation.

Batch cultivation

In the batch experiment a suspension adapted hybridoma cell line, 6.G4.B4.1.5 producing monoclonal antibodies against human IgG4, was cultivated in a total volume of 100 litres. The inoculation was performed in a volume of 50 l at a starting density of 0.4*10+5 cells/ml. The cells were cultivated in Dulbecco's Modified Eagle Medium (DMEM) supplemented with sodium bicarbonate, 0.5% foetal calf serum (FCS), penicillin-G and streptomycin.

The initial setpoint DO was 40% at 1 atm. After 23 hours, 50 litres of media was added to give the final volume of 100 l and after 44 hours the setpoint DO was lowered from 40% to 10%. The pH setpoint was 7.3 and the temperature setpoint was 37°C.

Continuous cultivation

In the continuous experiment a hybridoma suspension cell, 1.5293.A57.3.4 producing monoclonal antibodies against Gonococcus (Neisseria gonorrhoeae), was cultivated. The cultivation started as a batch culture and starting density was 0.2*10+5 cells/ml. The cells were cultivated in the same medium as in the batch experiment and total cultivation volume was 100 litres.

Continuous cultivation began after 71 hours of cultivation and the flow rate was set to 50 l/day. After 265 hours this flow rate was lowered to 25 l/day. The following setpoints were set: DO to 10%, pH to 7.3 and temperature to 37°C.

In both cultivations measured values for DO, pH and temperature were logged and saved in datafiles. Cell density and antibody concentration in the fermenter were determined by cell counting and nephelometer measurements, respectively, on samples taken offline and saved in separate datafiles.

RESULTS

The batch cultivation was terminated after 140 hours when less than 0.4*10+5 viable cells/ml remained.

The continuous cultivation was terminated after 600 hours. The hybridoma cell used in this culture was a higher producer (MAB/cell) than the hybridoma cell used in the batch culture.

Figures 3 and 4 show how online measured cultivation parameters have been displayed, together with offline measured production data on a graphical screen. Figures 3 and 4 were printed out directly from the multicolour screen, but on a one colour printer.

The setpoints for DO, pH and temperature were determined from previous results obtained in our laboratory (1).

DISCUSSION AND CONCLUSION

The pH control and the temperature control worked well and the deviations from chosen setpoints were low, as seen in figures 3 and 4. In the continuous cultivation the measured value for pH varied +/- 0.04 pH-units from setpoint and the measured value for temperature varied +/- 0.1'C from setpoint.

The control of dissolved oxygen in the reactor was more difficult to perform and the deviations from chosen setpoints were bigger, about +/- 5 %-units DO as shown in figures 3 and 4. The difficulties depended mainly on:

1) that the gas exchange was performed by surface aeration which gave a long deadtime before any response was seen
2) an increasing cell density changed the demand on the control action during the cultivation.

To improve the DO control an autotuner can be integrated in the block with the user algorithm for adaptive control in the DO control loop, illustrated in figure 2. The autotuner registers the control action of the PID-controller and monitors the effect it will have on the process. The autotuner will then calculate and set new PID-parameters for the PID-controller.

In conclusion, some of the advantages with the new process control system are:

* All processes can be monitored and controlled from one control station, which reduces the required manpower. During weekends the system can be operated from home with the help of a telephone modem and a terminal.

* The plant management system ALCOM makes it easier to perform the optimization of culture parameters for maximal cell growth and/or maximal antibody production. As previous cultivations can be displayed during or after a running cultivation, comparisons between different cultivations are easy.

* Cascade control loops and user algorithms for feedforward and sequence control can be input into the controller software. This provides a fast response to parameter regulation and can anticipate and compensate for expected disturbances.

* Several fermenters can be controlled at the same time with one process control system. This will mean a less expensive system than using dedicated controllers for every reactor. A one-controller system will however decrease the safety margin.

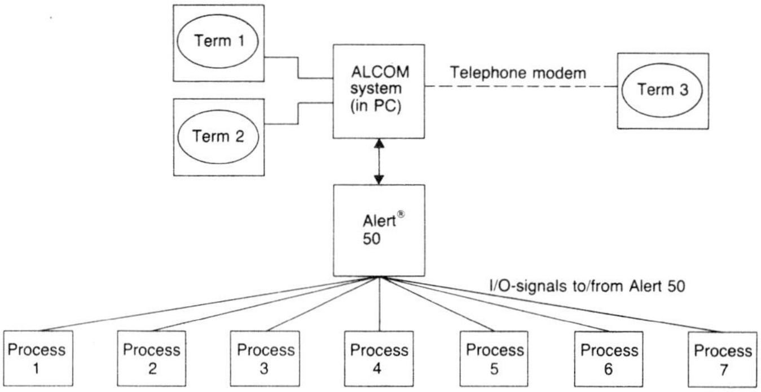

Figure 1 "Schematic example of the process control system installation"

The plant management system ALCOM can be defined as a communication system that is able to collect process control data, evaluate and present information, and perform supervisory control to optimize whole processes. Three terminals can be connected to the personal computer DEC380 loaded with ALCOM.

The controller Alert 50 can perform feedback control with PID regulation and the feedback loops can easily be programmed into cascade loops. The system also allows the user to input user algorithms into the controller for feedforward and adaptive control to obtain a complete control strategy.

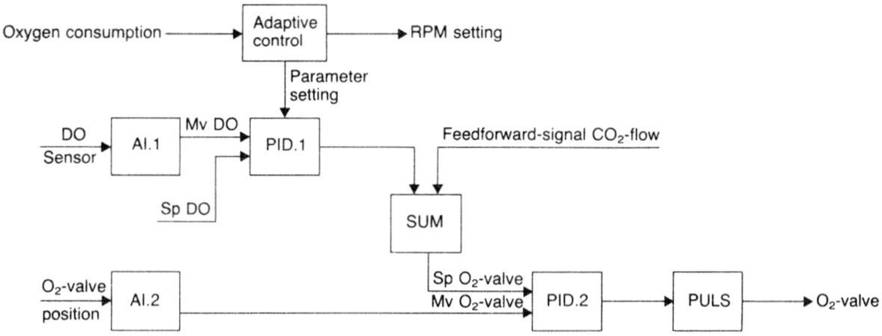

Figure 2 "DO control loop configuration"

Each square represents one functional block with the following function:

* Analog Input block "AI" reads an input from the process and converts it as required.
* PID-controller block "PID" is a proportional, integral and derivative controller.
* Sum block "SUM" adds input signals.
* Pulse outblock "PULS" converts an analog signal to an on/off output signal.

PID-controller "PID.1" compares the measured value (Mv) DO in the fermenter with the setpoint (Sp) DO. The output signal from "PID.1" is added to a feedforward signal from the pH control loop (to compensate for the CO2-flow into the reactor and the consequent dilution effect it will have on the oxygen level). This addition takes place in the "SUM" block.

The resulting output signal from the "SUM"-block, setting the percentage valve opening of the oxygen valve, is setpoint to PID-controller "PID.2". "PID.2" compares the actual position of the O2-valve with the setpoint. The output signal from "PID.2" is input signal to the following "PULS"-block that gives pulses to the valve to change its position and therefore the incoming quantity of oxygen.

When cell density increases a modified response pattern is needed. A user algorithm for adaptive control has been designed to set new PID-parameters for the "PID.1"-controller to improve control action. The adaptive controller also has the possibility to manipulate the stirring speed setpoint if the mass transfer of oxygen is too low.

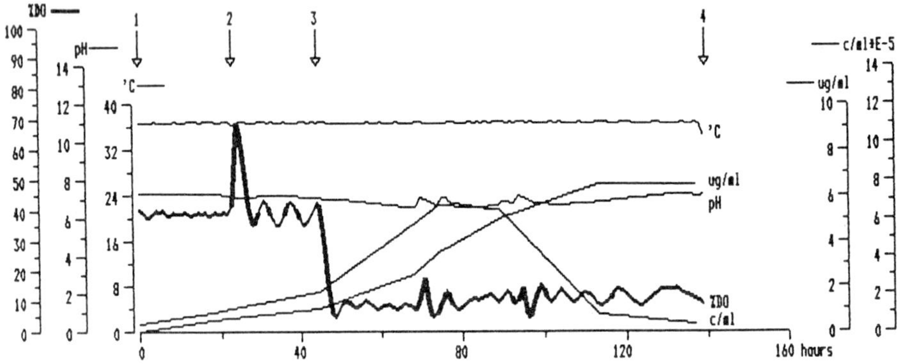

Figure 3 "Batch cultivation"

Events

1: Cultivation started in a volume of 50 litres
2: 50 litres of media added
3: Setpoint DO changed from 40% to 10%
4: Cultivation terminated after 140 hours

Data

Cell line	:	6.G4.B4.1.5 producing monoclonal antibodies against human IgG4
Medium	:	DMEM, NaHCO3, 0.5% FCS, penicillin-G, streptomycin
Starting cell density	:	0.4*10+5 cells/ml
Max cell density reached	:	6.7*10+5 cells/ml after 79 hours
Cultivation volume	:	100 litres
Setpoint DO (% air sat.)	:	40%, after 44 hours 10%, at 1 atm
Setpoint pH	:	7.3
Setpoint temperature	:	37°C

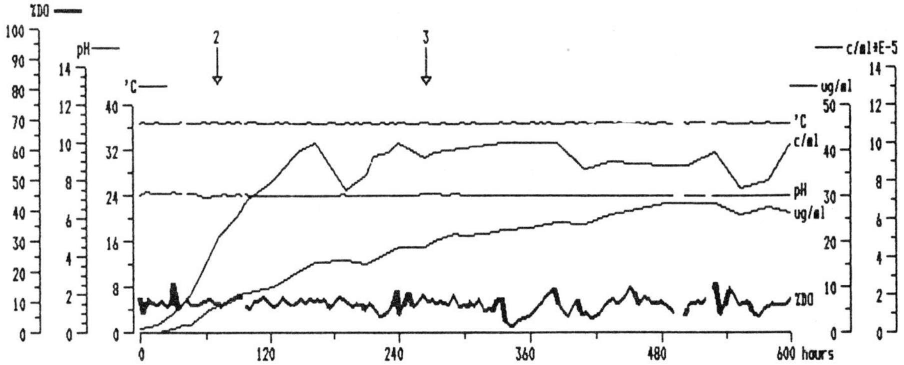

Figure 4 "Continuous cultivation"

Events

1: Cultivation started as a batch culture in a volume of 100 litres
2: Continuous cultivation started, flow rate F=50 l/day
3: Flow rate changed, F=25 l/day
4: Cultivation terminated after 600 hours

Data

Cell line	: 1.5293.A57.3.4 producing monoclonal antibodies against Gonococcus (Neisseria gonorrhoeae)
Medium	: DMEM, NaHCO3, 0.5% FCS, penicillin-G, streptomycin
Starting cell density	: 0.2*10+5 cells/ml
Max cell density reached	: 10.0*10+5 cells/ml
Cultivation volume	: 100 litres
Setpoint DO (% air sat.)	: 10% at 1 atm
Setpoint pH	: 7.3
Setpoint temperature	: 37°C

* Further installations of new reactors or new sensors for existing reactors are possible without disturbing already connected processes.

REFERENCES

1. Wergeland,I. and Wallberg,C. Continuous hybridoma cultures for the production of monoclonal antibodies: a factorial design experiment for the process optimization. Presented at ESACT april 1987.

Discussion

H. KATINGER — This is an upstream session and I want to give some personal views of this session. I was waiting for the most important points and all the speakers switched to defined media - but they never defined it. Speakers gave results but as you didn't know what was limiting they were not too meaningful. Yield of cells and product per unit medium volume gives us the concentration of product, not the fermenter. We need a strategy to perfuse knowing the limiting factor.

W-S. HU — It is not easy to define what is exactly limiting. There are so many nutrients and some can substitute for others. We have accumulated a lot of data from 15000 hours of culture and we conclude that there is no single limiting factor: I do not blame todays speakers because the animal cell systems are very complicated and need a lot of further work. Microorganisms must be environmentally efficient to survive but animal cells have a different strategy - they are used to interacting in vivo. They don't have to use nutrients efficiently - we are trying to force them to do this.

R. ARATHOON — There is one nutrient that can be used as a rate-limiting substrate and that is cystine. There is an absolute requirement with no alternative source. The problem is that under certain conditions it can limit product formation but not growth.

J. FEDER — You can make anything you want limiting. Glucose can be limiting, we even had mention of a proline limited system (Tolbert). So you can difine your growth-limiting medium.

There has been a lot of discussion on substrates - encapsulation, microcarriers etc. Traditionally we have been happy with a substrate if it gave a high plating efficiency. What I want to do is raise the flag 'on the effect of substrate on product expression levels'. An example, mammary cells on styrene do not express proteins (eg, casein) unless the plate is coated with collagen.

R. SPIER — I agree but you do not have to raise the flag as it is well recognised. B. Griffiths has shown the effect of substrate on Herpes virus, and parvovirus expression is also dependent upon the nature of the substrate. Unravelling the interaction is something which needs sorting out.

A. KADOURI — I agree with the previous speakers, you have to adjust the substrate to the biological needs of your project.

A way of determining whether cells are growing or non-growing by metabolic measurements is made by different patterns in quiescent and growing cells. Some quiescent cells produce high levels of lactic acid with no growth.

A. HANDA — Has anyone related these shear experiments to impellers, and in particular the zone in which the impeller is based? My feeling is that at high impeller speeds you are creating a vortex and entraining your gas-liquid interface into the fermenter. It would be interesting if you could dissociate the effects of gas bubbles at the interface and the shearing effect of the impeller.

B. GRIFFITHS

I wish to make a general comment as a result of certain comments I have heard on the papers in this upstream session. There are obviously groups representing two objectives. One is pursuing the challenge of growing cells at high density, to test model systems, to study cell behaviour at high density which is going to be different to that at low density - J. Feder mentioned the problem of diminishing returns. The other group wish to apply available cultures immediately for production use. They look at some of these experimental systems - see the complexity, the lack of scale up potential and possible licencing problems and show impatience at this work. This group says why all this effort for 10^8 cells/ml when 10^7 cells/ml works well. I sympathise with this view having tried all sorts of substrates over the years such as polystyrene turnings, glass wool, foams, to try and improve cell density yet, like the quotation used by J. Petriciarni, "I have gone round the full circle and come back with a bit more understanding and returned to my original system". Incidentally, this is the glass bead system and we found that 3 mm beads traditionally used to maximise surface area gave us lower cell densities per unit volume than 5 mm beads ie, to maximise cell performance we had to reduce the intensity of the system. To conclude, different people want different things out of their systems and that we can learn a lot from each other when using different approaches and end-points.

Section 5
Downstream unit processes

39 Design criteria for the purification of biopolymers for the pharmaceutical industry

S. Johansson, H. Johansson and D. Low

INTRODUCTION

Liquid chromatography utilizing different physical and chemical properties such as molecular weight, charge, isoelectric point, biospecific interactions or hydrophobic character of proteins, has for many years been the leading research technique for the purification of proteins.

At industrial scale the technique has not been widely spread since not many proteins have been produced in particulary large volumes. The outstanding exception is the purification of insulin that has developed into a world wide ton-scale per year business since 1971 (1).

With the commercialization of recombinant DNA-technology and the general trend towards native pharmaceuticals that has appeared during the past five years, liquid chromatography has become a crucial technique for the separation of proteins in pharmaceutical industries. Irrespective of the upstream source of the protein mixture, a micro-organism, an animal cell culture, tissue extracts, the technique for achieving the product quality needed today is liquid chromatography.

This paper will discuss how a research method for purification of a protein can be transferred to a well designed industrial process. The different chromatography techniques used in downstream processes with characteristics and applicability are presented. Starting by defining the design data that has to be established by the researchers, the paper will discuss what the implications of these are on method development, system design and choice of equipment in terms of reliability and reproducibility.

CHOICE OF METHODS

The scale-up of a chromatography process has shown to be fairly straightforward concerning the chromatographic performance itself (2). Therefore the process development work can generally be performed at small laboratory scale without causing major problems when scaling up. All of the methods commonly used in the laboratory for protein purification can be scaled up to large scale, however, not all of them are equally suitable for large scale use. The range of methods are given in Table 1, together with comments on their applicability. Methods based on adsorption, such as affinity-, ion exchange-, and hydrophobic interation-chromatography are more suitable for handling large volumes of dilute sample than partition methods such as gel filtration, where the sample volume is limited to a fraction of the total column volume.

PURIFICATION SEQUENCE

Since some techniques are better suited to the early stages of a purification whilst others are better suited to the final stages, it follows that it is important to consider the overall sequence of steps. This is perhaps best illustrated by looking at how the nature of the separation problem varies throughout the purification scheme.

At the early stages of a purification the chief problem is recovering the product of interest from a large volume of dilute sample. The product may be less than 1 % of total protein present. It is therefore desirable to use methods which are capable of handling large volumes of sample and which can show a concentrating or enrichment effect such as ion exchange or affinity chromatography. Hydrophobic interaction chromatography is also useful in the early stages, especially if the sample is in a high salt concentration after a precipitation step.

In the case where product has been concentrated and is present in roughly equal proportions to its contaminants, some form of fractionation step is required. This step need not remove all contaminants, and is an intermediate step. Again, yield is important, and the degree of purification can be optimized by careful manipulation of elution conditions. Sample volumes are still likely to be high and methods based on adsoprtion are favoured.

The final stages in a process can be of two types, the removal of low concentration contaminants or the removal of polymers.

Separation can be achieved by selective adsorption of the contaminant or the product. Selective adsorption of the product at this stage is likely to require a larger column.

Many biological products have a tendency to exist in polymeric forms. If these are to be removed then gel filtration can be used

Table 1. Chromatographic methods for large scale purifications of biological molecules

Characteristic	Separation method	Characteristics	Applicability
Size	Gel filtration	Resolution is moderate in the fractionation mode, but excellent for desalting. Speed is low for fractionation (cycle time 8h) but very fast for desalting (approx. 30 min). Capacity is limited by sample volume.	Fractionation by gel is best suited to the final stages of large scale purification and removing critical contaminants. Desalting can be used at any stage, particularly to link two steps by buffer exchange.
Charge	Ion exchange	Resolution is usually high. Speed is very fast when using the correct matrix. Capacity is very high. Sample volume is not limiting.	Ion exchange is most suitable during the early stages of a separation, when large quantities of material are to be handled.
Isoelectric point	Chromatofocusing	Resolution is very high Speed is high. Capacity is very high. Sample volume is limited by column size.	Chromatofocusing is more suitable towards the end of a purification.
Hydrophobic nature	Hydrophobic interaction	Resolution is good. Speed can be very fast. Capacity is high. Sample volume is not limiting.	HIC is suitable at any stage in a separation, particularly when the sample is at high ionic strength, e.g. after precipitation, ion exchange or affinity.
	Reverse phase	Resolution is very high. Speed is very fast. Capacity is high.	RPC is most appropriate at the final stages when purifying low molecular weight substances.
Biological affinity	Affinity chromatography	Resolution is excellent. Speed is very fast. Capacity is very high. Sample volume is not limiting.	Affinity is suitable at any stage, particularly when a low concentration of sample is present in a large volume together with a lot of contaminants.

as a final purification step. Gel filtration is slower than other techniques used in process chromatography and can only handle limited volumes (5-10 % of V_t) of sample per cycle, so it only be used at the end of a purification process if required.

DESIGN DATA

When doing the process development as many of the large scale premises as possible should be involved and a proper laboratory scale process development should therefore lead to a number of design data for each of the chromatographic steps in the purification sequence. The established parameters and data are of two different kinds.

A. Method design data.

Information related to the chemistry of the separation.

- Gel media type
- Capacity in grams per litre of column volume
- Column height
- Flow rate (linear flow rate in cm per hour)
- Buffer types and allowed variations in pH and conductivity
- Recovery based on a well defined product quality assay
- Cleaning procedures

B. System design data

Information necessary to design the system and to choose equipment.

- Back pressure created in the column
- Acceptable materials in contact with the liquid
- Protein sensitivity for shear forces
- Buffer consumption
- Product volume
- Number of product fractions to be collected
- Monitor principles
- Feedstock and buffer storage stability
- Process time

At laboratory scale the biochemists tend to focus on the method parameters under A. However, when a large scale process is under development all of the parameters have to be taken under consideration.
The above lists cover the data needed to establish a functional flow sheet and to dimension pumps, valves, the column and tanks. From this information it is also possible to calculate the

approximate space needed for the system installation. Even initial investment and production costs can be calculated at this stage.

REPRODUCIBILITY AND RELIABILITY

Reproducibility and reliability are two key words in process chromatography and they are both affected by the process development and system design.

The reproducibility of a chromatographic process is strongly related to variations in a limited number of process parameters. The most important are; separation media quality, cleaning procedures, feedstock and buffer quality.

To convert the laboratory results into a reliable production unit the critical parts of the process must be defined. This simply means to investigate what can go wrong in the process and in the equipment. What are the consequences if problems arise and how can the process be designed to compensate for this?

IMPLICATIONS OF METHOD DESIGN DATA

Choice of gel media

There are a number of special factors which have to be considered when choosing a separation media for large scale use. Firstly, the support must be compatible with biological materials in order to get the highest possible yields. Secondly, supports for production purposes must have excellent physical stability, in order to withstand the high flow rates used in large scale chromatography. Thirdly, supports must have excellent chemical stability, so that they can withstand the rigorous cleaning and sanitation procedures used in the pharmaceutical industry for the safe and hygienic production of therapeutic agents. Finally, the support must be made according to specific safety standards (e.g. GMP), must be available in sufficient quantities and above all must be made to a consistent quality.

Buffer conditions

It is important to determine the allowed tolerances in pH and conductivity to get reproducible separation performance. If pH and/or conductivity are out of the allowed range it normally means a substantial decrease in purification effect and these parameters need to be continously monitored and the monitors should be

equipped with alarm functions.
The buffers must be defined within certain limits during the process development. An example of how this can be defined is given in Table 2. The table shows the permissible variations of the buffer conditions in a process for production of Human Serum Albumin (3).

Feedstock conditions

Just as for the buffers the feedstock has an interval in terms of pH, conductivity and concentration that is permissible in order not to affect the purification adversely. Conductivity and pH should therefore be continously monitored whilst the concentration preferably is analysed separately.
The feedstock quality is of great importance since a clean and well-defined sample will decrease the number of re-packings needed and minimizes the influence of variations in separation media quality and in the column packing. If the feedstock quality does vary substantially from batch to batch, the chromatographic process cannot produce a product with the same quality from batch to batch. In such circumstances a small scale laboratory separation might give the information needed to run a full scale process to give even product quality.

Cleaning procedures

Chromatography processes are not usually run under completely sterile conditions. The main reason for this is that the product coming out from the purification step goes through additional process steps such as final formulation and bottling where the sterility is complete. Another reason is because of the problems involved with sterilizing the packed chromatographic bed. However, there is a need for keeping good process hygiene in the chromatography and there are two main goals with process hygiene routines in chromatography. These are; to prevent bacterial growth in the process system and to keep the endotoxin level in the product down.

Good process hygiene to prevent bacterial growth should involve such aspects as; fresh, sterile filtered buffers, hygienic design of equipment, Cleaning In Place (C.I.P.) methods compatible with gel media and system, batch filtered process solutions and short process times.

The C.I.P. routine preferred in modern process chromatography is often NaOH washing of the complete system. Other methods are autoclaving and the use of bacteriostatic additives. Advantages and disadvantages of the different methods are given in Table 3. NaOH also recovers precipitated material from the chromatographic

TABLE 2

Permissible variations of pH and ionic strength (I) or molarity in buffers used in the different chromatographic steps.

Step	Correct		Range	
	pH	I	pH	I
Buffer exchange	7.0	0.025	Not critical	0.023 - 0.027
Anion exchange				
- Buffer 1	5.20	0.025	5.10 - 5.25	0.023 - 0.027
- Buffer 2	4.50	0.025	4.40 - 4.60	0.025 - 0.030
- Buffer 3	4.00	0.15	Not critical	Not critical
Cation exchange				
- Buffer 1	4.50	0.025	4.30 - 4.70	0.023 - 0.027
- Buffer 2	5.50	0.11	5.45 - 5.55	0.11 - 0.12
- Buffer 3	8.00	0.40	Not critical	Not critical
Gel filtraiton		0.05 M NaCl	Not critical	Maximum 0.02

media, e.g. lipids, proteins and therefore contributes to a longer life time for the separation media.

IMPLICATIONS OF SYSTEM DESIGN DATA

In the previous section we discussed how some of the method parameters effect the process and how to design the methods and system to compensate for a negative contribution. The system design data and the control of these will also affect the reliability and the reproducibility of the process. The negative contribution can be compensated for by a proper system design and choice of equipment.

Air in the column

Air that is transported into the column affects the flow profile over the column area and hence will affects the separation. Air-bubbles may not have been completely excluded when the system was first installed or it might have been sucked into the system during a run, on the low-pressure side of the pump. To avoid this from happening the system must be equipped with manual valves and connections that allow for emptying the system from air easily. An air-trap installed between the pump and the column prevents air from reaching the column. An air-sensor for continous monitoring of the liquid stream should also be installed between the pump and the column. Air-sensors based on capacitance are available for detecting air-bubbles in waterstreams.

Volumes in feedstock tanks

Since the biological raw material is often very expensive it is, despite the risk for getting air into the tubing, important to utilize 100 % of the material in the feedstock tank. This can be achieved by an air-sensor on the tubing leading from the storage tank instead of a level sensor in the tank which always leaves a certain volume un-utilized.

Volumes in product tanks

Over-filling of the product tank is monitored with level sensors on the product vessel. Since chromatography is not a very time critical process these events can be handled by simply pausing the process whilst changing/emptying the product vessel, whereafter the purification is continued where it was paused.

Flow stabilities

Since the flow rate is a crucial process parameters for both productivity and product quality, proper monitoring of the flow rate is important. The origin of exceptional flow rates can vary and the most common are listed below.

* Clogging of in-line filters.
* Clogging the column.
* Malfunction of the pump and other equipment such as valves causing flow restrictions.
* Leakages, can lead to a too low flow rate through the column.

Over-pressure

Over-pressure is often related to the same factors as those causing flow rate disturbances. The main reason for over-pressure in chromatography is clogging or ageing of the chromatography media. Other reasons involve clogging of in-line filters and malfunction of valves or other flow restricting equipment.

The pressure should be continuously monitored to avoid damaging the equipment or the separation media. The pressure sensor should therefore be equipped with an alarm function. Long term monitoring of the pressure in the system also gives information on the condition of the chromatographic bed.

With the above listed process parameters properly monitored and controlled with alarm functions, the chromatographic process can be considered as a reliable unit operation. The variation in these parameters stems from poor equipment performance and in the coming section the equipment and its performance is discussed.

The column

The most crucial part of a chromatographic process installation is of course the column and its packing. The column performance is continuously monitored by the separation. It is therefore continuously under control. Once the column is packed and checked for even packing, it can be seen as a reliable part of the system until external factors effect it. These included parameters such as, flow rate, pH and conductivity, i.e. the packed column has a defined chemical and mechanical resistance which must not be exceeded.

Another important parameter for the column performance is the quality of the biological feedstock. It is not uncommon that the feedstock contains material that binds very strongly to the separation media leading to a reduction of the separation performance. Therefore comparison of the separation profile between consequtive runs no. 1, no. 10, no. 50 etc., should be a routine for determining the overall condition of the column including chemical, mechanical and contamination effects.

The product monitors

The second most important equipment for successful process chromatography is probably the monitor used for monitoring the separation. Normally this is a UV-monitor for continuous measurement of the A^{280} of the liquid stream coming out of the column. Malfunction of this monitor can turn out to be extremely expensive since the monitor signal is normally used to control the product stream between waste and product vessel.

The problems with UV-monitoring in process scale involve aspects such as ageing of the UV lamp and quantitatively measuring at high absorbances. The ageing of the lamp should preferably be continuously monitored to avoid errors caused by a low signal strength from the monitor. Typical life lengths of lamps in conventional UV-monitors are 8 000 hours.

The pumps

The pump in a chromatograhic system effects not only the flow rate and pressure in the system but also the overall productivity of active substances. The pumping principle in many cases needs to be "gentle" to avoid denaturing the sample. For example, gear pumps have been shown to cause product stability problems in the production of HSA (Eriksson (4)), whilst in the purification of Monoclonal anatibodies has shown to be less crucial (5). Recommended pump types involve peristaltic pumps, membrane pumps or lobe rotor pumps. These have all been installed in a number of plants and have proved to give good performance in chromatographic processes (6).

Valves

The positioning of automated valves in the system needs to be monitored to assure that the flow is taking the right path. This is especially important for the valves used for collection of the product.

The reliability of a chromatographic process system can be summarized by Table 4.

TABLE 3

Autoclaving	Bacteriostatics	NaOH
+ gives sterility - no cleaning effect - does not destroy endotoxins - not possible with packed columns	+ gives a low bacterial level or sterility + only possible with some affinity media - does not remove endotoxins - no cleaning effect - risk of traces in the product	+ sterilizes complete column systems + good clening effect + destroys pyrogens + cannot contaminate the product - careful handling

TABLE 4

Crucial parameters	Precautions	Alarm function
pH	Continuous monitoring, defined limits	X
Conductivity	Continuous monitoring, defined limits	X
Feedstock concentration	Off-line analysis, defined limits	
Air presence in the column	Continuous monitoring, in-line air-trap	X
Tank levels		X
Flow rate	Continuous monitoring	X
Pressure	Continuous monitoring	X

Crucial equipment	Precautions	Alarm function
The column	Operate within given specifications In-line filter Feedstock preparation	X
The product monitors	Signal level control Off-line calibrations	
The pump	Continuous monitoring of flow rate and pressure	X
The valves	Continuous monitoring of position	X

Time planning

When planning the production it is important not to consider each step in isolation, but rather the total process. Practical aspects such as how many steps can be done in a single working day and how steps can be linked together are frequently important. When transferring the procedure into production, the production manager must consider production time, capacity and batch sizes for each step and dimension the steps with the overall process in mind.

BIOCOMPATIBILITY

The overall recovery of the biological activity available in the feedstock is essential to get good process economy. To design for biocompatibility means that biological degradation during the process must be minimized. This takes a deep knowledge of the involved proteins, and is where the biochemical science really meets the engineering science.

Achieving high recoveries effects the whole process design, involving buffer choice, concentration of the feedstock, chromatographic technique, choice of separation media etc. That is, already in the initial choice of process principles the biochemist must select conditions from his knowledge of protein properties and behaviour which ensure a high recovery in process scale.

Aspects like what material are allowed to be in contact with the proteins will affect the choice of equipment all through the process. The presence of proteolytic activity in the protein solution will affect the process in terms of storage times, column size, pumping capacity and tank volumes.

REFERENCES

1. Janson, J.-C., 1971, J. Agr. Food Chem., 19. 581.

2. Berglöf, J.H., and Cooney, J.M., 1983, "Scale up of laboratory systems for large scale chromatography", poster presentation at Biotech '83, London, England.

3. Eriksson, S., and Curling, J.M., 1983, J. Appl. Biochem., 5, 282

4. Eriksson, S., Pharmacia Process Development Dept., Work from Pharmacia AB, Uppsala, Sweden.

5. Malm, B., Phgarmacia Process Development Dept., Work from Pharmacia AB, Uppsala, Sweden.

6. Lindquist, L.-O., Pharmacia Process Engineering Dept., Work from Pharmacia AB, Uppsala, Sweden.

Discussion

B. MAIORELLA — You indicated early on a disappointment that reverse phase chromatography is not used more. It is very appropriate for bacterial products but in the case of mammalian cell products expressed in the native form wouldn't you be concerned with denaturation of the product using reverse-phase.

JOHANSSON — I did not exclude reverse-phase chromatography but up to now it has not been a key technique for us.

MAIORELLA — One problem is that if the probe is in an organic solvent environment the biological structure of the probe could be changed and it could effect recovery of your activity.

40 Use of a dynamic filtration method for separation of animal cells

E. Rebsamen, W. Goldinger, W. Scheirer, O-W. Merton and G. E. Pàlfi

Abstract:

For solving the problem of concentrating and washing of viable cell suspensions a dynamic filter device has been evaluated. Results show that concentrating of cells under sterile conditions can be performed. With a laboratory scale filter (Sulzer-Biodruckfilter BDF-01) a 9-fold concentration was achieved, the filtration speed was about 15 l/h. The viability did not change significantly. Details of experiments are given and advantages are discussed.

1. Introduction

To reach cell counts of more than 2×10^6 per ml in fermentation processes of suspension cells is still a great problem. Up to nowadays, only little progress has been achieved. The Group of Tolbert et al. (1981) and Tolbert and Feder (1983) described a large-scale rotating filter perfusion system for high-density growth of mammalian suspension cultures based on the use of 1 um microporous porcelain filters. With this system, it was possible to reach cell densities of about 10×10^6 per ml, with the advantage of a better product titer.

All classical filtration systems in a cross-flow-configuration show the following disadvantages:

- After a short use the membranes are fouled.
- The obligatory used pumps damage the cells.
- The shearing forces at the spacers of the filtration moduls harm the cells.

Conventional separators lead to cell damage, special constructions for cell culture cannot be used for recycling the cells. In practice up to now no real perfusion process with cell recycling or cell concentration or washing of cells was possible.

Therefore we tested the 'Dynamische Biofilter BDF-01', made by Sulzer, for the following characteristics: Influence on the cell viability under certain performance parameters, the sterility of the system and the specific filtration effect.

. Materials and methods

One mouse-mouse-hybridoma cell line and one human breast carcinoma cell line, SKBR 5, adapted for the growth in aggregates in suspension were used for the experiments. Both were cultivated in DMEM-H21-10% FCS (GIBCO) and passaged twice a week.

The Sulzer-Biodruckfilter BDF-01 (fig. 1) was adjusted to the following conditions: The system was temperated at 25°C, the pump lift of the suspension pump was varied between 10 and 100% and of the concentrate pump was varied between 0 and 50%. The rotation speed of the pumps varied between 100 and 450 rpm and of the rotor of the dynamic filter between 500 and 2000 rpm. The filtration area was 0.033 m^2 the filtration material WO2.

The technical data like pressure, flow of concentrate, and flow of filtrate were registrated. Only the inner filtration area was employed during the experiments.

The cell count and the viability were estimated in a haemocytometer after staining with trypan-blue.

3. Principle of function

The function of the Sulzer-Biodruckfilter is shown in fig. 1 and is described shortly: The suspension (8) to be filtered is pumped continuously into the pressurized ring-gap of the rotating filter. The ensuing Taylor vortices flow across the filter surfaces of the inner and outer cylinders.(9,10). The concentrate (11) is withdrawn

Fig. 1a : Sectional drawing of the "Sulzer Biodruckfilter BDF-01."

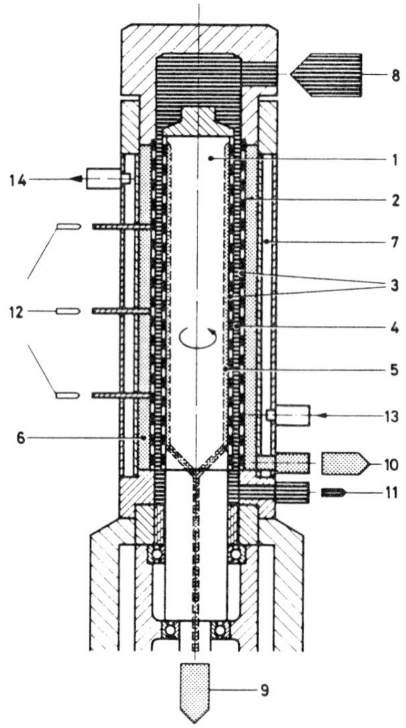

Fig. 1b : Application for cell- and enzyme recycling

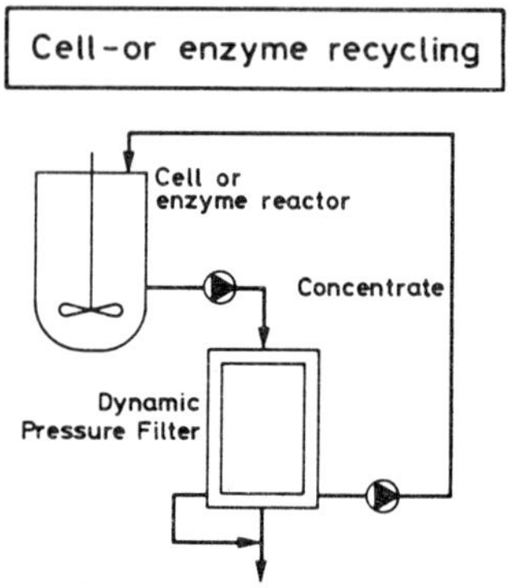

1 Inner cylinder
2 Outer cylinder
3 Filter surfaces
4 Filter chamber (ring-gap)
5/6 Filtrate chambers of inner and outer cylinder
7 Cooling jacket
8 Suspension
9 Filtrate, inner cylinder
10 Filtrate, outer cylinder
11 Concentrate
12 Washing solution/eluent
13/14 Cooling water

from the ring-gap by a pump which maintains the pressure in the filter chamber. The filtrate (9,10) escapes through the membranes.

4. Results and discussion

In a first set of experiments 1860 ml of the mouse-mouse-hybridoma cell suspension were pumped through the Sulzer-Biodruckfilter, to test the influence of the rotation speed of the pumps and the rotor on the cell viability. The cell concentrate and filtrate were pumped back to the rest of the cell suspension. The details and results are shown in table 1 and fig. 2. It is evident, that the variation of the rotation speed of the pumps between 100 and 450 rpm has no negative effect on the viability of the cell suspension. The increase of the rotation speed and the decrease of the pump lift of the concentrate pump increased the pressure, the flow of the concentrate and the flow of the filtrate only, but had no negative influence on the viability. The rotation speed of the rotor of 500 rpm (Taylor number of 350) had no negative influence on the cells. The increase via 1000 rpm (Taylor number of 700) up to 2000 rpm lowered the viability of the suspension and the concentrate significantly. It looks like, that the Taylor number of 700 is already the upper limit for the use of this system in animal cell culture.

After the 135th minute one liter of filtrate was removed. During the time of 240 minutes the cells passed through the filtration unit for more than 40 times. Finally, starting with a cell density of 500000 per ml a final cell count of more than 2.5×10^6 per ml could be achieved.

For testing the performance under sterile conditions a cell suspension of SKBR 5-cells was taken and concentrated in the Biodruckfilter. The results are shown in table 2.

4.5 l of the cell suspension were concentrated with the sterilized apparatus. The input cell density was about 1×10^6 cells per ml, the cell count of the concentrate was 4.1×10^6 cells per ml after the filtration time of 0,3 h. The viability did not change. Although the concentration factor of the suspension was 9-fold, only a cell concentration of about 4.5-fold could be achieved. The reason for

Tab. 1. Influence of different performance parameters of the "Sulzer Biodruckfilter BDF-01" on the cell count and the viability

Time (min)	Rotation speed of rotor (rpm)	Rotation speed of the pump (rpm)	Pumplift (%)		Pressure (bar)	Flow of concentrate (l/h)	Flow of filtrate (l/h)	Total cell count (10E3/ml)		Viability (%)	
			Suspension	Concentrate				Suspension	Concentrate	Control	Concentrate
000	500	100	100	50	0.05			555	555	74	74
020	500	100	100	50	0.025	1.86	1.71	396	665	72	74
045	503	101	100	50	0.030	1.98	1.71	385	745	68	73
075	503	200	100	50	0.040	3.80	3.35	475	610	78	71
105	503	300	100	50		5.82	5.25	490	800	70	75
135	503	301	100	10	0.095	1.34	9.75	210	1735	60	73
1. ltr. of filtrate was taken away											
165	503	301	100	10	0.155	1.31	9.75	315	2100	67	76
195	503	400	100	10	0.215	2.10	13.20	235	2525	72	69
225	503	450	100	10	0.235	2.31	14.70	175	2165	71	62
240	503	450	100	05	0.225	1.56	15.15	325	2320	68	68
270	1000	450	100	05	0.305	1.59	15.30	095	2245	58	53
300	2000	450	100	05		1.71	15.15	065	1.805	46	13

Fig.2.: Influence of the rotation speeds of the pumps and the rotor on the total cell count and the viability of the cells in the original and concentration suspension see also table 1

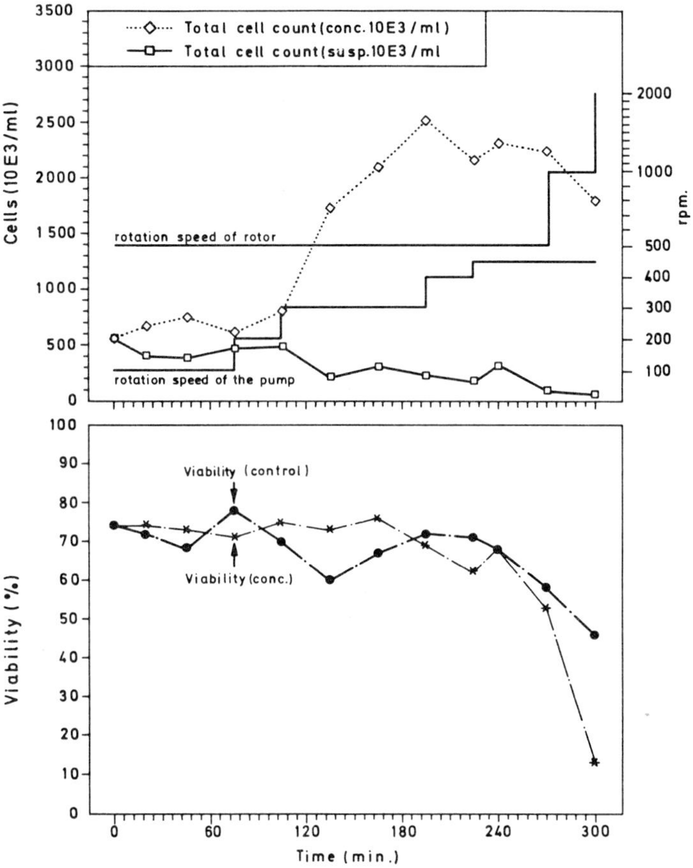

Tab. 2. Performance parameters of the "Sulzer Biodruckfilter" under sterile conditions.
A: Filtration
B: Cell - Cultivation

	Time (h)	Cells/ml (10E6)	Viability (%)	Concentration factor		Loss of cells(%)	Total cell count	Total volume(l)
				supernatant	cells			
A	0	1	100	1	1	0	4.5E9	4.5
A	0.3	4.1	100	9	4.5	50	2.05E9	0.5
B	0.31	3.0	100				2.05E9	0.7
B	21	3.7	100				2.59E9	0.7

this high cell loss is caused by the remain of cells in the filtration unit, because it was not washed afterwards. 200 ml of fresh DMEM-H21-10% FCS were added to the cell concentrate of 0.5 l and cultivated for 21 h. During this time, there was an increase in cell density from 3×10^6 cells per ml to $3{,}7 \times 10^6$ cells per ml. The viability rested at 100%. This result indicates too, that this construction does not contain any toxic material.

5. Conclusion

As the results show, the Sulzer-Biodruckfilter BDF-01 can be used for the separation and washing of animal cells under sterile conditions and shows the advantage of the possible recycling of the cells to create a perfusion system similar to that previously published by Tolbert et al. (1981) and Tolbert and Feder (1983). The main advantage, in comparison to the published systems is the ring-gap in combination with Taylor vortices which causes a longer half life time of the filtration unit.

If a miniaturization of this system is possible it can be used as on-line sampling system with cell recycling for the fermentation industry. This would solve a lot of sampling problems.

6. References

Tolbert, W.R., Feder, J., Kimes, R.C. (1981) Large-scale rotating filter perfusion system for high-density growth of mammalian suspension cultures. In Vitro 17, 885-890

Tolbert, W.R., Feder, J. (1983) Large-scale cell culture technology. In: Annual Reports on Fermentation Processes. Vol.6 (ed.: Tsao, G.T.), Academic Press, New York, pp. 35-74

This paper was presented at the Poster-Session of the 7th ESACT-Meeting held on September 30th to October 4th, 1985 in Baden, Austria.

41 Continuous culture of murine hybridomas with integrated recovery of monoclonal antibodies

J. Rudge, M. A. Desai, S. A. Shojaosadaty and A. Lyddiatt

ABSTRACT

A continuous culture apparatus with integrated product recovery is described for the laboratory scale production of monoclonal antibodies. Comparisons are presented of continuous and batch immunopurification using immunoadsorbents based upon agarose and kieselguhr-agarose composite materials. Consideration is given to the effect of cell debris upon adsorbent solid phases, the influence of proteases upon product integrity, and the use of rapid affinity-HPLC for quantitative monitoring of product.

INTRODUCTION

A growing need exists for the production of monoclonal antibodies (MCAB) having purity specifications and product requirements constrained to different degrees by applications as varied as immunopurification, diagnostic analysis and human or vetinary therapy (1). Commercial production has succeeded in rodent ascites, but has been increasingly superceded, for reasons of scale, quality control and economics, by *in vitro* culture of hybridomas. Monoclonal antibodies are currently produced in commercial quantities (1-1000 g) using dispersed or encapsulated hybridoma cells grown in stirred batch or air-lift fermentors, perfused tanks or membrane limited reactors on scales up to, and exceeding, 1000 litres (2). Despite such advances in process development, there remains a requirement for significant production on laboratory scales for MCAB purified to specificities governed by applications in research and development, or special commercial operations. We report the development of a continuous culture system with integrated product recovery having economic and practical advantages over discrete fermentation and product recovery operations as conventionally practiced.

METHODS

Batch cell culture

TBC-3 murine hybridomas used in this study were derived from WC2 parent cell line (courtesy of Immunology Dept. Birmingham University) and produced MCAB specific for the Cγ2 region of human IgG (4). Hybridomas were grown in RPMI 1640 medium (Gibco, Paisley) supplemented with 600 mg/l glutamine, 2 mg/ml sodium bicarbonate, 0.0004% (v/v) 2-mercaptoethanol, 2 ppm antifoam c (Sigma Chemicals, Poole), 50 IU/ml penicillin, 50 μg/ml streptomycin (Gibco) and 5% foetal calf serum (BDS, Birmingham). Stock cultures (10 ml) were maintained in standard plastic cell culture flasks (Nunc, Gibco). Batch cultures were expanded by dilution up to 2 litre volumes in round flasks agitated by means of a magnetic stirrer in a 5% CO_2 incubator at 36°C. Viable and non-viable cells were estimated by dye-exclusion using erythrosin B (Sigma).

Analytical methods

MCAB was quantified by a sandwich-type ELISA (3), and by affinity-HPLC (8). Proteolytic activity at neutral pH (7.5) was assayed according to the method of Tomarelli (9), calibrated against a trypsin standard curve, and expressed as trypsin equivalent units (specific activity 10000 BAEE units per mg). Acidic protease activity (pH 3.2) was assayed by the method of Kassell and Meitner (5), calibrated against a pepsin standard curve, and expressed as pepsin equivalent units (specific activity 1 Anson unit per g). Purity of MCAB was determined by SDS polyacrylamide gel electrophoresis according to the method of Laemmli (6)

Continuous cell culture

Cells were maintained in growth media as above in a round flat-bottomed flask (working volume 730 ml) with ports for medium addition and removal, air venting, headspace gassing, straight tube sparging, sample removal, and pH and dissolved oxygen probes (see Figure 1). Temperature (36°C) was controlled by immersion in a water bath and agitation provided

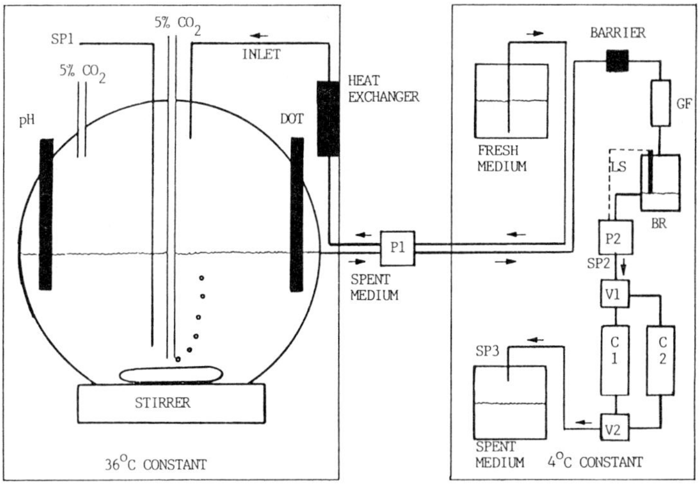

Figure 1. Schematic diagram of apparatus for mammalian cell culture with integrated product recovery.

SP1–SP3; sampling points, P1,P2; peristaltic pumps, V1,V2; switching valves C1,C2; immunoadsorbent columns, BR; buffer reservoir, LS; level sensor, GF; glass wool filter.

by a magnetically driven stirrer (60 rpm). Medium dilution was performed at 28 ml/hr and spent medium passed through a bacterial barrier before clarification and fractionation at 4°C (see later). The vessel was inoculated with cells at a density of 1 x 10^5/ml and the medium was gassed initially via the headspace with sterilized 5% CO_2:95% air at 100 ml/min. When the cell density exceeded 4 x 10^5/ml, the medium was sparged at 35 ml/min through a 6 mm diameter tube to minimise bubble damage to cells. Cell densities were maintained at 7 x 10^5/ml for productive cultures.

On-line product recovery

In order to maintain dilution rates and ensure constant contact of outlet flowstreams with on-line immunoadsorbents, a number of design features were incorporated into the apparatus. The working volume of the culture vessel was accurately maintained by means of a weir with the spent medium outlet having a larger inner diameter than the inlet (1.6 mm cf 0.8 mm). As a consequence, a buffer reservoir (volume 50 ml) with a side arm was required to collect the discontinuous flow of spent medium. A level sensor activated a pump placed between reservoir and immunoadsorbent when spent medium rose above the outlet side arm, thereby protecting the adsorbent column against air inclusion. Cells and cellular debris in spent medium which promote fouling and blinding of affinity adsorbents were removed by the application of the buffer reservoir as a settling tank and/or the use of a glass wool filter (1.0x3.0 cm) placed upstream of the buffer reservoir. Since both devices for cell separation were sited downstream of the bacterial barrier, they could readily be maintained or replaced without compromise to the sterility of the continuous culture vessel.

Human IgG, the specific antigen for TBC-3 MCAB, was purified to molecular homogeneity by DEAE ion exchange chromatography (3) from outdated human blood (courtesy of the Blood Transfusion Service, Birmingham). Purified IgG was immobilised on Macrosorb K4AX (Sterling Organics) and Sepharose CL-4B (Pharmacia) previously activated using the CNBr method of March *et al.* (7), to yield a final ligand density of 4.5 mg/ml as judged by spectrophotometric analysis of the protein reaction. Columns (1.6x8.5 cm; 15 ml matrix) of immobilised ligand were prepared in phosphate buffered saline at pH 7.5 (PBS), and spent medium was loaded at 15 cm/hr. Pre- and post-column concentrations of MCAB were monitored, and when MCAB breakthrough exceeded 10% of starting concentrations, a fresh column was brought on-line and the saturated column removed for further processing. Saturated columns were washed with three column volumes of PBS at 30 cm/hr. The MCAB was eluted at 15 cm/hr with 3 M KSCN in PBS, and desalted on a Sephadex G15 column (5.0x72.0 cm) at 15 cm/hr. Purified MCAB was stored frozen at -20°C.

RESULTS

Batch and continuous culture

A schematic diagram of the continuous culture vessel with on-line purification is shown in Figure 1. In order to limit the degredation of sensitive medium components such as glutamine, fresh medium was stored at 4°C and equilibrated at 36°C in a coiled heat-exchanger before addition to the culture vessel. Spent medium was removed through a horizontal side arm on the culture vessel to offset cell accumulation and settling.

A typical batch growth profile is shown in Figure 2 which indicates that cells attain maximal growth betwen days 3-4. Dead cells increase in number toward day 6 with concomitant increases in the concentration of extracellular MCAB and acidic protease. No neutral protease activity was detected in batch culture. A typical profile for cell growth and MCAB production in continuous culture is displayed in Figure 3. Viable cell concentrations were maintained at approximately 7×10^5/ml in the culture vessel, whilst the concentrations of acid protease (expressed in μg/ml pepsin equivalents) and MCAB remain constant in the culture vessel. In Figure 3, the temporal decline in cell numbers at days 20 and 34 were promoted by operational difficulties with gas sparging and medium input respectively. In both cases, MCAB productivity was readily restored and the culture actively maintained for a total of 48 days.

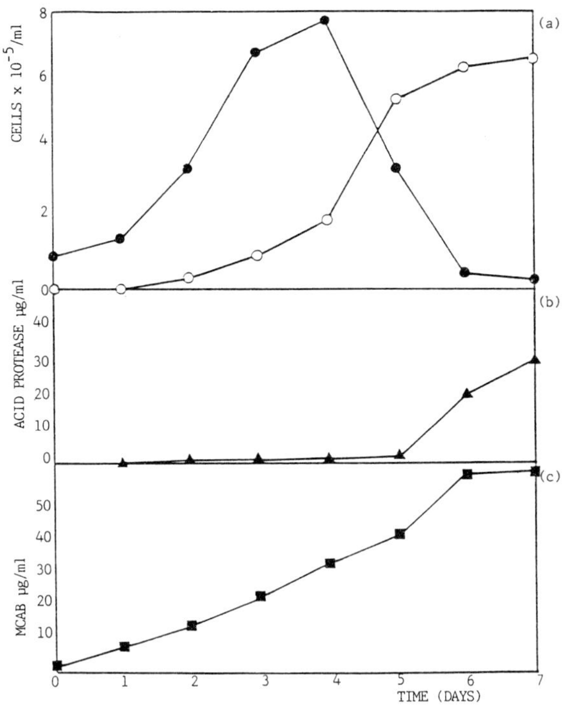

Figure 2. Batch culture of TBC-3 murine hybridomas.

Cells were grown in 2 litre batches and monitored for cells, acid protease and MCAB as described in Methods.

(a) Live cells (●) and dead cells (○) determined from dye exclusion studies.

(b) Acid protease (▲) expressed as equivalent units of pepsin determined from standard plots.

(c) MCAB concentration (■) determined by ELISA.

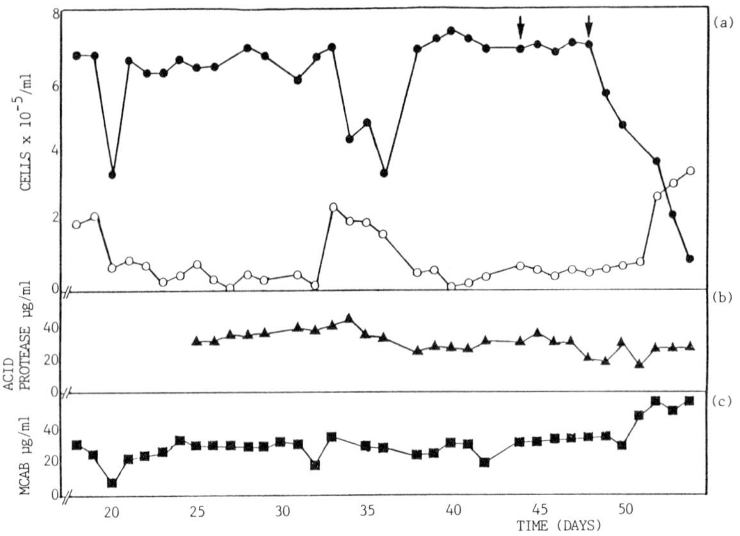

Figure 3. Continuous culture of TBC-3 murine hybridomas.

Culture and analyses conducted as described in Methods.

(a) Live (●) and dead (○) cells in culture vessel. ↓↓ indicates period (days 45-49) of integrated product recovery recorded in Figure 4.

(b) Acid protease concentrations in culture vessel (▲) expressed as equivalent mass units of pepsin determined from standard plots.

(c) Extracellular concentration of MCAB (■) in culture vessel determined by ELISA.

Spent medium clarification

It was essential to reduce the solids loading of culture output prior to MCAB adsorption on immunoadsorbents. Methods exploiting a settling vessel and/or a glass wool filter proved effective in the removal of the majority of cells and debris, but total clarification requires the design of on-line minicentrifuges or cyclones considered inappropriate to an economic, laboratory operation. Glass wool filters, replaced at 2-4 day intervals along with immunoadsorbent columns, were better suited to continuous MCAB recovery than the cell settler when used alone. Particulates which evaded the filters blinded Sepharose immunoadsorbents (1.6x8.5 cm) promoting a resistance to flow (15 cm/hr) equivalent to a back pressure of 500 mbar in less than 24 hours. Macrosorb immunoadsorbents operated in similar conditions were unaffected by particulates and displayed negligible resistance to media flow over 7 day periods.

MCAB purification

Figure 4 depicts the concentration of acidic proteases and MCAB product in samples taken from both culture vessel and upstream of immunoadsorbent columns operated in a 4 day period of continuous recovery (see Figure 1; days 45-49). It is clear that acid protease and MCAB was released from autolysed cells trapped in cell settlers or glass wool filters. In addition, Macrosorb immunoadsorbents were capable of greater than 90% MCAB recovery during 48 hour operations.

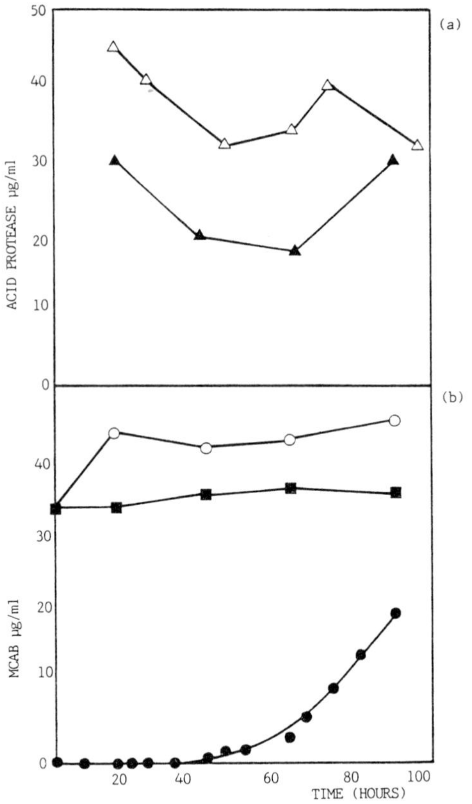

Figure 4. On-line recovery of monoclonal antibodies from continuous culture.

Culture and analyses were conducted as described in Methods. Samples taken at points SP1-SP3 indicated in Figure 1. Productivity is associated with day 45-49 in Figure 3.

(a) Acidic protease concentration in samples taken from culture vessel (▲) and pre-column (△) and expressed as equivalent units of pepsin determined from standard plots.

(b) MCAB concentration in samples taken from culture

Analysis of MCAB purified by immunoadsorption in batch operations, or on-line processes integrated with continuous cell culture, was undertaken by SDS-polyacrylamide gel electrophoresis under non-reducing conditions. Gel photographs (Figure 5) indicate the relative concentration of products and impurities (approximately 1:50) in clarified cell culture medium, and the high degrees of purification possible by immunoadsorption. It is also clear that the protease detectable in batch and continuous culture has no deleterious effect upon the molecular integrity of the product, even over time scales of experiments exceeding 170 hours.

DISCUSSION

A simple, and readily-constructed apparatus for the continuous culture of murine hybridomas with on-line MCAB recovery has been described. Purified MCAB was found identical to that purified from batch culture in terms of biological quality and product yield. As indicated in Table 1, significant savings may be achieved in labour costs when using continuous culture. Continuous culture reduced the time taken to produce 1 g of MCAB from 52 to 33 days through the elimination of work-up procedures associated with batch culture. Costs of MCAB production would be further reduced if the concentrations of foetal calf serum were reduced from 5%, or eliminated from serum-based media. For example, 75% of the costs of media and consumables quoted in Table 1 for continuous culture were accounted for by foetal calf serum alone.

vessel (■), pre-column (○) and post-column (●) and determined by ELISA and affinity-HPLC.

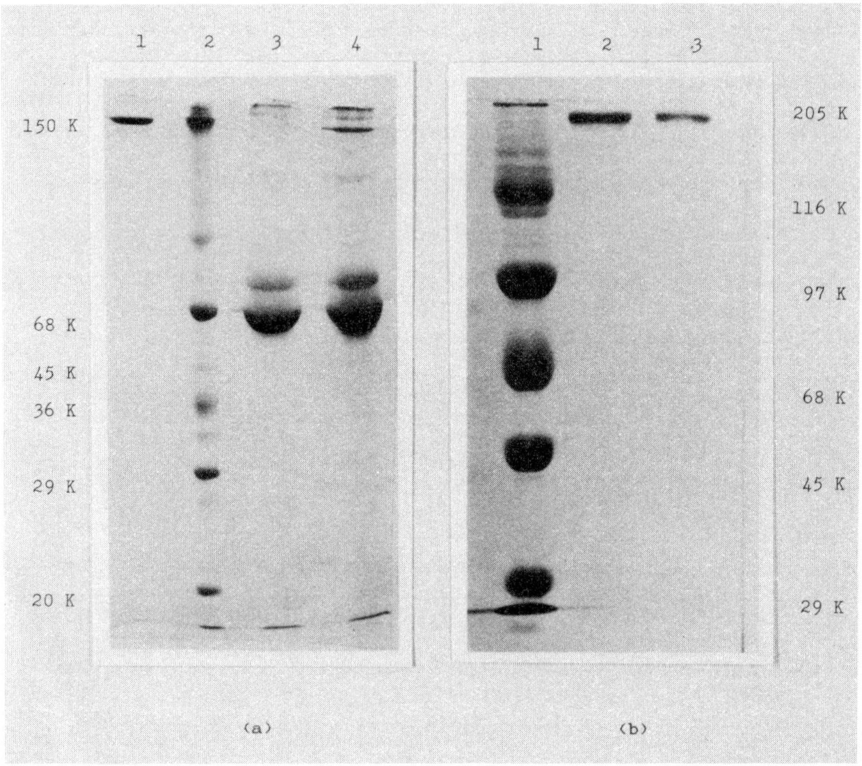

Figure 5. SDS-polyacrylamide gel electrophoresis of stages of immunopurification of monoclonal antibodies from batch and continuous cultures.

10% acrylamide gels were performed by the method of Laemmli (6) under non-reduced conditions. Protein visualised by staining with Coomassie blue R-250.

(a) Purification of MCAB from batch culture using off-line immunoadsorption on human IgG-Sepharose-CL 4B (3).

Track 1: Purified MCAB.

Track 2: Standard proteins including human IgG (150 K), albumin (68 K), ovalbumin (45 K), glyceraldehyde-3-phosphate dehydrogenase (36 K), carbonic anhydrase (29 K) and trypsin inhibitor (20 K).

Track 3: Culture medium in immunoadsorbent effluent stream.

Track 4: MCAB containing culture medium.

(b) Purification of MCAB from batch culture using off-line purification and from continuous culture using on-line purification.

Track 1: Standard proteins including myosin subunit (205 K), β-galactosidase subunit (116 K), phosphorylase B subunit (97 K), albumin (68 K), ovalbumin (45 K) and carbonic anhydrase (29 k).

Track 2: MCAB purified off-line.

Track 3: MCAB purified on-line.

The production of acid protease as an inevitable process of cell death in both batch and continuous culture has been highlighted in this study. However, there appears to be no compromise to integrated product recovery from continuous culture even where times of exposure of imunoadsorbent and bound product to protease exceeded 170 hours. This is clearly because the pH of spent medium (≃6.8) was higher than the pH optimum of the acid protease (pH 3-4), and because of the presence of high concentrations of serum proteins (≃2.5 mg/ml) relative to MCAB (30-45 μg/ml). Preliminary studies of repetitive use of immunoadsorbents suggest that these conditions similarly protect affinity ligands. However, acid proteases may pose a threat to product or ligand integrity in cultures using low protein media, or recovery operations dependent upon product adsorption to protein-A solid phases with subsequent desorption at pH 2.0.

The fouling promoted by cellular debris deposited in packed-bed adsorbents is an important consideration for the design of integrated MCAB recovery. Incomplete removal of debris, inevitable in such a simple apparatus as that reported here, places a time constraint upon the maintenance of culture dilution rates and the efficient operation of on-line immunoadsorbent columns. In the present study, resistance to flow was monitored as a function of changes in back pressure in an agarose-based immunoadsorbent (Sepharose), and a kieselguhr-agarose composite material (Macrosorb). Preliminary results have shown that the agarose column fouled within 24 hours. Whilst similar columns of composite materials show no significant resistance to flow after

170 hours operation on-line to continuous culture. In addition, the composite material, consisting of a rigid, porous kieselguhr superstructure filled with 4% cross-linked agarose, is resistant to the radial and axial compression problems which can compromise agarose materials at high flow rates. In the context of this study, composite materials may be washed and re-equilibrated more rapidly than agarose materials during off-line processing and re-equilibration.

Monitoring of MCAB product at various points within the continuous culture system may be routinely performed off-line by conventional ELISA methodologies (3). Recent work in this laboratory (8) has developed affinity-HPLC using columns of human IgG-silica (5-20 μm; 60-4000 Å pores) to quantitatively estimate concentrations of MCAB in cell cultures. Direct comparison of HPLC determination with standard ELISA indicates the HPLC to be more reliable and more sensitive across broad ranges of MCAB concentration (ng-mg). The HPLC system has the advantage of analysis times of 10-15 minutes, as compared to 4-5 hours for ELISA, which offers possibilities for dynamic control of mammalian cell culture and product recovery. Using a computer and controllable switching valves, it will be possible to automatically sample and monitor MCAB concentrations in the culture vessel and medium flow upstream and downstream of on-line affinity columns. MCAB concentration data, logged as integrals of UV absorbance specifically eluted during HPLC analyses, will be used to control the exchange of saturated for unsaturated immunoadsorbents when MCAB breakthrough exceeds values deemed inefficient for optimal recovery (eg. 10% load concentration).

Such an exchange would automatically trigger the processing and desorption of MCAB. Affinity-HPLC has potential for the monitoring and control of column desorption and regeneration, and also the dynamic control of continuous cultures (eg. dilution rate, gassing etc.).

In conclusion, the production of MCAB on a laboratory scale by continuous culture with integrated recovery shows distinct economic and practical advantages over conventional batch production. Intensive manual operations are reduced and MCAB may be produced more cheaply than in batch operations without compromise to product quantities and qualities. Automation of continuous culture and on-line recovery exploiting data gathered from rapid affinity-HPLC analyses promises further advantage. Whilst not rivalling current intensive approaches to commercial production of MCAB, the method offers an attractive vehicle for routine production of high quality material suited to research and diagnostic use.

ACKNOWLEDGEMENTS

JR and MAD would like to thank the Science and Engineering Research Council for financial support.

REFERENCES

1. Adamson, S.R. and Schmidli, B. (1986). Industrial mammalian cell culture. *Can. J. Chem. Eng.* **64**, 531-539.

2. Birch, J.R., Thompson, P.W., Lambert, K. and Boraston, R. (1984). The large scale cultivation of hybridoma cells producing monoclonal antibodies. *Presented at the American Chemical Society, Philadelphia.*

3. Desai, M.A. (1987). Affinity chromatography in downstream processing. *M.Phil. Thesis, University of Birmingham.*

4. Jefferis, R., Lowe, J., Ling, N.R., Porter, P. and Senior, S. (1982). Immunogenic and antigenic epitopes of immunoglobulins. 1. Cross-reactivity of murine monoclonal antibodies to human IgG with the immunoglobulins of certain animal species. *Immunology*, **45**, 71-77.

5. Kassell, B. and Meitner, P.A. (1970). Bovine pepsinogen and pepsin. *In Methods in Enzymology (Perlmann, G.E. and Lorand, L. eds), Academic Press, New York and London.* **19**, pp 336-347.

6. Laemmli, U.K. (1970). Cleavage of structural protein during the assembly of the head of bacteriophage T4. *Nature*, **227**, 680-685.

7. March, S.C., Parikh, I. and Cuatrecasas, P. (1974). A simplified method for cyanogen bromide activation of agarose for affinity chromatography. *Anal. Biochem.* **60**, 149-152.

8. Shojaosadaty, S.A. and Lyddiatt, A. *Manuscript in preparation.*

9. Tomarelli, R.M. (1949). The use of azoalbumin as a substrate in the colorimetric determination of peptic and tryptic activity. *J. Lab. Clin. Med.* **34**, 428-431.

TABLE 1. Economic comparison for the production of 1 g of monoclonal antibody.

ITEM	Batch culture[1]	Continuous culture
Total time taken (Days)[1]	52	33
Media and consumables[2] (including plastics etc)	176	185
Cost of man hours[3]	150	40
TOTAL COST (£):	326	225

1. Batch culture values based on maximal performance using typical laboratory based culture facilities, i.e. one CO_2 incubator containing three stirring positions (6 litres/week).
2. Assuming an average of 50 μg/ml MCAB from batch culture, and an average of 45 μg/ml MCAB from continuous culture.
3. Man hours arbitrarily costed at £10.00 per hour. Man hours represent the time taken for vessel washing, autoclaving, media preparation and sterilisation, cell counting, cell maintenance and MCAB purification.

Discussion

W. SCHEIRER A main problem in continuous purification processes in the pilot plant is the impossiblity of sterilising the downstream equipment. Can you give any information on the pyrogen content of your preparations?

RUDGE — We have not measured pyrogen levels basically because we are interested in the MCAb itself. The system is not used for materials that are to be used in vivo.

B. GRIFFITHS — The main difference in your cost figures was man hours. I am wondering how useful it is to compare processes on this basis as the man hour figure is arrived at by actually adding up the hands-on time. Now if staff were fully occupied all the time it would be meaningful but in practice this is difficult to achieve as jobs can't be sequenced that well. Thus saving in man hours might not satisfy the accountants if it just means the staff have extra time for coffee breaks etc. What are other peoples views on the validity of using man hours to compare the cost-effectiveness of different systems?

RUDGE — I only counted the time that people were actively involved and not the time waiting for things to happen eg, I did not count autoclaving as 3 hours but as half an hour.

GRIFFITHS — This is the point I am trying to make - did the staff have useful occupation whilst waiting for things to happen?

A. MILLER — I have been struck by the lack of probes to measure all the parameters you discussed, and lack of progress in developing them. Are you aware of FTIR, (Fully Transformed Infra Red Spectroscopy) which might register the infra-red spectra of individual molecules, store yet produce a meaningful approach for safety assessment.

GRIFFITHS — Does the use of primary cells constitute a saving in animals?

GOLDBERG — It cuts down the exposure of animals to pain.

Section 6
Vaccine products

42 Improved production of Marek's disease vaccine with the multi-tube technology

R. Dugré, M. Corbeil and M. Boulay

ABSTRACT

A process for the production of Marek's disease vaccine using an improved roller bottle system, the multiple-tube (TM) technology, is described. The very high productivity of this technology combined to its simplicity and reliability are such that it must be considered as one of the most outstanding technology for this type of production.

INTRODUCTION

Marek's disease vaccine, a very important poultry vaccine, is produced using anchorage dependent cells namely, chicken embryo fibroblasts (CEF). Microcarriers have been shown to be a suitable system for production of this vaccine (3). However, roller bottles are still the most widely used technology for this type of production. This generally means multiple labor-intensive manipulations and increased risk of contamination if hundreds roller bottles are used.

Some years ago, an improved roller bottle technology called the multi-tube system (TM) was described for the mass production of anchorage dependent cells and viruses (1,2). Increased growth surface area was obtained by the insertion of glass tubes (spaced by silicone rings) in the roller bottle (fig. 1). Culture medium was at that time recirculated from the TM bottle to a reservoir containing a supplementary volume of medium.

We now present data concerning the industrial application of a more simplified TM technology to the production of Marek's disease vaccine (MDV) using the live herpesvirus of turkey (HVT), a cell-associated virus. This innovated process resulted from a combination of different factors such as the increased growth surface area (15 000 cm²) of the TM roller bottle, the design of an improved tissue culture medium, the use of silicone tubing for the proper oxygenation of the culture medium and the elaboration of a protocole of production providing optimization of cell and virus growth.

MATERIALS AND METHODS

CELLS

Primary chicken embryo fibroblasts (CEF) cells are prepared by enzymatic dissociation of 11 day old embryonated specific pathogen free eggs (SPAFAS Inc., Connécticut USA).

VIRUS

Strain FC 126 of the herpesvirus of turkey was obtained many years ago from Dr. B.R. Burmester and used in production since that time.

MEDIUM

The medium used for the propagation of the cell and the virus has been specially developed in our laboratory based on the characteristics of the TM technology, the basal medium being the 199 with Hank's salts (GIBCO, Grand Island, New-York USA). Only three percent of foetal bovine serum (FBS) was used in the medium for the production of the vaccine.

CELL AND VIRUS CULTIVATION

The primary CEF are mixed with the HVT at the appropriate multiplicity of infection and both are inoculated simultaneously to the TM bottles. After an incubation period of three days at 37°C, the infected cells are harvested with proteolytic enzymes, concentrated and processed for freezing or freeze-drying. Before lyophilisation, the cells are disrupted to free the virus.

VACCINE TITRATION

Vaccines were titered in the quality control laboratory of the Institut Armand-Frappier. Twenty-four hours primary CEF seeded in small Petri dishes are inoculated with the proper dilution of the vaccine, left at 25°C on a rocker platform for an adsorption period of 45 minutes and incubated at 37°C for five days with the appropriate medium. The PFU are then counted using a standard inverted microscope.

RESULTS AND DISCUSSION

IMPROVEMENT OF THE TM TECHNOLOGY

The original TM system consisted of a TM roller bottle connected to a reservoir of culture medium that was recirculated continuously through a peristaltic pump. Oxygenation of the medium was done during that recirculation. Elimination of the reservoir appeared preferable if many TM bottles were going to be used in production.

The design of an improved culture medium using only 3% FBS, its recirculation through a silicone tubing for oxygenation and the determination of the best procedure for cell and virus growth resulted in the complete elimination of the reservoir and, consequently, in a tremendous simplification of the whole process. The construction of a rotative apparatus for the simultaneous use of 20 TM bottles and the recirculation of the medium through the silicone tubing completed the engineering work (fig. 2).

QUALITY OF THE VACCINE PRODUCED

The TM technology has been used for production of the two types of HVT vaccine: the cell-associated one that is kept frozen in liquid nitrogen and the cell-free vaccine that is freeze-dried and kept at 4°C. This technology has been used during a period of five years although not necessarily on an annual basis. More than 200 lots of vaccine have been produced totally. Thirty TM bottles were used for each lot and two lots per week could be manufactured. During that time, there has been no major break or failure of

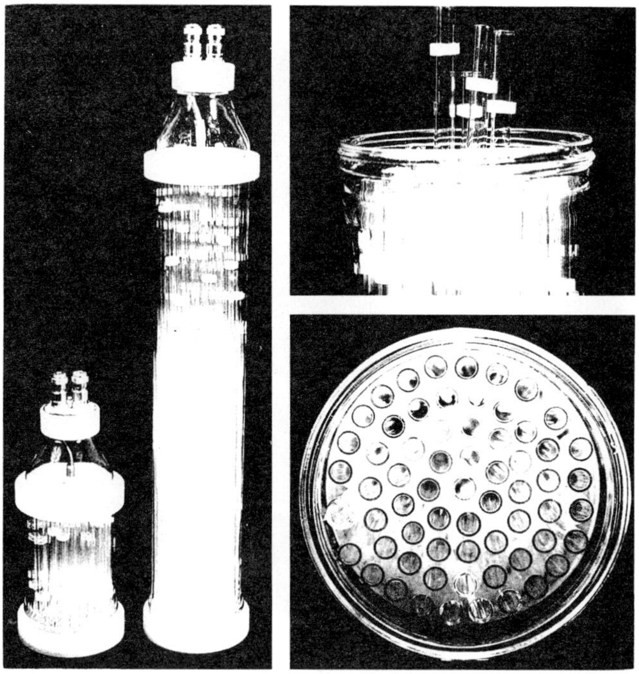

Fig. 1 Improved TM roller bottle designed in collaboration with Bellco Glass Inc. (Vineland, NJ)

Fig. 2 Rotative apparatus developed for the simultaneous use of 20 TM roller bottles with recirculation of the culture medium through silicone tubing.

TABLE I

PRODUCTION OF HVT VACCINE (1000 DOSES/AMPOULE)

FROZEN VACCINE (CA): 28 CONSECUTIVE LOTS*

100%	OF LOTS	WITH	TITRE	≥	6,000	PFU/BIRD DOSE
80%	OF "	"	"	≥	8,000	"
60%	OF "	"	"	≥	10,000	"

* 1250 AMPOULES/LOT

TABLE II

PRODUCTION OF HVT VACCINE (1000 DOSES/VIAL)

FREEZE-DRIED VACCINE (CF): 18 CONSECUTIVE LOTS*

100%	OF LOTS	WITH	TITRE	≥	10,000	PFU/BIRD DOSE
75%	OF "	"	"	≥	15,000	"
65%	OF "	"	"	≥	25,000	"
45%	OF "	"	"	≥	30,000	"

* 2100 VIALS/LOT

TABLE III

CELL ASSOCIATED MD VACCINE

(FROZEN)

MEDIUM [1]	CONSECUTIVE LOTS (30 TM/lot)	NUMBER OF [2] VACCINE CELLS / AMPOULE	NUMBER OF [2] PFU / DOSE	NUMBER OF [2] AMPOULE / LOT
A	16	29.5×10^6	9 500	1 250 [3]
B	11	24.0×10^6	11 000	1 500

1 Medium A: Original culture medium developed for the TM technology

Medium B: Recently improved tissue culture medium

2 Average number

3 1000 doses/AMPOULE

TABLE IV

CELL-FREE MD VACCINE

(LYOPHILIZED)

MEDIUM [1]	CONSECUTIVE LOTS (30 TM/lot)	NUMBER OF SONICATED [2] VACCINE CELLS / VIAL	NUMBER OF [2] PFU / DOSE	NUMBER OF [2] VIALS / LOT
A	12	16.6×10^6	15 000	2 100 [3]
B	9	16.5×10^6	21 000	2 220

1 Medium A: Original culture medium developed for the TM technology

Medium B: Recently improved tissue culture medium

2 Average number

3 1000 doses/vial

the equipment and only 6 lots (3%) were contaminated, most of them during the first year of introduction of the technology. Reliability of this technology has been now well demonstrated.

During these five years, two specific periods could be identified: a first one when the culture medium originally developped for the TM technology was used and more recently, a second one when a more improved medium has been developed and used in production.

Tables I and II are showing examples of the very high quality of the vaccine manufactured with our technology. High infectious titres were obtained with both the cell-associated and the cell-free HVT vaccine. One must remember that, generally, only 2000 PFU are given as a dose of vaccine.

Use of the recently improved medium for the production of the cell-associated vaccine increased the percentage of infected cells from 32% to 46% of the total number of cells filled in the ampoule (Table III); 24 x 10^6 cells/ ampoule are giving an average of 11 000 PFU/bird dose for a production of 1 500 ampoules per lot, the maximum number of PFU that we could get being 15 000 PFU/dose. Data presented in table II concern the cell-free vaccine. It is also showing an increase of the virus titre when the improved medium was used with an average of number of 21 000 PFU/dose, the highest scores being at more than 30 000 PFU/dose.

CONCLUSION

The data presented in this paper are undoubtedly showing the high productivity of the TM technology when applied to the production of the HVT vaccine against Marek's disease, one of the major poultry disease in the world if proper vaccination is not used. If one considers that the usual dose of vaccine given to one chicken is 2 000 PFU for both the cell-associated and the cell-free vaccine, it could be concluded that thirty TM roller bottles could produce more than 7.5 x 10^6 doses of cell-associated HVT vaccine and more than 20 x 10^6 doses of cell-free vaccine. The fact that it is using reusable glass roller bottles is also of considerable importance particularly for the developing countries who would be interested in producing their own Marek's disease vaccine without having to import microcarriers or even plastic bottles . The high productivity of the TM technology, its reliability and its simplicity mean that it must be definitely considered as one of the most outstanding technology for that particular type of production.

Discussion

KADOURI — You showed an EM of the mumps virus and the change in shape from spherical to a filamentous structure. My impression is that we were looking at RNA.

DUGRE — We first of all thought that but are now sure it is the virus.

DEKORT — Do you really think the virus changes shape?

DUGRE — Yes, it is not that extraordinary because if you look at some vaccines in which viruses have been passaged (eg, influenza in eggs) then you see such morphological changes.

43 Procedure for the adaption of the new mumps vaccine strain 'rubini' to human diploid cells

R. Glück

Abstract

Commercially available mumps vaccines or combined measles- + mumps- + rubella vaccines are generally safe, immunogenic and protective against the clinical diseases. However the acceptance rate of these vaccines by the general practioners and by the public in most parts of Europe is not very high. One of the reasons for this low acceptance rate is the fear of side-reactions. Physicians and a proportion of the population may be scared by the contraindications, such as the hypersensitivity to eggs and antibiotics - mentioned in the "physicians circular" and "patient product information".

Safe rubella and measles vaccines cultivated on human diploid cells already exist. A mumps vaccine produced in the same cells would be a safe immunogen and, in addition, could be used to prepare an improved trivalent measles-, mumps-, rubella vaccine.

The following communication details the development of such a new vaccine strain. The initial isolation of the so called "Rubini" strain, its attenuation in "spf" hen eggs and subsequent attenuation in human diploid cells (MRC-5) are described. Some data concerning the manufacturing process and the cultivation conditions are also given.

Finally it can be demonstrated that this new vaccine strain combines high safety with excellent immunogenicity.

Introduction

A major strategy in the development of viral vaccines is the adaptation of the virus to a different host in order to alter the nature of the host-virus interaction. For mumps virus, a member of the paramyxovirus group, this has commonly involved adaptation of human isolates to the chick embryo (1-3). While the first vaccines were crude and more or less purified preparations of infected chick amniotic and allantoic fluids, later vaccines were developed using chick embryo cell cultures as substrate (2,3). Almost only the latter type of mumps vaccine is used today for human immunization. Commercially available mumps vaccines or combined measles- + mumps- + rubella vaccines are generally safe, immunogenic and protective against the clinical diseases, but the acceptance rate of these vaccines by the general practioners and the public in most parts of Europe is not very high. One of the reasons for this low acceptance rate is the fear of side reactions. Physicians and a proportion of the population may be scared by the contraindications, such as the hypersensitivity to eggs and antibiotics mentioned in the physicians circular and patient product information. Centers for Disease Control are also warning that those people with a history of anaphylactic reaction following ingestion of eggs or a receipt of neomycin are at risk when vaccinated with today's combined measles-mumps-rubella vaccines (4,5). Safe rubella and measles vaccines cultivated on human diploid cells already exist. A mumps vaccine produced in the same cells would be a safe immunogen and, in addition, would be used to prepare an improved trivalent measles-mumps-rubella vaccine.

The purpose for the attenuation of the new mumps vaccine Strain "Rubini" was the adaptation of the wild virus to chick embryos. This adaptation of the virus then lead to the restriction of its ability to replicate in vitro in cells derived from human tissue. The attenuated strain then had to be once more adapted to MRC-5 cells to initiate productive infection without reversion to virulence.

Attenuation procedure of mumps virus

The mumps virus was obtained from the urine of a mumps patient at the end of viremia and purified and concentrated by differential centrifugation and ultracentrifugation for 4 hours at 25,000 x g (Fig. 1). The fraction containing the mumps virus was assayed by the hemagglutination test. The virus containing fraction was diluted in Eagle's basal medium (EBM) + 0.4% human serum albumin and aliquoted.

One aliquot was further processed as follows: A freshly confluent human diploid cell monolayer (Wi-38) was inoculated with the mumps suspension and, after addition of EBM + 10% fetal calf serum, was incubated at 37°C. The cell monolayer was harvested after 7 days. The resulting harvest was passaged a further time on Wi-38 cells. The so enriched wild virus was purified and concentrated by sucrose density gradient ultracentrifugation at 100,000 x g.

Subsequently three passages were carried out with the peak fraction at a density of 1.185 kg/l in specific pathogen free chicken eggs. The following different inoculation and harvesting techniques were employed in each case:

a) Inoculation into and harvesting from the allantoic cavity,

b) Inoculation into and harvesting from the amniotic cavity and

c) Inoculation into the amniotic cavity and harvesting from the allantoic cavity or amniotic cavity.

The eggs were incubated at between 32°C and 35°C. The viruses obtained after these three egg passages with the different inoculation/harvest variants described above were then pooled.

Ten further egg passages were carried out with this pool of virus by inoculation into the amniotic cavity and harvesting from the allantoic fluid. The temperature was maintained at about 32°C. The mumps antigen titer was in each case determined by the hemmaglutination test.

This was followed by four rapid passages on human diploid cells (MRC-5) at about 30°C (about 7 days for each passage). After the fourth passage mumps antigen was detectable by the hemagglutination test. Seven further passages on MRC-5 cells at 35°C (10 - 20 days for each passage) provided a first attenuated seed virus with a titer of about $10^{5.5}$ PFU/ml. But the production procedure for this seed virus was still too complicated: The scaling up procedure had to be performed from the beginning with mumps infected cells which restricted the employment of known large scale techniques.
Four further passages in MRC-5 cells including two plaque-purification steps by endpoint dilution cloning passages yielded a master seed Rubini virus. The last four passages were carried out at 35°C and the inoculation procedure was consequently done by infection of confluent tissue monolayers.

STRAIN HISTORY OF RUBINI MUMPS

Isolation from Urine

↓

Purification

↓

2 Passages in Wi-38 Cells (37°C)

↓

Concentration (UZ)

↓

3 Passages in SPF Hen Eggs (32°C)
(Allantois - Amnion - Combined)

↓

10 Passages in SPF Hen Eggs (30°C)
Inoculation in Amnionic Cavity
Harvest from Allantoic Cavity

↓

"REBA 31"
Titer: 7.6/ml

↓

4 Rapid Passages in MRC-5 Cells ("Blind") (30°C)
7 Days for each Passage

↓

7 Passages in MRC-5 Cells (35°C)
10 - 20 Days for each Passage

↓

2 Endpoint Dilution Cloning Passages (35°C)

↓

2 Passages in Confluent Monolayers of MRC-5 (35°C)

↓

Master seed Rubini MUP18
Titer: 6.4/ml

Fig. 1 Procedure of attenuation.

Titers were then in a range of 10^6 to 10^7 PFU per ml when assayed by a rapid immunofluorescence test in Vero cells.

Production procedure

A large scale production procedure of the Rubini mumps virus on MRC-5 cells can easily be done according to the requirements of WHO and FDA (6,7). The MRC-5 cells are at a PD between 30 and 40. Freshly confluent tissue monolayers in roller bottles or other monolayer tissue culture systems for large scale production are infected with the seed virus from the production seed at a dose of 10^7 PFU per squaremeter. After seven days the harvests can be performed according to the known requirements (6,7). Figure 2 shows the kinetics of virus growth and the slow cytopathogenic effect on human diploid cells in roller bottles of 1750 cm^2. The data of two consecutive lots show the consistency of virus production.

The cytopathic signs of Rubini virus in culture are similar to other paramyxoviruses by the typical formation of syncytia. Mumps viruses manifest diverse host-virus interactions ranging from acute, highly cytopathic infection to persistent non-cytopathic infection (8). The nature of the interaction depends on both, host cell and infecting virus. Studies with chick adapted vaccine strains have demonstrated that they can no longer be propagated in certain cells, particularly human diploid cells (9,10,11). As shown in previous publications the Rubini virus had lost its capacity to multiply in human cells (MRC-5) after 13 consecutive passages in embryonated eggs (11,12). In spite of the fact that the cytopathogenicity of mumps virus for a given in vitro cell system may be determined by the "adaptive status" of the virus, the Rubini virus combines high infectivity in human cells with lowest pathogenicity in human beings. The high infectivity and productivity is characterized by the altered cytopathogenicity and propensity toward persistent infection.

Characterization and safety tests

The first stage of development of a new vaccine is complicated and requires extensive laboratory and animal studies. Before clinical studies could be started in man the new vaccine strain Rubini had to undergo numerous investigations in the laboratory.

The primary virus seed lot was identified by neutralization tests in Vero cells. The Rubini virus strain could be distinguished from other vaccine strains by the size of plaques in these cells: After three weeks incubation at 35°C the Jeryl Lynn and Urabe vaccine strains yielded plaques of about 3 mm and 2 mm diameter respectively; the Rubini virus produced plaques of no more than about one mm under same conditions.

It has been shown by appropriate tests (6,7) that the primary and secondary seed lot was free from demonstrable adventitious microbial agents including retroviruses.

S.c. and i.c. inoculation of high doses of Rubini virus in baby and adult monkeys did not provoke any side effects. Furthermore all monkeys showed seroconversion after s.c. or i.m. inoculation.

In addition investigations by the electron microscope have shown that the Rubini mumps strain differed from the wild virus by its morphological structure (Fig. 3 and 4). Rubini viruses show filamentous structure compared to the spherical structure of mumps wild virus.

Clinical studies

The first investigations which have been carried out on 150 human subjects indicated that inoculation with the vaccine was well tolerated. No one

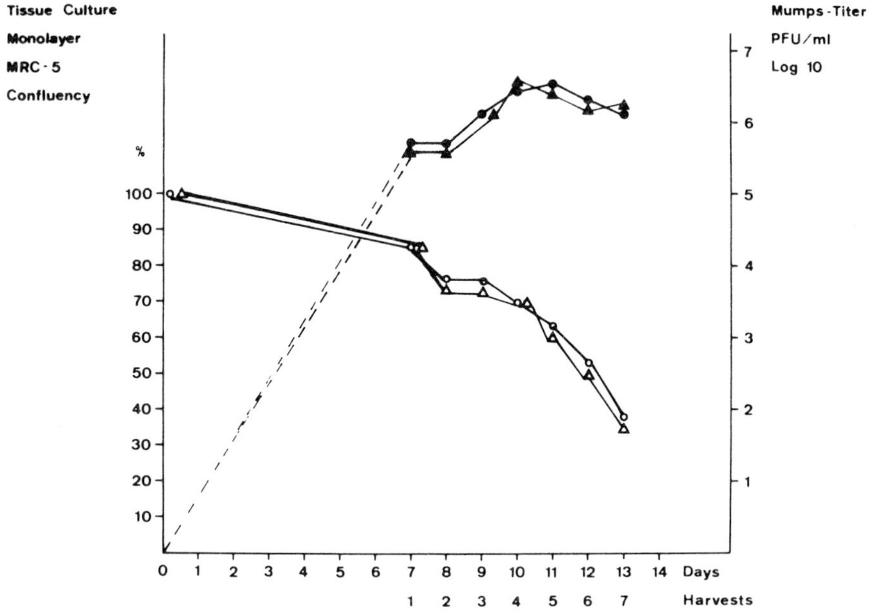

Fig. 2 Kinetics of Rubini virus production and monolayer confluency of MRC-5 cells. Curves of two consecutive lots demonstrate consistency of manufacture.

Titers of harvests of Lot A

Titers of harvests of Lot B

Monolayer confluency of Lot A

Monolayer confluency of Lot B

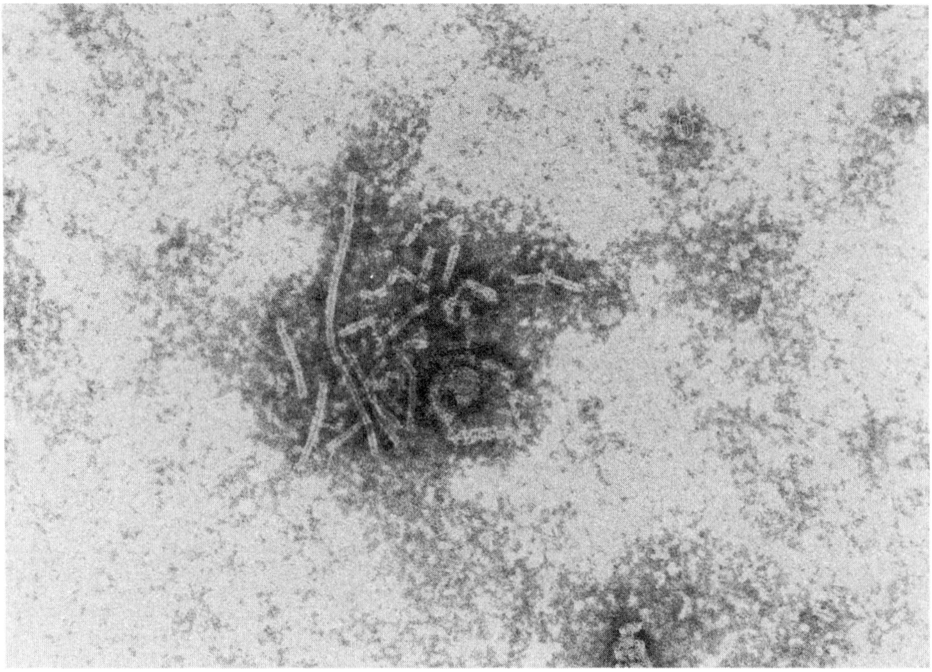

Fig. 3 Electron micrograph of Rubini mumps virus, showing the envelope and the hollow helical strands forming the nucleocapsid. The particles show filamentous structure. Viruses embedded in phosphotungstate. x 70,000.

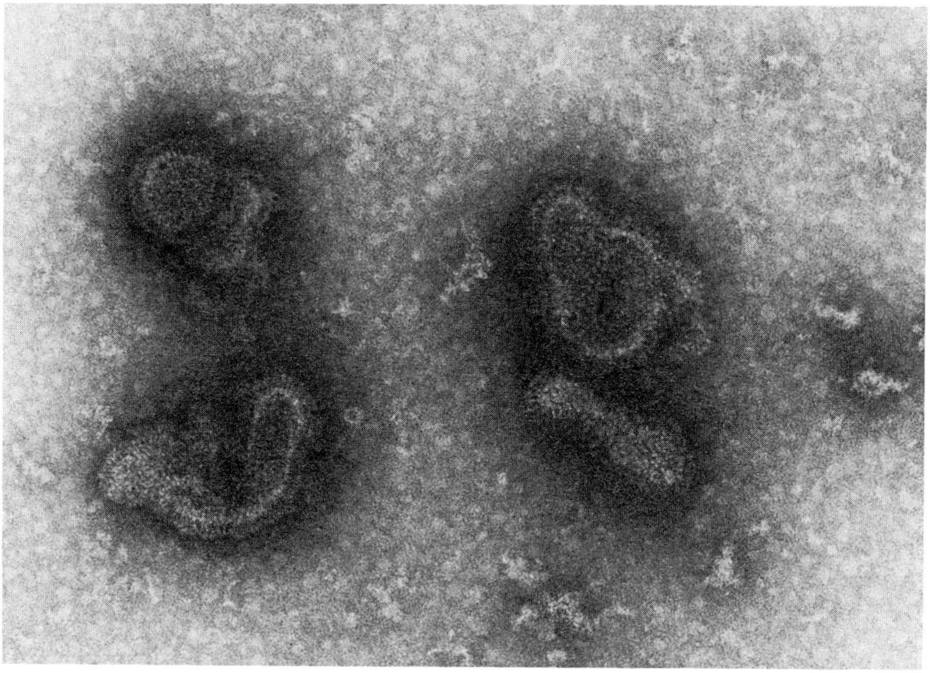

Fig. 4 Electron micrograph of wild mumps virus. The envelope and peripheral short projections are apparent. The particles show spherical structure. Viruses embedded in phosphotungstate. x 200,000.

Triviraten Berna Lot No.	Mumps Content Log_{10} PFU/dose	Vaccinees	Age years	Mumps seroconversion %
8401	5.0	60	1	95
8605	4.5	39	1-4	98
8605	4.5	28	1	70
8605	4.5	14	1	100
8605	4.5	148	7-14	95
8614	5.5	15	1	100
8613	4.3	15	1	93
Total	–	319	1-14	94

Table 1 Clinical evaluation of the mumps vaccine strain Rubini combined with EZ 19 measles and RA 27/3 rubella in four different lots of Triviraten Berna in children.

developed a fever and there were local reactions in only five per cent of all vaccinees. Critical organs such as the parotid glands or testicles showed absolutely no swelling, inflammation or any painful reactions.

The efficacy of the new vaccine was tested in combination with the EZ-19 measles and the rubella RA 27/3 vaccine strains, all viruses being grown in MRC-5 human diploid cells. As shown in Tab. 1, four different lots containing between $10^{4.3}$ and $10^{5.5}$ PFU per dose of Rubini mumps viruses were used for the different studies. The average seroconversion rate as determined by immunofluorescence test (13) for mumps in about 300 children aged between 15 months and 14 years was 94%. Up to now about 100,000 vaccine doses containing the Rubini mumps strain have been vaccinated.

Conclusions

Adaptation of Rubini mumps virus away from the human subject into the chick and back into human diploid cells gave rise to a strain with markedly altered cytopathogenicity for mammalian cells. Altered cytopathogenicity and propensity toward persistent infection are the properties that make the adapted virus strain successful as vaccine capable of eliciting long lasting immunity.

In addition current mumps vaccines may cause allergic reactions in vaccinees who are sensitive to avian proteins and to antibiotics. In other cases, there may be initial sensitizations with the possibility of attendant sequelae. These risks are avoided with the new vaccine.

References

1. Enders, J.F., Levens, J.H., Stokes, Jr.J., Maris, E.P. and Berenberg, W. (1946). Attenuation of virulence with retention of antigenicity of mumps virus after passage in the embryonated egg. J. Immunol. 54, 283.
2. Buynak, E.B. and Hilleman, M.R. (1966). Live attenuated mumps virus vaccine. 1. Vaccine development. Proc. Soc. Exp. Biol. Med. 123, 768.
3. Yamanishi, K., Hosai, H., Ueda, S., Takahashi, M. and Ohuno, Y. (1970). Studies on live attenuated mumps virus vaccine. II. Biological characteristics of the strains adapted to the ammiotic and chorioallantoic cavity of developing chick embryos. Biken J. 13, 157.
4. Immunization Practices Advisory Committee (1984). Recommendations of the Immunization Practices Advisory Committee. Ed. U.S. Department of Health and Human Services. Centers for Disease Control. Atlanta, Ge. USA.
5. Herman, J.J., Radin, R. and Schneiderman, R. (1983). Allergic reactions to measles (rubeola) vaccine in patients hypersensitive to egg protein. J. Pediatrics, 102, 2, 196.
6. WHO Expert Committee on Biological Standardization (1978). Technical Report Series No. 610, Twenty-eighth Report. Requirements for rubella vaccine (live), 54.
7. Food and Drugs (1986). Code of federal regulations. Vol. 21, Part 630, Subpart F.
8. Rima, B.K. and Martin, S.J. (1976). Persistent infection of tissue culture cells by RNA viruses. Med. Microbiol. Immunol. 162, 89.
9. Henle, G. and Deinhardt, F. (1955). Propagation and primary isolation of mumps virus in tissue cultures. Proc. Soc. Exp. Biol. Med. 89, 556.
10. Mc. Carthy, M., Wolinsky, J.S. and Lazzarini, R.A. (1981). A persistent infection of Vero cells by egg-adapted mumps virus. Virology 114, 343.
11. Glück, R., Hoskins, J.M., Wegmann, A., Just, M. and Germanier, R. (1986).

Rubini, a new live attenuated mumps vaccine virus strain for human diploid cells. Develop. biol. Standard., 65, 29.

12. Glück, R., Just, M., Berger, R. and Wegmann, A. (1986). Ein neuer attenuierter Humandiploidzell-Impfstoff gegen Mumps. Jahresvers. Schweiz. Ges. Pädiatrie, Appenzell, 6.-7. Juni 1986, 70.

13. Just, M., Berger, R., Glück, R. and Wegmann, A. (1985). Feldversuch mit einer neuartigen Humandiploidzellvakzine (HDCV) gegen Masern, Mumps und Röteln. Schweiz. Med. Wschr. 115, 1727.

44 Immuno detection of flaviviruses by monoclonal antibodies directed against West Nile virus specific and flavivirus group epitopes

N. Epstein, B. Lachmi, S. Lehrer, D. Katz, S. Cohen and U. Olshevsky

ABSTRACT

Eleven monoclonal antibodies (McAbs) directed against West Nile (WN) virus, were characterized for WN virus specificity and flavivirus group specificity. Six McAbs were WN virus specific and five McAbs were potentially flavivirus group specific as determined by haemagglutination inhibition test (HI) with WN virus and 17D (the vaccine strain of yellow fever virus) antigens. Immunoglobulin fractions from three McAbs of each group were purified by ion exchange or by affinity chromatography. The purified immunoglobulins were conjugated to horse-raddish peroxidase (HRP). McAbs and HRP conjugates were used for the detection of viruses using a two step (2.5 hours) and one step (1.5 hours) enzyme linked immunosorbent assay (ELISA). The detection limit of all assays ranged from 1×10^5 to 1×10^6 PFU/ml.

INTRODUCTION

The flavivirus group comprises of 4 major subgroups (Table 1), consisting of 50-70 serologically related human or veterinary pathogenes, causing many serious illnesses (1,2,3). Most flaviviruses are transmitted to vertebrate hosts by blood-sucking arthropods (mosquitoes or ticks), although some evidently lack an arthropod vector. The viruses consist of an isometric ribonucleoprotein core of about 35 nm in diameter surrounded by a lipid envelope.

Among other flaviviruses, particularly noteworthy is WN virus because of its occurrence in 3 continents, Africa, Asia and Europe. This virus has been isolated from mosquitoes, birds and humans. Its infection is characterized by diverse clinical manifestations ranging from frequent inapparent forms to severe encephalitis. In the Near East, the virus has been shown to produce a clinical picture closely resembling dengue. The diseases, in various parts of the world, may occur as sporadic cases or epidemic outbrakes. WN disease is endemic in Israel with an antibody prevalence of 38-60% (4,5,6,7,8).

The aim of this work was to employ McAb to construct sensitive reproducible and rapid assays for detection of WN virus specific antigens (Ags) and similar assays for the detection of flavivirus common group Ags.

MATERIALS AND METHODS

Production of monoclonal anti WN virus Antibodies

Monoclonal anti WN virus Antibodies (Abs) were produced as described elsewhere (9). In brief, BALB/C mice were immunized with WN virus. Spleen cells from immunized mice were fused with x63 NSI/1-Ag-4 (NSI) myeloma cells according to the procedure described by Köhler and Milstein (10). In order to obtain large quantities of the relevant McAbs, hybridoma cells were injected into pristane (2,6,10,14 tetramethylpentodecane) primed mice, to produce McAbs containing ascitic fluids.

Evaluation of McAbs to WN viruses by ELISA

Ninety-six well polystyrene microtiter plates (Dynatech) were coated with an optimal concentration of WN virions (150 μl/well) in glycine buffer pH 9.6 overnight at 4°C. After washing 3 times with TST buffer (Tris-HCl 0.05M, pH 7.6, NaCl 0.15M Tween-20 0.05%) wells were incubated with 100 μl McAbs to WN virus for 1 hour at 37°C, and then rinsed 3 times with TST and incubated with 100 μl rabbit anti mouse IgG conjugated to alkaline phosphatase (Sigma). Following 1 hour incubation at 37°C plates were rinsed 3 times with TST and incubated with 100 µl of the substrate p-nitro-phenyl phosphate (1mg/ml Sigma). The coloured product was monitored using an automatic ELISA reader (Dynatech) at 405 nm wavelength. For negative control, assay was performed under the same conditions using plates coated with sindbis (SIN) virus.

Purification of monoclonal antibodies

Ascitic fluid containing the McAbs were diluted with PBS to a final protein concentration of 10 mg/ml. McAbs were precipitated with an equal volume of 100% saturated ammonium sulfate, cooled at 4°C for one hour, and centrifuged for 15 minutes at 10000 rpm. The pellet was dissolved in PBS pH 7.4 and dialyzed against PBS.

The IgG_1 and IgM immunoglobulin fractions, containing the anti WN virus activity, were isolated employing DEAE-Sephacel (Pharmacia), according to Lems Van Kan et al. (11). Fractions were eluted using the following phosphate buffers, added sequentially 0.0175 M pH 6.8, 0.08 M pH 6.6 and 0.3 M pH 6.5. Each eluted peak was tested for anti-WN virus reactivity by ELISA as described above.

The IgG_2a and IgG_3 immunoglobulin fractions, containing the anti-WN virus activity, were isolated employing protein A-Sepharose CL-4B (Pharmacia) according to a modified method by Ey et al. (12,13). Briefly, ascitic fluids were dialyzed against 0.05 M Tris HCl, 0.15 M NaCl, 3 mM NaN_3, pH 8.0, and were loaded on the column in the same buffer. Elution of specific Ig subclasses was accomplished using a linear pH gradient with a starting buffer of 0.1 M Tris HCl, 0.15 M NaCl, 3 mM NaN_3, pH 7.0, and a second elution buffer of 0.1 M acetate, 0.15 M NaCl, 3 mM NaN_3, pH 3.0. Fractions from individual peaks were pooled, concentrated and assayed for anti-WN virus activity, by ELISA.

Preparation of horseradish peroxidase conjugated monoclonal anti-WN virus antibodies.

HRP was conjugated to the McAbs, according to Nakane and Kawaoi (14,15) as modified by Boorsma (16): HRP (4 mg, Sigma, purification grade RZ-3) was dissolved in 1 ml H_2O and incubated with 0.2 ml of 0.1 M sodium periodate for 20 minutes at room temperature followed by a 5 minutes incubation with 75 µl of ethylene glycol. The activated enzyme mixture was applied to the Sephadex G-25 column (fine, 1.5 x 24 cm) pre-washed with 1 mM acetate buffer, pH 4.4. The brown eluant (~2 ml) was adjusted to pH 9.5 (using 20 µl of 0.2 M carbonate buffer, pH 9.5), and mixed with 8 mg of purified IgG (in 0.01 M carbonate buffer, pH 9.5). After two hours at room temperature, the formed compound (Aldimine) was incubated for two more hours at 4°C with 0.1 ml of 4 mg/ml freshly dissolved sodium borohydride in H_2O, to form a stable conjugate. Excess reactants were removed by dialysis against 50 mM Tris-HCl pH 7.6 containing 0.15 M NaCl (TS buffer). Schematic illustration of HRP-IgG conjugation is shown in Fig. 1.

Two-steps, ELISA for WN virus detection

The assay was conducted as follows: Polyvinylchloride (PVC) 96-well microtiter plates (Dynatech) were coated with 100 µl/well of 20 µg/ml solution of monoclonal anti WN virus immunoglobulin fraction in TS buffer, by incubation for 3 hours at 37°C. The wells were rinsed three times with TS buffer and incubated overnight at 4°C with TS buffer containing 5% BSA. After three more washes with TS containing 0.05% Tween 20 (TST), plates were incubated for 1 h at 37°C with 100 µl /well of virions at concentrations of 1 x 10^4 - 5 x 10^7 PFU/ml (in TST buffer containing 0.5% BSA). Following three washes with TST 100 µl of monoclonal anti WN virus Ab-HRP conjugates diluted in 10% normal ascitic fluid were added to each well. After 1 hour incubation at 37°C and three washes with TST, plates were incubated for 20 minutes at room temperature with the substrate 2'2 azinobis-3-ethylbenz-thiazoline sulfonic acid (ABTS) + 0.003% H_2O_2. Optical density (OD) was determined at 405 nm using an automatic ELISA reader

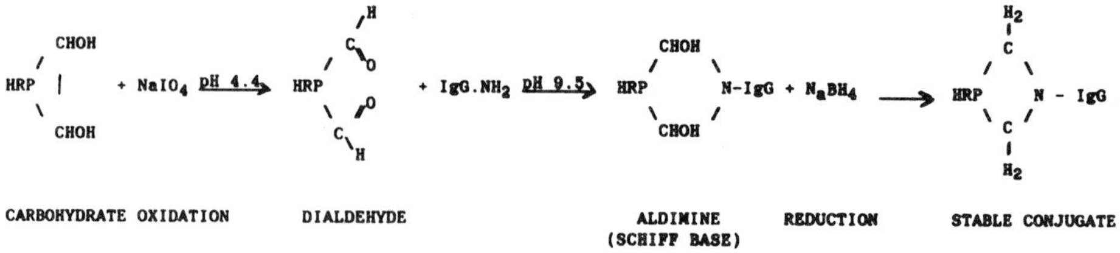

Fig. 1.: Schematic illustration of HRP-IgG conjugation.

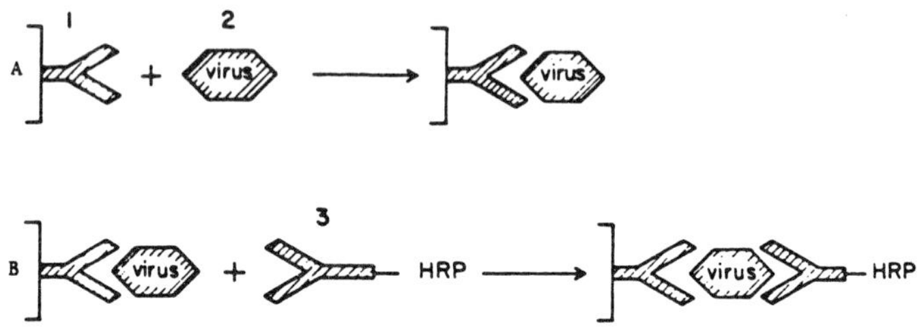

C SUBSTRATE ADDITION

Fig. 2.: Detection of Flavivirus-Group Ags: 2-steps assay.

A. Incubation (1 hour); B. Incubation (1 hour); C. Enzyme reaction (20 minutes) H_2O_2-ABTS $\xrightarrow{HRP}$ coloured product. 1. McAb WN adsorbed to PVC microplate wells. 2. WN or 17D antigens. 3. HRP-labelled anti WN McAb

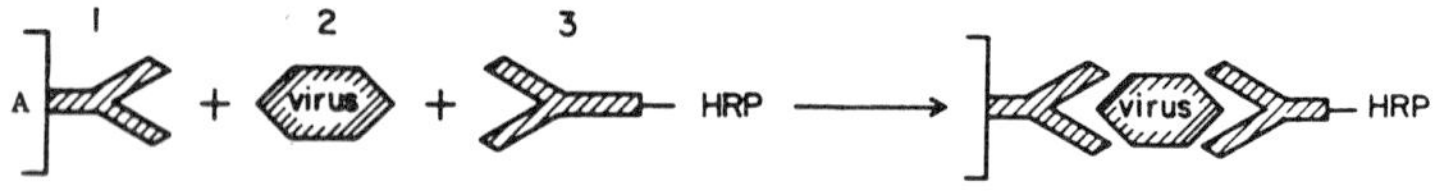

B SUBSTRATE ADDITION

Fig. 3.: Detection of Flavivirus-Group Ags: 1-step assay.

A. Incubation (1 hour); B. Enzyme reaction (20 minutes) H_2O_2-ABTS $\xrightarrow{HRP}$ coloured product. (1-3 as in Fig. 2).

(Dynatech). A schematic illustration of the two step test is shown in Fig. 2.

One-step, ELISA for WN virus detection

The assay was conducted similarly to the two steps assay described above, except for the following changes: 50 μl of the viral Ag were added to each coated well followed immediately by the addition of 50 μl of the conjugate, (skipping the washing step). The starting concentration of the antigens and conjugates were 2 fold higher than those in of the two-step assay. A schematic illustration of the one step test is shown in Fig. 3.

RESULTS AND DISCUSSION

Development of McAbs to WN virus

Eleven McAbs directed against WN virus were generated. All McAbs were directed against V3, the major envelope protein of the virus as was evaluated by gel electrophoresis (9). The McAbs were tested for:

1. Titer by ELISA.
2. Titer by in vitro neutralization test (NT).
3. Titer by heamagglutination inhibition test (HI).
4. Titer by immunofluorescence assay (IFA).
5. Reactivity with WN-envelope protein (by immunoblots).
6. Cross reactivity (in HI and IFA) with 17D virus (the vaccine strain of yellow fever) a member of another flavivirus subgroup.
7. Immunoglobulin classes and sub-classes (by Ouchterlony).

Table 2 summarizes the characteristics of the 11 McAbs. The five McAbs presumably directed against the flavivirus group determinants are marked with an astrix. Typical titration curves of one anti WN-specific McAbs (5F10/1E4) and one anti flavivirus group McAb (1B4/2A7), are shown in Fig. 4. Both McAbs reach very similar titers ($>1:10^6$).

According to the NT test with WN virus and HI tests with WN virus and 17D Ags, we were able to demonstrate the existence of at least 5

Table 1.

MAJOR FLAVIVIRUSES SUBGROUPS*

Subgroup	Virus	Abbreviations
TBE Complex	Tick-borne encephalitis Russian spring and summer encephalitis	TBE RSSE
YF group	Yellow fever	YF
JE group	Japanese B encephalitis Kunjin West Nile St. Louis encephalitis Murray Valley encephalitis	JE KUN WN SLE MVE
DEN groups	Dengue types 1-4	DEN

*According to Straus and Straus (1).

Table 2.

CHARACTERIZATION OF WN MONOCLONAL ANTIBODIES

No.	Parameter tested / McAb	Immunoblot Reactivity with WN-E Protein	Ig Class	Titer ($Dilution^{-1}$) as determined by: ELISA	NT	HI		** IFA reactivity	
						WN	17D	WN	17D
1	7H7/1D6	-	IgG_{2a}	10^6	200	40000	40	+	-
2	7D10/1A4	+	IgG_3	$2.5.10^6$	12800	40000	80	+	-
3	[5F10/1E4]	+++	IgG_{2a}	$>10^6$	6400	40000	320	+	-
4	[2F2/2E6]	+++	IgG_{2a}	$>10^6$	25000	40000	80-160	+	-
5	2D8/2B10*	-	IgG_3	$6.4.10^4$	100	5000	10000	+	+
6	[1B4/2A7] *	+	IgG_{2a}	$>10^6$	200	40000	20000	+	+
7	2G4/2E7 *	+	IgG_1	$>10^6$	<100	20000	20000	+	+
8	[1C9/1B2] *	+	IgG_{2a}	10^6	<100	40000	20000	+	+
9	2A8/2H2 *	++	IgG_{2a}	10^6	<100	40000	20000	+	+
10	7G5/2B6	+	IgG_{2a}	10^6	<100	80	<20	+	-
11	7H2/1F3	ND	IgM	$2.5.10^6$	<100	640	ND	+	-

* McAbs directed to flavivirus group epitope.
** +; - Positive or negative reaction.

[ ] McAbs chosen for ELISA.

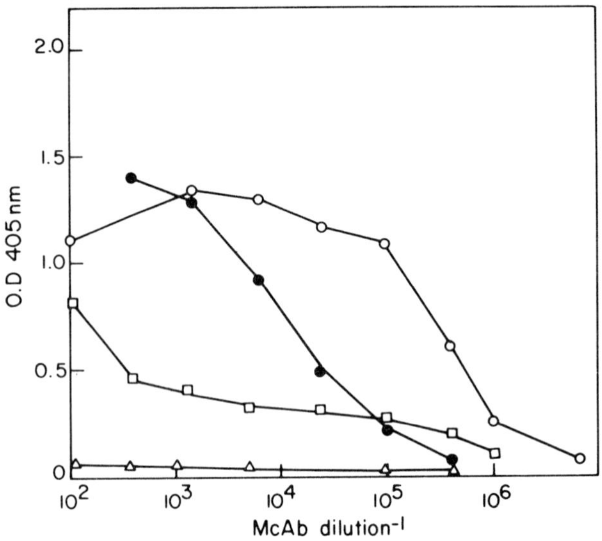

Fig. 4.: Titration curves of McAbs using WN virus adsorbed to polystyrene plates. The following Abs were added: □ -1B4/2A7; ○ -5F10/1E4; ● -mouse polyclonal anti WN Abs. The second Ab was rabbit anti mouse-alkaline phosphatase conjugate. △,▲ -SIN virus adsorbed plates, for 1B4-2A7 and 5F10/1E4 respectively

topographically distinct epitopes on V3 protein as follows:

1. Flaviviruses cross reactive (by HI) and non-neutralizing Abs. (Table 2, Nos. 7,8,9).
2. Flaviviruses cross reactive (by HI) and low neutralizing Abs. (Table 2, Nos. 5,6).
3. WN specific (by HI) and low neutralizing Ab (Table 2, No. 1).
4. WN specific (by HI) and high neutralizing Abs (Table 2, Nos. 2,3,4).
5. WN specific by ELISA (low HI and non neutralizing Abs). (Table 2, Nos. 10,11).

Use of McAbs for the detection of viral antigens

Six McAb, three directed against WN virus specific Ag (5F10/1E4, 2F2/2E6, 7D10/1A4) and three against the putative flaviviruses group epitopes (1C9/1B2, 1B4/2A7 and 2A8/2H2) were studied in immunoenzyme assays. The purified immunoglobulin fractions were conjugated to HRP as described in "Materials and Methods". Optimal conjugate dilutions were chosen for ELISA based on its titration curves by ELISA on polystyrene plates adsorbed with WN virions (Fig. 5).

ELISA for WN Ags detection using all combinations of conjugated and unconjugated McAbs were constructed.

For the detection of group specific Ags, the combination of 1C9/1B2 as the catching McAb and $1B4/2A7^{HRP}$ conjugate as the second Ab was found to be optimal for both WN and 17D viruses (Fig. 6). As shown in Fig. 7, in both one-step and two-step assays, the detection limit was similar ($\sim 10^6$ PFU/ml). This means that the test can be accomplished within 1.5 hours with the same sensitivity as within 2.5 hours.

By using the same catching McAb (1C9/1B2) and changing the conjugated McAb to $5F10/1E4^{HRP}$ (directed against WN specific Ags), the above described test can be used for WN specific assays. As shown in Fig. 8, the WN specific assay (1C9/1E4 - $WN\text{-}5F10/1E4^{HRP}$) is more sensitive (detects 10^5 PFU/ml) than the group specific assay

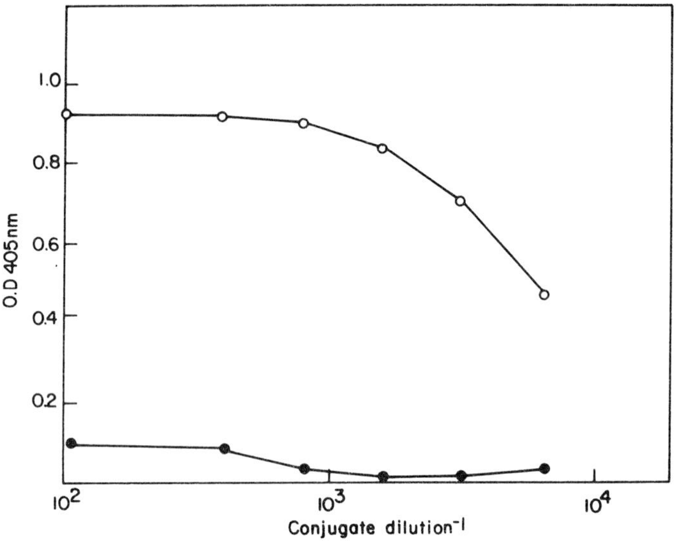

Fig. 5.: Titration curves of $1B4/2A7^{HRP}$.

O -Polystyrene plates adsorbed with WN; ● - plates adsorbed with SIN virus.

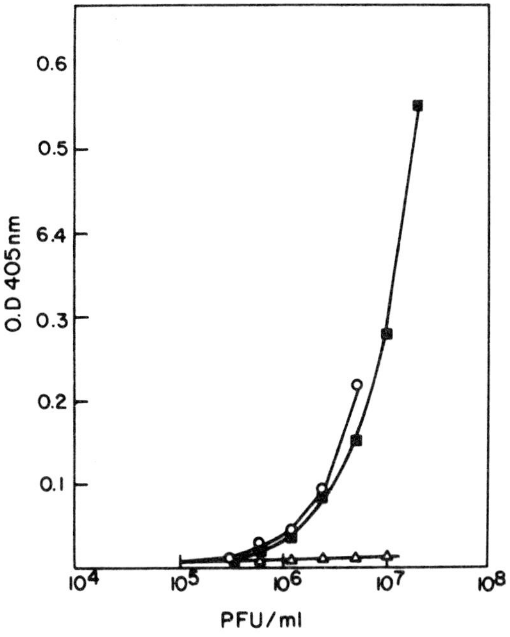

Fig. 6.: Detection of Flavivirus group specific Ags - 1 step ELISA

PVC plates were adsorbed with McAb 1C9/1B2 diluted 1:2000 in glycine buffer pH 9.6. After incubation with various concentrations of virus, the second McAb 1B4/2A7^{HRP} was added (diluted 1:2000 in PBS + 0.5% BSA).

■ - WN virus. O - 17D virus. Δ - SIN virus.

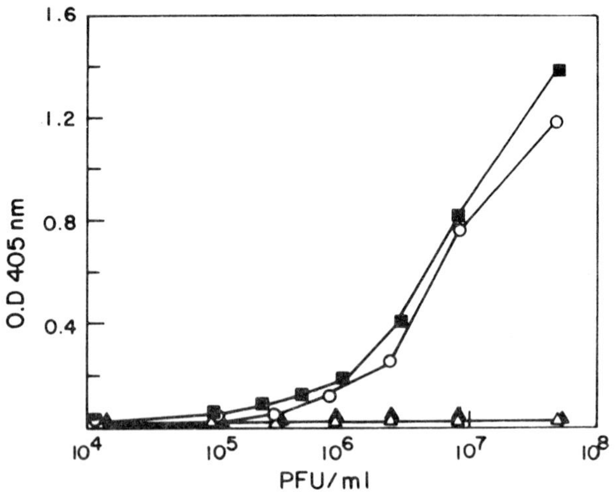

Fig. 7.: Detection of flavivirus group specific Ags by 1 and 2 step ELISA.

PVC plates were incubated with 1C9/1B2 McAb as described in Materials and Methods.

■ - 2 steps assay - Following incubation with WN virus, plates were washed and incubated with the second Ab 1B4/2A7^{HRP}.

○ - 1 step assay - WN and the second Ab were added sequencially. △ 1 and ▲ 2 step assays were performed with SIN virus.

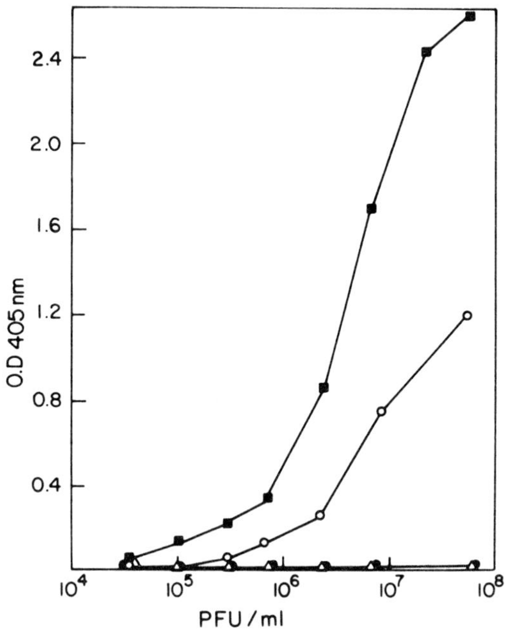

Fig. 8.: WN virus specific and flavivirus group 1 step ELISA

Virus and conjugates were added sequentially to PVC plates adsorbed with 1C9/1B2:

■ - WN virus and 5F10/1E4^{HRP} ; O - WN virus and 1B4/2A7^{HRP};
Δ - SIN virus and 5F10/1E4^{HRP}; ● - SIN virus and 1B4/2A7^{HRP}

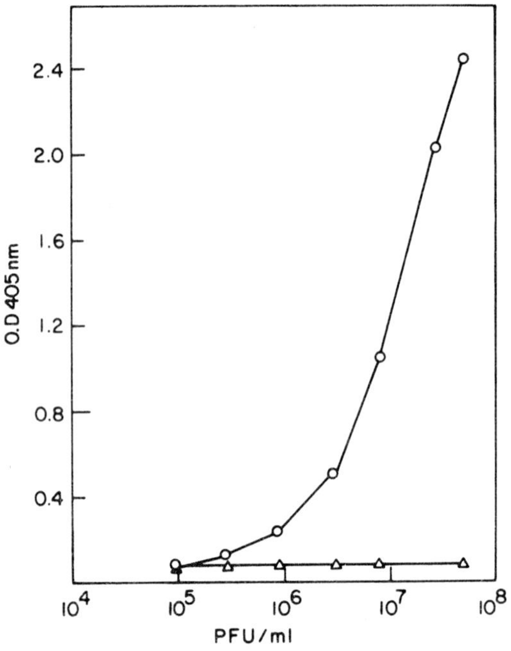

Fig. 9.: WN virus specific 1 step ELISA

O - WN virus and HRP conjugated McAb 5F10/1E4^{HRP} were added sequentially to PVC plates adsorbed with 2F2/2E6. Δ - Assay was performed under the same conditions with SIN virus.

(1C9/1B2-WN-1B4/2E6^{HRP}) which detects 10^6 PFU/ml.

When specific WN Ags were detected by 2 McAbs specific to the virus (2F2/1E4 - WN - 5F10/1E4^{HRP}) (Fig. 9) the test was less sensitive (1 x 10^6 PFU/ml) than the former combination (1C9/1B2 - WN - 5F10/1E4^{HRP}, 10^5 PFU/ml, Fig. 8). This difference might be explained by the fact that 1C9/1B2 and 5F10/1E4^{HRP} are directed against common and specific epitopes respectively, while 2F2/1E4 and 5F10/1E4^{HRP} are both directed against specific epitope(s).

In conclusion, eleven McAbs were generated against WN virus. Six were WN virus specific and five were potentially flavivirus group specific. Sensitive ELISAs were constructed with selected McAbs to specifically detect WN virus or the flavivirus group (Ag). One McAb 1C9/1B2 served as a universal catching Ab for both types of Ags. In the WN virus specific assay, a specific McAb-HRP conjugate was used (5F10/1E4^{HRP}), and for the detection of the flavivirus group antigen, a group specific McAb-HRP conjugate (1B4/2E6^{HRP}) was used. The ELISA carried in two forms: a two step assay (2.5 hours) and a faster, one-step assay (1.5 hours). The detection limit of all assays ranged from 1 x 10^5 to 1 x 10^6 PFU/ml.

ACKNOWLEDGEMENTS

The authors would like to thank Ms. Nita Ben-David for her assistance in preparing this manuscript.

REFERENCES

1. Strauss J.H. and Strauss E.G. (1985). Antigenic Structure of Togaviruses. In: Immunochemistry of Viruses. The Basis for Serodiagnosis and Vaccines. Elsevier Publisher BV. (eds.) M.H.V van Regenmortel and A. R. Neurath. p. 407-424.

2. Porterfield J.S. (1980). In the Togaviruses. (ed) R.W. Schlesinger. Academic Press, New York. pp. 13-46.

3. Mathews R.E.F. (1979). Third Report of the International Committee on Toxinomy of Viruses. Intervirology 12: 129-296.

4. Feinstein S., Akov Y., Lachmi B., Lehrer S., Rannon L. and Katz D. (1985). J. Med. Virol. 17: 63-72.

5. Goldblum N. (1954). Amer. J. Hyg. 59: 89-103.

6. Melnick J.L., Paul J.R., Riordan J.T., Barnett V.N., Goldblum N., and Labin B. (1951). Proc. Soc. exp. Biol. Med. 77: 661-665.

7. Panthier R., Hannoun C., Peytent D. and Mouchet J. Ann. Inst. Pasteur 115: 435-445.

8. Work T.H. (1971). Amer. J. Trop. Med. & Hyg. 20: 169-186.

9. Lachmi B., Epstein N. and Olshevsky U. (In preparation).

10. Köhler G. and Milstein C. (1975). Nature 256: 495-498.

11. Lems Van Kan P., Verspaget H.W. and Pena A.S. (1983). J. Immunol. Meth. 57: 51-57.

12. Ey P.L., Prowse I.S.J., Jenkin C.R. (1978). Immunochemistry 15: 429-436.

13. Epstein N., Miller L.H., Koushel I.C., Iroka J.V., Rener J., Howard R.J., Asofsky R., Aikawa M. and Hess R.L. (1981). J. Immunol. 127: 212-217.

14. Nakane E.K. and Kawaoi A. (1974). J. Histochem. Cytochem. 22: 1084.

15. Kawaoi A. and Makane P.K. (1972). Fed. Proc. 32: 840 (Abstract).

16. Boorsma D.M., Streefbork J.G. and Kors N. (1976). J. Histochem. Cytochem. 24: 1017.

45 A simple ELISA for the estimation of antibodies to equine herpes virus (EHV–1) and its application to the quality control of vaccines

M. Ackermann, S. Grüninger, L. Bruckner, H. K. Müller, R. Zutter and U. Kihm

ABSTRACT

An ELISA for the estimation of the EHV-1 antibody concentration in guinea pig- and horse sera was evaluated. The antigen preparation is based on the accumulation of structural viral proteins in the cytoplasm of infected cells and on the integrity of the cell membranes of infected cells in spite of the cytopathic effect (CPE). Thus, infected and mock infected cells may be grown under optimal nutritional conditions. Proteins added to the growth medium, e.g. fetal calf serum, that might interfere by competition during the coating of the antigen to the ELISA plates, can easily be removed by washings of the harvested cells. The cells are then treated with nonionic detergents under hypotonic conditions in order to disrupt the cell membranes but not the nuclear membranes. Nuclei are removed by centrifugation. The cytoplasm of infected and mockinfected cells, respectively, is coated alternatively to ELISA plates. Antibodies to EHV-1 are detected by antispecies antibodies and an avidin-biotin enhanced peroxidase system. The ELISA reaction is determined by the difference of the optical densities (Delta OD at 405 nm) of the wells containing infected and mock infected antigen.

This ELISA was applied for the quality control of EHV-1 vaccines. Guinea pigs were vaccinated with various dilutions of commercial as well as experimental EHV-1 vaccines. The induced antibody response correlated well with the amount of antigen used for vaccination.
The ELISA was adapted to the detection of EHV-1 antibodies in horse sera. The sensitivity and specificity of this test was equal to the serumneutralization test.

INTRODUCTION

The two most common and significant causes of viral respiratory disease in horses are equine influenza A and equine herpesvirus type 1 (EHV-1, equine rhinopneumonitis virus, equine abortion virus) (10). Horses are commonly immunized against the two infections with vaccines containing inactivated influenza A/equi-1 and A/equi-2 alone (a) or in combination with inactivated EHV-1. Such vaccines are licensed in Switzerland as well as monovalent EHV-1 vaccines. (The problems of EHV-1 immunizations have been discussed extensively elsewhere (6).)
The quality of equine influenza vaccines is tested according to the European Pharmacopoeia by the immunization of guinea pigs followed by the determination of antibody titers three weeks post vaccination (9).
In accordance to this test method, we have immunized guinea pigs with experimental and commercially available EHV-1 vaccines.
An ELISA was developed for the determination of EHV-1 antibodies in the sera of guinea pigs. The aim was to estimate the antigenic content of inactivated EHV-1 vaccines.
Cytoplasma of EHV-1 infected and mock infected cells, respecti-

(a) Footnote: Equine influenza vaccines are usually combined with Tetanus anatoxin.

vely, was used for the preparation of the ELISA antigen, according to the observation that glycoproteins specified by ALPHA herpes viruses are accumulated in the cytoplasma and expressed on the surfaces of infected cells (2,3,4,12,13). These glycoproteins are important elicitors of the host immune reponse and of viral pathogenicity (14). It was, for this reason, likely that the new ELISA correlated with the sero neutralization test (SNT).

In this paper, we describe the evaluation of an EHV-1 ELISA and its application for the detection of antibodies to EHV-1 in guinea pig- and horse sera.

MATERIALS AND METHODS

CELL CULTURE AND VIRUS PROPAGATION

Equine herpes virus strain Kentucky D was propagated in the RK 13 rabbit kidney cell line (7,8). The cells were grown in Eagle's MEM supplemented with 10% FCS and with 100 U of penicillin and 100 µg of streptomycin sulfate/ml.

SERONEUTRALIZATION

Neutralizing antibodies in guinea pig- and horse sera were evaluated by the standard microtiter technique (11).
100 TCID 50 of EHV-1 (Kentucky D) were incubated with the sera for one hour at 37°C before the inoculation of RK 13 cells.

HORSE SERA

In Austria, sera were collected from mares, foals, and recently imported horses (n=99). In Switzerland, sera were collected from young army horses (Trainbundespferde, 4-6 years of age). These draught horses are local Swiss breed. Among the Swiss animals (n=130) were 60 mares.

IMMUNIZATION OF GUINEA PIGS

Guinea pigs were vaccinated twice with experimental EHV-1 vaccines and with commercially available EHV-1 vaccines according to the European Pharmacopoeia (Monograph: Vaccinum influenzae equi inactivatum)(9).

Five guinea pigs were used per preparation. Blood samples were taken immediately before the first vaccination, three weeks later (immediately before the second vaccination), and finally three weeks after the second vaccination.

PREPARATION OF ELISA ANTIGEN

Confluent monolayers of RK 13 cells, grown in roller bottles (RB), were washed twice with phosphate buffered saline (PBS) and incubated with 10 ml of EHV-1 virus (m.o.i. = 0.1) at 37°C (AG+). For the preparation of Mock antigen (AG-), cells were incubated with the same amount of growth medium under the same conditions. After two hours, the inocula were replaced by growth medium containing 2% FCS. 36 hours post infection (p.i.), the cytopathic effect (CPE) in the infected cells was approx. 100%. AG+ as well as AG- cells were scraped into the supernate with a rubber policeman and further processed separately but simultaneously. After 10 minutes of incubation on ice, the cells were sedimented by low speed centrifugation at 4°C for 15 minutes (360xG). The pelleted cells were vortexed, resuspended in 50 ml PBS per RB, and sedimented again. This procedure was repeated twice. The washed cells were then resuspended in 5 ml of icecold hypothonic phosphate buffer, containing TPCK, TLCK, and detergent as described elsewhere (1). The suspension was incubated on ice and the disruption of the cell membranes was checked by the observation of samples in the phase contrast light microscope. As soon as approx. 90% of cell membranes were disrupted, the intact nuclei were pelleted by centrifugation (15 minutes, 1'000xG). The supernate (cytoplasmic fraction) was used as antigen (AG+ and AG-) in the ELISA.

COATING OF EHV-1 ELISA ANTIGEN

Appropriate amounts (as indicated in RESULTS) of AG+ and AG-, were diluted in 0.1 M Na Bicarbonate buffer, pH 9.5 and coated overnight (4°C) to Microtiter ELISA plates (M 129 A, Dynatech Produkte AG, Kloten, Switzerland). The rows were coated alternatively with AG+ and AG-. The coated plates were dessicated, sealed in plastic bags and stored at 4°C until use.

DETECTION OF EHV-1 ANTIBODIES IN SERA BY ELISA

Washing and dilution solutions, containing PBS and Tween and the preparation of ABTS substrate solution have been described extensively elsewhere (5). The dessicated ELISA plates were washed with dilution buffer before being incubated with diluted (1:100) guinea pig- or horse sera (as indicated in RESULTS) for one hour at 37°C. Then the plates were washed three times. VECTASTAIN ABC kits (Vector Lab. Inc., Burlingame, CA, 94010, USA) were used in order to visualize the antigen-antibody reaction. Antispecies-Biotin (at a dilution of 1:10'000) was added to the ELISA plates and incubated for 30 minutes at 37°C. 1 drop of Avidin solution and 1 drop of biotin-peroxidase was added to 10 ml of PBS (ABC solution) and incubated separately at 37°C. The ELISA plates were washed three times and incubated with ABC solution for 30 minutes before the ABTS substrate was added. After 15 minutes of reaction, the optical densities (OD 405 nm) of the substrate solutions were evaluated on a Titertek Multiskan photometer and transferred on line to a SORD M68 computer as described elsewhere (3). Absolute OD or DELTA OD (difference between AG+ and AG- wells) values were used as indicated.

RESULTS AND DISCUSSION

EVALUATION OF THE OPTIMUM ANTIGENCONCENTRATION

In order to evaluate the optimal concentration of AG+ and AG-, respectively, to detect EHV-1 antibodies in guinea pig- and horse sera, serial dilutions of cytoplasmic antigen were coated to the microtiter plates. The ELISA reactions (OD 405) of three different guinea pig sera with these antigens are illustrated in Fig. 1. The sera stemmed from the same guinea pig and were sampled before the first vaccination (NEG SERUM -), three weeks after the first vaccination (POS SERUM +), and three weeks after the second vaccination (POS SERUM +++) with a commercially available EHV-1 vaccine. As demonstrated in the figure by the reactions of the two posivive sera in the POS AG (AG+) wells, the specific reaction was approx. constant until an AG dilution of 1:1'000. At higher AG dilutions, the reaction was less intensive at the same time of incubation. Interestingly, a decreasing reaction of the POS SERA as well as of the NEG SERA with NEG AG (AG-) could be observed until an AG dilution of approx. 1:1'000. In contrast to this observation was the increasing reaction of NEG SERUM with increasing AG+ concentrations.

At AG dilutions of approx. 1:500, the reaction of the NEG SERUM was equal in AG+ as well as in AG- wells. On the other hand, a difference of 0.2 OD still remained in the reaction of the POS SERUM + (weak positive) with AG+ and AG-, respectively. For this reason, the optimal antigen concentration for the detection of guinea pig antibodies to EHV-1, was considered to be 1:500.

The same experiment was conducted using three different horse sera (considered negative, weakly positive, and positive, as determined by seroneutralization). The optimal AG dilution in the horse system was 1:300 (data not shown).

The decreasing reaction of POS SERUM with serial dilutions of AG- could be explained by the dilution of nonspecific binding sites contained in the non infected cytoplasma of RK 13 cells. The increasing reaction of the NEG SERUM with AG- might be due to the dilution of highly specific VIRAL binding sites present in the EHV-1 infected cytoplasma.

EVALUATION OF THE AMOUNT OF EHV-1 ANTIGEN IN VACCINES

The guinea pig EHV-1 ELISA was used for the determination of EHV-

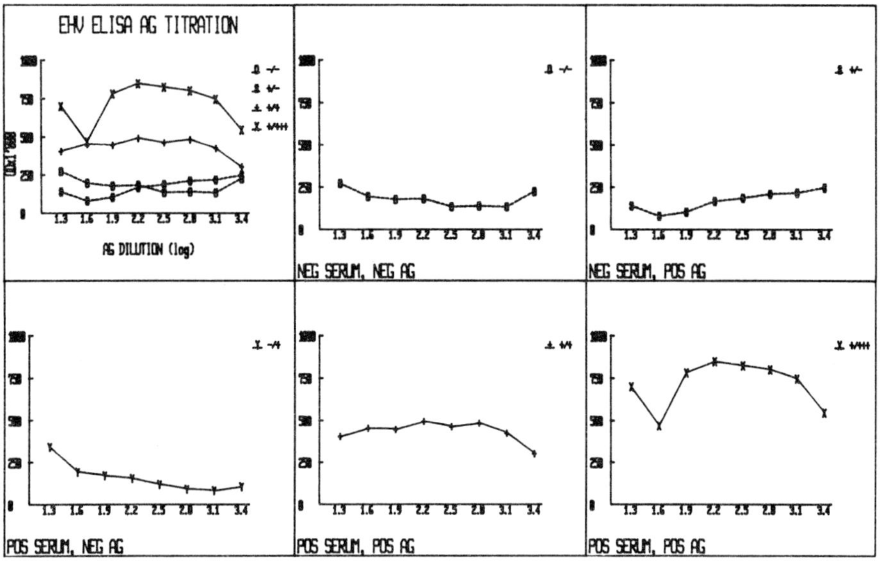

FIG. 1 ELISA reactions of guinea pig sera are shown.

Top row: Summary of the figure and reactions of negative sera.

First from left : Summary. Each serum is shown separately in the other parts of the figure.

Second from left: Reaction of preimmune serum with serial dilutions of mock infected antigen. (AG/serum = -/-).

Third from left : Reaction of preimmune serum with serial dilutions of EHV-1 antigen (AG/serum = +/-).

Bottom row: Reactions of positive sera.

First from left : Reaction of postvaccinational serum with serial dilutions of mock antigen (AG/serum = -/+). The serum was collected three weeks after the first vaccination.

Second from left: Reaction of the same serum with serial dilutions of EHV-1 antigen (AG/serum = +/+).

Third from left : Reaction of a further postvaccinational serum with serial dilutions of EHV-1 antigen (AG/serum = +/+++). The serum was collected threde weeks after the

1 antibodies in guinea pig sera, before and after vaccinations with serial dilutions of EHV-1 antigen. Four groups of five guinea pigs were vaccinated twice at three weeks intervals. Five guinea pigs received each one dose of commercial available EHV-1 vaccine (VAC 170). Each guinea pig of the remaining 15 received 1 ml of serumfree supernate of EHV-1 infected cells. The BPL inactivated virus was diluted in PBS in a manner to vaccinate groups of five guinea pigs each with 10*8, 10*6, and 10*4 TCID 50 of EHV-1, respectively. Sera sampled before the first vaccination, three weeks later, and three weeks after the second vaccination, were subjected individually to the EHV-1 SNT and to the EHV-1 guinea pig ELISA. The results are summarized in Fig. 2. The median values of reciprocal SNT titers as well as ELISA reactions (DELTA OD) correlated very well to the amount of antigen contained in the vaccine used.

In visual readings of the ELISA reactions, DELTA OD of approx. 0.2 is necessary to give a clear cut positive result. In the groups vaccinated with 10*8 and 10*6 TCID 50 of EHV-1, more than three of the five individual sera reacted stronger than DELTA OD 0.2. In the group vaccinated with 10*4 TCID 50, the visible reaction was not clear cut. The median value of the five sera was less than 0.2 DELTA OD but still clearly different to the reaction of the preimmune sera. Interestingly, the preimmune sera reacted stronger with AG- than with Ag+. These sera reveiled for this reason negative DELTA ODs. The commercial vaccine VAC 170 induced reactions similar to those due to the highest concentration of virus in the experimental vaccines. The median values of the ELISA reactions correlated obviously very well to the titers of neutralizing antibodies in the guinea pig sera. We conclude for this reason that the guinea pig EHV-1 ELISA might be very helpful for the determination of the antigen concentration in EHV-1 vaccines.

second vaccination.

x-axis: serial dilutions of antigen.

y-axis: OD 405 as determined 15 minutes after the addition of substrate.

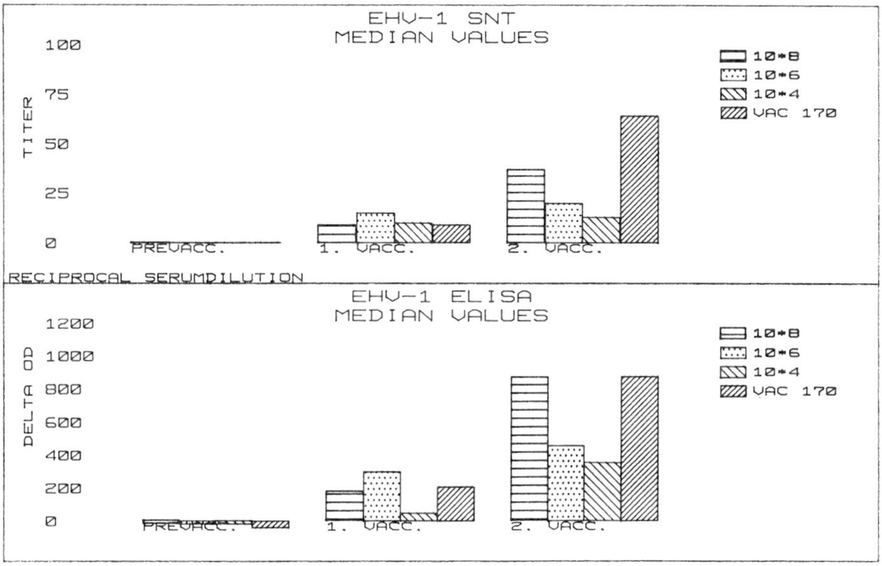

Fig. 2 Serologic reaction of guinea pigs before and following immunization with EHV-1 vaccines.

The symbols on the top of the right side indicate the vaccine used for immunization of the respective group of guinea pigs (see text).

Top : Course of median values of neutralizing antibody titers in guinea pig sera.

Bottom: Course of median values of ELISA reactions (DELTA OD) in guinea pig sera.

The sera were collected before the first vaccination (prevacc.), three weeks after the first (1. vacc.), and three weeks after the second (2. vacc.) vaccination.

APPLICATION OF THE EHV-1 ELISA FOR THE DETECTION OF EHV-1 ANTIBODIES IN HORSE SERA

In order to determine the value of a vaccine, it is necessary to demonstrate that vaccinated animals are, in contrast to not vaccinated controls, protected in challenge experiments. Such protection has to be shown in the animal species for which the vaccine is intended. With EHV-1 vaccines, even such complex aspects as the protection of mares from abortion have to be taken into account (6).

In order to correlate protection with the appearance and concentration of EHV-1 antibodies in horse sera, we have adapted our EHV-1 ELISA to horse sera. 99 horse sera, sampled in Austria, and 130 horse sera, collected in Switzerland, were subjected to SNT as well as to EHV-1 ELISA. The results are summarized in Tab. 1. Almost all of these sera contained antibodies to EHV-1. Apparently, it is difficult to find EHV-1 seronegative horses. This has to be considered in the planned challenge/protection experiments. The ELISA was equally suitable for the detection of EHV-1 antibodies as the SNT. Among the positive sera, we found three sera with high titers of neutralizing antibodies (1:160 and more) which reacted minimally in the ELISA (DELTA OD approx. 0.1). Interestingly, these sera were collected from mares. We are currently evaluating, if these reactions are due to crossreacting antibodies to EHV-3 (coital exanthema).

CONCLUSIONS

An ELISA for the detection of EHV-1 antibodies in guinea pig- and in horse sera was established. This ELISA, using a simple preparation of EHV-1 infected and mock infected cytoplasma as antigen, was equally sensitive and specific as the SNT. The ELISA is applied to estimate the potency of EHV-1 vaccines by the determination of EHV-1 antibodies in correlation to the antigenic content of the vaccines.

Tab. 1 PREVALENCE OF EHV-1 ANTIBODIES IN THE SERA OF AUSTRIAN AND SWISS HORSES

ELISA \ SNT	+	-
> 0.2	201*	1
0-0.2	17	4
< 0	0	6

SNT + = neutralizing antibodies detected in the horse sera as indicated

SNT - = no neutralizing antibodies detected

ELISA = Reaction of horse sera as determined by DELTA OD (Materials and Methods)

* = Number of sera with reaction as indicated

The guinea pigs are immunized according to the requirments for the control of equine influenza vaccines. For this reason, it is possible to use the same guinea pig sera for the control of the EHV-1 component as well as the influenza components of combined vaccines.

REFERENCES

1. Ackermann, M., D.K. Braun, L. Pereira, and B. Roizman. Characterization of herpes simplex virus 1 ALPHA proteins 0, 4, and 27 with monoclonal antibodies. J. Virol. 52: 108-118. 1984.

2. Ackermann, M., R. Longnecker, B. Roizman, and L. Pereira. Identification, properties, and gene location of a novel glycoprotein specified by herpes simplex virus 1. Virology 150: 207-220. 1986.

3. Ackermann, M., A.E. Metzler, H. McDonagh, L. Bruckner, H.K. Müller und U. Kihm. Stellen nichtbovine Paarhufer ein IBR-Virus-Reservoir dar? I. BHV-1 und CapHV-1-Infektions- und Reaktivierungsversuche an Ziegen, Virustyp-Spezifität der humoralen Antikörper und Charakterisierung der viralen Antigene. Schweiz. Arch. Tierheilk. 128: 557-573. 1986.

4. Ackermann, M., H. Ludwig, F. Conraths, G. Pauli, and L. Pereira. Conserved antigenic and functional domains on glycoprotein B of herpes simplex virus 1 and bovine herpesvirus 2. Submitted to J. Virol.

5. Bommeli, W., U. Kihm, M. Lazarowicz, and F. Steck. Rapid detection of antibodies to infectious bovine rhinotracheitis (IBR) virus by micro enzyme linked immuno-sorbent assay (Micro-

ELISA).
Proc. 2nd Int. symposium of vet. lab. diagnosticians. Lucerne (CH) II: 235-239. 1980.

6. Bürki, F.
Infections of the reproductive tract of equines.
9th International symposium of the word association of veterinary microbbiologists, immunologists, and specialists in infectious diseases: 111-117. Perugia, Italy, 1986.

7. Doll, E.R. and M.E. Wallace.
Cultivation of equine abortion and equine influenza viruses on the chorioallantoic membrane of chicken embryos.
Cornell Vet. 44: 453-461. 1954.

8. Engels, M., N. Nowotny, A.E. Metzler, R. Wyler, and F. Bürki.
Genomic and antigenic comparison of an equine herpesvirus 1 (EHV-1) isolate from the 1983 Lippizan abortion storm with EHV-1 reference strains.
Microbiologica 9: 221-234. 1986.

9. European Pharmacopoeia.
Vaccinum influenzae equi inactivatum.
2nd edition, part II-6: 249. 1983.

10. Jolly, P.D., Zhen Fang Fu, and A.J. Robinson.
Viruses associated with respiratory disease of horses in New Zealand: an update.
N.Z. vet. J. 34: 46-50. 1986.

11. Lazarowicz, M., F. Steck, M. Ackermann und U. Kihm.
Prüfung von zwei Impfstoffen gegen die Infektiöse Bovine Rhinotracheitis.
Schweiz. Arch. Tierheilk. 125: 797-808. 1983.

12. Roizman, B. and W. Batterson.
Herpesviruses and their replication.
in Virology. Eds. B.N. Fields et al.
Chapter 25: 497-526. 1985 Raven Press, New York.

13. Spear, P.G.
Glycoproteins specified by herpes simplex viruses.
in The Herpesviruses. Vol. 3.
B. Roizman ed. 1984 Plenum Press, New York.

14. Spear, P.G.
Antigenic structure of herpes simplex viruses.
in Immunochemistry of viruses. The basis for serodiagnosis and vaccines. Eds. M.H.V. van Regenmortel and A.R. Neurath.
Chapter 23: 425-446. 1985 Elsevier Science Publishers B.V.

ACKNOWLEDGEMENTS

Austrian horse sera were kindly supplied by Prof. Dr. F. Bürki, who also tested these sera for antibodies to EHV-1 in SNT and complement fixation.

The authors wish to acknowledge with thanks the help of Heather McDonagh in the initial EHV-1 ELISA experiments., R. Tschudin for supplying cells, and Mimi Specklin for helping to design the figures and the poster.

46 The production and application of an oil adjuvant vaccine against foot-and-mouth disease in cattle

H. Bahnemann, J. Mesquita, V. Astudillo and F. Dora

Abstract

To better protect cattle populations against foot-and-mouth disease under field conditions of South America a vaccine is needed which induces a longer lasting immunity than the standard FMD vaccine with aluminum hydroxide saponine adjuvant. Studies on an oil adjuvant FMD vaccine for cattle were begun at the PAFMDC in 1968 and this vaccine is now produced at the institute on a pilot scale level. The present production procedure is discussed in some detail, from the antigen production in BHK cells through inactivation and emulsification. The oil adjuvant consists of the mineral oil Marcol 52 and the emulsifier Montanide 888. Studies of the duration of immunity show that revaccinated adult cattle are protected against challenge for at least twelve months. In two large scale experimental field applications of this vaccine in Argentina and Brasil the FMD morbidity was reduced to one eighth of that in cattle vaccinated in the same area with the standard aluminum hydroxide vaccine. The vaccine prepared by the Center is at present applied by eight countries in South America in selected areas of government FMD control programs. For the last two years in the State of Rio Grande do Sul in Brasil over half the cattle population, more than seven and nine million animals, has been vaccinated with oil adjuvant FMD vaccines. The incidence of FMD in this state was reduced to a very low level, with no outbreaks in the oil adjuvant FMD vaccine application area.

Introduction

The preparation of vaccines against foot-and-mouth disease (FMD) improved considerably during the last 20 years in the antigen part of the vaccine. The antigen is now largely produced in cell culture and inactivated by first order reaction inactivants. The standard vaccine however still uses aluminum hydroxide and saponine as adjuvants. This so-called aluminum hydroxide saponine vaccine induces in cattle an immunity for a limited time of about four months only.

For government FMD control programs in South America and other parts of the world with similar field conditions it is quite difficult to vaccinate the cattle population in all areas every four months in order to ensure a continuous protection. A vaccine which could induce a longer lasting immunity would therefore be of great importance for the control of FMD. For this reason the Pan American FMD Center (PAFMDC) began in 1968 laboratory studies for the preparation of an oil adjuvant vaccine for cattle and in 1972 various projects for field application of this vaccine (13,16).

This work has led to a pilot scale production of an oil adjuvant FMD vaccine by the PAFMDC and an application of this vaccine by many countries in South America in selected areas of their FMD control programs.

VACCINE PRODUCTION

The main steps of the production of an oil adjuvant FMD vaccine at the PAFMDC are presented in Tables 1 to 3. More details of the production are given below and some important considerations will be discussed.

Antigen Production

Media Preparation

Serum is an important component of the cell culture medium and should be prepared with outmost care. It is of great importance to collect the blood under as clean conditions as possible in order to keep bacterial contamination and the presence of bacterial proteases at a minimum. Bovine blood is collected at a local slaughterhouse and at collection 10^5 I.U. Penicillin and 0.15 g Neomycin per liter of blood are added. The serum is allowed to drain from the coagulum overnight at 4°C and then processed immediately. All serum used in cell and virus production is treated with polyethylene glycol (PEG) 8000 (1,11,21) in the following way: to 100 l of clarified serum 16 l of a 50% solution of PEG in physiological saline adjusted to pH 7.9 with Tris is added. This mixture is agitated vigorously with a Vibromixer for 90 minutes, the precipitate is allowed to settle for 30 minutes and the supernate is clarified again by centrifugation. The PEG treated serum is added to the medium at 8%. If viral contamination is a problem the processed and filtered serum can be treated with BEI (10,20).

The cell culture medium is prepared with deionized water from commercially obtained Glasgow Eagle medium powder without glutamine. The glutamine is added at preparation. After addition of the serum the pH of the medium is adjusted to 7.4 and it is filtered through Millipore cartridge filters (CP19, CWSS and Durapore-TP 0.22µm).

The virus production medium is prepared with Glasgow Eagle medium, 70% of volume, and a 0.2 M solution of TRIS, 30% of volume. PEG serum is added to 1%. The pH of the medium is adjusted with HCl to 7.5 to 7.6. Filtration is through the same Millipore cartridge filter combination.

Cell production

The stock production of cells is done in suspension in a fermentor, from which cells are drawn for culture on substrate or in suspension.

The cells grown on substrate are seeded in 2 l Baxter type bottles at 40×10^6 cells in 200 ml of growth medium. After 72 hrs the cell concentration reaches about 600×10^6 cells.

The cells grown in suspension are seeded at 0.5×10^6 cells per ml and reach about 1.2 to 1.5×10^6 cells per ml in 24 hrs.

The cells grown in medium with PEG serum are mainly monodispersed in suspension and form a smooth multilayer on the glas surface, while the cell cultures grown in non-treated serum had many small and larger clumps of cells and grew also on the glas surface in clumps. The reason for this may be the removal of antibodies against BHK cellular proteins from the serum with the PEG treatment. All cattle vaccinated against FMD in Brasil are vaccinated with vaccines produced in BHK cells and their serum therefore is likely to contain antibodies against BHK cellular proteins.

Virus Production

The cell cultures are infected with a low passage (3-5 passages in BHK cells from bovine epithelium virus) seed virus produced in substrate grown cell cultures with a minimum titer of $10^{7.5}$ PFU/ml or an average titer of around 10^8 PFU/ml. This seed virus is added at 2 ml per l of virus medium, which at a cell concentration of 3 x 10^6 cells per ml gives a MOI of 0.02 to 0.06.

For bottles the same volume of medium as in the growth phase, 200 ml, is used which gives the desired cell concentration of 3 x 10^6 cells per ml. With the suspension cultures the cells are allowed to sediment in order to remove the growth medium supernate and the cells are then resuspended in a smaller volume of virus medium to the same concentration of 3 x 10^6 cells per ml.

The virus suspension is harvested when the cell count has dropped to less than 10% of the initial cell concentration, perhaps around 5%. With cells grown in PEG serum the virus production proceeds at a faster rate and the harvest time could be reduced by about 30%.

Virus Purification

The harvested virus suspension is cooled to 4°C, 2% of chloroform is added with agitation and the virus suspension is then passed through an in-line homogenizer in order to obtain a very intimate mixture of the cellular proteins with the chloroform. The precipitate is allowed to settle overnight and the supernate is passed through a continuous flow centrifuge for removal of residual precipitate.

Virus Inactivation

Virus inactivation is effected with bromoethylenimine, BEI (9). For this purpose a standard 0.1 M solution of BEI is used. The preparation of this solution is not difficult. The basic reaction is that bromoethylamine HBr (BEA) under alkaline conditions forms BEI by cyclisation. This reaction causes a drop in the pH of the solution from about 12.5 to between 8.5 and 9. A simple method is used to check on the cyclisation and the resulting pH drop. A pH indicator, ß naphthol violet, is added at 0.5 ml (of a 1% aqueous stock solution) per liter of the NaOH solution. After addition of BEA the color of the solution changes from violet to orange as the cyclisation progresses. If there is no color change the BEI preparation is not used for inactivation.

The 0.1 M BEI preparation contains only 0.5% BEI. It is not too dangerous if it is handled with reasonable care. More concentrated solutions of BEI can be prepared. However the following consideration should be taken into account. The dissolution of NaOH in water is exothermic and raises the temperature of the solution. With the 0.175 N NaOH solution this temperature rise is negligable. However with a 2.0 N NaOH solution the temperature goes up to between 60 and 70°C. The boiling point of BEI is at 52°C and if the temperature is not brought to a lower point, much of the BEI will go into the vapor phase. Therefore the preparation of higher concentrations of the BEI solution is more dangerous and must be controlled very carefully and be handled with proper precautions. One FMD vaccine production laboratory in South America, which used a 1 M BEI preparation, had considerable trouble with irregular inactivation. This was resolved with the use of a 0.1 M preparation.

Initially the inactivation was done at 37°C with 1.5 mM BEI. Unexplained low values for the antigenic mass in some virus preparations led to a study comparing this method with an inactivation with 3 mM at 26°C for the same period of time, 24 hrs. The results are presented in Table 4. The losses in antigenic mass were considerable with the A type and C type viruses and smaller with the O type virus. The standard procedure of inactivation therefore is now 3 mM BEI at 26°C for 24 hrs. The inactivation rate is determined for each virus suspension from five samples taken at hourly intervals, 0 to 4 hrs. The decrease of viral infectivity at this temperature and concentration of BEI is around 0.9 to 1 log PFU per hr.

The inactivation process is done under slow stirring in two different vessels, in order to avoid reinfection from any non-inactivated virus in dead spaces in valves, tubes or the vessel wall above the liquid level. After the 4 hour inactivation sample is taken from the first tank, in which the inactivation was begun, the virus suspension under inactivation is transferred to another tank in which the inactivation is continued to the end of the inactivation period.

At the end of the inactivation time the virus suspension is tested for innocuity. This is done by inoculation of two BHK roller bottles with 15 ml samples. The observation period is for 48 hrs. and two further blind passages are done. The supernate of the last passage is tested by CF for presence of viral antigen.

Vaccine Preparation

Formulation

The preparation of the antigen mixture is based on the antigenic mass determinations of each viral antigen batch. The minimum antigen quantity for each virus type per dose is given in Table 2. To the antigen mixture Na-thiosulfate (1 M solution) is added to a final concentration of 3 mM for neutralization of residual BEI and the pH is adjusted to 7.8 to 7.9 with glycine buffer. This antigen mixture is controlled by CF determination and an innocuity test in cell cultures.

To the mineral oil Marcol 52 the emulsifier Montanide 888 (3) is added at 10% and the oil-emulsifier mixture is filtered through the same Millipore cartridge filter combination as used in media filtration.

Emulsification

A volume of 1000 liter of the antigen mixture is added under agitation to an equal amount of the oil-emulsifier. From this premixture batches of 1000 liter are passed through two parallel in-line homogenizers developed by the Center.

In-process Controls

The in-process controls used in the preparation of the oil adjuvanted FMD vaccine are indicated in Table 3. One of the important controls is the determination of the quantity and quality of the antigen produced. The infectivity or CF titers give only an approximate indication. Therefore the quantity of the antigen is controlled by determination of the mass of the 140S antigen through density gradient centrifugation in CsCl (14) and the quality by study of the integrity of the viral polypeptides with PAGE. There are however indications that cleavage of the polypeptides may not necessarily indicate a poor immunogen. We have

TABLE 1

PRODUCTION OF AN OIL ADJUVANT FMD VACCINE
AT THE PAN AMERICAN FMD CENTER

1. Production of viral antigen

1.1 Cell production

Medium	Glasgow Eagle	commercial powder prepn. for 500 l
Serum	bovine serum	treated with PEG
Cells	BHK 21	substrate and adapted to suspension
Cultivation	on substrate	2 l roller bottles
	in suspension	fermentors up to 600 l

1.2 Virus production

Seed virus	low pass. (3-5) in cells on substrate	MOI = 0.02-0.06 PFU/cell
Cell concentr.	3 x 10^6/ml	Medium with 0.06 M Tris 1% serum PEG
Incubation time	Virus O	10 - 12 hrs.
	Virus A	16 - 20 hrs.
	Virus C	16 - 20 hrs.
Purification	2% chloroform	in-line homogenizer
Clarification	Centrifugation	400 l/hr. at 7680 G

1.3 Virus inactivation

0.1 M Prepn. BEI	3 mM BEI 26°C for 24 hrs.	two tanks, 4 + 20 hrs.
Inactivation rate	0.9 - 1 log/hr.	PFU titration
Neutralization residual BEI	Na-thiosulfate 3 mM	at time of vaccine formulation

TABLE 2

PRODUCTION OF AN OIL ADJUVANT FMD VACCINE
AT THE PAN AMERICAN FMD CENTER

2. Preparation of vaccine
Primary water-in-oil emulsion for i.m. application in cattle

2.1 Formulation

Antigen	50%	bi- or tetravalent	unconcentrated
		Type O_1	≥ 3.0 µg/dose
		Type A_{24} biv.	≥ 1.5 µg/dose
		Type A_{24} ttv.	≥ 1.0 µg/dose
		Type A_V ttv.	≥ 1.0 µg/dose
		Type C_3 ttv.	≥ 1.0 µg/dose
		Total (biv. O,A)	≥ 4.5 µg/dose
		Total (ttv. O,2xA,C)	≥ 6.0 µg/dose
Adjuvant	50%	mineral oil 9 parts	Marcol 52
		emulsifier 1 part	Montanide 888

2.2 Emulsification

In-line emulsifier	3 consec. passes	determin. of resistance

TABLE 3

PRODUCTION OF AN OIL ADJUVANT FMD VACCINE
AT THE PAN AMERICAN FMD CENTER

3. Controls

3.1 Antigen production

Cell production	Serum	growth promoting cap. 3 passages
Virus production	Titer	PFU (in IBRS-2) and CF
Virus inactivation	Inactivaht	Cyclization (ß N.viol.) inactiv. rate (5 points)
	Innocuity	Cell culture (3 pass.)
Viral antigen	140S mass	CsCl gradient
	Integrity of vir. peptide	PAGE

3.2 Vaccine formulation

Antigens	Antig. mixture	CF, Innocuity
Emulsification	Resistance	> 100 Megaohm
Emulsion	Stability	Centrifug. 60'/3000 rpm Temperat. 37°/15 days

3.3 Vaccine potency
(each valency tested)

Guinea pigs (orientat. only)	PD_{50} (0.25 ml)	> 25
Cattle 16-20 animals	EPP (MPI)	> 75 (low.lim. of 95% confid. level)

seen FMDV A_{24} antigens from another laboratory with completely cleaved polypeptides which in an oil adjuvant vaccine gave good protection upon challenge. There are probably different sites for cleavage and the site may be important for determination of immunogenicity of the antigen.

The degree of emulsification of the oil adjuvant FMD vaccine is measured by determination of the resistance of the emulsion. After each passage of the oil adjuvant vaccine through the emulsifier the resistance is determined with a locally made Ohmmeter. The first passage usually gives a value of about 30 Megaohm, the second passage then about 60 Megaohm and the third passage more than 100 Megaohm. This is the standard value of the vaccine.

POTENCY CONTROLS

The recommendations of the PAFMDC for potency control of oil adjuvant FMD vaccines are that in direct challenge tests in adult cattle at least 12 of 16 vaccinated animals should be protected against generalization (15). The potency can also be tested by an antibody assay and determination of the Expected Percentage of Protection (EPP). The mean EPP value (lower limit at 95% confidence level) should be no less than 75 (25).

Each batch of vaccine prepared at the Center is tested after bottling for innocuity in cell cultures and for potency in guinea pigs and cattle. In guinea pigs the DPC_{50} is determined by injection of 0.25 ml of vaccine diluted with active diluent and challenged with the homologous vaccine virus strain. The DPC_{50} is calculated for the injected vaccine volume, i.e. 0.25 ml. However a clear relation between the DPC_{50} value and protection in cattle has not been established and the guinea pig test is therefore considered more to be a screening test. It should be used as a definite test only if a test in cattle cannot be done.

Potency in cattle is tested at the PAFMDC by vaccination of 16 to 20 animals fully susceptible to FMD and through antibody assays with their sera taken at 21 to 28 days after vaccination. Neutralizing antibodies are determined by virus neutralization tests in microplates of IB-RS 2 cell cultures (micro neutralization titer - MNT), or by mouse protection tests in newborn mice (mouse protection index - MPI). Studies done at the PAFMDC (18) showed that MNT values of $\geq$ 3.0 and MPI values of $\geq$ 2.5 indicate a probability of protection against challenge of > 90%. The MPI values are used to calculate the mean EPP values (18,23,24). For vaccines prepared with the antigen quantities given in Table 2 the mean EPP for each virus type in the vaccine is a minimum of 80 but mostly falls between 90 and 95.

Government control laboratories in Argentina, Brasil and Uruguay have tested the oil adjuvant FMD vaccine of the PAFMDC in direct challenge tests in cattle for potency, stability and duration of immunity.

The PAFMDC tested the potency of an oil adjuvant FMD vaccine after storage at -70°C for five years. The potency of the vaccine was not reduced after storage for this time (26). This is an important observation for vaccine banks of disease-free countries and also allows the use of this vaccine as a reference vaccine for an extended period of time.

IMMUNOGENICITY

Experimental Studies

The immunogenicity of oil adjuvant FMD vaccines as well as the duration of immunity induced by these vaccines in young and adult cattle at primovaccination and at revaccination was studied in extensive laboratory experiments at the PAFMDC and field trials (2,5,6,7,13,19). Some of the results obtained for primovaccination and revaccination are given in Table 5. The serum antibody titers after primovaccination indicate a protection for about six months while upon revaccination the antibody titers indicate a protection for twelve months or more. A study of the oil adjuvant vaccine in Colombian cattle gave similar results (22). The results of the neutralizing antibody studies were confirmed in a challenge experiment with cattle vaccinated three times at six month intervals (19), where 29 out of 30 animals were protected against generalized infection at 13 months after the last vaccination, Table 6.

The recommendation of the PAFMDC for application of the oil adjuvant vaccine is that young animals under two years of age should be revaccinated at six month intervals while for animals more than two years old and previously immunized repeatedly one annual revaccination is sufficient.

Field Studies

All laboratory and field studies of the immunogenicity of an oil adjuvant FMD vaccine at the PAFMDC as well as in other countries, like Argentina, Colombia and Uruguay indicated that this type of vaccine induced high levels of antibody to FMD virus and better protection against infection for longer periods of time than the conventional aluminum hydroxide saponine FMD vaccine. Therefore two countries in South America, Argentina and Brasil, were interested in testing the oil adjuvant FMD vaccine in large scale field application programs in comparison to the standard aluminum hydroxide saponine vaccine.

In Argentina about 68.000 cattle were vaccinated from 1977 to 1981 in Hipólito Yrigoyen, Province of Buenos Aires, with oil adjuvant FMD vaccine and the incidence of FMD in these animals compared with that in a cattle population in the same area vaccinated with commercial aluminum hydroxide saponine FMD vaccine. C.H. Caggiano and collaborators reported on this experiment in 1982 (12).

Recently F. Dora and collaborators in Brasil published a report comparing the incidence of FMD during an epidemic of FMD in Bagé, R.S., in 1980 in two cattle populations vaccinated with aluminum hydroxide and oil adjuvant FMD vaccine respectively (17). The results of the publications from Argentina and Brasil are summarized in Table 7.

In the Argentinian experiment the number of animals vaccinated with oil adjuvant vaccine is conderably smaller than in the control group. But it can clearly be seen that the disease incidence is greatly reduced in the oil adjuvant vaccine group. The Brasilian experiment has comparable cattle populations for each vaccine type. The morbidity rates per thousand animals in both experiments are surprisingly similar for the aluminum hydroxide saponine vaccine and the oil adjuvant vaccine groups, 55 and 7 in Argentina and 45 and 6 in Brasil. In both countries the incidence of FMD in the oil adjuvant vaccine area was about eight times lower than in the area of the conventional aluminum hydroxide saponine vaccine application.

More detailed data are available for the Brasilian experiment. For animals primovaccinated with the oil adjuvant vaccine the morbidity

TABLE 4

ANTIGENIC MASS OF FMDV BEFORE AND AFTER INACTIVATION WITH BEI AT 37°C, 1.5 mM AND 26°C, 3mM FOR 24 HRS.

Virus Type	Temperature of inactiv.	140S µg/ml* before	after	Loss (in %)
O_1	37°C	3.59	2.77	23
	26°C	3.60	3.19	11
A_{24}	37°C	4.15	1.49	64
	26°C	2.90	2.57	11
A_V	37°C	2.20	1.26	43
	26°C	2.19	1.91	13
C_3	37°C	3.12	1.62	48
	26°C	2.85	2.68	6

* average of two antigen batches

TABLE 5

Mean micro neutralization titers of groups of twelve cattle primovaccinated and revaccinated once at six months after primovaccination with oil adjuvant FMD vaccine

Months p.vac.	Primovaccinated MNT O_1	A_{24}	Months p.rev.	Revaccinated MNT O_1	A_{24}
0	<1.0	<1.0	0	2.5±0.5	2.8±0.6
3	3.1±0.5	2.7±0.6	3	3.3±0.3	3.4±0.2
6	2.9±0.6	2.7±0.6	6	3.1±0.4	3.3±0.2
12	2.0±0.3	2.1±0.5	12	3.2±0.4	3.2±0.4
18	2.1±0.5	2.0±0.5	24	2.9±0.5	2.9±0.4

Modified from Auge and Gomes (6)

TABLE 6

Protection of cattle vaccinated three times at six month intervals with oil adjuvant FMD vaccine after exposure to FMD virus O_1 Campos 13 months after the last vaccination

MNT	Animals	Protected	Not Protected
≥ 3.0	26	26	0
2.0-2.9	4	3	1
3.3*	30	29	1

* = arithmetic mean of the group
Modified from Gomes and Astudillo (18)

TABLE 8

Application of oil adjuvant FMD vaccine and FMD outbreaks in the State of Rio Grande do Sul, Brasil, 1980 to 1986

Year[a]	Doses applied oil adj. vacc.	Outbreaks of FMD in R.S.[c]	O. ad. vacc. Outbreaks[c]	appl. area % of total
1980	240.616	3.926	1.546	39
1981	605.094	579	210	36
1982	632.624	128	39	30
1983	699.156	50	5	10
1984	2.702.184	88	7	8
1985	7.722.607	41	0	0
1986	9.400.000[b]	28	0	0

[a]The cattle population in these years was between 12 and 13 million animals
[b]estimated
[c]Data from Astudillo and Dora (4)

rate per thousand was 14 while for revaccinated animals it was only two. The morbidity for revaccinated animals therefore was more than twenty times lower than in the control group. One can only agree with Dora and his collaborators, that if all the animals in that particular area had been vaccinated with an oil adjuvant vaccine, the incidence of the disease would have been insignificant.

LARGE SCALE APPLICATION

Brasil

The excellent experience with the application of an oil adjuvant FMD vaccine in the epidemic in Bagé, R.S. in 1980 led the Veterinary Service of the State of Rio Grande do Sul to begin a program of systematic and controlled vaccination of the cattle population with this type of vaccine. The vaccination program was begun in the endemic area in the southwestern part of the state. The geographical coverage was increased gradually depending mainly on the availability of the oil adjuvant vaccine. In 1986 the program covered about 75% of the cattle population of the state. The vaccine for this program is supplied by the PAFMDC, a Federal Government FMD vaccine production laboratory in Campinas, S.P, a State Government FMD vaccine production laboratory in Porto Alegre, R.S., and a small amount by a private vaccine production laboratory. Recently other private vaccine production laboratories in Brasil have obtained a production license for an oil adjuvant FMD vaccine and the commercial production of this type of vaccine may therefore increase considerably in the near future.

The execution of the program in Rio Grande do Sul is based on administrative districts. The state veterinary service begins the introduction of the program into a new district by organizing meetings with local state officials, with private veterinarians and with the local farmers, where the purpose of the program and operative and administrative aspects are explained and also a demonstrative vaccination is done on a farm with a small number of cattle (30 to 60 animals). The farmers then receive letters informing them of the planned vaccination date. The supply of the necessary vaccine has been previously arranged through the local farm cooperative. The vaccination then will be supervised by staff from the state veterinary service.

Implementation of the vaccination program begins always in October and November with vaccination of all cattle including calves. This is followed five to six months later by another vaccination of all animals. Three to four months later only calves are vaccinated. The normal vaccination cycle then continues with the vaccination of all animals in February and March and vaccination of calves only three months later.

Figure 1 shows the area in the state covered by the vaccination program in 1986 and the planned expansion for 1987. Table 8 gives the number of doses of oil adjuvant FMD vaccine applied in the State of Rio Grande do Sul from 1980 to 1986 and the corresponding number of outbreaks of FMD in the state and in the oil adjuvant vaccine application area. Figure 2 shows the increase of the percentage of the cattle population vaccinated with oil adjuvant vaccine and the decrease of the percentage of FMD outbreaks in the area of application of this vaccine in comparison to all outbreaks in the state. The incidence of FMD in the state has been reduced drastically and the oil adjuvant vaccine application area has not had a single case of FMD for the last two years. The area of the field testing of this vaccine in Bagé in 1980, where application of the oil adjuvant FMD vaccine continued, has not had FMD for six years now. It therefore can be expected that with a complete coverage of the state with this program FMD will be eliminated in R.S.

FIGURE 1

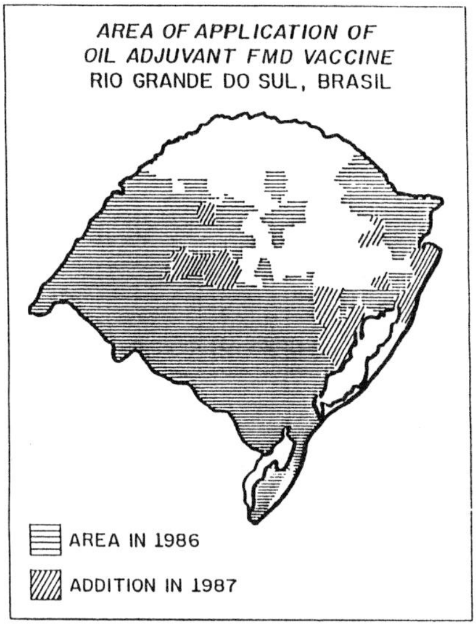

FIGURE 2

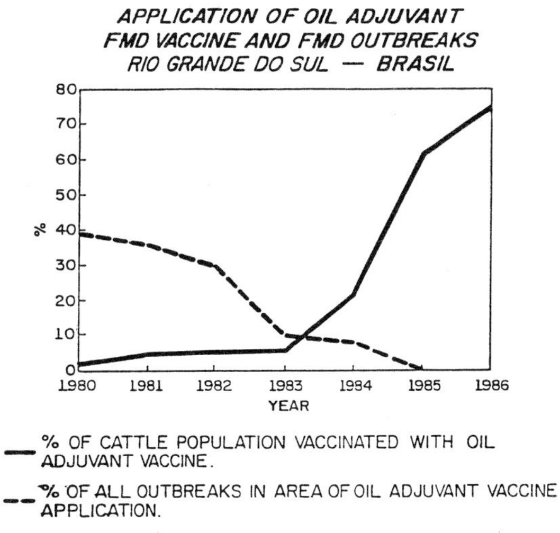

Other Countries

Government field projects with application of oil adjuvant FMD vaccine are in operation in Argentina, Brasil, Bolivia, Ecuador, Paraguay, Peru, Uruguay and Venezuela. These programs have been quite successful. So far the vaccine for these projects has been supplied by the PAFMDC. However the production capacity of the PAFMDC is limited and therefore Argentina and Venezuela are in the process of building FMD vaccine production laboratories for this type of vaccine. In Colombia a vaccine production laboratory has prepared and tested a number of preproduction lots and in Paraguay a commercial laboratory may begin production soon.

OUTLOOK

So far about 40 million doses of oil adjuvant FMD vaccine have been applied in South America and it is very likely that the production volume will soon increase considerably. The value of this vaccine for FMD control programs has been amply demonstrated in the State of Rio Grande do Sul in Brasil. However it should be emphasized that the success of the use of this vaccine is also due to the very well organized execution of the program and a controlled application of the vaccine. A systematic and massive vaccination of the cattle population with oil adjuvant FMD vaccine mainly in endemic areas (8) should have a profound effect on the epidemiological situation of FMD in South America.

REFERENCES

1. Abaracon, D. and Giacometti, H., Bol. CPFA 21-22:49-53, 1976
2. Abaracon, D. et al., Bol. CPFA 37-38:45-47, 1980
3. Abaracon, D. et al., Bol. CPFA 45-46:55-57, 1982
4. Astudillo, V. and Dora, F., 17. Conf. FMD Comm. OIE, 287-319, 1986
5. Auge, P. et al., Bol. CPFA 26:23-25, 1977
6. Auge, P. and Gomes, I., Bol. CPFA:27-28, 55-60, 1977
7. Auge, P. et al., Bol. CPFA 37-38:39-40, 1980
8. Auge, P., Bol CPFA 45-46:33-42, 1982
9. Bahnemann, H., Arch. Virol. 47:47-56, 1975
10. Bahnemann, H., J. Clin. Microbiol. 3:209-210, 1976
11. Barteling, S., Bull. Off. Intern. Epiz. 81:1243-1254, 1974
12. Caggiano, C.H. et al., 16. Conf. FMD Comm. OIE, 223-241, 1982
13. Casas Olascoaga, R., Bull. Off. Intern. Epiz. 89:1015-1054, 1978
14. Centeno, E. and Obiaga, J., First Intern. Conf. Impact Vir. Dis. Developm. Latin America, Rio de Janeiro, 1982
15. C.P.F.A., Serie de Manuales Tecnicos No. 2, 1980
16. C.P.F.A., Evaluacion de proyectos de vacunacion con vacuna oleosa, 1981
17. Dora, F. et al., Bol. CPFA 49-50:11-17, 1984
18. Gomes, I. and Astudillo, V., Bol. CPFA 17-18:9-15, 1975
19. Gomes, I. et al., Bol. CPFA 37-38:31-35, 1980
20. Heuschele, W., Art to Science 3:2, 1984
21. Mesquita, J. and Vieira, A., Bol. CPFA 35-36:64, 1976
22. Rocha, J. et al., Bol. CPFA 47-48:45-54, 1983
23. Sutmöller, P. et al., Bol. CPFA 39-40:37-42, 1980
24. Sutmöller, P. et al., Mtg. Techn. Comt., Europ. FMD Comm. FAO, 1983
25. Sutmöller, P. et al., Bol. CPFA 49-50:31-34, 1984
26. Sutmöller, P. et al., Mtg. Techn. Comt., Europ. FMD Comm. FAO, 1985

Section 7
Immune system products

47 Application of large scale perfusion systems for pharmaceutical production from mammalian cells

W. R. Tolbert, W. R. Srigley and C. P. Prior

ABSTRACT

APPLICATION OF LARGE SCALE PERFUSION SYSTEMS
FOR PHARMACEUTICAL PRODUCTION
FROM MAMMALIAN CELLS

W. R. Tolbert, W. R. Srigley and C. Prior
Invitron Corporation, St. Louis, MO 63134, USA

Two agitated perfusion systems have been used to grow attached and non-attached mammalian cells at perfusion rates between 50 and 150 liters/day. An immobolized non-agitated perfusion system has been used for extended production of secreted products from both cell types at similar rates. Products have included human and mouse monoclonals and human proteins from normal, transformed and recombinant cell lines. Multiple grams/day production has been achieved per reactor with antibiotic free reduced serum medium. Down stream protein purification emphasizes large scale systems to quickly and gently isolate desired molecules without harsh conditions that could reduce biological activity. Overall yield in the range of 50-70% is common. All operations are conducted under cGMP's in a facility specifically designed for pharmaceutical production. Double HEPA filtered air at up to 50 changes per hour is used in critical areas with seven direct digitally controlled pressure zones to provide both containment and an aseptic environment.

INTRODUCTION

This paper will report on the application of the perfusion technology described earlier in the meeting by Dr. Joseph Feder. This perfusion technology has been the basis for construction of a pharmaceutical manufacturing facility for the production of monoclonal antibodies, tissue plasminogen activator and other proteins by large scale mammalian cell culture.

In a sense I will describe a new 31,000 sq. ft. bioreactor (Figure 1) with a theoretical capacity of 1.3 million liters/year. The individual production operating systems are intimately connected to the manufacturing facility and both the facility and systems should be viewed as an integrated whole. The main objective of this facility is to make human injectable drugs,and therefore, considerable effort was extended to produce what we feel is a "state of the art" pharmaceutical facility built around this perfusion technology to be operated in full compliance to the U.S. Food and Drug Administration current Good Manufacturing Practices.

Dr. Elaine Esper, Director of the FDA, Office of Biologics Research and Review, noted in a Food and Drug Law Institute Meeting in Washington D.C. on December 9, 1986, that the agency has several concerns related to the alteration of molecular structure in bioengineered products. These concerns were related to the heterogenerity of amino terminal sequences, alterations due to non-glycosolation, and alterations in protein folding and disulfide bridge formation in mammalian proteins produced in

Figure 1: Invitron Corporation Manufacturing Facility in Berkeley, Missouri.

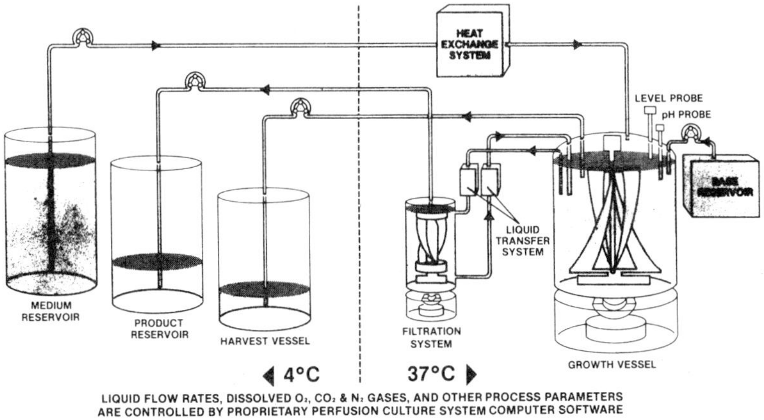

Figure 2: PERFUSION CULTURE SYSTEM: Liquid flow rates, dissolved O_2, CO_2, and N_2 gases, and other process parameters are controlled by proprietary perfusion culture system computer software.

bacteria such as E.coli (1).

As production of biopharmaceuticals in mammalian cells may answer many of these concerns, there seems to be an important place in the manufacture of the increasing number of new biological drugs for large scale mammalian cell culture technology. In the production of any drug product, compliance to extensive government regulations is necessary so that the consuming public is ensured of obtaining safe and effective products made in a reproducible and controlled manner.

Pharmaceutical products from mammalian cell culture must meet these same standards. This requires an integration of the cell culture technology used, the documentation and quality assurance methods to ensure product safety and efficacy, and the facility that provides the aseptic environment for manufacturing.

PERFUSION CULTURE TECHNOLOGY

The perfusion culture systems used in the pharmaceutical facility have been previously described (2,3,4). Figure 2 shows a diagram of the system in current use for growth of anchorage independent suspension cultures. This system allows high density cell growth of mammalian cells under conditions that maintain a high cell viability and stabilized cellular environment.

Perfusion of fresh media through the reactor is maintained at a rate proportional to the number of cells in the growth vessel until the desired cell concentration is attained. At this time a cell harvest is initiated to maintain that cell density and concomitantly a constant perfusion rate necessary for continued growth and health of the cells. The cell density at which this harvest begins is routinely 10 times the maximum concentration seen in batch production systems of the same cell. After establishing these conditions, production can continue from the same culture for months at a time.

For growth of attached cells a modification of the system is used as shown in Figure 3. This essentially allows cells attached to microcarriers to fall back into the agitated region of the culture vessel and not be transferred to the somewhat higher shear region of the filtration system.

Both of these systems provide the means for high density cell growth and production of secreted protein products. Cell associated products can be obtained from the cell harvest vessel in the earlier system or by collection of cells attached to the microcarriers in the latter system.

Figure 4 shows operation of a 100 liter perfusion system in one of the 37°C warm rooms at the plant. A process computer monitors and controls necessary culture parameters, gas flows, and liquid flows to the reactor through the adjacent reactor control cart. The process computer itself is located in a separate computer room outside of the contained production area.

The third type of perfusion system used in the plant is diagramed in Figure 5. This Static Maintenance Reactor (SMR) allows long-term maintenance of cells at very high density (approaching one tenth of tissue density) continually producing secreted protein products in very low or no serum conditions. Cells grown in one of the two above systems are concentrated, mixed with a matrix material and used to inoculate this reactor, such that one tenth of the reactor working volume is cells and nine tenths matrix material. By reducing serum or other proliferation stimulating components, cells are maintained for long periods of time efficiently producing secreted proteins.

Figure 6 shows the production scale system which can be inoculated with between one and two liters of packed cells. Table 1 compares production of several products in these three

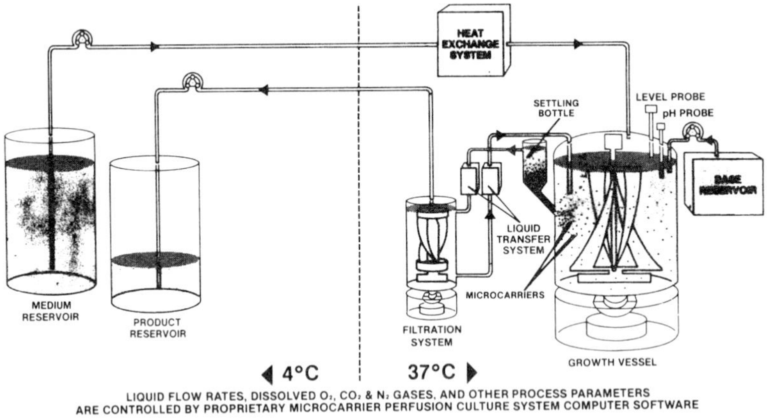

Figure 3: MICROCARRIER REACTOR SYSTEM: Liquid flow rates, dissolved O_2, CO_2 and N_2 gases, and other process parameters are controlled by proprietary microcarrier perfusion culture system computer software.

Figure 4: Operation of a 100L perfusion system has equivalent production capacity of thousands of liters of conventional batch suspension system. A typical perfusion culture run averages 90 days.

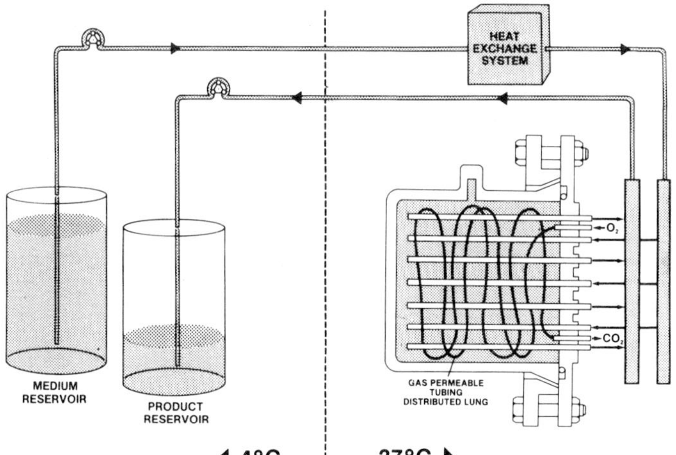

Figure 5: STATIC MAINTENANCE REACTOR: Liquid flow rates, dissolved O_2, CO_2 and N_2 gases, and other process parameters are controlled by proprietary static maintenance reactor computer software.

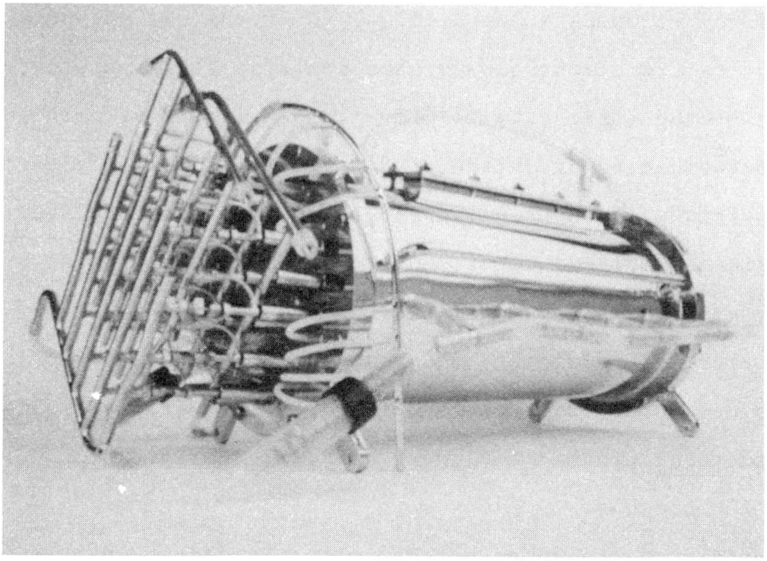

Figure 6: The Static Maintenance Reactor daily produces gram quantities of medically important proteins. A typical SMR run averages 90 days.

perfusion systems to parameters in small scale batch systems.

Because of the organization of the production facility, fresh media being perfused into the reactor systems and product contained in cell conditioned media is maintained at $4^{o}C$ in a cold room adjacent to the $37^{o}C$ warm room. This means that proteolytic enzyme activity which could cause significant protein product degradation can be dramatically reduced in the perfusion systems as compared to normal batch operation. Residence time of product in the perfusion growth systems is one to two days, and in the SMR may be as little as three hours compared to many days or weeks in conventional batch fermentation systems.

As production can be maintained for months from any of these systems, total annual productivity is greatly enhanced. One SMR run extended for four months with greater daily product output at the end of the run than at the beginning.

Another very important feature of this production method is the ability to operate completely without the use of antibiotics. Antibiotics are almost always used in large scale applications because of the difficulty in preventing a single microorganism from contaminating an entire production run. This major economic risk increases exponentially with the size of the culture. As the perfusion systems described are modular, operated at high density and so relatively small, they can be used without antibiotics. In addition, the production reactors, proprietary aseptic connection systems, and the facility were specifically designed to maintain the highest level of sterility for our production process.

The disadvantages associated with the use of antibiotics impact negatively the safety of the product. Antibiotic use encourages the growth of low levels of resistant organisms which are difficult to detect and may result in adulterated, unsafe drug products. Antibiotic use also tends to permit less than optimum operational standards which could lead to other types of contamination or loss of product integrity.

The absence of antibiotics guarantees very high quality in all aspects of the production process. Therefore, we feel that these substances should not be used in pharmaceutical systems.

LARGE SCALE PROCESS PURIFICATION

Perfusion systems have major advantages for down stream process purification. As mentioned above, serum proteins can be reduced or even eliminated, thereby removing the major background contamination from which the protein product must be isolated. Perfusion systems allow product to be collected at $4^{\circ}C$ and processed on a continual basis, thereby also reducing the opportunity for product degradation.

When handling large volumes of cell conditioned medium containing a protein product in dilute solution, the first step is a fifty to one hundred fold concentration. This is most often accomplished by ultrafiltration systems such as the plate and frame system shown in Figure 7.

For many proteins it is possible to establish conditions under which the ultrafiltration step can be eliminated and the conditioned media, directly or with some adjustment of pH or salt concentration, flowed over a highly cross linked ion exchange resin for a combination of concentration and initial purification. These resins allow extremely high flow rates of hundreds of liters per hour and have a high capacity for binding proteins. Elimination of the ultrafiltration step also reduces purification losses and damage to fragile proteins. Figure 8 shows use of such an ion exchange system.

Additional purification steps are defined by the physical

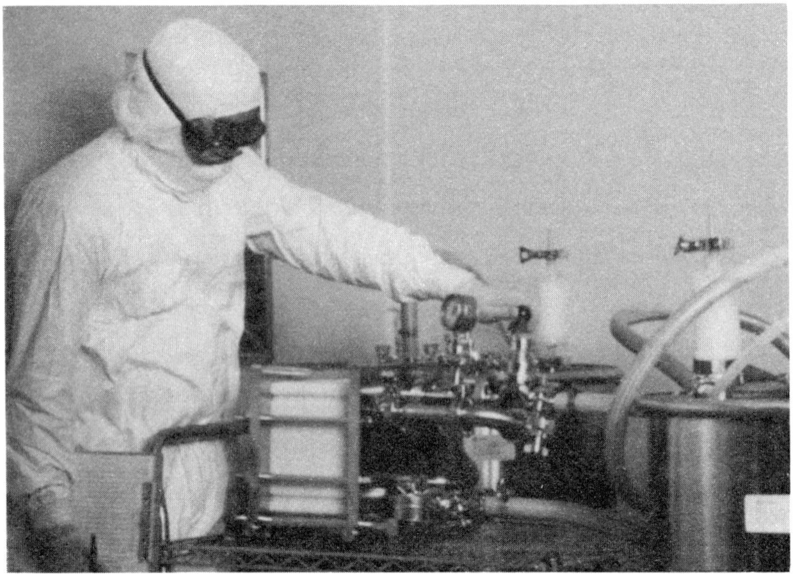

Figure 7: Aseptic concentration of product from conditioned cell culture media.

Figure 8: Chromatographic purification of product is conducted within the contained production facility.

characteristics of the desired protein and the fold purification necessary to achieve isolation. If very high fold purification is required, immunoaffinity columns can be used with monoclonal antibodies having a specificity against the desired protein.

Gel filtration chromatography provides an excellent finishing step in any pharmaceutical protein purification process. Acting as a molecular sieve, gel filtration removes potentially toxic residual chemicals and potentially antigentic large protein aggregates. In addition, gel filtration allows the purified protein to be formulated into a buffer system that confers the maximum solubility, biological stability, and suitability for pharmaceutical use.

Another important aspect of this protein purification scheme is the relative low cost of the chromatography materials. In making a pharmaceutical product, one of the major concerns is the segregation of any drug product from any other product that may be made in the same facility. With inexpensive chromatography methods, all resins and other materials can be completely replaced for each separate product. All columns and equipment are completely disassembled and cleaned with a validated cleaning procedure before being reassembled for use in another system. Figures 9 and 10 show purification of an IgG monoclonal antibody.

QUALITY CONTROL

The production of drug products by any means requires strict adherence to established pharmaceutical manufacturing principles and practices. The fact that the products of cell culture are often complex biomolecules whose efficacy may depend upon them retaining their native structure, and whose safety depends on freedom from extraneous contaminants only makes this all the more essential. These standards apply early in the design of

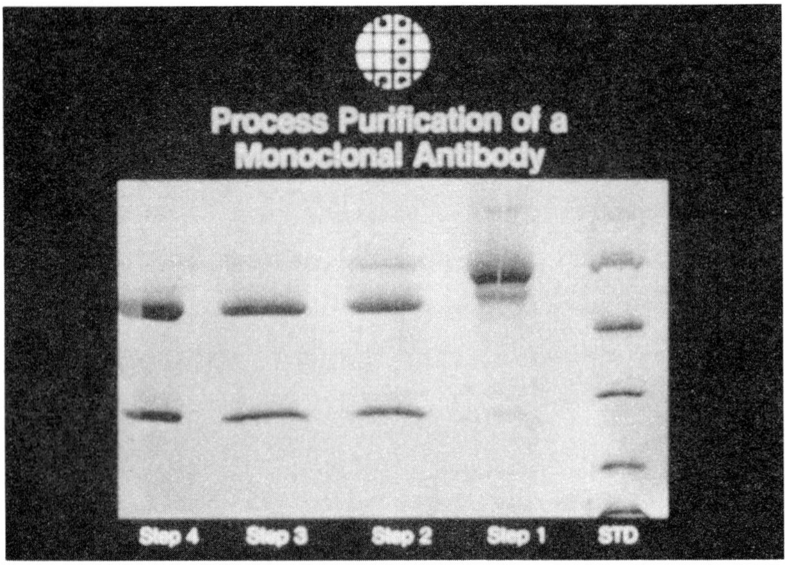

Figure 9: Gel Electrophoresis Pattern of steps in purification of an IgG monoclonal antibody. Lane 1 contains molecular weight standards. Lane 2 is conditioned medium starting material. Lane 3 is elution from first ion exchange column. Lane 4 is concentrated material before gel filtration step. Lane 5 is formulated product in bulk form ready for filling.

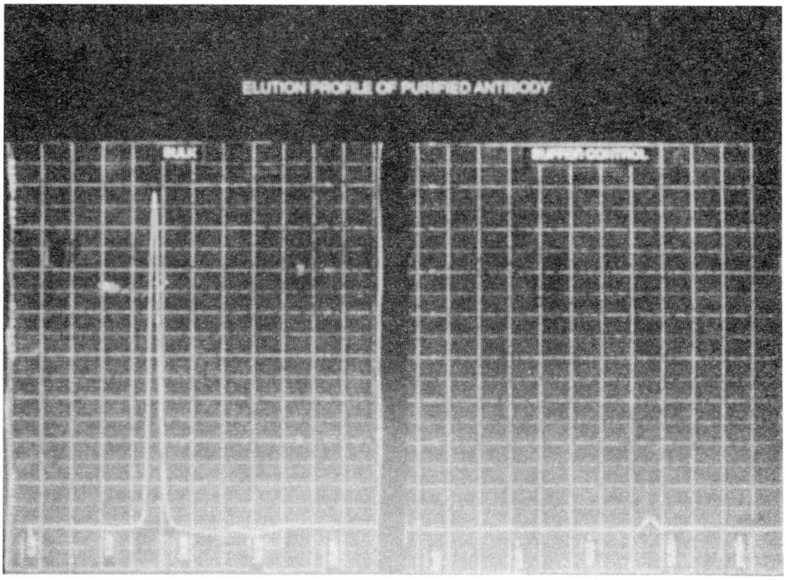

Figure 10: HPLC pattern of purified monoclonal antibody showing symmetrical peak and very high purity. Also shown is a buffer control.

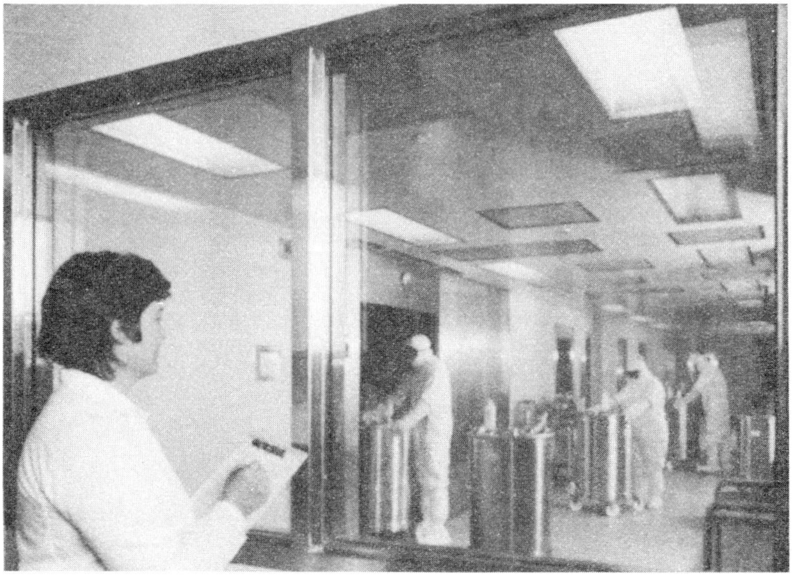

Figure 11: Quality Assurance person observes the unloading of autoclave-sterilized media containers.

equipment and extend to the operation of the production facility.

Adequate isolation and spatial separation should exist in the production area to preclude mix-up or cross-contamination of cells and products. Wall, floor and ceiling surfaces should be constructed of materials which are durable, smooth, non-absorbent, and able to withstand repeated cleaning and sanitization. Air to the production areas should be filtered and supplied at a rate which provides individual rooms with the pressurization necessary to achieve containment and aseptic operation. The number of air handlers used should be sufficient to permit appropriate isolation of critical production and quality control areas. Steam used to sterilize process or product contact surfaces should be saturated, clean and free of boiler additives. Process gas lines should be cleaned prior to use and constructed of non-additive materials.

The process water system should be of as high quality as possible. Water is the single largest component of cell culture-derived product, and high quality water is very difficult to obtain. A good pre-treatment system, followed by a multi-effect still which feeds water to a 316L stainless steel tank can consistently produce water which meets the U.S.P. monograph for Water for Injection. Holding the water at 80°C and circulating it through a properly installed and maintained stainless steel loop can generally assure that the water which is supplied to the production area is consistently of excellent quality.

Documentation of all aspects of the production is basic to a pharmaceutical operation. Cell culture is a developing science. Nevertheless, its use as a vehicle for the manufacture of pharmaceutical products requires that the process be described in detailed procedures. These procedures should include the cleaning and set-up of equipment, sterilization, cell culture , inoculations, media preparation, culture maintenance, collection and concentration of conditioned media, product purification, and the disposal of cultures and media. Procedures for testing, raw materials receiving, storage, facility maintenance and so forth are also necessary, but are typically more easily reduced to written form.

The design of a documentation system suitable for recording the critical steps of a cell culture process in a manner which complies with cGMP's present unique challenges. The variability of cells from different species, or in some cases, different clones of the same cell line, makes the advance preparation of a uniform batch record applicable to all cells virtually impossible. At the same time, the prospect of attempting to write a customized master formula and batch record for each individual cell line becomes formidable if any more than a few cell lines are involved. In a similar vein, while the natural variability of biological systems makes it likely that acceptable differences in the production process might be encountered from batch to batch, reasonable process control requires that some boundaries be established which prescribe the allowable latitude for process variability.

Training of employees in the proper execution of these procedures is essential and deviations from accepted procedures must be minimized and each instance documented.

Consistency in the manufacture of product is only possible to the extent that the characteristics of the raw materials used in the process are known and maintained within defined limits. Internal testing of purchased materials is an essential element of quality control. Of no less importance, however, is an active vender communication program. This would include vendor audits, discussions regarding sampling plans and notification of changes in vendor production operations. Quarantine and release of incoming material insures that required specifications are met before material is used for production.

A well-designed, in process testing program provides an understanding of the process which is often impossible to obtain from final product testing. Validation of both the process and the facility are required for pharmaceutical manufacture. Since the protein products of cell culture are complex biomolecules, efficacy depends upon their retaining their native structure, and safety depends on freedom from extraneous contaminants. This makes the manufacturing task more difficult than more traditional drug production.

Prior to the mid-70's, the method by which the quality of drug products was proven was primarily end product testing. The idea was that if a final lot of product passed a set of release tests, it was suitable for commercial distribution. In 1976, the FDA published the "Proposed Good Manufacturing Practices for Large Volume Parenterals." These proposed regulations set forth a concept which has dramatically changed the way that the quality of drug products is defined. In essence, the Agency has said that simply meeting final product specifications was no longer a sufficient basis on which to conclude that a lot of product was able to be released. In addition, the manufacturer must also be able to prove that the production and control systems used to make, hold and test the product were operating within established limits during the time the product was manufactured, held and tested in the facility. The basis for such proof is validation - proving that a system, process or test consistently works the way it was intended to work.

This entails defining the critical aspects of the facility

and equipment design and construction, verifying that they were adhered to, establishing that the plant and production systems function within acceptable specifications and documenting the results (see Figure 8). It also means that once things are working in a controlled fashion to yield the desired results, changes must be evaluated to ascertain that they do not adversely affect the product or production systems. The supposition is that if you have demonstrated that the plant and production systems are consistently capable of producing acceptable product, as long as no changes are made to those systems, it is likely that you will continue to produce acceptable product. End item testing then becomes a confirmation of that success, rather than a determination of it. This is an extension of the principle that quality can't be tested into a product; quality is either there or not there as a consequence of the manufacturing process. Validation is the means by which you insure that the quality is there.

CONCLUSION

The Manufacture of pharmaceutical products by cell culture involves the use of complex biological processes and subtle biochemical operations. That these processes must be carried out in compliance with the FDA's current Good Manufacturing Practices, and in accordance with the drug industry's high standards for process validation, makes the task even more challenging.

In the case of cell culture, a perfusion culture system was described which has been used successfully to manufacture safe and efficacious biomolecules on an industrial scale. Principles of protein purification typical to the manufacture of cell-derived products have been reviewed with emphasis on those

TABLE I

PRODUCTIVITY COMPARISON

Cell Type	Small Batch Vessels Product Concentration (mg/liter)	Small Batch Vessels Serum Concentration (%)	Large Perfusion Systems Reactor* Type	Product Concentration (mg/liter)	Serum Concentration (%)	Total Volume (liter)	Total Product (gm)
Hybridomas							
Line A	50	6	PC	80	1	3400	270
Line B	10	10	PC	32	10	2100	67
Line C	10	6	PC	16	2	3300	52.5
Line D	20	6	PC	36.5	2	2600	95
R-DNA							
Line E	10	7.5	PC	13	5	18500	240
Line F	0.2	10	MR	1.0	0	680	0.66
Line G	30	10	SMR	47	0	1400	66.2
Line H	10	5	SMR	12	0	8500	100

*PC Perfusion Culture System
MR Microcarrier Reactor
SMR Static Maintenance Reactor

factors which are critical to product safety. The importance of Quality Control in its role as a means of establishing and maintaining high operation standards for the facility and the process has been described.

REFERENCES

1. FDC Reports Prescription and OTC Pharmaceuticals, Vol 48, No 50, 8-9 (1986).
2. Feder, J. and Tolbert, W.R. The large-sclace cultivation of mammalian cells, Sci. Am. 248, 24-31 (1983).
3. Tolbert, W.R., Lewis, C.Jr., White, P.J. and Feder, J. Perfustion culture systems for production of mammalian cell biomolecules in Large-Scale Mammalian Cell Culture (J. Feder and W.R. Tolbert, eds), Academic Press, New York, NY, pp. 97-123 (1985).
4. Tolbert, W.R. and Feder, J. Perfusion systems for plant scale mammalian cell culture production, Pharm. Eng. 5, 23-27, 49 (1985).

48 The organotype culture model of human cancer cells: the effect of human recombinant interferons

R. Beaupain, M-C. Martyré and E. Falcoff

ABSTRACT

The organotypic culture method of human tumor cells may be a useful tool for assessing the effects of cytostatic agents on tumor growth and survival. Several monolayer cell lines have been adapted to this three-dimensional culture method : MCF-7 breast cancer cells, an ovary tumor line and A549 type II alveolar carcinoma cells. The extremely low traumatizing culture conditions (absence of enzymatic cell dissociation, growth measurements without handling the nodules) offer to injured cells a maximal survival chance, close to the *in vivo* situation.

The effects on A549 nodules of recombinant human interferon (IFN)-α2, -β and -γ and mixtures of -α2 and -γ were investigated. Changes of medium containing 10, 10^2, 10^3 or 10^4 U/ml of IFNs were carried out 3 times a week for 6 to 8 weeks. Low doses of the 3 IFNS stimulated growth of 25 % of the nodules (a 1.7 fold increase for the quickest growing cultures). No growth inhibition was observed after a treatment with IFN-β, even after the highest doses. Treatment with 10^4 U/ml of IFN-α2 or -γ resulted in a 1.9 to 2.3 fold inhibition respectively and necrosis and regression of the nodules occured. The same result was obtained with a mixture of IFN-α2 (10^3 U/ml) plus IFN-γ (10^3 U/ml) and in this case, nodule regression occured earlier. These data demonstrate that IFN-α2 and -γ cooperate to potentiate the IFN antiproliferative activity.

The organotypic culture of human tumor cells is a continuous culture system where cancer cells are growing on top of a semi-solid medium in a three-dimensional manner. This method has been developed by E.T. WOLFF and E.M. WOLFF (19) and more recently adapted for *in vitro* bioassay purposes (1,5,2,3,11). The organotypic cancer nodules, issued from monolayer cultures, are subcultivated without enzymatic cell dissociation procedures : the nodules are cut with microsurgical scissors and the fragments are transferred on a fresh medium. So, damage to the cell membrane in this culture method, is highly reduced. In order to perform the bioassay procedures, the cancer nodules are laid on Millipore filters which replace the semi-solid agar substratum. Medium changes can hence be done without handling the individual nodules and culture trauma to the cells is further more reduced. Growth evolution can be followed for long periods of time (2-3 months or even longer) and this makes it possible to study regeneration and regrowth processes after the different antitumoral treatments.

Another characteristic of the cultured organotypic nodules is the evidence of some cell differentiation. Nodules issued from the monolayer A549 lung cancer line (9) develop alveolar structures which are filled with mucus. This mucus appeared to be the pulmonary surfactant (1,5). Organotypic nodules of two other cell lines will be presented here : MCF-7 breast carcinoma cells (16) and ovary cancer cells (OD262) from a patient of the Institut Curie and established as a cell line by H. MAGDELENAT (Institut Curie, Section de Biologie, Paris, France).

This three-dimensional tumor model has first been used to test the effects of X-rays and cisplatinum on nodule growth (1). Combined treatments with cisplatinum (CDDP), 5-fluorouracile (5-Fu) and X-rays were applied (2) as well as a combination of X-rays and misonidazole (miso), a radiosensitizer used in therapeutics (5). In both 5-Fu and Miso treated nodules, growth enhancement has been observed at low drug concentrations ; the higher doses being growth inhibitory. This dual effect has also been observed using interferons (IFNs) : actually, it has been observed that besides their antitumoral activities (7, 14,8,4), they can stimulate growth of malignant cells (10,6) ; these latter results were obtained with the tumor clonogenic assay (10).

In the present study, we compared first the effect of three human IFNs, IFN-α2, -β and -γ, all produced as highly purified recombinant IFNs, on A549 lung carcinoma nodules maintained in organotypic culture to see whether the inhibitory and stimulating effects could be found. Then combinations of IFN-α/β and -γ were tested. This could result in the association of complementary effects at the tumor target level. Such an association could provide a more efficient use of IFNs in human cancer therapy in abolishing the growth enhancing and/or toxicity effects.

MATERIAL AND METHODS

Organotypic culture

The organotypic nodules (A549, MCF-7 and OD262) originate from monolayer cells. They are obtained by scraping the cultured cells from the culture surface of plastic flasks, followed by a centrifugation step (about 500 g) during 20 min. The pellet is placed on top of a semi-solid medium previously prepared in a Petri dish (Falcon 1006) having a tight fitting lid. The semi-solid medium is obtained by mixing (in a 1 % ratio) 1 % bacto-agar (Difco) in distilled water with two fold-concentrated RPMI 1640 culture medium containing 20 % heat inactivated fetal bovine serum. The RPMI 1640 medium 2 X concentrated is supplemented with 9 mg glucose per ml. For the MCF-7 and OD nodules, insulin was added (0.08 I.U./ml in the 2 X medium). The nodules are subcultured every 10 days by dividing them in two (or more) pieces with iridectomy scissors and by placing them on a fresh medium previously prepared in the Petri dishes. 0.2 ml complete liquid RPMI 1640 X 1 was added to the cultures.

In order to do the different experiments, instead of putting the nodules on the semi-solid agar medium, they are placed on Millipore filters (HABPO 2500, pore size 0.45 or 0.8 um, cobalt sterilized), which have previously been placed on the semi-solid medium. Five days later, when wound-healing of the nodules is finished, the filters containing the nodules (generally eight nodules per filter) are transferred to new Petri dishes without the semi-solid medium. RPMI 1640 X 1 medium is added (1.3 or 1.5 ml). The nodules on the filter are individualized to enable the follow up of growth or degradation of each individual nodule. For sterility purposes, the filters are placed in new Petri dishes every 14 days.

Growth measurements

Every 5 to 10 days, a two-dimensional measurement of each nodule is made (the greatest diameter and a second measurement at a right angle to the first diameter). The size(s) of the nodule is expressed as a percentage S_t/S_0 where S_0 is the size before the treatment and S_t the size at time t during the treatment. The mean growth percentage is calculated and the regression lines are established and they gave the mean growth percentage per day. Student's t test, done for all treated nodules, was used to compare the growth percentage of treated and control nodules. The growth index was the ratio :

$$\frac{\text{growth \% of treated nodules}}{\text{growth \% of control nodules}}$$

The analysis of the combined effects, according to the definitions of Denz (6), was performed as follows :

Additive $GR_{A+B} = (GR_A) \times (GR_B)$

Synergistic $GR_{A+B} < (GR_A) \times (GR_B)$

GR = growth rate A,B = IFNs A + B = mixtures of IFNs

Culture of the regeneration nodules

The healthy and outgrowing zones of necrotic nodules which are regenerating after the different treatments can be excised and put on the semi-solid medium or on Millipore filters. The nodule growth measurements can be done as described.

Interferons

Three highly purified IFNs were studied : IFN-α 2 obtained from Shering-Plough (1 X 10^8 U/mg protein), IFN-β, from Cetus (1.3 X 10^8 U/mg protein) and IFN-γ, from Biogen (2 X 10^7 U/mg protein). The three IFNs were found to be electrophoretically pure. The doses of IFNs in the cultures were based on the antiviral titers, as determined by the cytopathic effect inhibition assay on human Wish cells, challenged with Vesicular Stomatitis Virus, as already described (18). All three IFNs were assayed simultaneously, and the titers were standardized to the N.I.H. human leukocyte IFN N° GA 902 530.

IFN treatment of A549 nodules

In order to perform the IFN treatments, the A549 nodules cultured on the semi-solid agar medium were transferred on the Millipore filters, as described. The medium with or without IFN was renewed three times per week. IFN-α 2 and -γ were used alone or in mixtures ; IFN-β was used alone. The concentrations were 10, 10^2, 10^3 and 10^4 U/ml, when IFNs were applied alone. For the mixtures, the following concentrations were used : -α 10^3 U/ml plus -γ 10, 10^2 or 10^3 U/ml and -γ 10^3 U/ml plus -α 10 or 10^2 and the control 2 X 10^3 U/ml for both IFN-α 2 and -γ.

RESULTS

The organotypic nodules

In Fig. 1 is shown a A549 alveolar II pulmonary tumor nodule _in situ_ grown on Millipore filter 10 days after subculturing, the "size" is 2.2 X 1.9 = 4.18 mm2 (magnification X 50). The A549 nodules present alveolar structures which are filled with mucus. This mucus is seen in Fig. 2A, where the histological structure of the A549 nodules is shown. The structure of MCF-7 breast cancer nodules is presented in Fig. 2B. Cord-like structures are seen, there is no mucus secretion and a tendancy to form secondary nodules. The ovary cancer nodules (OD262)(Fig. 2C) have a loose structure, an abundant mucus secretion and here is also a tendancy of secondary nodule formation. In Table I, is presented the growth percentage of the three types of nodules. MCF-7 appeared to be the quickest growing nodules and A549 the slowest.

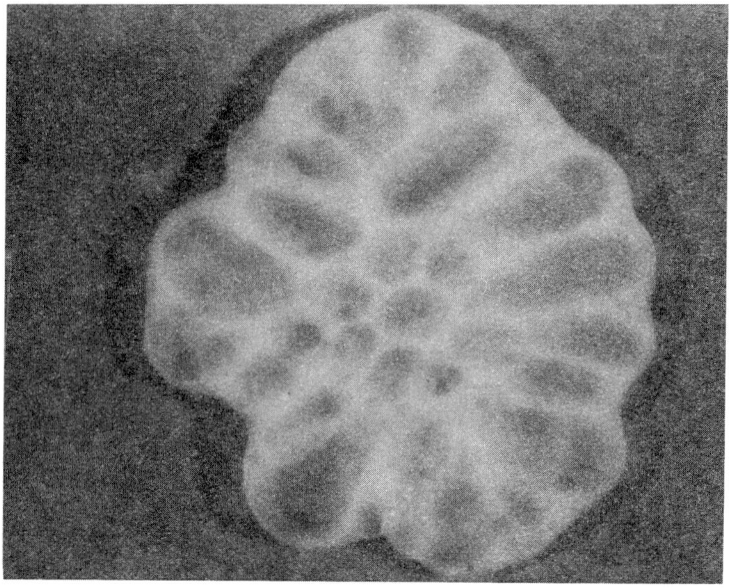

Fig. 1 A typical A 549 nodule in situ growing on a millipore filter in RPMI 1640 medium supplemented with 10% fetal bovine serum. " Size ": 2.2 x 1.9 mm = 4.18 mm^2 (5cm = 1mm). The translucent areas are the alveoli which are filled with mucus.

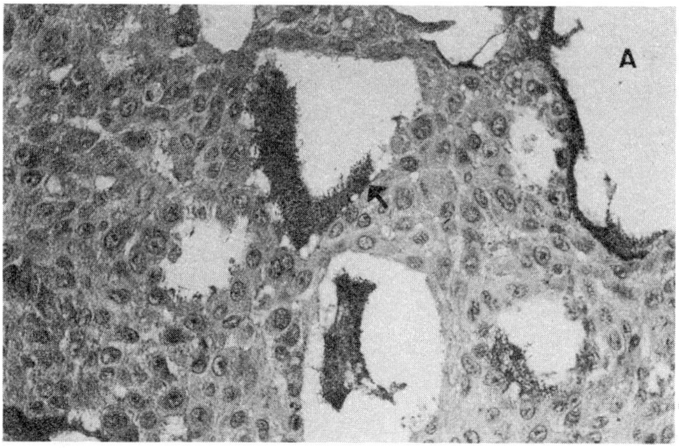

Fig. 2A A histological view of a A 549 pulmonary cancer nodule.
Note the mucus containing alveoli (arrow).

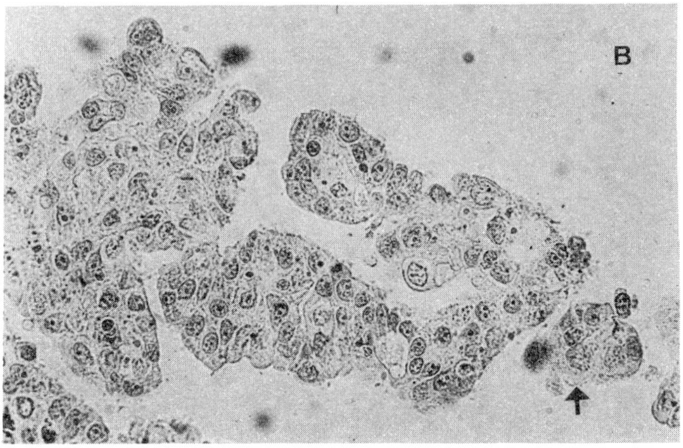

Fig. 2B A histological view of a MCF-7 breast cancer nodule showing cord-like structures and secondary nodule formation (arrow). No mucus secretion is present.

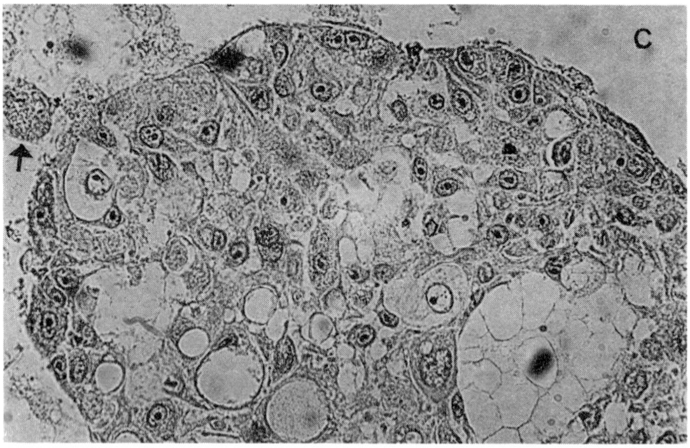

Fig. 2C A histological view of a OD 262 ovarian cancer nodule having a loose structure and an abundant mucus secretion. There is a tendency to form secondary nodules (arrow).

Table I

Comparison of growth percentage per day of A549, MCF-7 and OD262 nodules

	growth (% per day)
A549	6.86 ± 0.62
MCF-7	17.25 ± 2.86
OD262	13.21 ± 0.75

Table II

Comparison of growth percentage of A549 nodules during the different IFN treatments

IFN (U/ml)	Time (days) after treatment	Growth (%/day)	P
Control	0-65	5.62 ± 0.48	
α -2			
10	0-65	5.60 ± 0.48	N.S.
10^2	0-65	5.35 ± 0.35	N.S.
10^3	0-41	6.20 ± 0.41	N.S.
	41-65	0.66 ± 0.17	< 0.001
10^4	0-55	2.98 ± 0.31	< 0.001
	55-65	regression	
β			
10	0-65	5.21 ± 0.31	N.S.
10^2	0-65	5.57 ± 0.32	N.S.
10^3	0-65	5.09 ± 0.47	N.S.
10^4	0-65	4.95 ± 0.41	N.S.
γ			
10	0-65	5.61 ± 0.63	N.S.
10^2	0-65	5.92 ± 0.41	N.S.
10^3	0-65	3.73 ± 0.51	< 0.02
10^4	0-55	2.88 ± 0.27	< 0.001
	55-65	regression	

Table III

Comparison of growth percentage of A549 nodules in the absence of IFN after the different IFN treatments

IFN (U/ml)	Growth (%/day)	P
Control	6.06 ± 0.56	
α 2 10^2	5.06 ± 0.34	N.S.
α 2 10^4	4.16 ± 0.25	< 0.05
β 10^2	6.31 ± 0.79	N.S.
β 10^4	5.40 ± 0.32	N.S.
γ 10^2	5.78 ± 0.33	N.S.
γ 10^4	5.15 ± 0.31	N.S.

The treatment with one IFN

Growth patterns of IFN-α 2 treated nodules is represented in Fig. 3A and Table II. The controls had a growth rate of 5.62 $\pm$ 0.48 % per day. During a 10 and a 10^2 U/ml IFN-α 2 treatment, the growth rate was respectively 5.35 $\pm$ 0.35 % and 5.52 $\pm$ 0.49 % per day. A treatment with 10^3 U/ml of IFN-α 2 had no effect up to the 41st day : the nodules grew 6.20 $\pm$ 0.41 % per day. Later on, growth was inhibited 8.5 fold and nodules grew only 0.66 $\pm$ 0.17 per day. The nodules, however, kept healthy. A 10^4 U/ml dose of IFN-α 2 occasioned a 1.89 fold growth inhibition which was present from the begining of the treatment on ; nodules grew 2.98 $\pm$ 0.31 % per day. By the 55th day, necrosis appeared in the nodules and in 3 out of the treated 16 nodules, healthy alveolar structures remained present.

A similar effect was found with IFN-γ (Fig. 3B and Table II). No growth inhibition was observed during treatment with low doses. Growth rates of the 10 and 10^2 U/ml doses were respectively 5.61 $\pm$ 0.63 % and 5.92 $\pm$ 0.41 % per day. Growth was slowed down 1.51 times when the nodules were maintained in presence of the 10^3 U/ml dose and growth percentage was 3.73 $\pm$ 0.51 % per day. A greater effect was obtained with the 10^4 U/ml dose. A 2.30 fold growth inhibition was observed and all the nodules became highly necrotic from the 55th day of treatment on. The nodules tended then to desintegrate.

IFN-β had no effect on nodule growth. Daily growth rate for the 10, 10^2, 10^3 and 10^4 U/ml doses was respectively 5.21 $\pm$ 0.31, 5.57 $\pm$ 0.32, 5.09 $\pm$ 0.47 and 4.95 $\pm$ 0.41 % (Fig. 3C and Table II). The differences are not statistically significant for $p = 0.01$.

When studying the growth rates of the individual nodules, it became clear that some nodules grew faster than the fastest growing nodule of the untreated controls and this occured for the 3 IFNs used. Taking together all the treated nodules, 16 out of 192 (8 %) showed an enhance growth (from 1 % to 29 % stimulation) but a statistically significant difference with the fastest growing untreated control nodule was found in 2 nodules treated with 10^2 U/ml of IFN-α 2, in 1 nodule treated with 10^2 U/ml and 1 with 10^3 U/ml of IFN-β and in 3 nodules grown in presence of 10^2 U/ml dose of IFN-γ .

The IFN treatments were stopped when the nodules treated with highest concentrations became entirely necrotic. In order to verify if actually all the desintegrating nodules were dying, the filters containing the nodules were placed on top of a semi-solid medium to reduce the desintegrating effect of the wavelets of the liquid medium. In these culture conditions, all the necrotic nodules regenerated forming several extrusions of healthy tissue showing alveoli. The healthy regeneration zones of the IFN-α 2, -β and -γ treated nodules at the 10^2 and 10^4 U/ml doses were excised, placed on new filters and their growth was measured for another period of 70 days. The controls, which

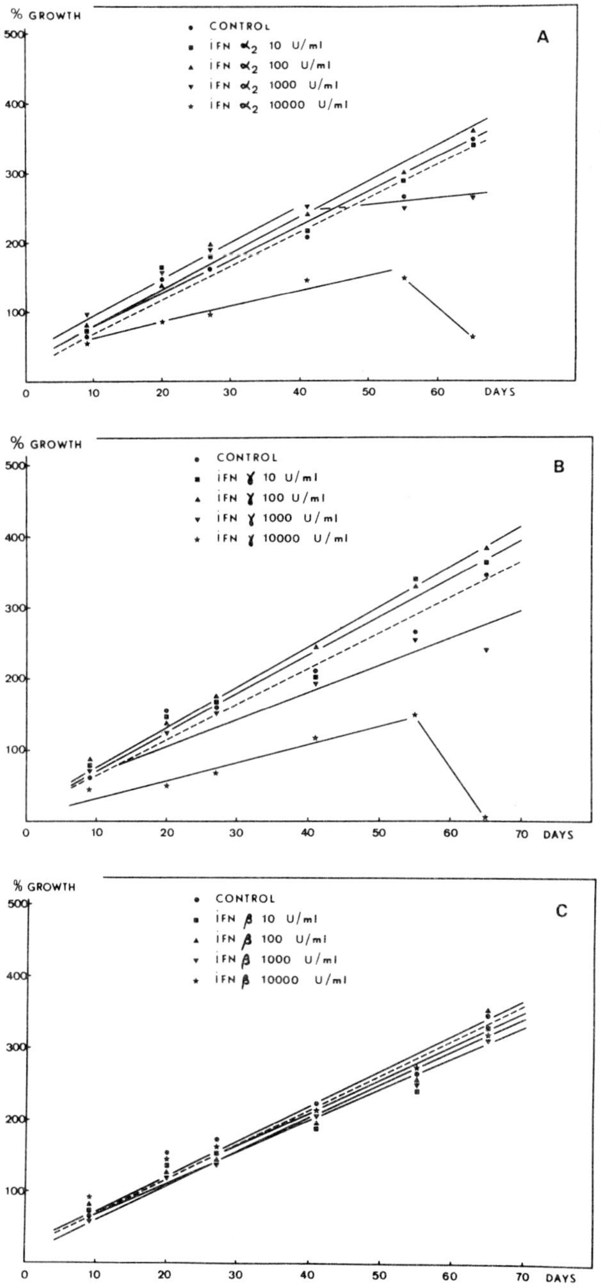

Fig. 3 Growth curves of A 549 nodules treated for 65 days with different doses of IFN-α_2(A), IFN-γ(B) and IFN-β(C).

were parts cutoff from the original untreated control nodules of the experimental series, grew 6.06 ± 0.56 per day. The 10^2 U/ml and 10^4 U/ml of IFN-α 2 treated nodules showed respectively a 1.23 and 1.43 fold growth inhibition with growth percentage of respectively 5.06 ± 0.34 and 4.26 ± 0.25 per day. Compared with the controls, growth rates of the IFN-β and -γ showed no significant difference. The growth percentage for the 10 and 10 U/ml doses of IFN-β were respectively 6.31 ± 0.79 and 5.40 ± 0.32 and for IFN-γ the values were 5.78 ± 0.32 and 5.15 ± 0.31 (Table III).

The treatments with mixtures of IFNs

To determine whether IFN-α 2 and -γ could interact in an additive or synergistic way in our tumor model, the IFNs were applied in combinations on A549 cells maintained in organotypic culture.

Figure 4A shows the effect of IFN-α 2 used at a concentration of 1,000 U/ml, alone or added with various concentrations of IFN-γ : 10, 100 and 1,000 U/ml. The controls had a growth rate of 6.8 ± 0.5 per day ; a treatment with 1,000 U/ml of IFN-α2 had no effect up to the 28th day with a growth of 6.6 0.8 %, afterwards the nodules grew only 0.34 ± 0.22 % per day, whereas treatment with the same dose of IFN-α2 plus 10, 100 and 1,000 U/ml IFN-γ resulted respectively in a growth rate of 6.42 ± 0.69, 2.95 ± 0.25 and 1.85 ± 0.6 % per day, thus demonstrating a dose-effect phenomenon as growth inhibition is well correlated with the IFN-γ concentrations. On the contrary, when used alone at doses of 10 and 100 U/ml, IFN-γ gave a growth rate of 9.92 ± 0.97 and 7.26 ± 0.67 % per day ; this indicates a stimulation growth effect of the low doses, effect specially marked with 10 U/ml (Table IV).

Moreover, the dramatic growth inhibition observed on the 35th day in the nodules treated with IFN-α 2 1,000 U/ml plus -γ 1,000 U/ml was accompanied by the complete disappearance of the healthy alveolar structure leading to a total necrosis.

Figure 4B shows the results obtained with IFN-α 2 or -γ used at a dose of 2,000 U/ml and compared to the combination using 1,000 U/ml of each IFN ; at day 21st, the growth rate per day was respectively of 6.70 ± 0.89 and 5.38 0.94 % for IFN-α 2 and -γ versus 1.96 ± 0.33 % for the nodules treated by the mixture, thus indicating an earlier effect of the combination ; between the 21st and the 45th day, the growth rate was no more than 0.54 ± 0.24 % for the IFN-α 2 and 1.03 ± 0.24 % for the IFN-γ with complete regression of the nodules treated by the mixture.

Finally, the effects obtained using 1,000 U/ml IFN-γ associated with 10 or 100 U/ml IFN-α 2 are reported in Fig. 4C, compared to those induced when IFNs were used separately at the same dose. In order to present a complete evaluation, 1,000 U/ml of IFN-α 2 were tested again when combined with 1,000 U/ml

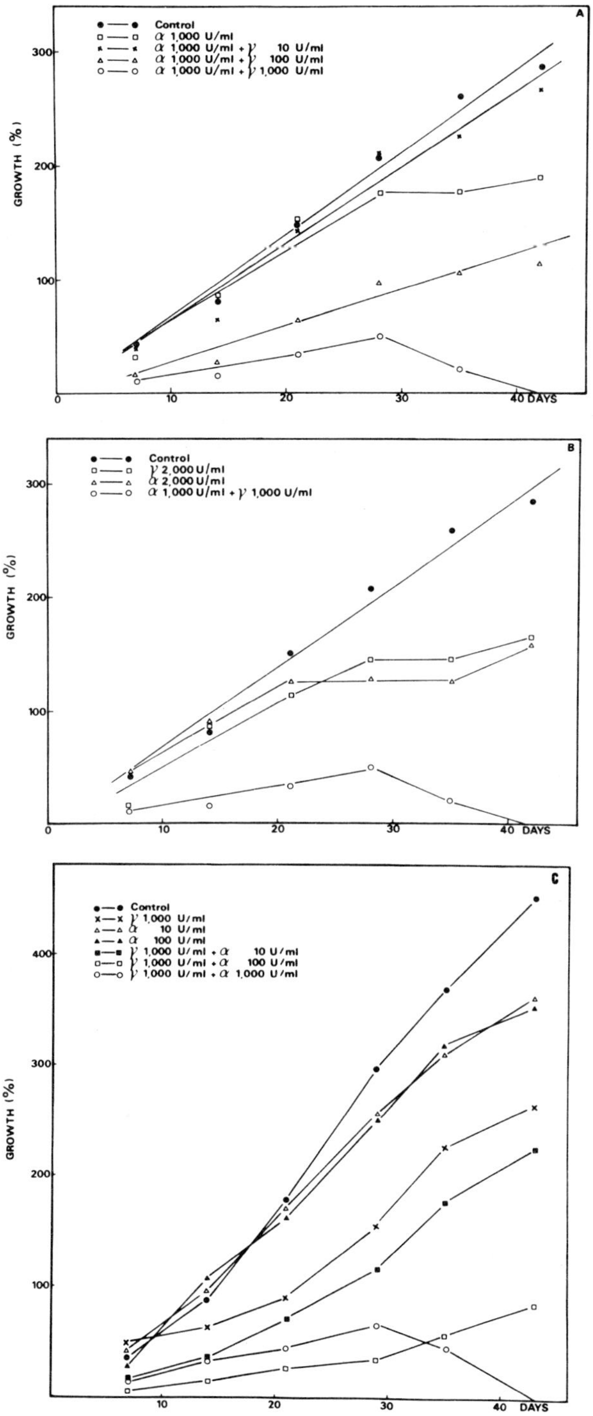

Fig. 4 Growth curves of A 549 nodules treated with IFN-α_2 and -γ used at different doses, separately or in combination.

of IFN-γ . In this experiment, the growth rate of the controls was of 10.05 ± 1.23 % per day ; no significant growth inhibition was observed in the nodules treated with 10 and 100 U/ml IFN-α 2 with a growth rate of respectively 8.48 ± 1.01 and 8.63 ± 0.77 % per day ; the growth percentage of IFN-γ treated nodules (1,000 U/ml) was of 5.55 ± 0.73 per day ; this percentage decreased to 5.23 ± 0.44 and 2.17 ± 0.28 when IFN-γ was combined with 10 and 100 U/ml IFN-α 2, the growth rate was of 2.45 ± 0.33 per day up to the 28th day, with regression between the 28th and the 45th day (Table IV).

The growth index was calculated for all series and the effects of the combinations were evaluated according to the definitions listed by Denz et al. (6). These evaluations were done from the data obtained on day 35, considering the state of necrosis of the nodules treated with 1,000 U/ml IFN-α 2 plus 1,000 U/ml IFN-γ . Additive or synergistic effects resulting from the use of mixtures of IFN-α 2 and -γ are reported in Table V.

The renewed treatments of regeneration nodules

Figure 5A shows the growth pattern of IFN-α 2 (10^4 U/ml) regeneration nodules. The controls (without a renewed treatment) grew 5.60 ± 2.08 % per day. The 10^2 U/ml and 10^3 U/ml treated nodules grew respectively 3.89 ± 1.16 and 3.49 ± 0.9 % per day, corresponding to a growth inhibition of 1.43 and 1.60. Mean growth rate of the IFN-β treated nodules was low : 3.49 ± 0.92 % per day showing a great dispersion in growth rates of the individual nodules (26 %). A renewed treatement with IFN-β 10^3 U/ml put the growth rate back to a higher level (5.34 ± 1.61 % per day). The same phenomenon was observed with IFN-γ (10^4 U/ml) treated regeneration nodules. Mean growth rate of the controls slowed down to 3.34 ± 0.70 % per day. Here again, a great diversity in growth rate of the nodules was observed. A 10^2 U/ml IFN-γ retreatment enhanced growth to 5.06 ± 1.42 % per day. A retreatment with 10^3 U/ml almost stopped growth (0.96 % per day) but it resumed at quite a high level from the 38th day on (4.78 ± 1.72).

DISCUSSION

In the present paper, we described a three dimensional tumor model for assessing the effects of antitumoral treatments. It has been shown that the histological aspects of the nodules of the three types of cancer are different. Specially, A549 alveolar II pulmonary tumor cells maintained typical characteristics of pneumocytes : the presence of multilamellar bodies. The organotypic method is not only interesting for the maintenance of the three di-

Table IV

Comparison of growth percentage of A549 nodules treated with IFN-α 2 and/or IFN-γ

IFN (U/ml)	Time (days) after treatment	Growth (%/day)	P
1. Fixed dose of IFN-α 2, variables doses of IFN-γ			
Control	0-45	6.8 ± 0.50	
α 2 (1,000)	0-28	6.6 ± 0.80	N.S.
	28-45	0.34 ± 0.22	< 0.001
α 2 (2,000)	0-21	6.70 ± 0.89	N.S.
	21-45	0.54 ± 0.15	< 0.001
γ (10)	0-45	9.92 ± 0.97	< 0.01
γ (100)	0-45	7.26 ± 0.67	N.S.
γ (1,000)	0-45	2.54 ± 0.49	< 0.001
γ (2,000)	0-21	5.38 ± 0.94	
	21-45	1.03 ± 0.24	< 0.001
α 2 (1,000) + γ (10)	0-45	6.42 ± 0.69	N.S.
α 2 (1,000) + γ (100)	0-45	2.95 ± 0.25	< 0.001
α 2 (1,000) + γ (1000)	0-28	2.47 ± 0.39	< 0.001
	28-45	regression	
2. Fixed dose of IFN-γ , variable doses of IFN-α 2			
Control	0-45	10.06 ± 1.23	
γ (1,000)	0-45	5.55 ± 0.73	< 0.01
α 2 (10)	0-45	8.49 ± 1.01	N.S.
α 2 (100)	0-45	8.63 ± 0.77	N.S.
γ (1,000) + α2 (10)	0-45	5.23 ± 0.44	< 0.001
γ (1,000) + α 2 (100)	0-45	2.17 ± 0.28	< 0.001
γ (1,000) + α 2 (1,000)	0-28	2.45 ± 0.33	< 0.001
	28-45	regression	

Table V

Potentiation of IFN antiproliferative activity by mixing IFN-α 2 and IFN-γ

IFN (U/ml)	Growth index*	Combination index°
No IFN	1	
α 2 (10)	0.84	
α 2 (100)	0.85	
α 2 (1,000)	0.75	
α 2 (2,000)	0.75	
γ (10)	1.33	
γ (100)	1.01	
γ (1,000)	0.55	
γ (2,000)	0.58	
α 2 (1,000) + γ (10)	0.86	Additive
α 2 (1,000) + γ (100)	0.53	Synergistic
α 2 (1,000) + γ (1,000)	0.15	Synergistic
γ (1,000) + α 2 (10)	0.48	Additive
γ (1,000) + α 2 (100)	0.15	Synergistic

$$* \text{ Ratio } \frac{\text{growth \% of IFN-treated cells}}{\text{growth \% of control cells}}$$

° Additive and synergistic effects of the combination were evaluated according to the definitions of Denz <u>et al</u>. (J. Interferon Res., 1985, **5**, 147-157).

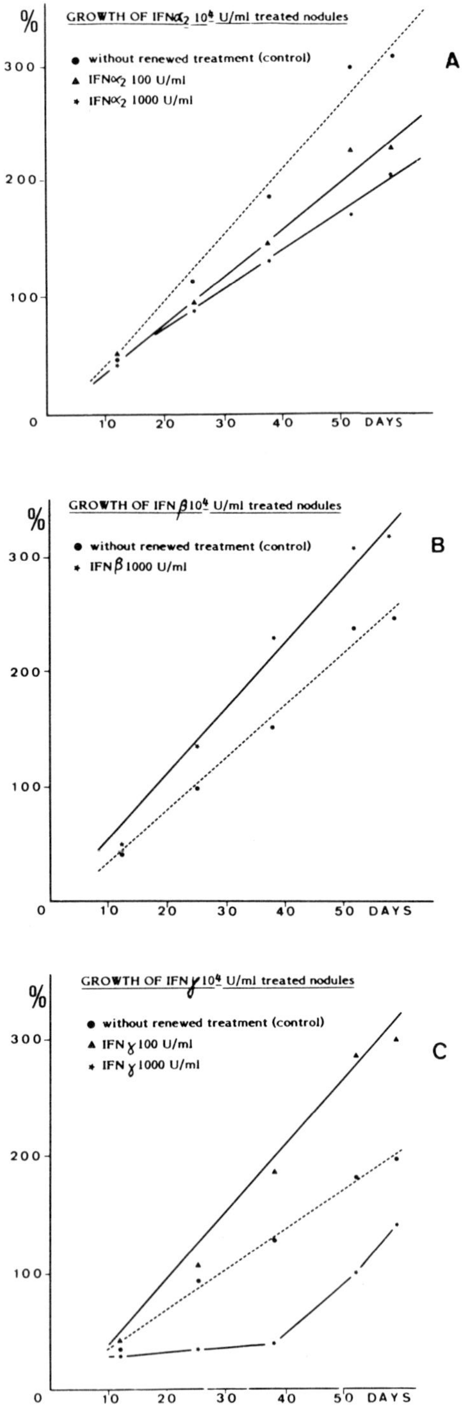

Fig. 5 Regrowth of IFN-α_2 (A), -β (B) or -γ (C) treated nodules after discontinuing the treatments with or without a renewed IFN-treatment.

mensional cell structure and of cell differentiation ; the possibility of keeping the nodules for long periods of time in low traumatizing culture conditions makes long term treatments possible and, this way, mimics the clinical long term treatments. This method enables also to study cells which resisted to the different antitumoral treatments.
Acquired great resistance or, on the contrary, enhanced sensitivity of these cells can easily be observed. This is demonstrated by the retreatment experiments (Fig. 5A,B,C). The IFN-α2 treated nodules appeared to be more sensitive to a renewed treatment than the untreated nodules. The IFN-β and -γ treated nodules reacted differently : here the growth stimulation phenomenon, already present at the initial treatments, was more clear-cut at the renewed treatments.

A549 nodules were particularly resistant to the three IFNs used. High doses (10^4 U/ml during 65 days) were necessary to induce cell necrosis which extended all over the nodules. However, these nodules regrew once the IFN treatment was discontinued. It is possible that a network of viable cells (undifferentiated or stem cells) remains and that their division is slowed down or blocked by the IFN treatment (17). Since no new differentiate cells are produced, the necrotic state of the nodule may be the result of the natural death of these cells. The fact that general nodule necrosis is developping only after about 50 days of treatment may support this hypothesis.

IFN-β was inactive and this may be related to the low level of IFN-γ receptors observed in A549 cells, which appeared to be lower than IFN-α2 rereceptors (12) in spite of the close relationship of IFN-α2 and IFN-γ binding sites (13,15).

Mixtures of IFN-α2 and -γ resulted in the potentiation of growth inhibition of A549 nodules. Such an association acts additively or synergistically on tumor growth inhibition according to the doses used. It should be emphasized that using 10^3 U/ml IFN-γ plus 10^2 U/ml IFN-α2 resulted in a greater growth inhibition than the combination of 10^3 U/ml IFN-α 2 plus 10^2 U/ml IFN-γ ; even, the growth inhibition was equivalent to that observed using 10^3 U/ml IFN-α 2 plus 10^3 U/ml IFN-γ. In the latter case, a 35 day treatment resulted in a complete growth inhibition and necrosis led to the disaggregation of the nodules on the 45th day. It should be noted that the combination of 10^3 U/ml IFN-α2 plus 10^3 U/ml IFN-γ was more effective than a 10^4 U/ml IFN-α2 or -γ treatment. Moreover, growth inhibition was greater and disaggregation of the nodules occured earlier.

The regeneration of tumor nodules after treatment with IFNs and the possibility to observe this clinical relevant phenomenon in the organotypic culture model emphasize the interest of the nodule model in the approach of human malignant pathology ; it provides a possibility for the setting up of the most efficient not aonly palliative but also curative therapeutic modalities.

ACKNOWLEDGEMENTS

We gratefully acknowledge Prof. E.T. WOLFF for his help and the Fondation Singer-Polignac for its generous grant.

REFERENCES

1. Beaupain, R., Baroche, C. and Lagarde D. (1983). Long term regeneration of cisplatinum and X-ray treated human tumor nodules in continuous organotypic culture. International Journal of Radiation Oncology, Biology and Physics, **9**, 707-712.

2. Beaupain, R. and Dionet, C. (1985). Effects of combined treatments of cis-diammine dichloroplatinum (II), 5-Fluorouracil and X-rays on growth of human cancer nodules maintained in continuous organotypic culture. Cancer Research, **45**, 3150-3154.

3. Beaupain, R., Billard, C. and Falcoff, E. (1986). Effects of human recombinant interferons -α 2, -β and -γ on growth and survival of human cancer nodules maintained in continuous organotypic culture. European Journal of Cancer and Clinical Oncology, **22**, 141-149.

4. Borden, E.C., Hogan, T.E. and Voelkel, J.C. (1982). Comparative antiproliferative activity in vitro of natural α- and γ-interferons for diploid and transformed stain cell assay. Journal of Interferon Research, **4**, 441-447.

5. Chevillard, S., Beaupain, R. and Coppey, J. (1985). Growth stimulation by misonidazole of lung carcinoma cells maintained in continuous organotypic and monolayer culture. Anticancer research, **5**, 241-248.

6. Denz, H., Lechleitner, M., Marth, C., Daxenbichler, G., Gastl, G. and Braunsteiner, H. (1985). Effect of human recombinant alpha-2 and gamma interferon on the growth of human cell lines from solid tumors and hematologic malignancies. Journal of Interferon Research, **5**, 147-157.

7. Gresser, I. (1977). On the varied biological effects of interferon. Cellular Immunology, **34**, 406-415.

8. Kirkwood, J.M. and Ernstoff, M.S. (1984). Interferons in the treatment of human cancer. Journal of Clinical Oncology, **2**, 336-352;

9. Lieber, M., Smith, B., Szakal, A., Nelson-Rees, W. and Todaro, G. (1976). Continuous tumor cell line from a human lung carcinoma with properties of type 2 alveolar epithelial cells. International Journal of Cancer, **17**, 62-70.

10. Ludwig, C.U., Durie, B.M., Salmon, S., and Moon, T. (1983). Tumor growth stimulation in vitro by interferons. European Journal of Cancer and Clinical Oncology, **19**, 1625-1632.

11. Martyré, M.C., Beaupain, R. and Falcoff, E. (1987). Potentiation of the antiproliferative activity by mixtures of human recombinant IFN-α 2 and -γ on growth of human cancer nodules maintained in continuous organotypic culture. European Journal of Cancer and Clinical Oncology, In press.

12. Merlin, G., Falcoff, E. and Aguet, M. (1985). ^{125}I-labelled human interferons alpha, beta and gamma : comparative receptor binding data. Journal of General Virology, **66**, 1149-1152.

13. Orchanski, P., Novick, D., Fischer, D.G. and Rubinstein, M. (1984). Type I and type II interferon receptors. Journal of Interferon Research, **4**, 274-282.

14. Quesada, J.R., Swanson, D.A., Trindade, A. and Gutterman, J.U. (1983). Renal cell carcinoma : antitumor effect of leukocyte interferon. Cancer Research, **43**, 940-947.

15. Sarkar, F.H. and Gupta, S.L. (1984). Receptor for human interferon : binding of ^{125}I-labeled recombinant human interferon to receptors of Wish cells. Proceedings of the National Academy of Science, USA, **81**, 5160-5164.

16. Soule, H.D., Vazquez, J., Long, A., Albert, S. and Brennan, M.A. (1973). A human cell line from a pleural effusion derived from a breast carcinoma. Journal of the National Cancer Institute, **51**, 1409-1412.

17. Stewart, W.E. and Blanchard, D.K. (1985). Modulation of immune functions

by interferons. In Immunity to Cancer, Academic Press New York, pp. 545-559.

18. Wietzerbin, J., Kolb, J.P., Senik, A. et al. (1984). Studies on purification of human gamma interferon : chromatographic behavior of accompanying IL-2 and B cell helper activity. Journal of Interferon Research, **4**, 141-152.

19. Wolff, E.T. and Wolff, E.M. (1975). Current research with organ culture of human tumors. In : Fogh, J. ed. Human Cancer Cells in vitro, Plenum Press London, pp. 521-571.

Discussion

ARATHOON Can you speculate as to why the synergistic effect of alpha and gamma (interferon end) was abolished when you put them in at the higher dose of 2000u

BEAUPAIN In the necrotic modules treated with 10,000 u/ml there are still living and viable cells and these viable cells are blocked in their cell dimension. There are no new differentiated cells produced. I have no explanation for the synergistic .

MEIGNIER Do you know the half-life of interferon in your culture conditions. As you change your medium three times a week does it mean that your nodules are exposed to an invariant level of interferon.

BEAUPAIN The level is stable. The interferon was still there two days after we added it. It was rather stable.

MEIGNIER Did you compare in vivo with in vitro treatment of the nodules.

BEAUPAIN No; it is possible to put the A549 nodules in nude mice but we have not yet done the experiment.

49 Immune interferon enhances the expression of MHC antigens in human placenta

B. Shomer, E. Gazit and V. Toder

Abstract .

The absence of the Major Histocompatibility Complex (MHC) antigens on villous trophoblast is postulated to be one of the main mechanisms enabeling the normal allografting of the embryo in-utero. The purpose of this study was to determine the effect of interferon (IFN) gamma on the expression of MHC antigens in villous trophoblast of human term placenta. Frozen sections of term placenta , fixed in cold acetone, were studied by indirect immunofluorescence , using monoclonal antibodies against class I antigens (W6/32- anti HLA A,B,C, and anti beta 2 microglobulin) and class II antigens (7.2- anti HLA D/DR). In untreated sections , syncytiotrophoblast was found inert, and moderate levels of expression were found on villous embryonic mesodermal core . In sections treated with IFN-gamma for 24-48 hr' , expression of class I antigens was enhanced in villous mesodermal core , but was not induced in syncytiotrophoblast . Expression of class II antigens was also enhanced in villous core , especially by cells morphologically identified as macrophages. Here too , syncytiotrophoblast remained inert . Enhancement of both classes of MHC antigens was also evident in basal decidua . These findings support our knowledge on IFN's capability of enhancing antigenic expression in different tissues , and show syncytiotrophoblast to be immunologically inert even after relatively high doses of IFN-gamma , what emphasizes our current concepts that the syncytiotrophoblast is an immunologically inert barrier between mother and conceptus.

Introduction.

The successfull grafting of allogeneic embryos in-utero remains a misterious phenomenon after two decades of intensive research. Since fetal syncytiotrophoblast (villous trophoblast) is practically the sole membrane directly bounding maternal blood and therefore directly bounding maternal immune factors, it may be concidered as constituting the immunological frontier between the fetus and it's mother. The lack of the Major Histocompatibility Complex (MHC) antigen expression upon syncytiotrophoblast is considered as one of the major mechanisms enabeling the normal fetal development (1). Indeed other placental tissues do express MHC antigens , mainly HLA A,B,C, and betta 2-microglobulin (beta 2-m) (3,5,6), but from the very first stages en-route delivery , no expression of MHC antigens can be detected on villous trophoblast (1). Late findings show that MHC gene expression is of the induced type (i.e. the gene is switched "off" or "on" according to the action of specific factors on inducing or suppressing elements associated with the gene). Various mediators were demonstrated as capable of causing suppression or induction / enhancement of MHC gene expression . The most potent mediator known , is immune (gamma)

Interferon (IFN) , which is reported as an inducer and an enhancer of MHC genes (2,7). Acting on the cell , IFN is a modifier, binding to a specific receptor, which activates several second messengers and causes MHC specific m-RNA transcription (7). In the present work , we bring evidence , that although IFN enhances and induces MHC expression in villous core tissue, it has no potency of inducing expression on villous trophoblast.

Materials and methods.

Tissues - Human term placentae were collected sterile from normal deliveries, and were kept at 4°C until proccessing . Human early (8-12 weeks) placentae were obtained from abortions of normal pregnancies , and were kept in cold Phosphate Buffered Saline (PBS -pH 7.4) until processing .
Interferon - Recombinant immune (gamma) Interferon was a generous gift of Dr. N, MOAV (Interyeda).
Antibodies - Antibodies used for this study were : W6/32 - Anti HLA A,B,C, shared determinants, Anti betta 2-microglobulin (Non heavy chain associated), 7.2- Anti HLA D/DR shared determinants, Fluorescein Isothiocyanate Conjugated Goat Anti Mouse (FITC-GAM- Tago inc. USA), and Mouse control ascites fluid (Bio-Makor , Israel).
Incubations - Incubations of thin slices of tissue were carried out in a complete culture medium (composed of RPMI-1640 medium containing 1% L-Glutamin , 1% Hepes buffer, Penicillin + Streptomycin 100 IU/ml each, 10% heat inactivated fetal calf serum and 0.1% mercaptoethanol), with or without 10^4 IU/ml of IFN , at 37°C in a 7% CO_2 incubator for 24 hours.
Indirect immunofluorescence - Tissue to be examined was snap-frozen in liquid Nitrogen, and sectioned in a cryostat (IEC-Demon, USA). Sections were fixed in cold aceton, and incubated with first antibody for 30 minutes at room temperature in a humid-camera, then washed three times in PBS . The second wash PBS contained 0.1% Agepon (Agfa Geveart - Germany) as a wetting agent. Incubation with FITC-GAM was then carried out for 30 min' in dark at 4°C . Sections were then washed, coverslipped by using 90% Glycerol in PBS, and examined on a LEITZ fluorescent microscope.

Results

A total of 5 term placentae were examined , placenta incubated without IFN , showed no expression at all (of any of the MHC antigens checked) on villous trophoblast , and moderate levels of all antigens in villous mesodermal core (table 1). in the same placenta incubated with 10^4 IU/ml of gamma IFN for 24 hours ,no induction of any of the antigens was observed on villous trophoblast. In villous mesodermal core, however, HLA A,B,C, and betta 2-m levels were higher , and class II level was very high as compared to tissue incubated without IFN. Amplification was normally expressed by means of fluorescence intensity , but in class II the total number of expressing cells was also larger in IFN treated tissue, what shows de-novo expression.
Early placenta is much less dense then term placenta, and intervillous tissue was normally seen. Morphological appearance of this tissue suggests of its being a maternal decidual contamination. A total of 6 early placentae were examined (table 2) . Here too, no expression of any of the MHC antigens was evident on villous trophoblast .Initial moderate levels of HLA A,B,C, and betta 2-m were observed, but they were somewhat lower then in term placenta. Following incubation with IFN , a significant intensification of the class I antigen levels was detected. Non of the early villous tissue expressed class II antigens, even after incubation with IFN . Intervillous tissue expressed initial low -to moderate levels of antigens , and enhancement of expression following IFN treatment was especially intence in class II.

Table 1 : MHC antigens expression on term placenta.

	villous trophoblast		villous core	
	---	IFN	---	IFN
HLA A,B,C,	--	--	+	++
betta 2-m	--	--	+	++
HLA D/DR	--	--	+	+++

Slices taken from the fetal side of term placenta were incubated with or without 10^4 IU/ml of IFN gamma . 24 hours later tissue was washed sectioned and labeled with monoclonal antibodies as mentioned in Matirials and Methods.
(--) no staining, (+) low density, (++) moderate density, (+++) high density.

Table 2. MHC antigen expression on early placenta.

	villous trophoblast		villous core		intervillous tissue	
	---	IFN	---	IFN	---	IFN
HLA A,B,C,	--	--	+/-	++	+/-	++
betta 2-m	--	--	+	++	+	++
HLA D/DR	--	--	--	--	+/++	+++

see footnote as for table 1.

Discussion .

Since villous placental tissue forms an interface between fetal and maternal immune systems, MHC antigen expression upon it is mostly interesting . Villous mesodermal core of term placenta appears to be positive for class I (HLA A,B,C, and betta 2-m) as well as for class II (HLA D/DR) histocompatibility antigens. Villous mesodermal core of early placenta ,on the other hand, appears negative for class II antigens. Being directly exposed to maternal blood throughout gestation , and thus playing a significant role in all feto-maternal interactions, the syncytiotrophoblast is characterized by it's apparent failure to express class I and class II MHC antigens in a form or density that can be identified immunohistologically (1). The reasons for failure of villous trophoblast to express MHC antigens in a normal manner are not known, although two basic possibilities are considered reasonable . A depletion or another alteration of the MHC gene might occur , or a down regulation of the above mentioned gene expression , by different mechanisms acting at the feto-maternal interface.

Exposure to IFN-gamma causes enhancement of HLA specific m-RNA , and cell surface expression of class I (11) and class II (10) antigens , as well as de-novo synthesis of these antigens by various cell lines (7). Based on these effects we examined the possibility of MHC gene depletion in syncytiotrophoblast. Data presented here clearly show that even large doses of immune IFN could not induce the expression of class I and class II antigens on syncytiotrophoblast surface. None of the monoclonal antibodies we have used could label IFN treated syncytiotrophoblast tissue , while significant enhancement of expression of class I antigens was evident in term as well as in early placental villous mesodermal core, and enhancement of class II antigens was observed in term placental villous core only . Generally ,the total expression of MHC antigens was higher in term placenta. Anyway , the failure of relatively high doses of immune IFN to activate MHC expression indicates a possibility of a certain alteration of this gene within syncytiotrophoblast , and a recent work has shown that HLA antigen expression by villous trophoblast is under transcriptional control, since this tissue fails to express detectable levels of class I antigen specific m-RNA (4). The exact nature of the MHC gene alteration in syncytiotrophoblast still have to be explored.

Conclusions .

The present work has shown that immune IFN is a potent enhancer of MHC antigen expression in placental tissue. Immune IFN did not induce expression on syncytiotrophoblast . Since we have not detected intracellular staining of MHC antigens in syncytiotrophoblast , it is not reasonable that disruption of expression is at the cellular transport level. IFN acts directly on the MHC gene therefore we suggest there is a certain alteration in MHC gene within syncytiotrophoblast.

Aknowledgements.

We are gratefull to Prof. L. NEBEL for his constant support , and to Dr. N. MOAV for her kind gift of gamma interferon.

This research was partly supported by the Fund for Basic Research administered by the Israeli Accademy of Sciences and Humanities, and M.& F. Ashkenasy Fund for Cancer Research.

REFERENCES.

1.) Faulk, W.P.,and McIntyre, J., (1983) Immunological studies on human trophoblast . markers , subsets and functions. Immunol. Rev. 75,139-175.

2.) Halloran, P.F., Autenried, P.,and Wadgymar, A., (1986) Regulation of HLA antigen expression in human kidney. Clinics in Immunol. and Allerg. 6,411-435.

3.) Johnson, P.M., (1984) Immunobiology of the human trophoblast. In: Immunological aspects of reproduction in mamals. Crighton P.B., Ed. Butterworths, London , 109-125.

4.) Kawata, M., Parues, J.K.,and Herzenberg, L.A., (1984) Transcriptional control of HLA A,B,C, antigens in human placental cytotrophoblast isolated using trophoblast and HLA specific monoclonal antibodies and the fluorescence-activated cell sorter. J. Exp. Med. 160,633-645.

5.) Loke, Y.W.,and Butterworth, H., (1987) Heterogeneity of human trophoblast populations . in: Immunoregulation and fetal survival. GILL T.J. III, Wegmann T.G., Eds. , Oxford Univers. press N.Y.

6.) Meziou, W., Chardon, P., Flechon, J.E., Kalil, J.,and Vaiman M., (1983) Expression of beta 2 microglobulin on preimplantation pig embryos. J. Reprod. Immunol. 5, 73-77.

7.) Rosa, F., Hatat, D., Abadie, A.,and Fellous M., (1985) Regulation of histocompatibility antigens by interferon Ann. Inst. Pasteur/Immunol. 136C, 103-112 .

8.) Simmons, R.L., (1969) Histoincompatibility and the survival of the fetus : current controversies. Transplant. Proc. 1, 47-56.

9.) Sunderland, C.A., Naiem, M., Mason, D.Y., Redman, C.W.G.,and Stirrat G.M., (1981) The expression of major histocompatibility antigens by human chorionic villy. J. Reprod. Immunol. 3, 323-332.

10.) Vizelizter, J.L., Perez, N., Areuzana-Seisdedos, F.,and Devos, R., (1984) Pure Interferon -gamma enhances class II antigens on human monocyte cell lines. Eur. J. Immunol. 14, 106-111 .

11.) Wallach, D., Fellous, M.,and Revel, M., (1982) Preferential effect of gamma interferon on the synthesis of HLA antigens and their m-RNAs in human cells . Nature 299, 833-835.

50 Gamma interferon and anti-thy-1 antibody induce specific killing activity in memory-like cytolytic hybridomas

T. Ozery and Y. Kaufmann

ABSTRACT

The T cell receptor (TcR) for antigen (Ag) is the cell surface molecular complex known to be involved in Ag-specific T cell activation. We have used memory-like cytotoxic T lymphocyte (CTL) hybridomas to examine whether CTL can be activated by alternative agents which interact with non-TcR surface molecules. Two inducers were found to be effective, an anti Thy-1 monoclonal antibody (mAb) G7 and gamma-interferon (IFN-γ). G7 mAb was a potent inducer of specific killing activity as well as IL-2 production, similar to the antigenic signal, suggesting that the Thy-1 activation pathyway is associated with elements of the TcR complex. IFN-γ effectively stimulated cytotoxicity without inducing IL-2 production indicating that the signal mechansim that mediate cytotoxicity is different from that trigger IL-2 production. Furthermore the results suggest that activation of the cytolytic program through the receptor to IFN-γ triggers a mechanism alternative to the TcR activation pathway.

INTRODUCTION

Cytotoxic T lymphocytes (CTL) are important in both cellular and humoral immune responses. They carry out direct effector functions such as cytolysis and mediate regulatory functions via secreted lymphokines. The physiological activation of CTL requires the triggering of the α/β heterodimer (Ti) of the T cell receptor after its recognition of antigen (Ag) (1). The Ti is part of the TcR molecular complex that contains also four distinct, noncovalently associated polypeptide chains of the T3 structure (2,3). The T3 tranduces the Ag signal from the Ti through the cell membrane (4). In addition to the T3-Ti complex, the T11 molecule (5) and the 44-KD polypeptide (6), identified on human T cells as well as Thy-1 (7,8), the TAP protein (9), and Ly-6 (10) described on murine T cells have been implicated in the T cell activation process since, mAb against all of these structures can be mitogenic. These cell surface molecules may function either to amplify or modify signals generated via the T3-Ti complex, or may serve as distinct alternative pathways for T cell activation. In normal T cells Ag-stimulation induces cell-surface receptors for IL-2 and cellular proliferation and maturation occurs as the result of binding of the secreted IL-2 to the newly induced receptors (11). In CTL clones maturation of cytolytic function occurs following Ag-induction of cell proliferation hence it is difficult to evaluate the contribution of the stimulating Ag to differentiation.

The use of memory-like CTL hybridomas (12) obviates this difficulty since the hybridomas are devoid of IL-2 receptors, do not respond to IL-2 (13), and activation of their cytolytic function occurs without induction of cell proliferation (14). We have previously shown that these CTL-hybridomas that had been derived from BALB/c anti EL4 ($H-2^d$ anti $H-2^b$) CTL, can be induced by Ag-presenting cells to lyse specific target cells (TC) as well as to secrete IL-2 (13,14). Hence the hybrid cells are reminiscent of memory CTL whose functional maturation is independent of IL-2-mediated cell proliferation.

We have examined here whether functional maturation of the hybridomas can be induced via non-TcR cell surface molecules such as Thy-1 and the receptor for gamma-interferon (IFN-γ). The activation pathways triggered in the hybridomas by these agents were compared to that triggered by the antigenic signal by following the induction of specific TC lysis as well as IL-2 secretion.

MATERIALS AND METHODS

CTL-hybridomas, tumor cells and CTLL-2 cells.

The generation, cloning and growth characteristics of the CTL hybridomas had been described (12,15). The hybrid cells and C57BL/6 thymoma EL4 cells were cultured in Dulbecco's modified Eagle's medium supplemented with 15% fetal calf serum (FCS) (DME/15 medium), antibiotics, Fungizone 250 μg/ml, 1mM Na-pyruvate, Tylocine 60 μg/ml (anti-PPLO reagent) and 2 mM L-glutamine, without supply of specific antigen or lymphokines. The IL-2 dependent cell line CTLL-2 was propagated in CTLL medium (RPMI 1640 containing 5% FCS and 50 μM 2-mercaptoethanol) supplemented with 30% (V/V) crude rat IL-2 (15).

Reagents

Mouse recombinant IFN-γ was obtained from Shionogi Research Laboratory, Osaka, Japan, and anti Thy-1 monoclonal antibody (mAb) G7 (ascitic fluid) from Dr. E. Shevach, NIH, Bethesda, Maryland, USA. Concanavalin A (Con A) was from Bio-Yeda, Rehovot, Israel. Neuraminidase from Behringwerke AG, Marburg, West Germany and Alpha methylmannoside (α-MM) from Sigma, St Louis, Mo. USA.

Stimulation of the hybridomas.

Hybrid cells were incubated with Con A, G7 mAb, IFN-γ or x-irradiated (4000 rads) EL4 cells for the indicated time in DME/15 medium containing 50μM 2-mercaptoethanol (activation medium).

Assays for proliferation, IL-2 and cytolytic activity.

Proliferation: hybrid cells, ($3x10^4$), were incubated in 0.2 ml activation medium with G7 mAb, Con A or x-irradiated EL4 cells, for 20 hr and harvested after a 4 hr terminal pulse with ^{3}H-thymidine. IL-2 was monitored by measuring the ability of conditioned media to support proliferation of the IL-2-dependent CTLL-2 cells; ($5x10^3$) CTLL-2 cells, in 0.1 ml CTLL medium were incubated with 0.1 ml aliquots of serially diluted hybridomas supernatants for 24 hr. CTLL-2 proliferation, monitored by ^{3}H-thymidine incorporation in triplicate cultures, was determined during the last 4 hr of incubation. Cytolytic activity of hybrid cells was assessed against the specific TC EL4. Hybrid cells ($1x10^4$) were incubated in triplicate with ($1x10^4$) neuraminidase-treated (12) ^{51}Cr-labelled EL4 TC in 0.15 ml standard medium containing RPMI, 10% FCS, in microtiter plates. The plates were centrifuged (200 x G, 5 min) and incubated 4 hr at 37 ^{o}C. Lytic assays were terminated by centrifugation (450 x G, 10 min) and 50% vol of the supernatants were removed and counted. Percent specific lysis was calculated as described previously (12).

RESULTS AND DISCUSSION

Induction of specific killing activity and IL-2 production by the anti-Thy-1 mAb G7.

The Thy-1 molecule is a 25000-30000 daltons cell-surface glycoprotein found on thymocytes, peripheral T cells, fibroblasts, epithelial cells and neurons in the mouse (16). Recently it has been shown that Ab. to some epitopes on the Thy-1 molecule are mitogenic for resting T cells and also induce lymphokine secretion from normal T cells and T cell hybridomas (7,8).

We asked whether such an anti-Thy-1 mAb is capable of inducing specific killing activity as well as IL-2 production in the memory-like CTL hybridomas. We have previously shown that these CTL hybridomas can be induced by $H-2D^b$ presenting cells or by a T cell mitogen (such as Con A or PHA) to lyse specific TC as well as to secrete IL-2 (Ref. 14,15 and fig. 1).

The effect of G7 anti-Thy-1 mAb (7) on the functional reactivities of several unrelated monoclonal hybridomas was analysed. The hybrid cells were incubated with either G7 Ab., with stimulating EL4 cells or Con A for 20 hrs. Following stimulation conditioned media were tested for IL-2 and the hybrid cells examined for their ability to lyse the specific ($H-2D^b$) TC EL4. The effect of the stimulants on hybrid-cell proliferation was examined by testing incorporation of ^3H thymidine into the hybrid cells during the last 4 hrs of stimulation. Fig. 1 shows that hybrid cells' proliferation was slightly blocked by G7 Ab. while killing activity was stimulated (the low levels of killing exhibited by non-stimulated hybridomas resulted from partial activation promoted by the EL4 TC during the lysis assay). IL-2 production was also induced by G7 mAb, however while low doses of G7 were sufficient to activate cytotoxicity, high doses were needed to trigger IL-2 secretion. The finding that induction of killing and IL-2 secretion occurs while cell proliferation is partially blocked, and that these cells do not express receptors for IL-2 (13), indicate that functional maturation of memory CTL is <u>in</u>dependent of IL-2-mediated cell proliferation. Activation of both functions, cytotoxicity and lymphokine production as well as the dependency of IL-2 induction on higher doses of stimulant, were also observed when activation was affected via the TcR complex by Ag or Con A (fig. 1). The similar pattern of hybridomas stimulation by Ag and G7 Ab. suggests that the activation

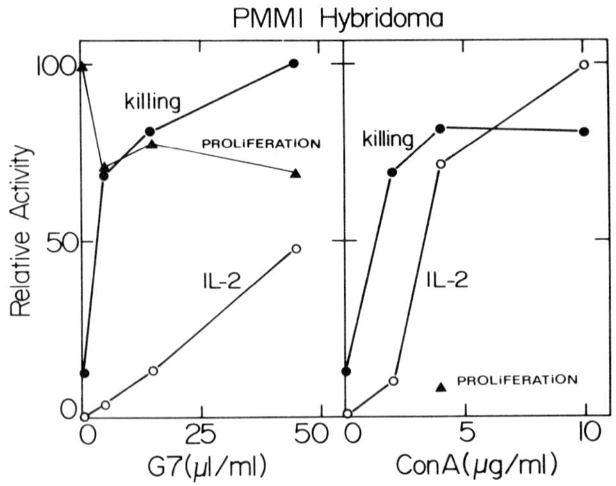

Figure 1: Effect of G7 mAb, Con A and Ag-presenting cells (EL4) on cytotoxicity and IL-2 secretion activity of the hybridomas.

PMM1 or $Md90^+$ hybrid cells were incubated for 20 hr at 37 °C with the indicated amounts of the stimulating agent: G7 mAb (ascitic fluid), Con A or irradiated EL4 cells. Conditioned media were assayed for IL-2 by testing their ability to promote proliferation of CTLL-2 cells. Maximal ^{3}H thymidine uptake by the test cells induced by 50% conditioned media from activated hybridomas, designated as 100 units was 102×10^3 cpm ($Md90^+$) and 37×10^3 cpm (PMM1). Hybrid cells were tested for killing activity after washing 3 times in standard medium. Con A was removed by incubating the hybrid cells in 0.1M αMM for 15 min at 37° and washing. Specific lysis of ^{51}Cr labelled EL4 TC was examined at an effector/TC (E:T) ratio of 1. Maximal lysis designated as 100 units was 18% ($Md90^+$) and 36% (PMM1). Spontaneous release of ^{51}Cr from TC was 10%. To measure proliferation, hybrid cells (3×10^4) were incubated for 20 hr in the presence of G7 mAb or Con A and harvested after 4 hr of terminal pulse with ^{3}H thymidine. ^{3}H thymidine uptake by $Md90^+$ and PMM1 hybridoma in absence of the inducing agent was 28×10^3 and 32×10^3 cpm respectively.

pathway triggered by Thy-1 molecules is associated with elements of the TcR complex-pathway. This conclusion is in agreement with the finding of Schmitt-Verhulst et al. that Ti variant of a CTL clone which expressed Thy-1, was unresponsive to Thy-1 mAb activation (17).

Immune interferon (IFN-γ) stimulates specific killing activity without inducing IL-2 production.

IFN-γ has an immunoregulatory role, it is produced by stimulated lymphocytes and influence a variety of cells including T, natural killers and B cells (Rev. in ref. 18). Because in normal lymphocytes heterogenous clones respond to IFN-γ it is difficult to directly determine the affected population. In IL-2-dependent T cell clones proliferation and differentiation are not readily separable. The use of the monoclonal CTL-hybridomas, whose activation occurs in absence of accessory cells and without induction of cell proliferation, allows direct analysis of the effect of IFN-γ on functional maturation of memory CTL. Fig. 2 shows that cytolytic activity of PMM1 and $Md90^+$ hybrid cells was stimulated by recombinant IFN-γ . Optimal induction was achieved by 40 u/ml. The efficiency of induction of killing activity in PMM1 and $Md90^+$ cells by IFN-γ , was comparable to that obtained with Ag-presenting cells (EL4) (fig. 2), while cytotoxicity of M8.23 hybrid cells was better stimualted by IFN-γ (Table I). However in contrast to Ag-mediated triggering which induced <u>both</u> cytotoxicity and IL-2 secretion IFN-γ did <u>not</u> induce secretion of IL-2 in any of the hybridomas. Higher doses of IFN-γ (up to 1000 u/ml, fig. 2) and longer incubation (up to 6 days, Table I) did not result in induction of IL-2 secretion. The fact that IFN-γ stimulates specific killing activity of the hybrid cells suggests that this biological agent directly triggers maturation of the

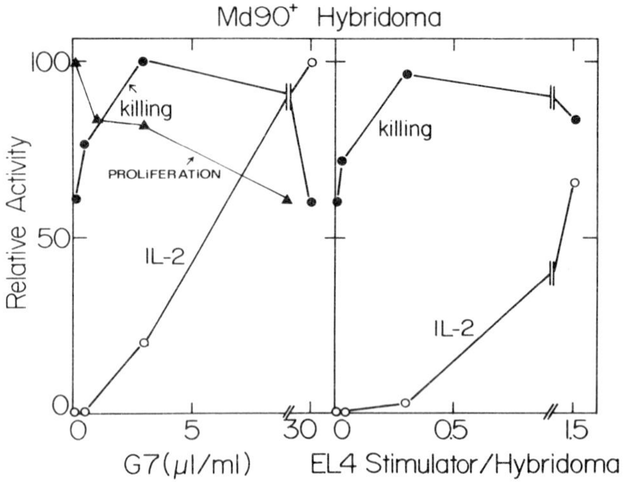

Figure 2: The effect of IFN-γ and antigen-presenting cells on stimulation of killing activity and IL-2 secretion by CTL hybridomas.

CTL hybridomas were incubated for 20 hr at 37 °C with IFN-γ or irradiated EL4 cells at hybridoma/EL4 ratio of 2. IL-2 was measured in the hybridomas' conditioned media as in fig. 1. Specific killing activity of freshly added EL4 TC was tested as in fig. 1. Spontaneous release of ^{51}Cr from TC was 10%.

Table I: IFN-γ does not induce IL-2 secretion in PMM1 and M8.23 hybridomas.

Hybridomas	Stimulation inducer	days	Specific lysis* % (±SD)	IL-2 secretion ^{3}H Thy uptake cpmX10^{-3} (±SD)
PMM1	-	-	12.5 ± 1.4	0.1 ± 0.0
	EL4#	1	47.3 ± 6.0	17.5 ± 6.0
	IFN-γ @	1	33.7 ± 3.7	0.1 ± 0.0
		3	23.5 ± 3.6	0.1 ± 0.0
		6	13.1 ± 1.6	0.0 ± 0.0
M8.23	-	-	3.9 ± 0.7	0.1 ± 0.0
	EL4#	1	12.7 ± 1.2	15.6 ± 1.0
	IFN-γ @	1	30.3 ± 1.2	1.1 ± 0.1
		4	15.6 ± 1.6	0.2 ± 0.0

* Killing assay was for 5 hr. Spontaneous release was 10%.

\# Stimulation with EL4 was at E/T of 2.

@ IFN-γ 200 u/ml was used.

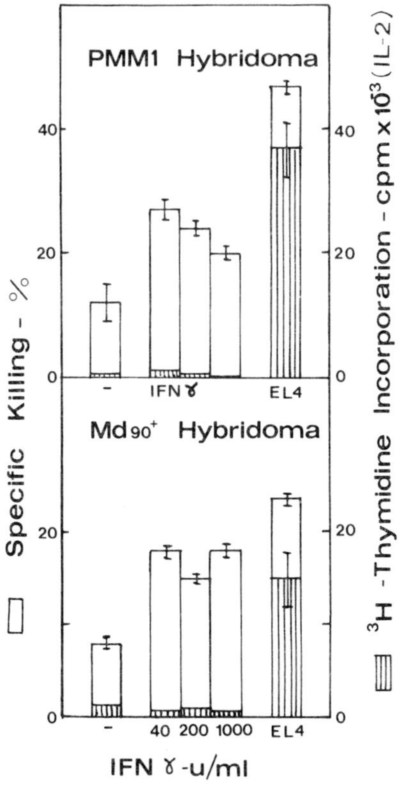

cytolytic function in memory CTL. That IL-2 production is not induced in spite of stimulation of killing activity, indicates that the signal mechanisms involved in CTL cytotoxicity and lymphokine production are not identical.

Taken together the data indicate that the induction of cytotoxicity in memory CTL can be mediated through three different cell surface molecules, the TcR complex, the Thy-1 molecule and the receptor for IFN-γ. In addition they suggest that Thy-1 signaling depends on the TcR activation pathway while IFN-γ seems to trigger the cells via an alternative pathway.

Acknowledgements

We thank Dr. E. Shevach for G7 Ab. and Dr. K. Sugita, Shionogy Research Lab. for IFN-γ. This work was supported by the fund for basic research administrated by the Israel Academy of Sciences and Humanities.

References

1. Kronenberg, M., Siu, G., Hood, E.L., and Shastri, N. (1986) The molecular genetics of the T cell antigen receptor and T-cell antigen recognition. Ann. Rev. Immunol.4, 529-591.
2. Samelson, L.E., Harford, J.B., and Klausner, R.D. (1985) Identification of the components of the murine T cell antigen receptor complex. Cell 43, 223-231.
3. Oettgen, H.C., Pettey, C.L., Maloy, W.L. and Terhorst, C. (1986) A T3-like protein complex associated with antigen receptor on murine T cells. Nature (London) 320, 272-275.
4. Weiss, A., Imboden, J., Shoback, D., and Stobo, J.D. (1984) Role of T3 surface molecules in human T cell activation; T3-dependent activation results in an increase in cytoplasmic free calcium. Proc. Natl. Acad. Sci. U.S.A 81, 4169-4173.
5. Meuer, S.C., Hussey, R.E., Fabbi, M., Fox, D., Acuto, O.,

Fitzgerald, K.A., Hodgdon, J.P., Protentis, J.P., Schlossman, S.F., and Reinherz, E.I. (1984) An alternative pathway of T-cell activation; a functional role for the 50Kd T11 sheep erythrocyte receptor protein. Cell 36,897-906.

6. Moretta, A., Pantaleo, G., Lopez-Botet, M., and Moretta, L. (1985) Involvement of T44 molecules in an antigen-independent pathway of T cell activation. Analysis of the correlations to the T cell antigen-receptor complex. J. Exp. Med. 162, 823-838.

7. Gunter, K.C., Malek, T.R., and Shevach, E.M. (1984) T-cell-activating properties of an anti-Thy-1 monoclonal antibody; possible analogy to OKT3/Leu 4. J. Exp. Med. 159, 716-730.

8. MacDonald, H.R., Bron, C., Rousseaux, M., Horvath, C., and Cerottini, J.C. (1985) Production and characterization of monoclonal anti-Thy-1 antibodies that stimulate lymphokine production by cytolytic T cell clones. Eur. J. Immunol. 15, 495-501.

9. Rock, K.L., Yeh, E.T.H., Gramm, C.F., Haber, S.I., Reiser, H. and Benacerraf, B. (1986) TAP, a novel T cell-activating protein involved in the stimulation of MHC restricted T lymphocytes. J. Exp. Med. 163, 315-333.

10. Malek, T.R., Ortega, G., Chan, C., Kroczek, R.A., and Shevach, E.M. (1986) Role of Ly-6 in lymphocyte activation. Induction of T cell activation by monoclonal anti-ly-6 antibodies. J. Exp. Med. 164, 709-722.

11. Meuer, S.C., Hussey, R.E., Cantrell, D.A., Hodgdon, J.C., Schlossman, S.F., Smith, K.A., and Reinherz, E.I. (1984) Triggering of the T3-Ti antigen receptor complex results in clonal T-cell proliferation through an interleukin-2 dependent autocrine pathway. Proc. Natl. Acad. Sci. U.S.A. 81, 1509-1513.

12. Kaufmann, Y., Berke, G., and Eshhar, Z. (1981) Cytotoxic T-lymphocyte hybridomas that mediate specific tumor-cell lysis in vitro. Proc. Natl. Acad. Sci. U.S.A. 73, 2502-2506.

13. Kaufmann, Y., Chang, A.E., Robb, R.J., and Rosenberg, S.A. (1984) Mechanism of action of cyclosporin A on antigen-stimulated T-cell hybridomas. J. Immunol. 133, 3107-3111.

14. Kaufmann, Y., Moscovich, M., Robb, R.J., Rosenberg, S.A., and Berke, G. (1985) Antigen/mitogen induced cytolytic activity and IL-2 secretion in memory-like CTL hybridomas. In Adv. Expl. Biol. Med. pp. 535-550. Plenum Press, New York.

15. Kaufmann, Y., and Berke, G. (1983) Monoclonal cytotoxic T lymphocyte hybridomas capable of specific killing activity antigenic responsiveness and inducible interleukin(s) secretion. J. Immunol. 131, 50-56.

16. Reif, A., and Allen, J.M.V. (1964) The AKR thymic antigen and its distribution in leukemias and nervous tissues. J. Exp. Med. 120, 413-433.

17. Schmitt-Verhulst, A.M., Guimezanes, A., Boyer, C., Poenie, M., Tsien, R., Buferene, M., Hua, C., and Leserman, L. (1987) Pleiotropic loss of activation pathways in a T cell receptor α-chain deletion variant of a cytolytic T-cell clone. Nature (London) 325, 628-631.

18. Trinchieri, G., and Perussia, B. (1985) Immune interferon; a pleiotropic lymphokine with multiple effects. Immunology Today 6, 131-135.

51 Development of a device and a double antibody microradioimmune assay for fast screening and sensitive quantitation of monoclonal antibodies

N. Epstein, G. Shavit and Z. Oril

ABSTRACT

Radioimmune assays (RIAs) are specific and quantitative assays. They find a wide use in screening cell cultures for production of relevant monoclonal antibodies (McAbs) where assays should be accurate, reproducible and rapid. The most common RIAs are of the solid phase type where antigens (Ags) in solution stick nonspecifically to the plastic. However, in cases where the Ag is expensive or not available in large quantities for adsorption purposes, it can be radiolabelled and bound to antibody (Ab) in solution. The complex Ab-Ag is precipitated with ammonium sulphate or polyethylene glycol. Traditionally the reaction is carried out in small plastic centrifuge tubes. This procedure is time consuming and is limited in the number of tests which can be handled at once. For the purpose of screening hundreds of hybridomas supernatants we have developed a microassay which is performed in wells of microtiter plates which are cut off mechanically and transferred to γ-counting tubes. A device was built for cutting out the individual cups of the flexible thin-walled microwelled plate, which is superpositioned on a complementary rigid microwell plate. The device consists of an electrically heated wire, which is passed at a, predetermined height, slightly below the upper edge of the flexible microwell plate. The cutting operation takes a few seconds and results in 96 individual cups, resting in the respective wells of a standard rigid microwell plate. The cups are then easily removed by pincers for radioactivity evaluation.

INTRODUCTION

RIAs are accurate, sensitive and rapid assays for Ab or Ag quantitation (1). RIAs have numerous applications, among others in screening cell cultures for production of (McAbs).

The most common RIAs are of the solid phase type (SP-RIA) where Ags (proteins and other macromolecules) in solution are non-specifically adsorbed to the microtiter plastic wells. The Ab sample to be determined is reacted with the adsorbed Ag followed by a reaction with a radiolabelled Ab (second Ab) to the tested Ab. The amount of Ab bound to a fixed quantity of Ag is determined by the amount of bound radioactivity of the second Ab after removal of unbound radioactive material.

In order to determine the amount of bound radioactivity, the 96 wells of the microtiter plate are cut off manually with scissors, and dropped into carrier tubes which are suitable for the gamma counter.

However, in cases where the Ag is expensive and not available in large quantities for adsorption purposes, it can be radiolabelled and bound to its Ab in solution. The Ab-Ag complex is reacted with a second Ab followed by precipitation with ammonium sulphate or polyethylene glycol (PEG). In conventional immunoprecipitation assays (IP-RIA) precipitates of immuno-complexes are generally prepared in small vials. Each vial is individually spun in a mini centrifuge, excess of Ag is aspirated and bound Ag in complex is counted for γ-radioactivity in each vial. This procedure is time consuming and limited in the number of samples which can be handled simultaneously (by the number of vial slots in the centrifuge, usually 12).

For the purpose of screening hundreds of hybridoma supernatants, and in order to overcome the above mentioned problems, we have developed a microassay performed in wells of a 96 well microtiter plates. Severing the wells of the soft plastic plates cannot be accomplished simply by cutting the microplate with scissors, because the Ab-Ag precipitate is likely to splash (since it is not adsorbed to the plastic), thereby contaminating the environment.

In order to facilitate the cutting of the individual cups of the flexible microtiter plate, we have designed a simple instrument, termed a microtiter plate cutter. This device can be applied in the above described assay systems.

1. SP-RIA where cutting out the 96 individual wells manually with scissors is time consuming, cumbersome and inconvenient.
2. Micro IP-RIA, which is otherwise prohibitive.

<u>Description of the microtiter plate cutter and mode of operation</u>

Microtiter plates are commercially available in standard sizes, 96 (8 x 12) individual wells. Such plates are either made of a rather rigid plastic, (as polystyrene) or of a thin flexible plastic, (as polyvinyl chloride (PVC)). The rim of the PVC is cut off with scissors and the rimless plate is superimposed on a rigid polystyrene plate and reasonably tightened (Figs. 1 and 2). The immunoprecipitation takes place in the superimposed flexible PVC plate. At the end of the incubation series, the plates are centrifuged for 20 minutes at 3000 rpm at 4°C. Supernatants are discarded from the wells by shaking off into a waste container followed by hitting the plates onto absorbent material on the workbench to remove any additional unbound material, until completely dry. The sandwiched plates are placed on the aluminium platform (1/4" thick) of the microtiter plate cutter. The instrument (Figs. 3 and 4; specifications - Table I), consists of an electrically heated wire fixed at a predetermined height, slightly below the upper edge of the flexible microwell plate, the cups of which are to be cut. During cutting, the microwell plates are moved towards and through the heated wire as demonstrated in Fig. 3.

The cutting operation is carried out within a few seconds and results in 96 individual cups resting in the respective wells of the rigid microwell plate (Fig. 5). The cups are then easily removed by pincers to gamma vials and counted in a gamma counter. As four microplates can be centrifuged simultaneously, one experiment results in up to 384 reaction cups.

<u>An example of micro-RIA application for screening of McAbs to the tumor marker carcinoembryonic antigen (CEA).</u>

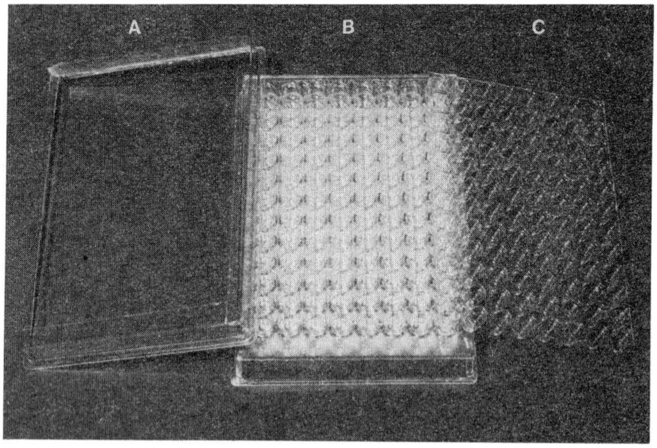

Fig. 1.: A PVC plate superpositioned on a polystyrene plate. (A) The removed rim. (B) Top view of the rimless PVC plate superpositioned on a polystyrene plate. (C) The rimless PVC plate.

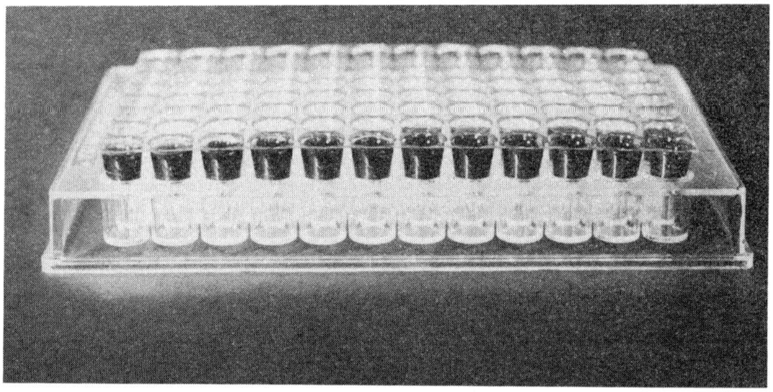

Fig. 2.: Side view of the upper flexible microtiter plate superimposed on the lower rigid plate.

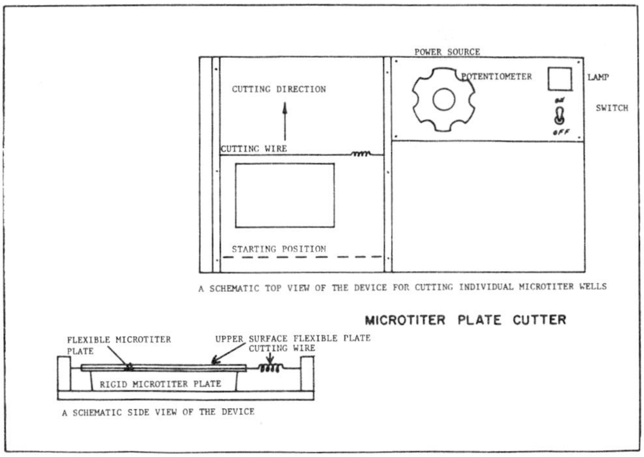

Fig. 3.: Schematic top (A) and side views (B) of the microtiter plate cutter.

Fig. 4.: The microplate cutter in operation.

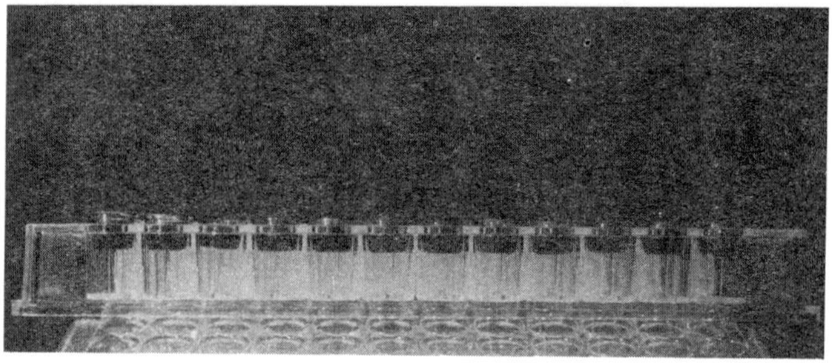

Fig. 5.: Side view of a microwell plate supporting cut off cups.

Initially the test used for monitoring McAbs to CEA was as follows:

Macro-IP in conical tubes. Mouse McAbs (200 μl) in 0.01M Na-phosphate buffer pH 8.0, containing 0.15M NaCl, 0.5% bovine serum albumin (BSA) and 0.01% NaN_3 were mixed in plastic conical tubes (Eppendorf), with 100 μl containing 50.000 cpm of ^{125}I-CEA at 37°C (specific activity of 30-60 μCi/μg). Each tube was mixed well by vortex. After one hour of incubation, the following reagents were added: 100 μl of 0.1M EDTA (adjusted to pH 8.0), 100 μl of 2% v/v normal mouse serum (carrier) and 100 μl of 1 mg/ml goat anti-mouse immunoglobulin (Lot No. G406 7 mg/ml Bio-Yeda), diluted in 0.01M Na-phosphate buffer pH 8.0 containing 0.15M NaCl 0.5% BSA, and 0.01% NaN_3. Each tube was well mixed by vortex and mixtures were incubated at 37°C. After one hour mixtures were further incubated with 700 μl of 25% PEG MW 6000 (Merck) in Na-phosphate buffer 0.1% BSA at 4°C on ice. After 30 minutes, tubes were centrifuged in Eppendorf centrifuge 5414 at 4°C for 15 minutes. Supernatants were carefully aspirated off, and the cover of each Eppendorf vial was cut to fit the gamma vials and radio activity was evaluated in a gamma counter (Gammamatic, Kontron).

Since the macro IP RIA is a time consuming, tedious and inefficient procedure, the assay was modified to the following micro-RIA (as was detailed in Ref. 2).

Microassay in microtiter plates

Diluted antibodies (40 μl) were incubated with 10 μl 12000 cpm of ^{125}I-CEA (under the same conditions and buffers as in the macroassay), in a flexible rimless microwell plate, superpositioned on a similar rigid microwell plate. The sandwiched plate was shaken in a micro-shaker (AM 69, Cooke) device. After one hour, 20 μl EDTA, 20 μl goat anti-mouse, and 20 μl normal mouse serum (all at concentrations and conditions as in the macroassay) were aded. One hour later, the mixtures were incubated with 80 μl, 25% PEG on ice. Plates were centrifuged for 20 minutes at 3000 rpm using the International centrifuge, and supernatants were thoroughly removed. The upper surface of the

Table I.

Specifications of microtiter plate cutter

Transformer	220V/9V AC (2 ampere)
Voltage	9V AC
Current	1.5 A
Resistence wire	Length 145 mm Diameter 0.4 mm resistance 11.2 Ω/meter
Alternating potentio-meter	0 - 100%

Table II.

Comparison between macro and micro RIAs

Macroassay	Microassay
Time consuming	Fast
Tedious	Convenient
Inefficient (Simultaneous handling of 12 reaction vials)	Efficient (Simultaneous handling of 384 wells)
Expensive (Reaction volume 1300 μl)	Economical (Reaction volume 190 μl)

flexible plate was removed by the heating wire resulting in 96 severed wells. The cut off cups were removed from the rigid plates by pincers and were placed in gamma vials and counted in the gamma counter. Results obtained in both methods were identical.

In summary, the advantages of the micro IP-RIA (facilitated by the use of the microtiter plate cutter) over the macro RIA are summarized in Table II.

ACKNOWLEDGEMENTS

The authors would like to thank Ms. Nita Ben-David for her assistance in preparing this manuscript and to Ms. Lilly Ashani for her photographic skills.

REFERENCES

1. Farr, RS: A quantitative immunochemical measure by the primary interaction between ^{125}I-BSA and Ab. J. Infect. Dis. 103: 239, 1958.

2. Epstein, N., Zandani, P., Inbar, I., Halperin, B., Cohen, S., Reuveny, S. and Ariel, N. Development of monoclonal antibodies directed against carcinoembryonic antigen (CEA) and their application for CEA detection. Develop. biol. Standard. 1987 (In press).
(A note to the Publisher: This paper was presented at this meeting, please add the relevant volume and page numbers).

52 Development of monoclonal antibodies directed against carcinoembryonic antigen (CEA) and their application for CEA detection

N. Epstein, R. Zandani, I. Inbar, B. Halperin, S. Cohen, S. Reuveny and N. Ariel

ABSTRACT

Six monoclonal anti CEA antibodies were developed. Two McAbs were derived against CEA purified from human metastases to the liver of primary adenocarcinoma. These McAbs were purified from ascites fluids using a protein A column and were characterized as IgG_2b. Four McAbs were derived from CEA purified from a human colon adenocarcinoma cell line. These McAbs were isolated from ascites fluids using DEAE-sephacel and were characterized as IgG_1. The dissociation constant of these McAbs averaged around 10^{-9}M. The purified immunoglobulin (Ig) fractions were conjugated to horseraddish peroxidase and applied in a two-site immunoassay for the detection of both metastatic and cell line derived CEA. An optimal Ab combination was found, facilitating detection of CEA at 3-60 ng CEA/ml. Their potential use in testing clinical specimens will be evaluated. Two of the McAbs were found to differentiate between malignant (colon carcinoma) and normal tissue in immunohistopathological staining of fresh frozen tissues. The same two McAbs, when tested as tagging Abs in a clinical assay, were found to give dose response curve identical to those obtained using the commercial reagent.

INTRODUCTION

Carcinoembryonic antigen (CEA) is a tumor associated glycoprotein, first described by Gold and Freedman (1) as a marker for colonic cancer. Although originally thought to be specific for human colon adenocarcinoma, it is now well known that CEA is associated with a range of other carcinomas (2) and various nonmalignant inflammatory diseases (3). CEA is a glycoprotein with an apparent molecular weight of 200000 daltons, of which about 40-60% are carbohydrate moieties. It is an important clinical marker of malignant transformation. In current clinical practice, CEA serum levels are used as criteria for prognosis of several malignancies (4), for monitoring the success of surgery or chemotherapy and for early detection of recurrencies (5). In the absence of any known function of the CEA molecule, immunoreactivity, the best defined property of the molecule, is used for its determination.

There are several areas in which antibodies (either polyclonal or monoclonal) to CEA are being used in the management of cancerous patients. These include: serum assays to monitor tumor burden (4,5); immunohistopathological analyses of tissue samples (6) (to detect or further characterize tumor cells) and in situ detection of carcinomatous lesions (at primary sites, regional lymph nodes or distal sites) using radiolabelled antibody (7). Moreover, the potential use of antibodies to CEA as therapeutical reagents is currently being contemplated (8,9).

A major drawback of using polyclonal anti-CEA antibodies for the above mentioned purposes has been their cross reactivity with various normal adult tissues such as spleen, lung colon, polymorphonuclear leukocytes, lymphocytes and red blood cells (10). Thus CEA was defined as a family of isoantigens which have multiple antigenic determinants, some of which may be common to all isoantigens, whereas others may be antigenically distinct. Therefore, monoclonal antibodies with restricted CEA specificity have the potential of greatly improving the reliability, specificity and reproducibility of the above mentioned applications. In this paper we describe the production and characterization of monoclonal anti-CEA antibodies. The potential use of these antibodies for CEA purification, CEA monitoring and immunohistochemical studies is demonstrated.

MATERIALS AND METHODS

Immunization schedules

Two groups of 6 to 8 weeks old female BALB/C mice (Charles River, U.K.) were immunized with the following antigens:

Group A - with purified CEA of human hepatic metastases of pri--mary adenocarcinoma of the colon (11) (Table 1, Schedule A).

Group B - with CEA (crude or pure) isolated from a human CEA secreting cell line (HuCCl) (Table 1, Schedule B) (12,13).

Myeloma culture

Myeloma cells (x63 NSI/1-Ag-4 - NSI), originated in BALB/C mice, were grown in DMEM with high glucose (4.5 gr/liter - Biological Industries, Beit Haemek) and with 2 mM glutamine (Biological Industries), nonessential amino acids 0.1 mM (Gibco), penicillin 100 U/ml, streptomycin 100 μg/ml and 15-20% inactivated bovine fetal serum (= complete DMEM). Cells were occasionally cultured in a medium containing 8 azaguanine (30 μg/ml), to eliminate growth of the HAT resistant myeloma cells.

Hybridization and cell cultivation

Hybridization was performed according to modification of the procedure described by Köhler and Milstein (14) and Galfre et al. (15).

In brief, spleen cells of the mice exhibiting the highest antibody titer were fused with the NSI myeloma cells (at a ratio of 1:5-10), by exposure of the mixed cells to 35% polyethyleneglycol for 5 minutes (PEG, 1000 daltons - Serva). The mixture of the cells was dispersed in a selective medium comprised of super-DMEM and HAT solution, and distributed into microculture plates (96 wells) at a concentration of 5×10^5 cells/ml, using normal mouse spleen cells (at a concentration of 5×10^4 cells/ml) as feeder cells.

Super-DMEM: complete DMEM, as described for myeloma culture, supplemented with 10% NCTC (Whittaker Bioproducts), 1 mM oxaloacetic acid (Sigma), 0.2 U/ml insulin (Sigma), 1 mM sodium pyruvate (Biological Industries). Super-DMEM/HAT: Super-DMEM containing 10^{-4}M hypoxanthine (Sigma), 4 x 10^{-7} M aminopterin, 1.6 x 10^{-5}M thymidine. The plates were maintained at 37°C in an humidified incubator with 7% CO_2 in air, and growth medium was changed twice a week.

Growing hybridomas were screened for secretion of anti-CEA activities, as described below. Positive colonies were cloned, expanded and injected into mice, to produce ascitic fluids. Cells (at a concentration of 5 x 10^6 cells/ml) were frozen at -180°C in liquid nitrogen, until further treatment.

Hybridoma growth in mice - McAbs isolation

BALB/C mice were injected intraperitoneally with 0.5 ml pristane (2,6,10,14-tetramethylpentadecane), an immunosuppressive substance, at least 7 days prior to an anticipated injection of hybridoma cells. Hybridomas (10^6-10^7 cells per mouse) were then injected into BALB/C mice intraperitoneally, where an antibody-rich ascites fluid develops. Seven to fourteen days after injecting the cells, the mice exhibit substantial abdominal swelling, and the ascitic fluid is removed into test tubes. Subsequently, the ascites are allowed to clot and are clarified from the fibrin, by filtration over glass wool.

Double antibody (Ab) RIA for anti-CEA antibody quantitation

For the purpose of screening supernatants of growing cell lines from microtiter wells after fusion procedures, we have modified the double antibody RIA previously described (13) as follows 40 µl diluted antibodies in TST buffer (Tris-HCl 0.05 M pH 7.6, NaCl 0.15 M, Tween-20, 0.05%) + 0.1% BSA or in 0.01M Na-phosphate buffer pH 8.0 containing 0.15 M NaCl + 0.5% BSA were incubated with 10 µl of 12000 cpm of ^{125}I-CEA at 37°C (specific activity of 30-60 µCi/µg, diluted in 0.01 M Na-phosphate buffer pH 8.0 containing 0.15 M NaCl + 0.5% BSA), in a flexible thin-walled microwell plate, the rim of which was cut off. The plate was superpositioned on a similar rigid microwell plate. One hour later, the following reagents were added: 20 µl of 1 mg/ml goat anti-mouse Ig (Lot No. G406 7 mg/ml - Bio-Yeda) diluted in 0.01 M Na-phosphate buffer pH 8.0, containing 0.5 M NaCl 0.5% BSA and 0.01% NaN_3; 20 µl of 2% normal serum (carrier, diluted in the same buffer); and 20 µl of 0.1 M EDTA in H_2O (adjusted to pH 8.0). One hour later, the mixtures were incubated with 80 µl of 25% polyethylene glycol (PEG) MW 6000 (Merck) in Na-phosphate buffer containing 0.1% BSA, at 4°C on ice. Plates were centrifuged for 20 minutes at 3000 rpm using the International centrifuge and supernatants were thoroughly removed. The upper surface of the flexible plate was removed by the heating wire of the microtiter plate cutter (16) resulting in 96 severed wells. The wells were counted in a gamma-counter (Gammamatic, Kontron).

Iodination of CEA

CEA (cell line) was labelled using a modification of the Greenwood-Hunter-Glover method (13,17).

The specific radioactivity of the CEA was determined before separation on the Sephadex column, by precipitation of the CEA (using 5% trichloroacetic acid) and determination of the fraction of radioactivity in the protein precipitate, ranging between 30-60 µCi/µg

Determination of the antibody affinity constant

Affinity constants of the McAbs were calculated from binding assays using the Scatchard equation (18). For the affinity measurements, a competition test was carried out in 60 µl reaction mixture, as follows: A constant amount of 40 µl diluted McAb (exhibiting 50% or less of binding radio labelled CEA in titration curves) was incubated with 50000 cpm of ^{125}I-CEA (in 10 µl), in the presence of increasing amounts of 10 µl unlabelled CEA, from 0.5 ng-1000 ng/well). The reaction was followed by the regular double Ab RIA procedure as described above. All dilutions were done in 0.01 M Na-phosphate buffer pH 8.0 containing 0.15 M NaCl, 0.5%/BSA and 0.01% NaN_3.

Purification of McAbs

McAbs were precipitated by ammonium sulphate as described in (19). The IgG immunoglobulin fraction containing the anti CEA activity was isolated employing DEAE-Sephacel (Pharmacia) according to Lems van-Kan et al. (20). The IgG_2b immunoglobulin fraction containing anti-CEA activity was isolated employing protein A-sepharose CL-4B (Pharmacia) (Fig. 1) according to a modified method of Ey et al. (22,23). Fractions from individual peaks were pooled, concentrated by ultra filtration (PM 30 membrane, Amicon) and assayed for anti-CEA activity by double Ab RIA.

Sodium-deodocyl sulphate polyacrylamide gel (SDS-PAGE) electrophoresis

Electrophoresis was performed on SDS-PAGE slab gel (10%) according to Laemmli (24).

Determination of immunoglobulin subclasses

Classes and subclasses of the various monoclonal antibodies were determined by polyethylene glycol Ouchterclony analysis (23) using purified class specific heavy chain reagents (goat anti-mouse IgM, IgG, IgG_{2a}, IgG_{2b}, IgG_3 and IgA).

RESULTS AND DISCUSSION

A set of six hybridomas directed against CEA have been generated from two fusions (described in part in Ref. 13 and "Materials and Methods"), using purified metastatic (13) and cell culture CEA (12,13) for immunization (Table 1). Hybridomas were initially selected for their ability to produce immunoglobulins reactive with labelled purified CEA preparations, determined in the double antibody RIA (see "Materials and Methods"). The hybridomas were injected into pristane primed mice for the induction of ascitic fluids, purified on Protein A-Sepharose and DEAE-sephacel and further characterized.

Table 2, describes the antibody origin, subclass and the titer of anti-CEA reactivity of each ascitic fluid.

The dissociation constants of the McAbs were determined by

TABLE 1.

Immunization protocol of mice for the production of monoclonal anti-CEA antibodies

Schedule A

Injection No.	Day	CEA (metastatic, μg)	Adjuvant	Route
1	0	100 crude	CFA	IP+FP
2	49	100 crude	InCFA	IP+IP
3	60	50 crude	InCFA	IP
4	78	70 pure		
Fusion	82			

Schedule B

Injection No.	Day	CEA (cell line, μg)	Adjuvant	Route
1	0	100 crude	Alum + B Petussis	FP+IP
2	24	50 pure	Alum	SC
3	120	150 pure	Alum	IP
4	121	150 pure	-	IV
Fusion	122			

IP - Intraperitoneal.
FP - Footpads.
IV - Intravenous.
SC - Subcutaneous.

Table 2.

Origin and subclasses of McAbs

Clone	Origin	Igb subclass	Anti-CEA titer (Ascitic fluid)
G10G11	Metastatic	IgG_2b	1:25000
H2G8	Metastatic	IgG_2b	1:25000
3C8	Cell line	IgG_1	1:100000
4G9	Cell line	IgG_1	1:25000
4C5	Cell line	IgG_1	1:100000
IE4	Cell line	IgG_1	1:100000

inhibition RIA. The values were calculated from Scatchard plots. By and large, the dissociation constant averaged around 10^{-9}M. Fig. 2 demonstrates a displacement curve of monoclonal antibody H2G8, insert is Scatchard plot.

The monoclonal antibodies, derived from mice immunized with metastatic CEA (13) were employed for immunoaffinity purification of cell culture CEA (13). Supernatants from a CEA cell line cultivated in tissue culture (29), were pooled. Following perchloric acid extraction, ultrafiltration and dialysis, CEA was purified to homogeneity employing the above monoclonal antibodies coupled to CNBr-sepharose 4B. As demonstrated in Fig. 2, ^{125}I-labeled purified cell culture CEA is homogeneous with an apparent MW of 200.000 (similar to metastatic CEA).

The evaluation of the suitability of the above McAbs for CEA detection, was tested in the following systems:

a. McAbs G10G11 and H2G8 were radiolabelled with ^{125}I and employed as detecting antibody in a commercial CEA-RIA, resulting in dose response curves identical to the ones obtained with the commercial reagent (Fig. 3, reproduced from Eiken).

b. In a two site CEA-EIA. In order to construct a CEA-ELISA, based on two monoclonal anti-CEA antibodies, pairs which recognize different epitopes (or possibly CEA specific repetetive epitopes) should be chosen. To assess the possibility of employing the above McAbs in a CEA EIA, each McAb was conjugated to HRP and tested for retaining its the ability to detect CEA (when using polyclonal Ab as the catching layer, results not shown). Each conjugate was tested with all McAbs as catching antibodies in a one step EIA (Fig. 4), where CEA sample and conjugated antibody were simultaneously included with the immobilized antibody. As can be seen from Fig. 5, McAb 3C8 and 4G9-HRP conjugate represent an optical combination for CEA detection at the above range (3-60 ng/ml in buffer) whereas 1E4 HRP is not suitable for this purpose. Each of the above mentioned McAbs, when tested with their homologous conjugate, resulted in a flat CEA standard curve; suggesting that McAbs 3C8 and 1E4 probably recognize the same

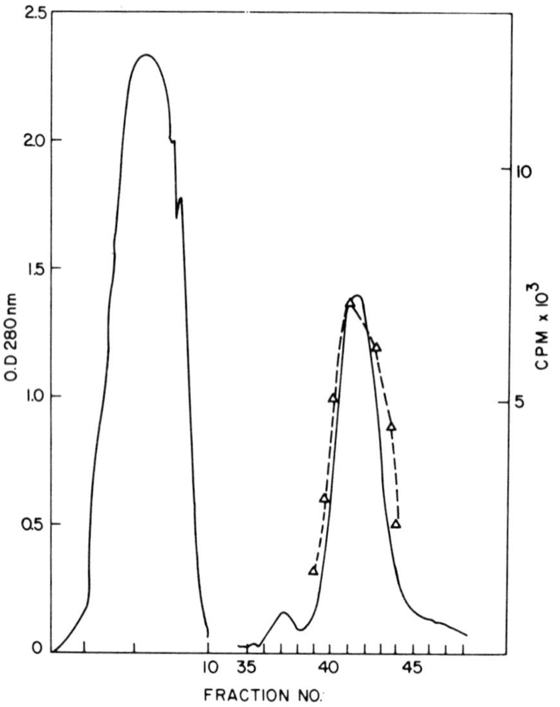

Fig. 1.: Purification profile of anti CEA McAbs (G10G11) on protein A sepharose column. Ascites fluids (partially purified by ammonium sulphate precipitation) was loaded on column equiliberated with 0.05M Tris HCl/0.15M NaCl/3mM NaN_3 pH 8. Elution of specific immunoglobulin subclasses was done as described by Epstein et al. (23). Protein was monitored at 280nm. Anti CEA activity (---) was determined by DAB RIA according to "Materials and Methods".

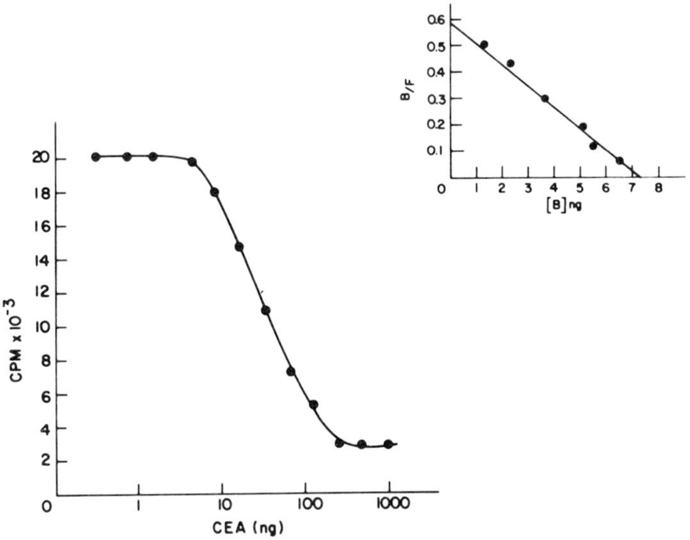

<u>Fig. 2.:</u> Displacement curve of McAb H2G8. Insert - a Scatchard plot. The calculatd dissociation constant Kd - 1 x 10^{-9}M.

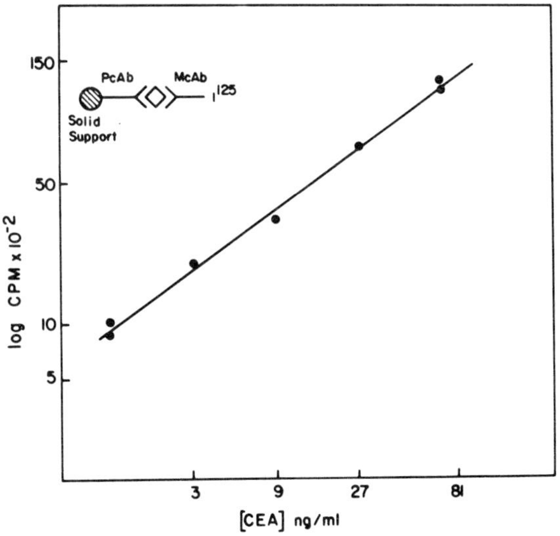

<u>Fig. 3.:</u> McAb G10G11 as tagging antibody in a commercial CEA-RIA (reproduced from Eiken). PcAb - Polyclonal anti CEA Abs. McAb - Monoclonal anti CEA Abs.

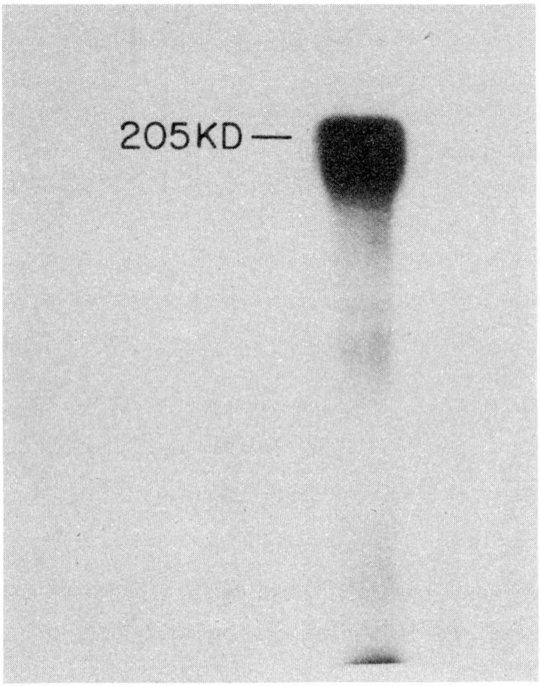

Fig. 4.: SDS PAGE of affinity column purified ^{125}I-CEA.

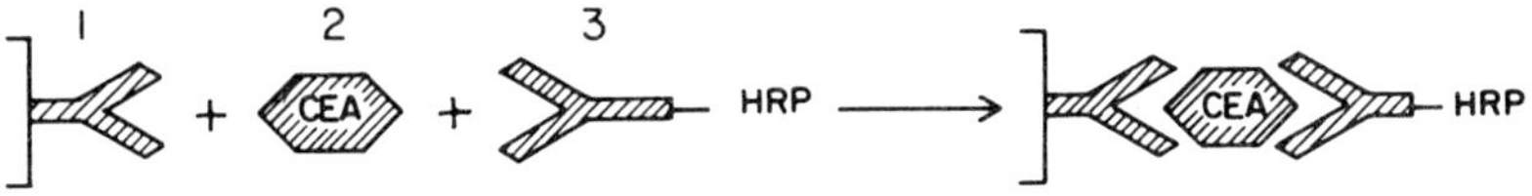

Fig. 5.: Scheme of 1 step CEA-EIA test. (1) McAb anti CEA adsorbed to PVC microwell. (2) CEA antigen. (3) McAb anti CEA enzyme labelled (horse-radish peroxidase).

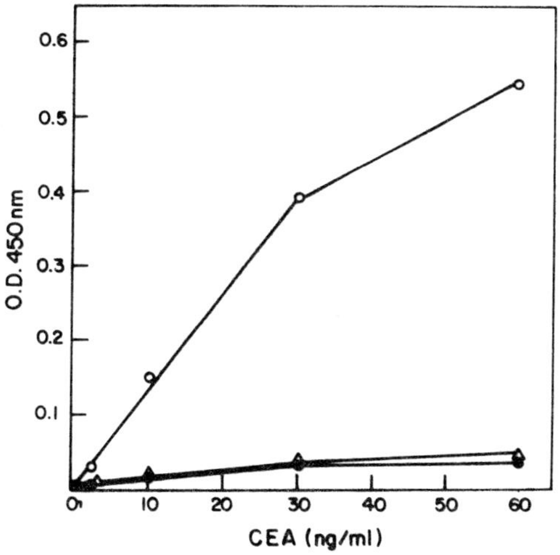

Fig. 6.: One step, two-site CEA-EIA (as described in "Materials and Methods".

O - 3C8-CEA-4G9 HRP; △ - 3C8-CEA-IE4 HRP; ● - 3C8-CEA-3C8 HRP.

epitope, whereas McAbs 3C8 and 4G9 may recognize dif ferent (or repetetive) epitopes of the CEA molecule. However, CEA levels in serum samples should be determined and results compared with those obtained from commercially available CEA kits.

c. It was shown in clinical studies (30) that in indirect immuno-fluorescent staining of freshly frozen tissues, McAbs G10G11 and H2G8 could discriminate between colon adenocarcinoma and normal adult tissues, in 19 out of 25 cases studied. (Normal tissues showed non or very little staining). Patients who had elevated serum CEA levels were positively stained, while patients whose tissues were negative upon staining had normal CEA serum levels.

Further studies (immunohistochemical staining and CEA serum quantitation) are required to ascertain the CEA specificity of the above McAbs and to determine whether the epitopes recognized by the above antibodies are unique to CEA or common to CEA-like antigens. Provided CEA specificity is demonstrated, the above antibodies have the potential to improve reliability, specificity and reproductivity of CEA detection.

ACKNOWLEDGEMENTS

The authors would like to thank Ms. Nita Ben-David for her assistance in preparing this manuscript.

REFERENCES

1. Gold P. and Freedman S.O. (1965). J. Exp. Med. 121: 439-462.

2. Fuks A., Shuster J. and Gold P. (1979) In: Cancer Markers: Diagnostic and Developmental Significance. S. Sell (ed.). Humana Press, Clifton, N.J. pp.315-327.

3. Reynoso G. and Keane M. (1979). In: Prognosis and Monitoring of Patients with Cancer. Pt. I. Herberman R.B. (ed.). Marcel Dekker, Inc. N.Y. pp.239-254.

4. NIH Concensus Report (Summary) (1981). Brit. Med. J. 287: 373-375.

5. Mackay R.R. (1979). In: Immunodiagnosis of Patients with Cancer. Pt. I. Herberman R.B. (ed.). Marcel Dekker Inc. N.Y. pp.255-264.

6. Yachi, A., Imai, K., Fugita, H., Moniya, Y., Tanda, M., Endo, T., Tsujusaki, M. and Kawaharada, M. (1984). J. Immunol. 132: 2998-3003.

7. Haskell C.M., Buchegger F., Schreyer, M., Carrel S. and Mach J.P. (1983). Cancer Res. 43: 3857-3864.

8. Griffin T.W., Haybes L.R. and DeMartino J.A. (1982). J. Natl. Cancer Inst. 69: 799.

9. Goldenberg D.M., Sharkey R.M., Primus F.J., Mizusawa C. and Hawthorne M.F. (1984). Proc. Natl. Acad. Sci. USA. 81: 560.

10. Muraro R., Wunderlich D., Thor A., Lundy P., Noguchi P., Cunningham R. and Schlom J. (1985) Cancer Res. 45: 5769-5780.

11. Krantz M.Y., Laferte S. and Ariel N. (1984). In: Methods in Enzymology. Vol. 84 Langone E.E. (ed.). Academic Prsss. pp.32-48.

12. Lazar A., Reuveny S., Geva J., Marcus D., Silberstein L., Ariel N., Epstein N., Altboum Z., Sinai J. and Mizrahi A. (1987) Develop. Biol. Standard. 66: 432-428.

13. Epstein N., Zandani R., Inbar I., Altboum Z., Lazar A., Reuveny S., Geva J., Marcus D., Mizrahi A. and Ariel N. (1987) Develop. biol. Stand. (S. Karger, Basel) 66: 429-437.

14. Köhler G. and Milstein C. (1975). Nature 256: 495-498.

15. Galfre G., House S.C., Milstein C., Butcher G.W. and Howard J.C. (1977). Nature 266: 495-498.

16. Epstein N., Shavit G. and Orli Z. (1987) Develop. biol. Stand. (In press).

(A note to the publisher: This paper is presented in this meeting, please add the relevant volume and page numbers).

17. Greenwood F.C. and Hunter W.M. (1963). Biochem. J. 89: 114-123.

18. Scatchard G. (1949). Proc. Natl. Acad. Sci. USA 51: 660-670.

19. Herde K. and Schwick H.G. (1973). In: Salt Fractionation of IgS. Handbook of Experimental Immunology. Weir D.M. (ed.) Vol. 1. "Immunochemistry", Oxford, Blackwell Scientific Publications, p.61.

20. Lems Van Kan P., Verspaget H.W. and Pena A.S. (1983). J. Immunol. Meth. 57: 51-57.

21. Epstein N., Lachmi B., Lerer S., Katz D., Cohen S. and Olshevsky U. (1987) Develop. biol. Stand. (In press).
(A note to the Publisher: This paper is presented in this meeting. Please add the relevant Vol. and page numbers).

22. Ey P.L., Prowse I.S.J., Jenkin C.R. (1978). Immunochemistry 15: 429-436.

23. Epstein N., Miller L.H., Kaushel I.C., Iroka J.U., Rener J., Howard R.J., Asofsky R., Aikawa M. and Hess R.L. (1981). J. Immunol. 127: 212-217.

24. Laemmli U.K. (1970). Nature 227: 680-689.

25. Ochterlony O. (1958). Prog. In Allergy, Kallos P. (ed.). S. Karger, Basel and New York, 5: 1-78.

26. Nakane E.K. and Kawaoi A. (1974). J. Histochem. Cytochem. 22: 1084.

27. Kawaoi A. and Makane P.K. (1972). Fed. Proc. 32: 840 (Abstract).

28. Boorsma D.M., Streefbork J.G. and Kors N. (1976). J. Histochem. Cytochem. 24: 1017.

29. Reuveny S., Lazar A., Mizrahi A., Marcus D., Shahar A. and

Epstein N. (1987). Develop. biol. Stand. 8 ESACT Meeting. (Submitted for Publication).
(A note to the Publisher: This paper is presented in this meeting. Please add the relevant Vol. and page numbers).

30. Zidon J., Bartal A.H., Epstein N., Ariel N., Hoffman A., Yagoda A., Steal S., Munichor M., Lichtig C., Hashmonai M. and Robinson E. (1986). Proc. 14th. Internat. Cancer Congr. Budapest, Hungary (Abstract).

53 Carcinoembryonic antigen (CEA) production in low protein growth medium

S. Reuveny, A. Lazar, A. Mizrahi, D. Marcus, A. Shahar and N. Epstein

ABSTRACT

A process for production of carcinoembryonic antigen (CEA) by a human adenocarcinoma cell line (HuCC1) propagated on microcarriers in serum-containing medium, was described in an earlier publication (1). The purpose of this study was to improve the existing process for CEA production. The goal was to develop an efficient culturing system for continuous production of large quantities of CEA in low protein growth medium.

A low protein growth medium was developed. This growth medium was based on RPMI-1640 medium supplemented with 10% bovine serum dialyzate, Primatone RL and Pluronic F-68. The protein content of this medium was 200ug/ml, as compared with 1600 ug/ml in medium containing 2% serum. The HuCC1 cells growing in this medium lose their ability to grow on culture surface (microcarriers or bottom of T-flask) and propagate in suspended aggregate culture as rounded hollow beads in which the cells grow in several layers on the outer surface of the bead while the center of the bead was filled with growth medium. The advantages of the new medium are: a. crude CEA with high specific activity is obtained; b. most of the proteins present in this medium can be easily removed by ultrafiltration during the downstream purification process; c. HuCC1 cells can grow in this medium without the need for microcarriers, thus simplifying the scaling-up process; and d. the medium is significantly cheaper than the former growth media, since expensive fetal bovine serum is not used.

The culture was scaled up to 15 l culture volume. Fifty percent of the culture supernatant (containing CEA) was harvested every 2-3 days, CEA titers ranged between 100-700 ng/ml. The specific activity of the CEA in the crude supernatant was about 0.75 ug CEA/mg protein, as compared with about 0.08 ug/mg protein in the serum-containing process.

For comparison, the HuCC1 cells were propagated in two additional culturing systems: Immobilized on polyurethane sponge matrix(2) and in an open perfusion culture in stirred reactor equipped with a decantating column.

The CEA containing culture supernatants obtained from these cultivation systems were further purified up to homogeneity. The efficiency of the purification process was 84% .

INTRODUCTION

Carcinoembryonic antigen (CEA) is a tumor associated glycoprotein present in the sera of patients with metastatic colonic cancer (3), with primary lesions of various cancers (4), or having various nonmalignant inflamatory diseases (5). In current clinical practice, CEA serum levels are used for monitoring cancer patients for response to therapy and early detection of recurrences (6).

The regular source of CEA is liver metastases of colon adenocarcinoma. The main problem in using this source as CEA standard is its non-reproducible nature and its limited availability. CEA produced in tissue culture may be an alternative source. The antigen can be obtained from this reproducible source in essentially unlimited amounts.

In a previous publication (1), we have described a method for large scale cultivation of a CEA-producing cell line (HuCC1) on cellulose-based microcarriers. The microcarriers allow the growth of the cells for extended periods in low-serum medium (2% fetal bovine serum); large amounts of CEA were obtained by periodical harvesting of the culture.

The main problems encountered during the operation of the above mentioned system were:

1. The specific activity of the crude CEA obtained in medium containing 2% fetal bovine serum, was very low (approximately 0.08 ug CEA/mg protein). Thus, the efficiency of the downstream purification process was very low (10-40%). Bovine serum albumin (BSA) which is present in high concentration in serm tend to adsorb on the CEA molecule and to interfere with the down stream purification process.

2. Harvesting of cells from the microcarrier surface during the scale-up process was problematic for two reasons: it increased the chances of contamination, and the efficiency of recovery of cells from the microcarrier surface was relatively low (approximately 50%).

The purpose of this work was to develop an efficient method for production of CEA, that will overcome the above mentioned problems.

A low protein growth medium was developed, in which the protein content was 200 ug/ml. HuCC1 cells grew in this medium in aggregate culture without need for microcarriers. Scaling-up of the culture was easily achieved by gradual addition of fresh growth medium. The specific activity of the CEA in the culture supernatant was about 0.75 ug CEA/mg protein.

MATERIALS AND METHODS

Cells: A CEA secreting human colon adenocarcinoma cell line

(HuCCl) was used throughout the whole work (7).

Media: Cells were propagated in RPMI-1640 (Gibco, Grand Island, N.Y.). Three media combinations were used, as described in Table I. Serum dialyzate was prepared as follows: Bovine serum was seperated from blood collected at slaughterhouses. The serum was dialyzed against phosphate buffered saline (PBS) (pH 7.2) in a tangential flow ultrafiltration system (Pellicon cassette, Millipore, Bedford, Mass.), using membranes of 100K, 30K and 10K molecular weight cutoff and a surface area of 15 square feet.

Cell propagation

Monolayer cultures: Cells are propagated in T-flasks as monolayers in RPMI 1640 medium plus 10% fetal bovine serum (SM) medium at 37°C. Confluent cell cultures are passaged by treatment with 0.05% trypsin-0.02% EDTA solution in PBS(pH 7.8). Cells are seeded at a concentration of $2X10^4$ cells/cm^2 vessel surface. Confluent cultures contain $15\text{-}35X10^4$ cells/cm^2.

Table I: Composition of growth media

Material	Dialyzed media (DM) (g/100 l)	Primatone pluronic dialyzed media (PPDM) (g/100 l)	Serum containing media (SM) (g/100 l)
RPMI-1640	1040	1040	1040
$NaHCO_3$	200	200	200
Neomycin sulfate	5	5	5
Glutamine	29.2	29.2	29.2
Glucose	150	150	150
Dialyzate	101	101	-
Pluronic F-68*	-	10	-
Primatone RL**	-	250	-
Fetal bovine serum	-	-	101
Protein concentration (ug/ml)	10	210	8000

* BASF-Wyandotte, Wyandotte, Mich.
** Humko-Sheffield, Norwick, N.Y.

Maintenance of monolayer cultures for long periods is achieved by using DM medium (replaced three times a week).

Aggregate cultures (semi-continuous harvesting): Cell inoculum from trypsinized monolayer culture is added to a spinner flask containing PPDM medium (Table I). The initial cell concentration is $2X10^5$ cells/ml. Scaling up of the culture is achieved simply by adding fresh PPDM medium every 2-3 days, to keep the cell concentration at about $5X10^5$ cells/ml. In this way, the cultures are scaled up to 15 liter volume. Cultures

Table 2: CEA Production Systems

Production system	Stirred reactor	Polyurethane sponge matrix	Open perfusion culture
Reactor volume (1)			
Total	18	1	5
Culture	15	0.3	4
Medium reservoir	-	0.7	-
Cell concentration ($\times 10^6$/ml)	0.6-1	3.3* 5 **	4
Dissolved oxygen (% air saturation)	30-80%	-	10%
Glucose consumption rate (mg/10^6cells/day)	1	0.68***	0.5
Lactate production rate (mg/10^6cells/day)	0.4	0.52***	0.22
Conversion of glucose to lactate (%)	20	38	23
CEA culture supernatant harvesting:	semicontinuous: 50% of culture every 2-3 days	semicontinuous: 100% of culture every 4 days	continuous: 35% of the culture vol/ day
CEA titer (ng/ml)	100-700	300-500	1100
Specific prod. rate (ng/106cells/day)	100-200	-	100
Productivity (mg CEA/1 culture/day)	0.08	0.33	0.38

* Estimated from CEA production rate.
** Estimated from glucose consumption rate.
***Cell concentration estimated at 3.3×10^6/ml.

ACKNOWLEDGEMENT

The authors whish to express their gratitude for the skillful technical assistance of Mrs. L. Silberstein.

are incubated at 37°C stirred at 50-100 RPM. Cultures larger than one liter in volume are aerated by sparging 5% CO_2 in air at a rate of 0.2 liter/minute. When the culture reaches a concentration of about $1X10^6$ cells/ml, CEA harvesting is initiated. Every 2-3 days, stirring of the culture is stopped for 1-2 hours. Aggregates are allowed to settle down and 50% of the culture supernatant (containing crude CEA) is harvested and fresh medium is added. This type of harvesting was carried out for a period of at least 4 months for each culture. At the one-liter scale, DM medium is used. In cultures larger than one liter (where gas sparging is used) PFDM medium is used.

Packed bed cultures: HuCCl cells were propagated immobilized on polyurethane sponge matrix. Detailed description of the system and culturing condition are given elsewhere (2).

Open perfusion system: HuCCl cell aggregate culture is transferred into a 5 liter CelliGen fermentor (16), (New Brunswick Scientific, Edison N.J.) at a concentration of $4X10^6$ cells/ml. Perfusion is achieved by pumping out culture supernatant through a column separator located above the liquid level (17). In this column aggregate settling rate exceed the upward flow rate and thus supernatant free of cell aggregate is being pumped out at a rate of 0.35 l/l culture/day. Fresh growth medium is added at the same rate. Dissolved oxygen was controlled at 10% air saturation and pH at 6.9 by mixing air, nitrogen and CO_2 into the gas stream entering the fermentor and by adding bicarbonate to the culture. Cell concentration was maintained at $4X10^6$/ml.

Cell counts: Cells in aggregate cultures were counted by the nuclei count method (8). Free-suspended cells were counted by the trypan blue dye exclusion method.

Protein, glucose and lactate assays: Protein was determined by the lowry method (9), using BSA as standard. Glucose levels were measured by the glucose oxidase colorimetric method (Sigma method # 510) and lactic acid levels by the lactate dehydrogenase method (Sigma method #826).

CEA assay: CEA titer was determined by ELISA-CEA, developed at our institute (10), or CEA Abbott kit (11).

RESULTS AND DISCUSSION

Development of low protein maintenance medium (DM)

Attempts were made to increase the specific activity of CEA in the culture supernatant, by reducing protein content in the growth medium.

A low protein medium was developed. This medium was based on addition of dialyzate of bovine serum, rather than fetal bovine serum, which contributes all the protein to the growth medium.

Bovine serum was prepared from blood collected at the slaughterhouse. This serum was dialyzed against PBS, using pilot scale ultrafiltration systems of 100K, 30K and 10K cut off membranes.

We have found that RPMI-1640 medium supplemented with 10% 30K dialyzate (DM) can be used as maintenance media for CEA production. Media containing 10K dialyzate did not maintain the cells, while in the 100K dialyzate most of the serum protein (BSA) were present.

HuCC1 monolayer cultures propagated in serum-containing medium were transferred directly to dialyzate medium. Supernatants from 150 cm^2 T-flasks were harvested three times a week (50 ml each time) and fresh dialyzate medium was added. The cultures were maintained for at least three months. The total amount of the cells in the vessel was $20\text{-}50 \times 10^6$ ($1.5\text{-}3.5 \times 10^5$ cells/cm^2) and CEA titers were 100-700 ng/ml (i.e., about 20 ng CEA/10^5 cells/day). These values were similar to those obtained from the serum-containing process.

Spinner cultures were maintained using the same dialyzate medium for periods of at least four months (Figure 1). CEA titers in this experiment were about 400 ng/ml and cell concentration about 0.8×10^6 ml (i.e., approximately 20 ng CEA/10^5 cells/day).

The protein content of the 30K dialyzate was 70 ng/ml, thus, its contribution to the maintenance medium (DM), which contains 10% dialyzate, was 7 ng/ml (Table 1).

We believe that the 30K serum dialyzate contains small molecular weight growth factors, which support cell maintenance (12). The DM medium did not support significant cell growth, cell doubling times of 15-20 days were recorded, as compared with 3-5 days in the case of serum containing media.

Five hundred ml microcarrier spinner culture were initiated using RPMI-1640 medium + 10% fetal bovine serum. On the tenth day of growth, medium was changed to dialyzate medium. On the indicated days, 50% of the culture supernatant was harvested and fresh dialyzate medium was added. This culture was maintained for four months.

Serum proteins which are present in cell culture media protect the cells from mechanical damage, during agitation and, especially, during sparging of gases into the culture (13). We have found that HuCC1 microcarrier cell cultures, maintained in the low protein medium, are totally destroyed, as a result of gas sparging at a rate of one l/minute. Gas sparging is essential for supplying oxygen to the culture, in the large scale system.

Thus, the next step in our research was to develop a modified dialyzate medium, which will enable cell growth and will protect the cell from mechanical damage during gas sparging.

Development of low protein growth medium which protect cells from mechanical damage

Attempts were made to modify the dialyzate medium, in order to supply mechanical protection to the cell and to enable cell growth.

The concept was to enrich the dialyzate medium with a chemical polymer for mechanical protection and add protein hydrolyzate as nutritional element, in order to support cell growth (14,15). Several media combinations were tested (results not shown).

The best results were obtained using Pluronic F-68 (copolymer of polyoxypropylene polyoxyethylene) and an enzymatic digest of animal tissue (Primatone RL) in concentrations of 0.01% and 0.25%, respectively (PPDM medium).

The new PPDM medium supported HuCCl cell growth. Cell seeded at a concentration of 2×10^5 cells/ml, reached a concentration of 0.9×10^6 cells/ml in 5-8 days. These doubling times were similar to that obtained in the serum containing media. CEA titers achieved range between 100-700 ng/ml and the cell concentration between $0.6-1X10^6$ cells/ml (Fig. 2).

The protein concentration in the PPDM medium was 400 ng/ml. This value include 200 ng/ml contribution of peptides from the Primatone RL and only 7 ng/ml from the 30k dialyzate. CEA obtained in this medium had a specific activity of about 0.75 ng CEA/mg protein, a value similar to that obtained using the DM medium .

The molecular weight of the peptides in the Primatone RL is smaller than 1000 and that of the Pluronic F-68 is 800. So they can be easily removed from the supernatant by ultrafiltration during the CEA downstream purification process. Most of the foreign proteins in the culture supernatant are secreted by the cells or generated as a result of cell death or lysis in culture , and do not originate in the PPDM growth medium.

The HuCCl cells which grew in this low protein growth medium, lose their ability to attach to growth surface, in general, and particularly, to microcarriers, probably since the new medium does not contain the high molecular weight attachment factors. The HuCCl cells grew in the low protein medium (PPDM) as rounded hollow beads (Figs 3-5), in which the cells are located in several layers on the outer surface of the bead. These beads range in size from hundred microns to 1-2 millimeters. The cells in the beads were viable, active in mitosis and demonstrated some degree of differentiation (microvilli were generated and typical epithelial cell organization was observed), (Fig 5).

There was no need for use of microcarrier in the culture, since a culture of free floating cell beads (aggregates) was achieved. Moreover, it was observed that single cells, detached from the bead surface, lead to generation of new beads (Figure 4D), thus, scaling up of the culture volume was easily achieved by gradually adding fresh growth media.

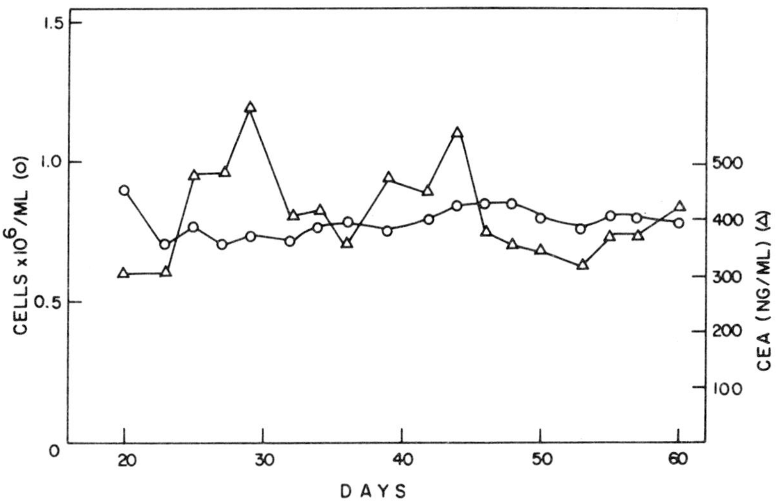

Figure 1: Cell maintenance and CEA production in spinner culture using dialyzate medium.

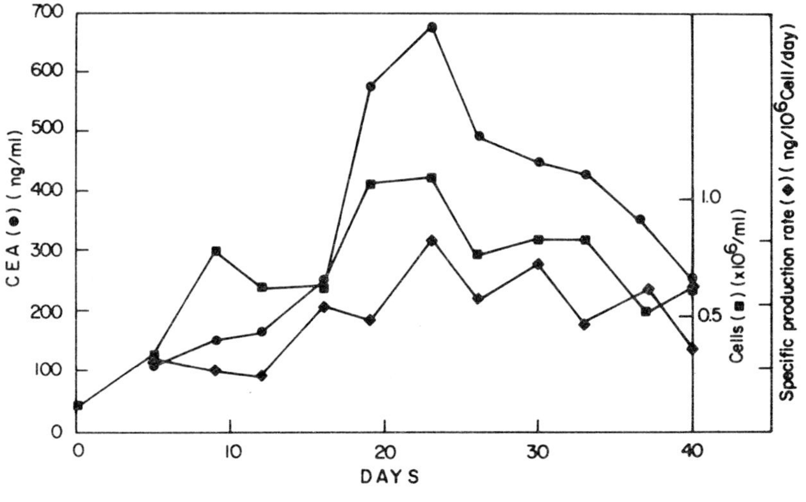

Figure 2: CEA production in a 15 liter spinner culture.

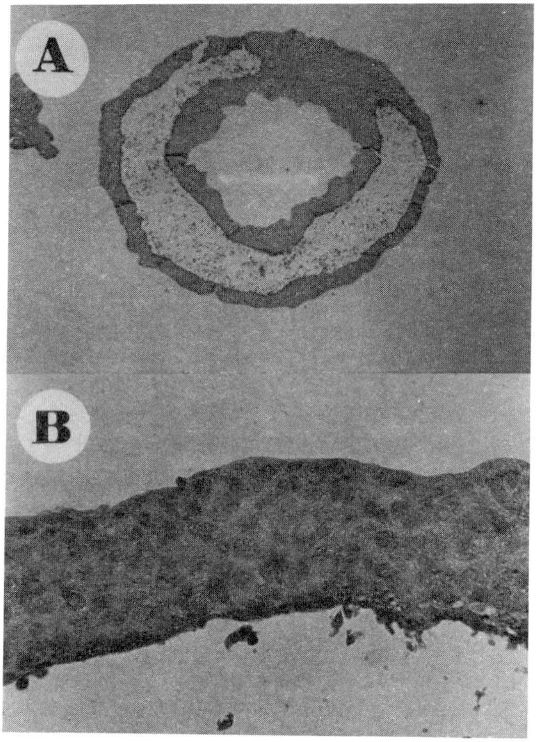

Figure 3: Hematoxylin-eosin light microscopy section of HuCCl beaded aggregate.

(A) One beaded aggregate; (B) The wall of one aggregate.

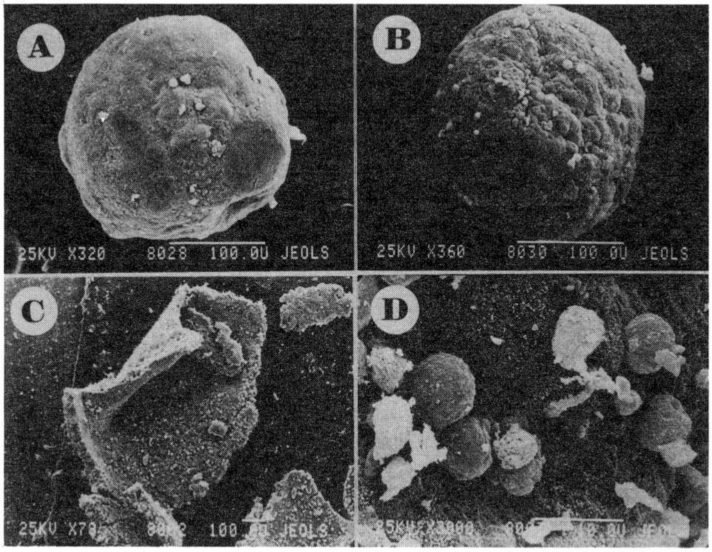

Figure 4: Scanning electron micrograph of HuCCl cell aggregates.

(A,B) Beaded aggregates; (C) The wall of the beaded aggregate composed of several layers of cells; (D) Rounded cells released from the aggregate surface.

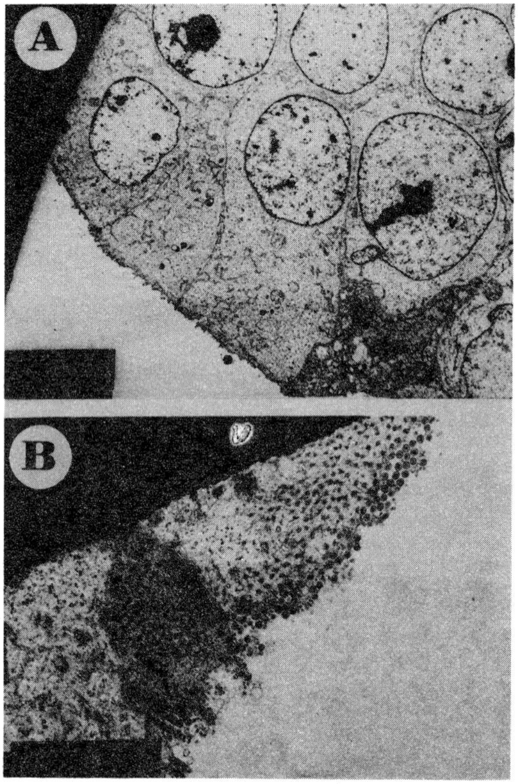

Figure 5: Transmission electron micrograph of HuCCl cells.

(A) Cross section of an aggregate; (B) Microvilli at the surface of the aggregate.

The main advantages of this process over the old CEA production process in serum containing media are:

a. Crude CEA with high specific activity is obtained (0.75 ng CEA/mg protein, as compared with 0.08).

b. All the media proteins, peptides and polymers are smaller in molecular weight than 30k and thus can be easily separated from the CEA by 100k ultrafiltration.

c. There is no BSA in the crude CEA preparation (BSA tend to absorb on the CEA molecule).

d. Microcarriers are not used in this process.

e. Scaling up is done easily by medium addition, without the need for cell trypsinization from the microcarriers surface. Thus reducing dramatically the chances of culture contamination.

f. The process is cheaper, since the expensive fetal bovine serum is not used.

Culture supernatants obtained, were purified to homogeneity by a three steps purification process (1), which included ultra-filtration, perchloric acid extraction and anti-CEA affinity chromatography. The efficiency of the purification is 85% and the purification factor is 1900.

CEA production in perfused sysyems.

Huccl cells were propagated in two types of perfusion systems: a.Polyurethane sponge packed bed reactor in which growth medium is recirculated continuously through the cell growth chamber(2). CEA is harvested semicontinuously by periodical change of the culture supernatant with fresh growth medium. b.Open perfusion culture using a special low shear stirred reactor(16 ,CelliGen New Brunswick Sci. Edison ,N.J.),eqipped with a decantating column(17). In this system steady state conditions were maintained as a result of the continuous addition of fresh growth medium and removal of culture supernatant.

Results obtained with these perfusion systems were compared to those obtained with the previously described system (Table 2).

The best results were obtained using the steady state open perfusion culture. Cells were maintained at high concentration ($4X10^6$/ml) under homogenious conditions, achieving high CEA titers (1.1ug/ml) and high productivity(0.38 mg/l culture/day). The polyurethane sponge matrix system achieved similar high productivity rates (0.33 mg/l culture/day).However this system operates under non-homoginuous conditions, representable culture samples can not be taken and nutrients and metabolites gradient are generated inside the growth chamber. Moreover, since there is channeling in the flow pattern of growth medium through the sponge bed , anaerobic pockets are generated leading to higher rates of glucose conversion to lactate (38% as compared with 20-23% in the other systems).

REFERENCES

1. Lazar, A., Reuveny, S., Geva, J., Marcus, D., Silberstein, L., Ariel, N., Epstein, N., Altboum, Z., Sinai, J. and Mizrahi, A. 1987. Develop. Biol. Standard. 66, 423-428.

2. Lazar, A., Reuveny, S., Mizrahi, A., Avtalion, M., Whiteside, J.P. and Spier, R.E. (1987). These proceedings.

3. Gold, P. and Freedman, S.O. 1965. J. Exp. Med., 121, 439-462.

4. Gold, P. and Shuster, J. 1980. Cancer Res., 40, 2973-2980.

5. Rogers, G.T. 1983. Biochim. Biophys. Acta., 645, 227-249.

6. Booth, S.N., Jamieson, G.C., King, J.P.C., Leonard, J., Oates, G.T. and Dykes, P.W. 1974. Brit. Med. J., 4, 183-187.

7. Yaniv, A., Altbaum, Z., Gazit, A., Bloch-Stacker, N. and Eylon, E. 1978. Exp. Cell. Biol., 46, 220-230.

8. Sanford, K.K., Earle, W.R., Evans, V.S., Waatz, J.K. and Shanon, J.E. 1951. J. Natl. Cancer Inst., 11, 773-795.

9. Lowry, O.H., Rosebrough, N.J., Farr, A.L. and Randall, R.J. 1951. J. Biol. Chem., 193, 265-275.

10. Epstein, N., Zandani, R., Inbar, I., Altboum, Z., Lazar, A., Reuveny, S., Geva, J., Marcus, D., Mizrahi, A. and Ariel, N. 1986. Develop. Biol. Standard. 66, 429-437.

11. Abott-CEA-EIA Monoclonal (1-Step). Immunological test system as an aid in the prognosis and management of cancer. Abbot Laboratories. Diagnostic Division, March, 1986.

12. Pledger , W.J., Estes, J.E., Howe, P.H. and Leof, E.B. 1984. In: Mammalian cell culture: The use of serum-free hormone-supplemented media. Mather, J.P., editor. Plenum Press, N.Y., pp. 1-15.

54 Production and purification of a bladder cancer associated membrane protein

B. Lundgren, I. Wergeland, S. Braesch-Anderssen, S. Paulie, H. Koho and P. Perlmann

ABSTRACT

The aim of the study was to produce an antigen, mainly confined to bladder carcinomas, by use of cell culture in a pilot scale fermenter. Human bladder cancer cells, TCCSuP, grown on Cytodex[R] 1 were scaled up by serial subcultivation to a final culture volume of 40 litres. The cell doubling time was reduced during the scale up process by replacing the NBS culture supplement with FBS. An increased cell growth rate was obtained in a specially designed pilot scale fermenter, as compared to growth rates obtained in spinner flasks, by carefully controlling the pH, dissolved oxygen and stirring speed. Following a cell-microcarrier separation step the antigen was purified from a cell membrane preparation. The monoclonal antibody, 7E9, which was earlier produced against TCCSuP cells, was used to define the antigen under study. The antigen was further used to obtain polyclonal antisera from immunised rabbits. This antisera is now being used in developing a clinical assay for the detection of bladder carcinoma antigen present in the urine of affected persons.

INTRODUCTION

Mouse monoclonal antibodies have been used to define five different antigens present in transitional cell carcinoma (TCC) of the urinary bladder. These antigens are present in the majority of TCC tissues (1.2.4.) and TCC cell lines (6), and are potentially useful as markers of human bladder cancer, and for developing clinical tests. Two of these antigens defined by the antibodies S2C6 and 7E9, have now been purified and characterized. The 7E9 antigen seems to be confined to TCC tissues and to a lesser degree in vessel endothelia (3). Neither of the two antigens have been found in normal bladder tissues and therefore both may be useful in diagnostic and/or prognostic tests of urine or bladder washings. Such tests are now in progress.

MATERIALS AND METHODS

Cells and culture media

TCCSuP cells were obtained from the Immunology department (10). They were cultured in medium 199 containing 5% (v/v) of newborn calf serum, 2 mM glutamine.

Microcarriers

Cytodex 1 was prepared as previously described (13) and microcarrier cultures contained a final concentration of 3 grams Cytodex/litre culture volume.

Cell culture

The initial cultures were contained in Techne microcarrier vessels at a volume of 1 l. The first scale up step consisted of 1x1 l spinner, the second of 4x1 l and the third of 8x1 l (Fig. 1). 50% of the culture medium was changed midway through each culutre. The stirring speed was continiuosly at 25 rpm from the start of the cultures. The fourth step was a pilot scale culture of 40 l which were contained in a specially designed, conical, stainless steel fermenter (150 litres total volume, supplied by Electrolux). The stirring speed was kept at 15-20 rpm. The temperature, dissolved oxygen level and pH were monitored and controlled throughout the culture period.

Subpassaging cell cultures

This method has earlier been described in detail (7).

Harvesting of cells

Cells were washed thoroughly in 0.9% NaCl to remove the serum, and then harvested by incubation in 1 mM EDTA in PBS. A nylon mesh was used to separate the cells from the microcarriers. Cells were then lysed in a N2-cell disruption bomb, and then the membrane fraction was isolated by centrifugation (3). This was finally dissolved in Tris-HCl pH 7.5 containing 0.2% Triton X-100, 1% 6-aminocaproic acid, 1 mM EDTA and 50 000 IU aprotinin.

Monoclonal antibodies

The production and specificities of the mouse monoclonal antibodies used in this investigation has been described (1,6,11). The monoclonal antibodies were purified on protein A-Sepharose^{R}.

ELISA

To estimate antigen quantities an inhibition of cell ELISA was primarily used (6) but also an inhibition of antigen ELISA (3).

Affinity chromatography

Protein A-Sepharose was washed with ten volumes of phosphate buffer (0.1 M, pH 8.1). 5-10 mg of monoclonal antibody in phosphate buffer was added for each ml of protein A-Sepharose. The column was washed with ten bed volumes of phosphate buffer, before the addition of 0.05% glutaraldehyde in the same buffer. The volume of the glutaraldehyde solution was such that it would surely pass the column within five minutes. A few ml of phosphate buffer were added and finally ten bed volumes of phosphate-buffered 2% glycine (Fig. 2). Elution from these columns was either with 3 ml 0.1 M glycine buffer, pH 2.8 or with 0.7% acetic acid, both containing 0.1% NP 40. Concentration of the eluates was performed on Minicon B15.

Characterization

The isoelectric point of the antigen was determined by isoelectric focusing, and the Mw by SDS-PAGE (12). Affinity for Con-A, sensitivity to pH, teperature and SDS were also evaluated (3).

RESULTS

Cell culture

In the first culture the cell doubling time was 69 hours. At the beginning of the second culture the doubling time increased to 108 hours. In order to reduce the cell doubling time we replaced the newborn calf serum with fetal calf serum. After the serum change the doubling time decreased to 70 and 73 hours in the third and fourth cultures respectively. Under more controlled conditions in the fermenter, the cell doubling time decreased to 40 hours (Fig 1). The final cell density in the 40 l culture was $8x10^5$ cells/ml.

Affinity chromatography

Treatment of the antibody-protein A-sepharose complex with 0.05% glutaraldehyde (Fig 2) resulted in material which could be used repeatedly without affecting its performance. Recoveries after acid elution of the affinity column were 60-80%. The amount of 7E9 antigen revovered from 20 mg of TCCSuP membrane proteins after affinity purification was approximately 10 ug of 7E9 antigen.

Characterization

The pI of 7E9 was fond to be 7.4. The antigen did not bind to Con A. It was sensitive to SDS, and to heating . Most of its activity was lost already after 5 min at 50°C. The antigen was stable for up to 24 h at pH 2.0 but not at pH 9.0. These data are summarized in Table 1.

Table 1 A summary of the properties of the 7E9 antigen

Tissue distribution	Bladder cancer and some blood vessel endothelia
Cell line used for isolation	TCCSuP (bladder cancer)
Size estimated by SDS-PAGE	23 kDa/29 kDa
pI	7.4
Affinity to ConA	-
Antigenic activity susceptible to :	
heat 70°C, 5 min	+
pH 2.0 , 24 h	-
SDS (0.1%), 24 h	+

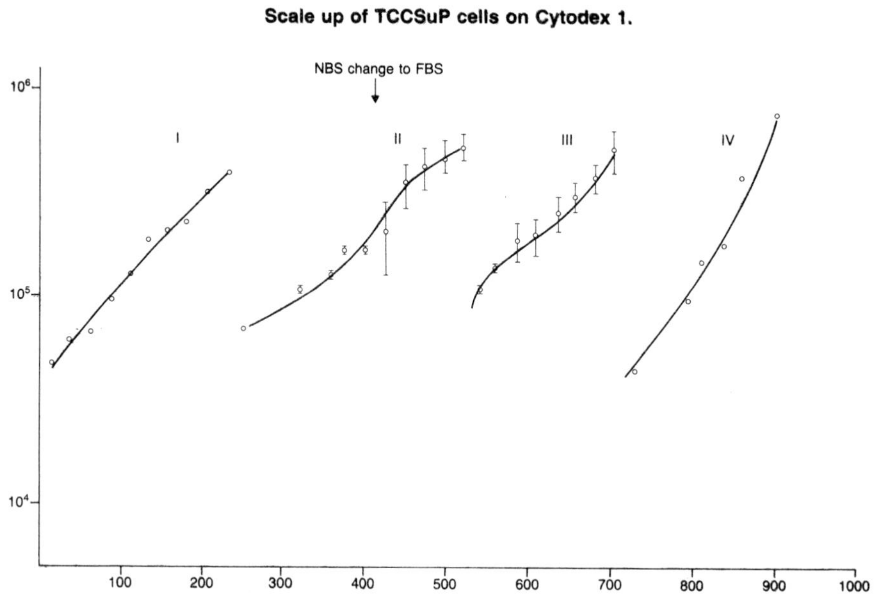

Fig. 1

SCALE UP OF TCCSuP CELLS ON CYTODEX 1.
TCCSuP cells were grown on Cytodex 1 (3 g/l) in DMEM containing 5% NBS. During the second culture the NBS was changed to FCS to reduce the cell doubling time.

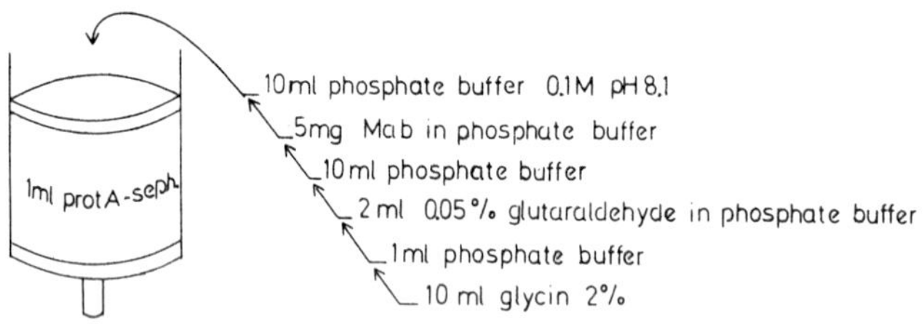

Fig. 2

Glutar aldehyde fortified coupling of monoclonal antibodies to protein A-Sepharose.

DISCUSSION

Our results indicate that the bladder associated antigens, reactive with the monoclonal antibodies S2C6 and 7E9 seem to be different from the antigens obtained by several other groups (5,8,9,14,15).These antigens also seem to have a different tissue distribution. No attempts to isolate and purify any of the TCC antigens refered to above have been reported in the literature. This might reflect the difficulties of processing these relatively hydrophobic membrane components. The cells from which S2C6 and 7E9 antigens were obtained can easily be grown in large scale cultures, and assures access to the quantities of antigen required for clinical testing. Polyclonal antisera against the two antigens were raised in rabbits. These sera are now being used in combination with monoclonal antibodies to develop a double determinant test primarily of urine but in the future also of serum.

REFERENCES

1 Ben-Aissa, H. Paulie, S. Koho, H. Biberfeld, P. Hansson, Y. Lundblad, M.L. Gustafsson, H. Jonsdottir, I. and Perlmann, P. (1985) Specificities and binding properties of 2 monoclonal antibodies against carcinoma cells of the huamn urinary bladder. Br. J. Cancer 52, 65.

2 Ben-Aissa, H., Paulie, S. Koho, H. Biberfeld, P. Hansson, Y. Braesch-Anderssen, S. Lagerkvist, M. Lundblad, H. Gustafsson, H. and Perlmann, P (1986) Monoclonal antibodies against carcinoma cells of the human urinary bladder: Immunohistochemical staining of tissues. Anticancer Res. 6, 165.

3 Braesch-Anderssen, S., Paulie, S., Koho, H., and Perlmann, P. (1986) Isoaltion and characterization of two bladder carcinoma associated antigens. J. Immun. Methods 94, 145-151.

4 Bubenik, J., Kieler, J. Perlmann, P. Paulie, S. Koho, H. Christensen, B. Dienstbier, Z. Koprikova, H. Pospisil. J. Pouckova, P. Novak, F. Dvorak, P. Lauerova, L. Kovarik, J. Iversen, H.G. Hou-Jensen, C. Rasmussen, F. Bubenikova, D. Jandlova, T. and Simova, J. (1985) Monoclonal antibodies against huamn urinary bladder carcinomas: Selectivity and utilization for gamma scintigraphy. Eur. J. Cancer Clin. Oncol. 21, 701.

5 Fradet, Y. Cardo, C. Thomson, T. Daly, M.E. Whitmore, W.F. Lloyd, K.O. Melamed, M.R. and Old, L.J. (1984) Cell surface antigens of human bladder cancer defined by mouse monoclonal antibodies. Proc. Natl. Acad. Sci. U.S.A. 81, 224.

6 Koho, H. Paulie, S. Ben-Aissa, H. Jonsdottir, I. Hansson, Y. Lundblad, M.L. and Perlmann, P. (1984) Monoclonal antibodies to antigens associated

with transitional cell carcinoma of the human urinary bladder. Cancer Immunol. Immunother. 17, 165.

7 Lindner, E. Arvidsson, A.C. Wergeland, I. Billig, D. (1987) Subpasssaging cells on microcarriers: The importance for scaling up to production. Develop. biol. Standard, 66, 299-305.

8 Masuko, T. Yagita, H. and Hashimoto, H. (1984) Monoclonal antibodies against cell surface antigens present on human urinary bladder cancer cells. Natl. Cancer. Inst. 72, 523.

9 Messing, E.M. Bubbers, I.E. Whitmore, K.E. DeKernion, J.B. Nestor, M.S. and Farney, J.L. (1984) Murine hybridoma antibodies against human transitional carcinoma-associated antigens. J. Urol. 132, 167.

10 Otoole, C. Tiptaft, R.C. Stevens, A. (1982) HLA antigen expression on urothelial cells: Detection by antibody-dependent cell-mediated cytotoxicity. Int. J. Cancer 29, 391.

11 Paulie, S. Ehlin-Henriksson, B. Mellstedt, H. Koho, H. Ben-Aissa, H. and Perlmann, P. (1985) A p50 surface antigen restricted to human urinary baldder carcinomas and B lymphocytes. Cancer Immunol. Immunother. 20, 23.

12 Paulie, S. Lundblad, M.L. Hansson, Y. Koho, H. Ben-Aissa, H. and Perlmann, P. (1984) Monoclonal antibodies to antigens associated with transitional cell carcinoma of the human urinary bladder. Cancer Immunol. Immunother. 17, 173.

13 Pharmacia Fine Chemicals (1981). Microcarrier cell culture principles and methods. Technical Book Series, Pharmacia, Uppsala, Sweden.

14 Starling, J.J Sieg, S.M. Beckett, M.L. Schellhammer, P.F. Ladaga, E.F. and Wright, G.L. (1982) Monoclonal antibodies to human prostate and bladder associated antigens. Cancer Res. 42, 3084.

15 Young, D.A. Prout, G.R. and Lin, C.W. (1985) Production and Characterization of mouse monoclonal antibodies to human bladder tumor associated antigens. Cancer Res. 45, 4439.

Section 8
Toxicity testing with animal cells

55 The use of animal cells for evaluation of toxicity, carcinogenesis and mutagenesis

A. M. Goldberg

ABSTRACT

Over the last few years there has been increased societal pressure on the one hand and self generated scientific pressure on the other to develop new and better *in vitro* techniques for the evaluation of safety of commercial products.

In vitro methodology addresses four major areas within toxicology:

- it provides the basic science of the discipline with new information,
- in acute toxicity testing, it can provide alternatives to current approaches,
- in the area of chronic toxicity testing, it will allow us to define the mechanisms associated with organ specific insults, and
- in the area of risk assessment, the combined knowledge obtained by superior methods will allow us to assess risk more accurately.

In vitro methodology is rapidly developing as a result of major advances that have occurred in cell and organ culture techniques, and in the ability to maintain the functional capacity of cells in culture.

It has become apparent that there are several areas where alternative tests can and should be developed. The underlying assumption in the development of these tests should be that the mechanism of toxicity is of primary importance and that the endpoint be clearly defined. Examples from studies dealing with inflammation assessment, hepatotoxicity and carcinogenicity will be highlighted as examples of the use of animal cells in *in vitro* toxicity.

INTRODUCTION

The use of cell, organ and tissue culture has become routine in elucidating our understanding of basic biology, physiology, among other disciplines. It can be amply demonstrated that cell, organ and tissue methodologies are the routine, not the exception, in the new biology laboratory. In toxicology, this is not yet the case. During the last few years there have been several approaches to the development of *in vitro* methods in toxicology, and overall these have increased the awareness of the scientific community for the need for this development and for the appropriateness of the endeavor.

This presentation will look at the reasons for developing *in vitro* techniques in toxicology. We will then explore several areas, by way of example, to demonstrate how *in vitro* methodology can be applied in the study of inflammation, hepatotoxicity and in the development of methods appropriate to study carcinogenesis and therapeutic approaches to cancer. Further, the products of biotechnology, human specific proteins, may become the necessity for the acceptance of *in vitro* methodology in toxicology.

REASONS FOR DEVELOPING *IN VITRO* TOXICOLOGICAL METHODS

One can identify three reasons for developing new *in vitro* techniques in toxicology [11,14,15,16]. These include: (1) scientific and academic reasons; (2) societal or humane reasons; and (3) economic considerations.

Scientific and Academic Reasons

The scientific and academic reasons for the development of alternatives (*in vitro* methodology) in toxicology testing are numerous and compelling. The LD50 and several other classical methodologies currently used in routine toxicity testing are more than 50 years old. Clearly we should be asking the question, can we do better?

There are more than 100,000 chemical formulations and more than a million new chemicals on the shelves of industry that have yet to be evaluated. Our current methodology is both time consuming and expensive. In the case of chronic studies, it may take us anywhere from three months to two years to complete a battery of *in vitro* tests. With the number of compounds and formulations currently available to us that have yet to be tested, and new ones coming on the market or in the chemist's laboratory, it is clear that we must develop new methodologies which are more predictive and shorter-term so that these compounds can be evaluated.

The advances in tissue culture methodology over the last few years have been astonishing [17]. With the development of defined media and defined matrix, the ability to clone cells, and new equipment available, cell culture has been taken from the exotic to the routine.

Lastly, a most compelling reason for the development of *in vitro* methodology may be the products of biotechnology. By this I mean species-specific proteins. As new entities are produced in the biotechnology laboratory, for example, human growth hormone, it is apparent that an appropriate way, and maybe the only way, to evaluate the toxicity, and in some cases the efficacy, of these compounds will be in cultures of human cells.

Societal and Humane Reasons

During the last several years, there has been an increased awareness and pressure on the part of the public to decrease or reduce the use of experimental animals in the areas of fundamental research, toxicology testing, and education. This movement has grown and tends to be focused around the issue of pain. The concerns expressed by the animal protection movement tend to be genuine and raise the significant issue of whether or not the scientific community is aware of solutions to scientific problems which may be based on methodology that is not yet widely accepted.

Both in Western Europe and the United States, there have been significant legislative activities developed to approach the issue of animal welfare. The issue of animal welfare, that is, housing and management of animal use, is essentially common for both animal protection groups and the scientific community. On the other hand, the issue of the use of animals in fundamental research is one that is not completely understood by the animal protection movement and not carefully defended on the part of the scientific community.

Economic Reasons

The cost of doing research and testing is very expensive. *In vitro* methodologies on a routine basis are considerably less expensive, both in terms of resources necessary and time for the study. With the availability of greater opportunities from the commercial sector, these costs will continue to decrease.

INFLAMMATION

Using examples from the work of Norman Dubin [7] at Hopkins and Vincent DeLeo [5] at Columbia, it is possible for one to develop an understanding of an approach currently being applied to the development of the study of inflammation. These authors, independently and in different systems, have approached this area by utilizing the concept that prostaglandin metabolism participates in the inflammatory process and that chemical substances interact with cellular systems to produce alterations in prostaglandin release and metabolism.

The data exemplifies the approach that by understanding the mechanism one will be able to develop an appropriate test. The data that has been developed over the last several years supports the concept that prostaglandin release can be selectively modified and is not the result of leakage or nonspecific effects on the cell or culture system, but is directly related to insult by the chemical agents.

The general construct of using prostaglandin metabolism as an indicator of inflammation in *in vitro* systems is further strengthened by the fact that two independent laboratories, using different cell types and different approaches, have developed data which are not only compatible but confirmatory and suggests that knowing and understanding the basic mechanism of prostaglandin metabolism, one can develop a test system which is quantifiable and predictive of toxicity.

HEPATOXICITY

Hepatoxicity provides the opportunity for studying simultaneously two distinct areas [6,9]. The first of these being hepatocyte toxicity and the second, the study of acute effects of compounds on living cells. The hepatocyte is a cell which has been used in fundamental aspects of toxicology for several years. Since hepatocytes have been a cell used by toxicologists, it was also one of the first cells chosen as an approach to acute toxicity studies. It provides the opportunity to use a mammalian cell. It also allows the study of physiological functions of the cell, such as the production of specific proteins which can then be used as specific markers for the assessment of toxicity. Traditional markers, such as membrane integrity or the leakage of soluble enzymes are far less sensitive.

A most intriguing feature at this early stage of understanding hepatotoxicity is the consistent observation that hepatocytes can undergo significant reductions in the rate of total protein synthesis, while at the same time, specific proteins synthesis can be markedly increased. This offers the opportunity to study specific function of the cell while being able to determine, with considerable sensitivity, specific alterations in overall cellular function.

CARCINOGENICITY

Many assays have been developed to assess the mutagenic and carcinogenetic potential of compounds [2]. The Ames test has become an early standard by which all other approaches were judged, yet the Ames test is clearly inadequate and insufficient to predict carcinogenicity. Considerable other approaches have been developed over the course of the last 15 or 20 years, both for the assessment of carcinogenicity of environmental chemicals and for understanding of the disease process and development of therapeutic approaches. I will not focus on what has been, but will attempt to examine two new approaches to the development of methods for the study and understanding of the process of carcinogenesis and for the development of therapeutic agents for cancer.

Metastatic Processes

A group of scientists, Drs. Kleinman, Martin, Iwomoto and Albini and their colleagues, have developed an *in vitro* assay to study the invasive behavior of tumor cells and to screen antimetastatic drugs and hormones [1,12]. The assay is based on the understanding of biological matrices and the invasiveness of metastatic cells. This system also permits the screening of drugs that can prevent metastasis. If these agents are developed they will be a most useful approach to cancer treatment. The assay developed by these authors is based on the ability of cells to penetrate the biological matrix. Additionally, this assay is predictive of estrogen dependent metastatic activity.

This assay, the "Madrigel" Invasion Assay, represents an approach to the use of tumor cells in identifying metastatic potential as well as drug susceptibility.

Development of Drugs Against Cancer

The headline, 23 December 1986, *New York Times* reads: "Quest for Cancer Drugs: U.S. Devises Major New Strategy. Screening Thousands of Substances: Standard Mouse Gives Way to Technology" [8]. The National Institutes of Cancer in the United States has indicated that an entire new approach to the way it has and will screen for cancer chemotherapeutic agents is under development [3,4]. The change has been necessitated by the slow progress made in developing therapeutic approaches to this disease. The new methodology will be based on tumor cells, human cells, as well as many other *in vitro* test methodologies. A first approach is to use a vital stain as a marker of functional cells. That is, living cells pick up stain while nonviable cells do not. Therefore, it is very possible to screen thousands of compounds rapidly using a combination of cells in culture coupled to computerized image analysis. It is anticipated in this program that more than 100 different types of human cancer cells will be evaluated.

CONCLUSION

In vitro methods in toxicity assessment, mutagenicity and carcinogenicity programs have just begun. Clearly, it is too early in the development of the program to identify at this time which cells, cell lines or cell types will be used in the future. However, what is clear is that they will be used. As recently as the mid 70s, there were few (Nardone [13], Goldberg [10]) that were considering *in vitro* methods in toxicology as viable approaches for the prediction of toxicity. Now less than seven years later, the possibility for routine *in vitro* methodology in toxicology is here.

REFERENCES

1 Albini, A., Graf, J., Kitten, G.T., Kleinman, H.K., Martin, G.R., Veillette, A. and Lippman, M.E. 17 -estradiol regulates and v-Ha-ras transfection constitutively enhances MCF7 breast cancer cell

interactions with basement membrane. Proc. Natl. Acad. Sci., USA, 83:8182-8186, November 1986 Cell Biology.

2 Board of Scientific Counselors National Toxicology Program. Report of the NTP Ad Hoc Panel on Chemical Carcinogenesis Testing and Evaluation, U.S. Department of Health and Human Services, Public Health Service, U.S. Government Printing Office, 1984.

3 Boyd, M.R., Shoemaker, R.H., Cragg, G.M. and Suffness, M. Future perspectives on the anticancer drug discovery program of the National Cancer Institute; potential new avenues for the investigation of marine biologicals. Presented at SeaPharm Conference on Pharmaceuticals and the Sea, October 24-25, 1985, Harbor Branch Foundation, Fort Pierce, Florida.

4 Boyd, M.R., Shoemaker, R.H., McLemore, T.L., Johnston, M.R., Alley, M.C., Schudiero, D.A., Monks, A., Fine, D.L., Mayo, J.G. and Chabner, B.A. New drug development. In: Thoracic Oncology, Part 7, Chapter 51, W.B. Saunders, 1987.

5 DeLeo, V.A., Harber, L.C., Kong, B., and DeSalva, S. Surfactant-induced delayed neurotoxicity. In, *In Vitro* Toxicology, A.G. Goldberg (Ed.), Alternative Methods in Toxicology Vol. 3, Mary Ann Liebert, Inc. 1985, pp.467-482.

6 Din, W.S. and Frazier, J.M. Protective effect of metallothionein on cadmium toxicity in isolated rat hepatocytes. Biochem J., 230:395-402, 1985.

7 Dubin, N.H. Prostaglandin production as an index of *in vitro* cytotoxicity. In, In Vitro Toxicology, A.G. Goldberg (Ed.), Alternative Methods in Toxicology Vol. 3, Mary Ann Liebert, Inc. 1985, pp.43-52.

8 Eckholm, E. Quest for cancer drugs: U.S. devises major new strategy. The New York Times, Tuesday, December 23, 1986, pp.C1 and C12.

9 Frazier, J.M. and Din, W.S. Specific protein synthesis as an indicator of hepatotoxicity. In, Safety Evaluation and Regulation of Chemicals 2, F. Homburger (Ed.), 2nd Int. Conf., Cambridge, Mass, 1983, Karger, Basel, 1985, pp.295-304.

10 Goldberg, A.M. Mechanisms of neurotoxicity as studied in tissue culture systems. In, Toxicology, 17, 1980, pp.201-208. Elsevier/North-Holland Scientific Publishers Ltd.

11 Goldberg, A.M. Approaches to the development of *in vitro* neurotoxicological methods. In, Proceedings of 1986 OHOLO Conference, Israel, in press.

12 Melchiori, A., Allavena, G., Bohm, J., Remy, W., Schmidt, J., Parodi, S., Santi, L. and Albini, A.

Interferons inhibit chemotaxis of transformed cells and their invasion of a reconstituted basement membrane. Unpublished paper.
13 Nardone, R. Summary and synthesis. In, In Vitro Toxicity Testing 1975-1976, J. Berkey and C. Sherrod (Eds.), Collected proceedings of the First and Second Working Conferences of Toxicity Testing In Vitro, 1975/1976, The Franklin Institute Press, Jefferson, Arkansas, November 1977, pp. 356-362.
14 Rowan, A.N. Scientific perspectives in the development of alternatives in toxicology. Comments on Toxicology, 1987. In press.
15 Rowan, A.N. and Goldberg, A.M. Perspectives on alternatives to current animal testing techniques in preclinical toxicology. In, Ann. Rev. Pharmacol. Toxicol, 1985, 25:225-47.
16 U.S. Congress, Office of Technology Assessment. Alternatives to Animal Use in Research Testing and Education, U.S. Government Printing Office, Washington, D.C., OTA-BA-273, February 1986.
17 U.S. Congress, Office of Technology Assessment. Ownership of Human Tissues and Cells, U.S. Government Printing Office, Washington, D.C., OTA-BA-337, March 1987.

Discussion

A. MILLER You only had 2-4 passages for human keratinocytes. In the epidermis you have at least 4 cell layers starting with a germinal base layer. Once a basal cell has divided it is committed to die after about 28 days. My question is do these basal cells have infinite life or not, as we are trying to grow these keratinocytes on microbeads to have a constant production of epithelial cells for grafting etc?

GOLDBERG I don't know what happens to the basal cells in the slide you saw. Boyce's group in California have cloned keratinocytes for many passages and got the skin to grow as a clone. Another group (at MIT) have a matrix in which the keratocytes invade and form a skin-like material.

H. MURAKAMI What is your definition of viability? Dye tests reveal that cells are alive even when they cannot divide.

GOLDBERG If a cell is exposed to a foreign compound it is very important to look at several parameters (1) a physiological or biochemical function (2) you want an end-point eg, an enzyme function (3) you may have a recovery process in some instances.

B. GRIFFITHS Looking into the future a bit the tissue or organ type system which you have described wants to be replaced with a cell system so that one can standardise. Can you see this coming and if so in what time-scale?

GOLDBERG I believe that the toxicologists submitting the data are going to wait until they are comfortable interpreting that data. As methodology becomes available so it will be assessed. There are two factors, the science and the pressures from society to decrease animal testing and yet produce a meaningful approach for safety assessment.

GRIFFITHS Does the use of primary cells constitute a saving in animals?

GOLDBERG It cuts down the exposure of animals to pain.

56 Development of a model for the study of toxicity and carcinogenicity in vitro

C. B. Seymour, C. Mothersill, A. Cusack and M. McDonnell

ABSTRACT

A method has been developed which allows small samples of a variety of human tissues to be grown in culture. Explants established from surgical specimens of breast, bladder, oesophagus, skin, thyroid, larynx and colon can be maintained for at least one month. Cells from the explant migrate initially and then begin to divide. Quantitative analysis of cell growth and growth rate is possible using image analysis and autoradiography. Cell to cell contact and ultrastructural organisation has also been examined.

Preliminary results indicate that carcinogens need to be applied at very specific concentrations in order to stimulate growth. Lower levels of carcinogen than those needed for growth stimulation have no apparent effect; at higher levels a toxic effect predominates.

This technique should prove useful in determining the threshold levels at which chemicals become carcinogenic.

INTRODUCTION

Studies of the effects of carcinogens on isolated systems *in vitro* are one of the main ways for obtaining mechanistic data concerning the precise steps involved in the transformation of the normal cell. One of the major problems with these studies is that very few systems exist which show quantifiable changes *in vitro* which are acceptable evidence of transformation (Borek & Sachs, 1966; DiPaolo *et al.*, 1969; Kakunaga, 1973; Reznikoff *et al.*, 1973) and virtually all of these are systems based on rodent cells of fibroblast origin and the properties are limited to some isolated colonies derived from cultures of established cell lines. Data concerning changes in epithelial, endothelial or mixed cell populations are scarce (Borek, 1972, 1983; Mothersill & Seymour, 1984; Mulcahy et al., 1984; Wigley et al., 1985) and data on human systems are again mainly limited to fibroblast studies (Leavitt & Kakunaga, 1980; Milo & DiPaolo, 1978; Seymour & Mothersill, 1987; Sheela & Kennedy, 1986).

The carcinogenic process in vivo is known to involve several stages (Bannasch, 1984; Berenblum & Haran, 1955; Maskens, 1987; Mottram, 1944) and is now thought to involve interactions between mesenchymal, ecto and endodermal elements (Castellot et al., 1980; Van den Hooff, 1984; Gillman et al. 1955). Clearly studies on isolated and undifferentiated cells are unlikely to provide precise and relevant data.

In the present paper the development of a mixed cell population model system is described. The tissue cultured is from human specimens of various organs resected during surgical procedures. The effect of known chemical carcinogens on the growth of these cultures is described.

METHODS

Surgical samples

Samples of human tissue were carefully dissected from resection samples obtained during surgery. Pieces of mucosa ~2.5 cm^2 were removed from an area remote from the tumour which had a normal appearance. Samples of tissues were placed in a fixative for histological examination while the main samples were placed in containers of complete growth medium, see Table 1. Samples were rushed to the tissue culture laboratory for processing. Generally less than one hour elapsed between removal of the sample and plating of the tissue in culture flasks.

Culture technique

On arrival in the laboratory the tissue was rapidly chopped using a sterile scissors and forceps to give pieces 2 - 3 mm^2. These were placed in a solution of 0.25% Trypsin in 10 ml Earle's Balanced Salt Solution (Gibco Biocult) to which 10 mg/ml collagenase Type IV (Sigma) had been added. The tissue was digested in this mixture for 30 min at 37°C. After this time the tissue was placed in a 90 mm sterile petri dish and suitable pieces were plated in 25 cm^2 tissue culture flasks (Lux) in 2 ml growth medium. Explants were manoeuvred to the middle of the flask and were carried gently to the incubator to prevent dislodgement. They were cultured at 37°C in a humidified atmosphere containing 5% CO_2 in air. Medium was not changed at any stage during the culture period. Approximately 100 explants of tissue could be routinely set up within one hour.

Assessment of response

The cultures were fixed and stained with Haematoxylin and Eosin or Periodic Acid Schiff Reagent to confirm the presence of epithelial cells at intervals described in the Results section. The area of flask covered by cells was measured using a macroscopic or microscopic grid, depending on the size involved. Fibroblastic areas were easily identifiable and were excluded from calculations. Results were expressed as percentage of control area. Results could be converted to give cell numbers by counting the number of cells in a representative number of grid squares and this was done where comparisons of normal response to carcinogens were being made.

Statistical analysis

At least 10 plates were used per dose point and results expressed as mean $\pm$ standard error. Regression lines where appropriate were drawn using least squares analysis.

Carcinogens were obtained from Sigma. They were dissolved in complete growth medium and diluted such that 0.1 ml of carcinogen solution could be added to the culture medium already in the flask to give the desired final concentration.

Autoradiographic analysis

Cultures were exposed to 0.05 uCi/ml ^3H thymdine for 24 hrs. They were then washed, treated with Trichloroacetic acid to remove non DNA bound thymdines and fixed with 70% ethanol. Cultures were then apposed to Hypertech ^3H autoradiography film (Amersham International UK) for 2 - 3 days. The film was then developed using Kodak X19 developer.

Electron microscopy

Cultures were fixed in 2.5% gluteraldehyde followed by post fixation in 1% Osmium tetroxide. They were then embedded in EPON 812. Vertical and horizontal sections were cut and examined using a transmission electron microscope.

RESULTS

Fig. 1 shows a typical growth curve for oesophageal explants cultured for 3 weeks. It is apparent that a considerable increase in cell number occurs over the culture period. The autoradiograph (Fig. 2a) confirms that most of this increase is due to growth of the cells, although an area of non dividing cells can be identified around the explant. Similar results can be observed for bladder urothelial cultures (Fig. 2b). The ultrastructural appearance of the cultures (Fig. 3) confirms that the outgrowth is mainly composed of epithelial cells since characteristic microvilli and desmosomes can be seen. Table 2 lists the average cell number at 2 weeks for a number of human tissues. Since the initial migration comprises about 5×10^3 cells, this means that even if each cell is capable of division, the culture goes through at least five cell divisions in two weeks.

In Fig. 4 the growth of oesophageal explants following exposure to increasing concentrations of Benzo(γ) pyrene is shown. Stimulation of growth is apparent at a concentration of 1 μg/ml. Similar results are shown in Figs. 5 and 6 for β-propriolactone and n-nitrosoethanolamine, two other agents suspected of causing oesophageal cancer in humans. Fig. 5 shows the effect of B-propriolactone (peak growth after exposure to 2.5 μg/ml) and Fig. 6 shows the effect of n-nitrosoethanolamine (peak growth after exposure to 0.05 mg/ml). It can be seen that each carcinogen causes a growth stimu-

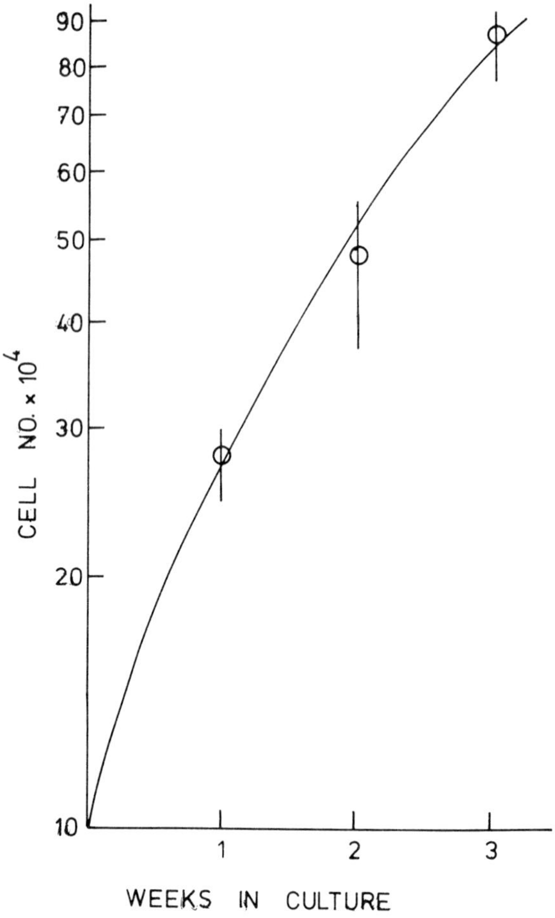

1. Growth curve for oesophageal explants cultured over a three week period.

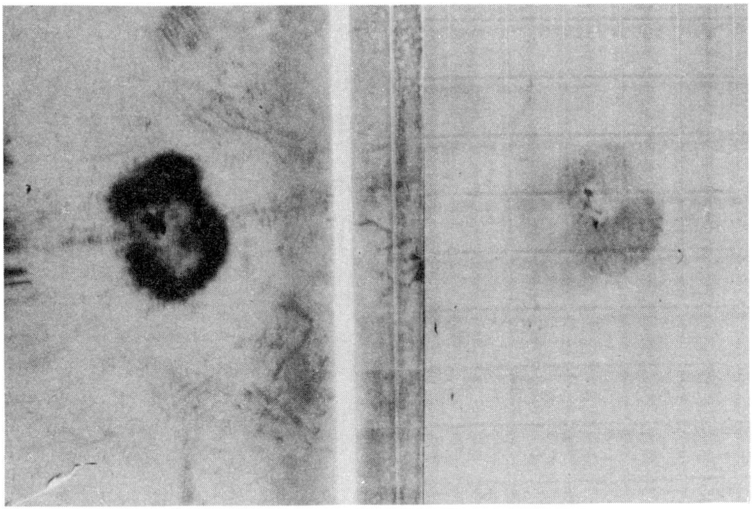

2a. Autoradiograph for an oesophageal explant after one week in culture.

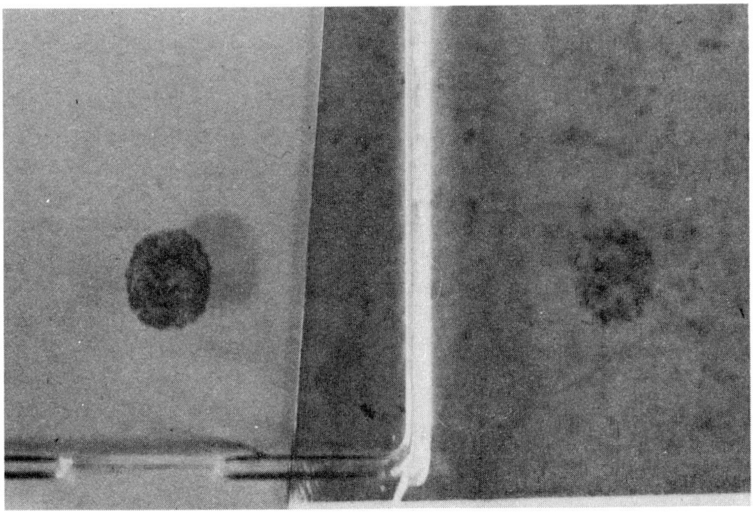

2b. Autoradiograph for a bladder urothelial explant after 1 week in culture.

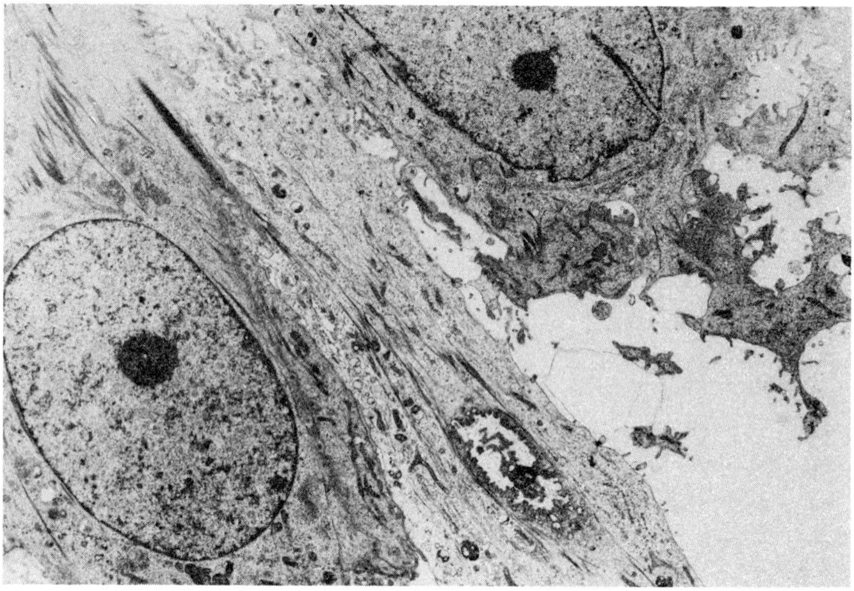

3a. Ultrastructural appearance of an oesophageal culture after 1 week.

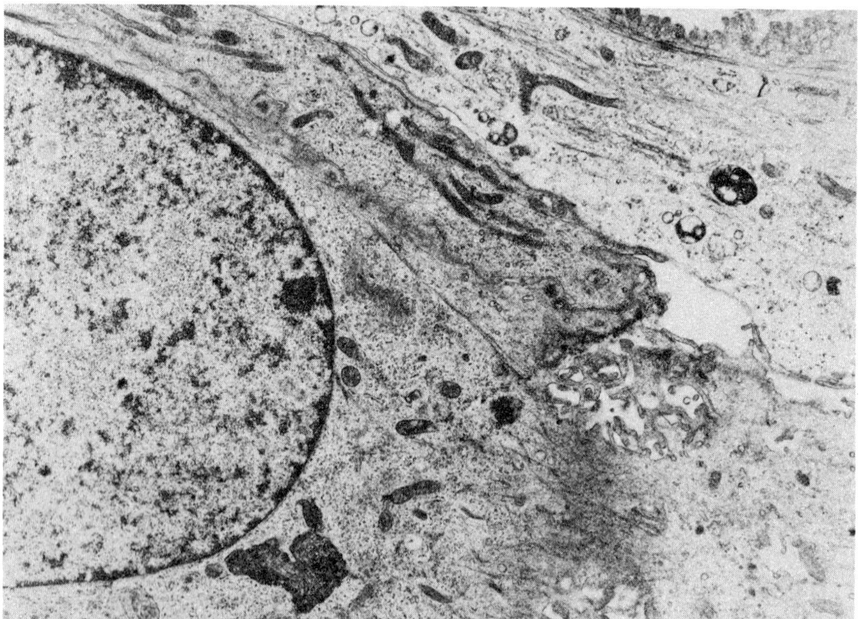

3b. Ultrastructural appearance of a bladder culture after 1 week.

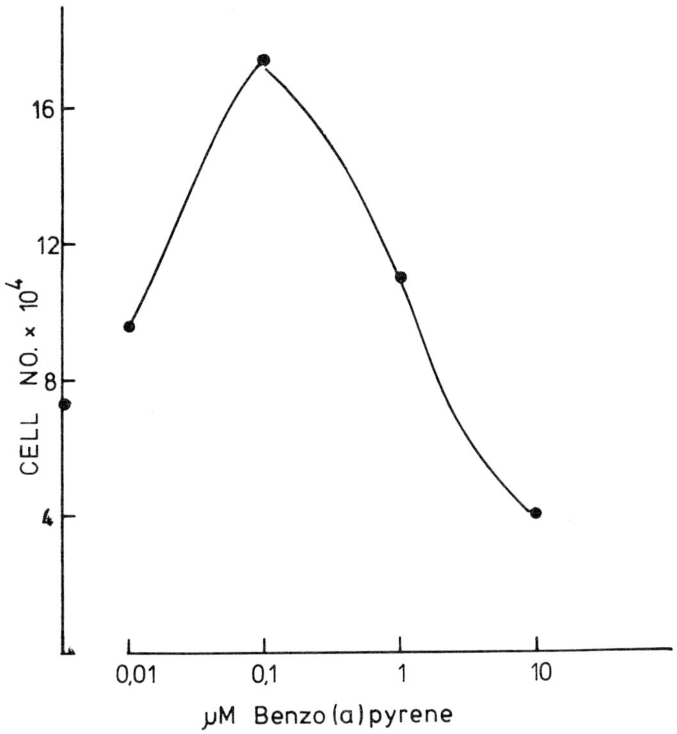

4. The growth of oesophageal explants exposed to increasing concentrations of Benzo(α)pyrene.

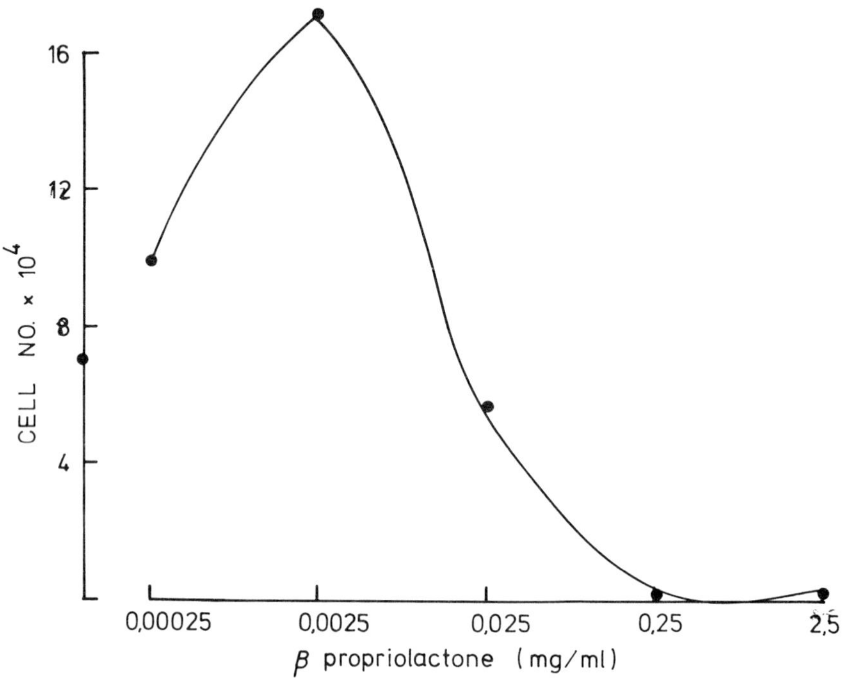

5. The growth of oesophageal explants exposed to increasing concentrations of β-propriolactone.

TABLE 1

COMPOSITION OF GROWTH MEDIUM FOR THE CULTURE OF HUMAN EXPLANTS

Basal medium (Earle's) with L-Glutamine	500 ml
Foetal calf serum	20% V/V
Hydrocortisone (Solucort, Glaxo)	0.1 ug/ml
Insulin (Neusilin, Wellcome Labs)	10 mIU/ml
Penicillin (Gibco Biocult)	20 IU/ml
Streptomycin (Gibco Biocult)	20 ug/ml
Fungizone (Gibco Biocult)	1 ug/ml
Gentamicin sulphate (Cidomycin, Wellcome Labs)	4 IU/ml
Mycostatin	50 IU/ml

TABLE 2

CELL NUMBERS OBSERVED AFTER TWO WEEKS IN CULTURE FOR HUMAN EXPLANT OUTGROWTHS. CELL NUMBERS HAVE BEEN CORRECTED FOR INITIAL MIGRATION.

Tissue of origin	Cell no. on Day 14
Breast	$1.6 \times 10^5 \pm 0.3 \times 10^5$
Oesophageal mucosa	$2.4 \times 10^5 \pm 0.2 \times 10^5$
Urothelium	$0.6 \times 10^5 \pm 0.04 \times 10^5$
Colonic mucosa	$0.09 \times 10^5 \pm 0.01 \times 10^5$
Larynx	$2.9 \times 10^5 \pm 0 2 \times 10^5$
Thyroid	$1.7 \times 10^5 \pm 0.2 \times 10^5$

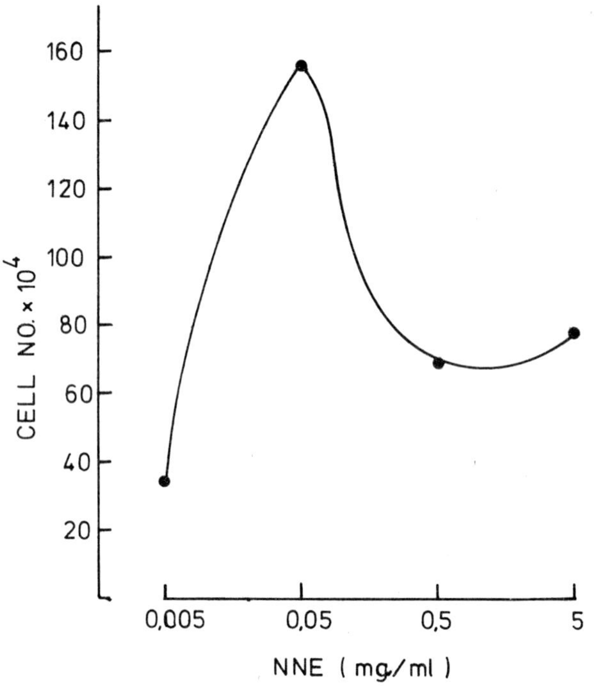

6. The growth of oesophageal explants exposed to increasing concentrations of n-nitrosoethanolamine.

lation at certain concentrations. Future investigations will examine the histochemistry and cell cycle kinetics of these stimulated cells.

DISCUSSION

The results indicate that using samples of a variety of human tissues obtained during routine surgical procedures it is possible to obtain viable growing cultures of epithelial cells. Success of the technique depends on using small amounts of medium and not changing it throughout the culture period. A similar adverse effect of medium changes was observed when culturing thyroid cells (O'Connor et al., 1980) where the effect was found to be due to a high requirement of thyroid epithelial cells for lactate (Mothersill et al., 1981). Preliminary experiments (unpublished) using conditioned medium obtained from other oesophageal cultures indicated that as was the case for thyroid cells, the life span of oesophageal explant cultures can be extended by this means.

Previous authors (Friedman et al., 1981) using explant cultures have found limited cell mgiration, which is followed by terminal differentiation of the epithelial cells and overgrowth of the culture by fibroblasts. In this study epithelial cell growth was confirmed by autoradiographic analysis. Presumably the lack of fibroblast growth is due to the lack of medium changes.

The experiments using carcinogens indicate that a stimulation of growth can occur at specific concentrations of each of the three carcinogens tested. High concentrations of the carcinogens were toxic. In the case of n-nitrosoethanolamine the enhancement of cell growth was very considerable. Other increases were more modest.

The results are interesting in the light of experimental results with animals which also indicate that precise dose combinations are required to produce tumours of various types without killing the animal (Hecht, 1985; Coggle, 1982). Other in vitro work using hamster embryo cells (Di Paolo et al., 1971, 1974) has also shown that the precise conditions of the experiment are important. They showed that greatly increased frequency of transformation occurred following pretreatment of cells with X-rays or alkylating agebts for 48 hours before exposure to a carcinogen (Benzo (α) pyrene). Longer or shorter times of pre-exposure did not increase the normal transforming effect of the carcinogen.

All the chemicals tested are known or suspected carcinogens for oesophageal mucosa in vivo (Hueper & Conway, 1964)

The technique should be useful for studying organ specific transformation by chemicals and other agents. Development of micro assay methods should allow biochemical and histochemical analysis of tumour related antigens or oncogene products in cells and media.

REFERENCES

1. Bannasch, P. (1984). J. Cancer Res. Clin. Oncol., 108, 11-22.
2. Berenblum, I. & Haran, N. (1955). Brit. J. Cancer, 9, 268-71.
3. Borek, C. (1972). Proc. Nat. Acad. Sci. USA, 69 (4), 956-9.
4. Borek, C. (1983). Ann. N.Y.Acad. Sci., 407, 284-90.
5. Borek, C. and Sachs, L. (1966). Nature, 210, 276-8.
6. Castellot Jr., J.J., Karnovsky, M.J. and Spiegelman, B.M. (1980). Proc. Nat. Acad. Sci. USA, 77, 6007.
7. Coggle, J.E. (1982). J. Soc. Radiol. Prot., 2 (1), 15-21.
8. Di Paolo, J.A., Donovan, P.J. and Nelson, R.L. (1969). J. Nat. Canc. Inst., 42, 867-74.
9. Di Paolo, J.A., Donovan, P.J. and Nelson, R.L. (1971). Proc. Nat. Acad. Sci. USA, 68, 1734-7.
10. Di Paolo, J.A., Donovan, P.J. and Casto, B.C. (1974). Chem. Biol. Interactions, 9, 351-4
11. Friedman, E.A., Higgins, P.J., Lipkin, M., Shinya, H. and Gelb, A.M. (1981). In Vitro, 17 (7), 632-44.
12. Gillman, T., Penn, J., Bronks, D. and Roux, M. (1955). Brit. J. Cancer, 9, 272-83.

13. Hecht, S.S. (1985). Clin. Physiol. Biochem, 3, 89-97.

14. Hueper, W.C. and Conway, W.D. (1964). Chemical Carcinogenesis and Cancers. C. Thomas, Illinois.

15. Kakunaga, T. (1973). Int. J. Cancer, 12, 463-73.

16. Leavitt, J. and Kakunaga, T. (1980). J. Biol. Chem., 255 (4), 1650-61.

17. Maskens, A.P. In: Concepts and Theories in Carcinogenesis. Maskens, A.P., Ebbesen, P. & Buruy, A. eds. Excerpta Medica, Amsterdam, New York, Oxford, 1987.

18. Milo, G.E. and Di Paolo, J.A. (1978). Nature, 275, 130-2.

19. Mothersill, C., Murphy, A., O'Connor, M., Seymour, C.B. and Malone, J.F. (1981). Cell Biol. Int. Reports, 5 (9), 877-86.

20. Mothersill, C. and Seymour, C.B. (1984). Cell Biol. Int. Reports, 8 (10), 887-96.

21. Mottram, J.C. (1944). J. Pathol. Bacteriol., 56, 181-7.

22. Mulcahy, R.T., Gould, M.N. and Clifton, K.H. (1984). Int. J. Radiat. Biol., 45 (5), 419-26.

23. O'Connor, M., Cullen, M.J. and Malone, J.F. (1980). Acta Endocrinol., 93, Suppl. 231.

24. Reznikoff, C.A., Bertram, J.S., Brankow, D.W. and Heidelberger, C. (1973). Cancer Research, 33, 3239-49.

25. Seymour, C.B. and Mothersill, C. (1987). Int. J. Radiat. Biol. (in press).

26. Sheela, S. and Kenneddy, A.R. (1986). Carcinogenesis, 7 (2), 201-5.

27. Van den Hooff, A. (1984). Perspectives in Biology and Medicine, 27 (4), 498-509.

28. Wigley, C.B., Trejdosiewicz, L.K., Southgate, J., Coventry, R. and Ozanne, B. (1985). J. Cell Physiol., 125, 156-65.

Discussion

I. YAMANE: How do you quantify cell numbers - with a haemocytometer? You sometimes get overgrowths of your target epithelial cells with fibroblasts. How often do you get this problem?

MOTHERSILL: If we use gentamicin we don't get fibroblast overgrowth.

Another trick is not to change the medium during the test.

We normally get an inch square of biopsy material which allows us to set up 106–200 explants which is sufficient for the test.

A. HANDA — You use serum levels up to 20%. Have you done cytotoxicity responses to serum?

MOTHERSILL — We get a very uniform response of our cells to serum so we don't find this a problem. We always get our serum from the same supplier and have no problems with batch variation.

HONDA — I am just a bit worried that you will find retrospectively that you should have done your work with serum-free medium.

MOTHERSILL — We are hoping to develop a serum-free method in time.

J. NORDENBERG — What substrate do you use for culturing your explants? Do you use extracellular matrix or fibronectin ?

MOTHERSILL — Nothing: We use a very small volume of medium to allow cells to attach.

NORDENBERG — What is the size of the tissue pieces after you have cut them up?

MOTHERSILL — Generally 2-3 mm^2.

NORDENBERG — We have to use single cells dispersed with collagenase for breast cancer specimens.

MOTHERSILL — The difficulty with breast cancer is to get enough epithelial cells out of the fat and fibre tissue.

General Discussion

K. JAMES — I would think that your assay techniques coupled with Flow Cytometry would give a tremendous amount of information. You could study about 1000 cells a second, there are markers for differentiation antigens, dedifferention antigens; markers for cell cycle and ploidy. Access to a cell sorter opens up a great potential for studying mutagenesis.

GOLDBERG Flow Cytometry has been recommended as a technique to be developed in this field and I agree that it will become a very important method in all areas of toxicology.

M. ACKERMAN Dr Katzir said we should superimpose problems to learn something so all week we have talked about growing cells and at the end of the week we talk about killing them. Superimposing these two problems should allow much to be learnt about both.

Using a drug to induce some product in the cell (eg, cadmium to induce metalathionine) we might find something common between growing and killing eg, if we could fuse the promotor of metalathionine to a product that would drive the cell to suicide it might be easy to see the effect of cadmium on the cell by killing it. Do you know of an enzyme that could be used in this way?

GOLDBERG I can't thing of an example but the principle of using custom designed cells is important, especially if you can induce some measurable protein that doesn't kill the cell.

H. MURAKAMI Is it possible to use cell lines to do this kind of job?

B. GRIFFITHS At the moment the answer is probably no - we need to design a cell line for the job.

Section 9
A glimpse of the future

57 The transformation of science fiction into technical reality by using animal cells in culture

R. E. Spier

1. Introduction.

The masters of science fiction of the late nineteenth and early twentieth centuries such as Jules Verne or H. G. Wells would, I imagine, smile ruefully to themselves when they view, from that distant shore, the practical manifestations of what they conjectured a mere 50-100 years ago. Manned spaceships, sattelite communications, television, Cray computers and a planet where infectious disease is on the defensive, (pace HIV), were the ideas of their fanciful books. Today they are so part of our lives that we take them for granted. But what of the science fiction of today which relates to our area of interest; animal cell biotechnology. My selection would include the following examples:

* Frank Herbert's "Dune" trilogy (in six volumes) has, as one of the themes running through it the regeneration of a whole human from a single living cell of that human, (Duncan Idaho). This feat is now commonplace for plant propagation and a small industry is capitalising on that ability, so the future may include an animal variant of this technique. (The fiction that Idaho was also capable of regenerating his memories will require

a little more development for its translation to practicality but it is for the animal cell biotechnologist to unravel the secret properties of the "axolotl" regenerating tanks).

* The instigation of human life in the test-tube (petri-dish) and the maintenace of that fertilised ovum for a period of two to three weeks through to the bastula phase in vitro is a discovery of recent times (Steptoe). It has also become commonplace for babies born after their 20th week gestation to be nurtured into fully developed and normal adults. That the 17 week gap between these two capabilities will narrow is inevitable. Whether it will come to the situation depicted in Aldous Huxley's "Brave New World" where humans were developed from fertilised ova through to independent existence in a "factory-type" environment remains for the animal cell biotechnologists of the future to determine.

* The relationship between the animate and the inanimate continues to become increasingly intimate. The signals of truncated nerves can be made to drive electro-mechanical devices in the form of artificial limbs. Why not then give a modicum of serious consideration to Terence Dick's creation, "The Dalek". This mechanical entity is directed by an on-board naked biological and cellular " brain".

* Finally, the pains and pleasures of the examination system may become a thing of the past were the extended (R.Ė.S.) thoughts of Clive Sinclair to become a reality. Into one cubic centimetre of silicon it is possible to encode a library's worth of information. Were this small entity to be implanted into the brain of a human in such a way that the information in the silicon could become available to the consciousness of the

bearer, then the end of exams, (and their marking), would be with us.

To achieve such ends we will have to adopt new ways of thinking. It is no longer viable to separate "science" from "techniques". Science encompasses,

* knowledge of the composition of our physical world, (analytic),
* understanding of how that physical world works, (analytic),
* the knowledge base required to make things, (synthetic).

Techniques on the other hand involve,

* science,
* art,
* skill,
* craft,
* capability.

It is the thesis of this author that it is only by the joint, interactive and synergistic application of science with techniques that the science fiction of today will become the reality of tomorrow.

That much progress has already been made as will be evident from the succeeding sections of this paper which will highlight the most recent developments and indicate how they are likely to contribute to the transformations depicted above.

2. Uses for Antibodies

The well known uses of antibodies are for passive immunisation, disease diagnosis and medical imaging, material purification and in the construction of the magic bullets of Paul Ehrlich. There could be other applications involving the use of monoclonal antibodies to so mark particular cells that they become the targets of the cytotoxic and natural killer cells of the body. Such targets could include those body cells whose specific function it is to store fat. Another would be those cells whose uncontrolled growth causes death. Or the monoclonal antibodies could delineate cells infected with the HIV virus series thus leading to their eventual demise.

There are presently in use two putative therapeutic techniques which seek to use antibodies for the elimination of particular cell types in vivo. On the one hand there is the anti-OKT3 monoclonal antibody which is effective in decreasing the immune response to transplanted kidneys by damaging the capability of the T-helper cells required to mount the immune response. This technique may also be applied in the case of bone marrow transplants often used in attempts to cure leukemia. On the other hand there could be beneficial effects resulting from the elimination of the effects of the T-suppressor cells so as to enable the T-helper cells to gain the upper hand in the removal of the excessive
number of lymophocytes generated in a leukemic individual.

Antibodies have beem made to the receptor sites of hormones and have been shown to cause effects which mimic the effects normally caused by the specific ligand. In this way it could become possible to specifically control the production of each or any of the hormones and their antagonists. This in turn could lead to the development of humans and animals in ways which could promote their well-being and/or productivity.

There are also some problems which have to be solved. For example, antiidiotypic antibodies secreted by human-human hybridomas, which could be used therapeutically to cure Hepatitis B or the infection caused by Cytomegalovirus, may generate an immune reaction to the part of the molecule which is foreign to the recipient; the antiidiotype paratope. Also making magic bullets out of the 'homing vehicle' of the monoclonal antibody and a warhead of a cytotoxin can also be problematic, as the cytotoxin may be immunogenic and the antibodies so raised would neutalise the beneficial effects. This would also occur were a toxin to which the body has already been exposed (diphtheria) to be used, as this would act as a restimulant to immunisation.

While it is possible to make antibodies in mouse cells which are chimeras of the human Fc portion with the mouse antigen combining section and thus overcome some of the foreigness of the wholly mouse antibody, there have been some objections raised (in Japan) to the other murine proteins present in the final product which would of themselves provoke an immune reaction. This would be less of a problem for the products of a human-human hybridoma cell line as the contaminating proteins would not be xenotypic. Yet it should be recognised that the homologous oncogenes of human cells probably (but not necessarily) present a greater hazard than those of murine origin: a hazard which reduces to vanishing proportions in view of the limits placed on the amount of such DNA allowable in the final product dose (10pg).

3. Animal Cells as Products.

Modified Autologous Cells (MAC's) are cells which have begun their lives in a particular individual and then removed from that individual for exogenous modification before they are reinserted. To achieve this effectively it is possible that the cells will have reproduced and will have been treated with various modifiers in the in vitro environment. A number of such systems are presently under examination or in some circumstances more freely available.

* The lymphocyte activated killer (LAK) cell system explored by Rosenberg and others involves the removal of the lymphocytes of a cancer patient and the treatment of those cells with concentrations of materials like interleukin 2 or interferon which would give rise to toxic symptoms were they administered to an individual directly. After exposure to such modifiers the cells are reinjected back into the patient with sub toxic doses of immunomodifiers. Experience has shown that circulating lymphocytes do not perform as well as lymphocytes which have been obtained from the tissues of the cancerous material itself. It is clear that there is much to learn about such techniques and in particular the way in which the cells which have been extracted can be both amplified and specifically modified before they are readmitted into the patient.

* Skin cells may be removed from undamaged parts of a patient suffering from burns and then amplified in the laboratory before they are returned to the patient in a way which will promote the healing of the extensive skin damage suffered. More recent work has shown that it may be possible to grow cells of one individual

and apply them to another after a period of in vitro cell culture. It is hypothesised that such culture conditions depress the expression of the MHC genes which give the cells their individual based uniqueness.

* There have been experiments on dogs which have demonstrated the possibility of grafting brain cells from one individual to another. Experiments have begun with humans on the replacement of those cells, which due to a biochemical defect, have resulted in a depletion of dihydroxyphenyl alanine (DOPA). According to a newspaper report embronic brain cells used for this purpose would migrate to the appropriate part of the brain, colonise that area and behave as if they were normal cells.

* The use of the cells of the islets of langerhans for the in situ generation of insulin has to be a technique which in the future will provide those individuals who need regular injections of insulin with an on-board method of providing that missing ingredient. There are two approaches. One of them is based on the use of the Damon encapcel process while the other is centred on the use of cells cultured in vitro prior to innoculation.

* Similar replacement/modification therapy is under investigation for the haematopoietic stem cell system. In this area the over production of lymphocytes or the under production of erythrocytes can lead to disease.

The growth of organs in culture and the production of highly defined cell cultures is leading to great strides in the improvement of our abilities to test drugs for both beneficial effects and for toxicity effects. Were it possible to so standardise animal cells so that they became like any other distributed reagent then again another major step forward will have been achieved.

4. Products from Recombinant Animal Cells.

A computer generated list of publications issued between 1983-1986 threw up 76 relevant references describing some 40 different products or expressed materials of which the most popular were:-

* Hepatitis B surface antigen	6 references
* Factor IX	6 references
* Interferon-gamma	5 references
* Growth Hormone	5 references
* Interleukin 2	4 references
* Tissue Plasminogen Activator	3 references
* Chimeric antibodies	3 references
* Interferon-beta	3 references

There is little doubt that the postranslational modifications of the native recombinant proteins is a key element in the final efficacy of the product based on expressed recombinant animal cells. The direct evidence that exists indicates that gamma-interferon produced in bacteria is less of a problem than glycosylated versions produced in yeast cells. Whereas there is some evidence that the glycosylated version of erythropoietin produced from recombinant animal cells has a life in the body some 100 times longer than its non-glycosylated bacterially expressed counterpart. However, there yet remains to be answered the question as to the factors which control the glycosylation of the native proteins. Such factors need not only include the source of the cell expressing the material but also the way that cell has been conditioned or handled in culture as well as the particular conditions of the culture at the time the protein is expressed.

5. Cell Technology.

The last 5 years have witnessed the transformation of the cell technology scene. From a base of monolayer cell culture in bottles, multiple plate and tube systems, packed static beds and microcarriers and the co-developed suspension cell culture systems based on stirred tanks and air-lifts there now has come into existance a number of cell generating and exploiting techniques which require the cells to be held in the bioreactor at local cell concentrations in excess of 10^8 cells/cc. Such bioreactors can be built around parallel planar membranes, hollow fibres, corruscated ceramic cartridges, hollow microspheres, beads of carbohydrate gels, collagen matix beads and many others. In such dense cell systems it would seem that the cells provide materials for one another which enables reactors to be operated with low concentration and less expensive proteineceous media.

However there is yet insufficient data to judge whether the specific yield of product/cell/unit time is radically increased in such systems. Similarly more data is needed to determine whether the concentration of material in the product stream, (not in the bioreactor and local to the cells), is greater in these perfusion reactors. In spite of such uncertainties there is adequate reason to believe that as the media into which the product is expressed is low in total protein the removal of product proteins from such feed streams is likely to be more efficient than would have been the case in more 'polluted' discharge streams.

Thus we still need to know how to run a bioreactor containing animal cells at concentrations approaching those found in animal tissues in vivo such that the supplied medium contains only those components required for product formation and for the maintenance of the production system. In addition to this there should be available an on-line self optimisation control and regulation system so that the system as a whole is operated at its most productive level.

6. The problems remaining.

Do lactic acid and ammonia build up to concentrations which are the cause of cell death or decreases in productivity ? How is it that some cell lines can 'tolerate' higher concentrations of both of these waste products than other cell lines ? What is the mechanism of toxicity and how can the effects of such toxicity be used to better control the state of differentiation of the cell and hence its overall productivity of a secreted material ? There is also the possibility that there are end product negative feed-back effects which operate in some cell systems and not in others.

On the physico-chemical side we know little of the effects on the cell of subjecting such life-forms to shear stresses nor do we fully understand the mechanism whereby the way in which a cell is attached to a surface controls its physiology and its ability to produce enzymes and viruses. Finally, while the work of Eagle in the 60's has shown that as the cell concentration in culture increased there is a sparing effect on the number and concentration of several of the monomeric materials needed in the formulation of the culture medium, there are few, as yet, indications as to how increases in the cell concentration in the bioreactor results in the ability of the cells to grow, divide and produce in a serum free and often a protein free environment.

7. Conclusion.

Cells can be grown and used in the in vitro environment at concentrations approaching those found in body tissues. From our knowledge of hormones and hormone receptors coupled with our ability to activate or mimic such materials using specific antibodies which can be produced ad libitum in culture, we are on the verge of controlling the way, and the extent, to which cells differentiate. Furthermore we have begun to appreciate that animal cells in culture may have interesting, surprising and probably useful properties when they are held in a non-homogeneous graded molecular fields.Surely this is looking like a serious approach to the foothills of those challenges set out by the science fiction writers mentioned in the introduction. In any event we have a long way to go. As the visions of the science fiction writers of the nineteenth century were realised in the techniques of our times so is it likely the the dreams of the science fiction writers of our day will have a realisation in the future.

Discussion

P. GRAY — We seem to have a pre-occupation with cell density. If we look at recombinant animal cells it is usual to get 20-50 ug of products per million cells. It is not difficult to get 10^7 cells/ml in a conventional reacter, maintained for weeks on end. This means a productivity of 200-500 mg/L/day which is good. So do we really need to go higher? I would have thought problems like medium optimisation rather than cell numbers should be studied.

SPIER We are looking for more than productivity. High cell density can get you into cheaper media faster but what will happen to specific productivity I do not know. You were defining productivity in terms of mg/ml which in fact, even in the densest system is determined by how much you harvest - and not on how concentrated your cells are. The idea is to give the environment in which the cells work best.

GRAY One of the problems in tissue culture is the desire to perpetuate this black art. In biotechnology the whole question of units, terms, dimensions etc is well worked out and documented. What we need is everybody to compare their systems in these terms (as laid down by UPAK) so we can get away from the magic box attitude.

Postscript

58 Terms and symbols for use in animal cell culture

P. P. Gray

There has been a rapid expansion in interest in recent years in animal cell culture. This interest has resulted from developments in the technology for growth and product formation by hydridomas and recombinant mammalian cells. The increased interest has led to the development of a range of novel bioreactors suitable for the growth of animal cells. Whereas traditionally animal cells were grown mainly in Roux flasks and roller bottles, they are now being grown in suspension cultures at scales up to 10,000 litres. In addition, novel forms of reactors involving membranes and other configurations designed for process intensification have been developed specifically for large scale animal cell growth.

The increased need for accurate kinetic data and for means of comparing efficiencies and productivities in different reactor configurations has emphasised the need for standardisation in the calculation and reporting of performance for animal cells in culture. Such problems are not new in biotechnology. In an attempt to minimise such problems with more traditional fermentations, the IUPAC Commission on Biotechnology published in 1982 a list of symbols with units recommended for use in describing fermentation parameters (1). These symbols are widely used in the biotechnology literature. It is felt that, with a few minor modifications, these units would be suitable for tissue culture work. A list of the more important symbols is shown in Table 1. Implementation of the use of a standard set of units would simplify considerably comparison of results appearing in different publications, and aid the evaluation of various bioreactor systems.

Looking at the symbols and units in greater detail:

<u>Cell concentration:</u>

IUPAC recommends that x be used as the symbol for biomass concentration (mass per unit volume). It is more common

TABLE 1

Description	Symbol	S.I.Units	Customary Units
Number concentration of cells	n	m^{-3}	mL^{-1}
Total number of cells	N	1	1
Substrate concentration mass or moles per unit volume	c_s	$kg.m^{-3}$	$mg.L^{-1}$, $mmol.L^{-1}$
Product concentration mass or moles per unit volume	c_p	$kg.m^{-3}$	$mg.L^{-1}$, $mmol.L^{-1}$
Inhibitor concentration mass or moles per unit volume	c_i	$kg.m^{-3}$	$mg.L^{-1}$, $mmol.L^{-1}$
Saturation constant as in growth rate expression $u = u_m c_s/(K_s + c_s)$	K_s	$kmol.m^{-3}$	$g.L^{-1}$, $mmol.L^{-1}$
Inhibition constant	K_i	$kg.m^{-3}$	$g.L^{-1}$
Yield, general mass ratio expressing output over input.	Y	1	1
Yield, growth mass ratio corrected for maintenance, where $\frac{1}{Y} = \frac{1}{Y_G} + \frac{m}{u}$	Y_G	1	1
Maintenance coefficient non-growth term associated with substrate consumption	m	s^{-1}	h^{-1}
Specific growth rate $u = (dn/dt)/n$	u	s^{-1}	h^{-1}, d^{-1}
Specific metabolic rate $q = (dc/dt)/n$ where c may be a substrate or product mass concentration. Subscripts may further define the rates, eg q_s, q_p, q_{O_2}, substrate uptake, product formation and oxygen uptake rates respectively.	q		$\mu gms.10^6 cells^{-1}.h^{-1}$

in tissue culture work for cell counts to be used, and accordingly the IUPAC symbol of n for cell number concentration is more appropriate. The symbol N for total cell numbers, eg total cells in a reactor is also suitable. If necessary these terms can be qualified by measurements of cell viability.

Concentrations of media components: The symbols $c_{s,p,i}$ should be used to denote concentrations (mass or moles per unit volume) of substrates product, inhibitor, respectively. $K_{s,i}$ to be used as saturation and inhibitor constants.

Yield terms: In spite of the fact that tissue culture media are complex, it is becoming increasingly common for yield terms to be calculated for various key ingredients. Given the relatively high maintenance energy requirements as a percentage of total substrate uptake, the two possible yield terms, Y and Y_G need to be carefully defined.

Specific metabolic rates: IUPAC recommends the use of $q_{s,p}$ to denote the specific rate of substrate uptake or product formation, defined as $q = (dc/dt)/n$. Common units for say q_p would be micrograms of product formed / 10^6 cells/hr.

The specific rate of product formation is an important variable for comparing productivities under different environmental conditions or with different reactor configurations. Where sufficient data points exist, it can be calculated and plotted as a function of time. Alternatively it can be averaged over a period of time or a whole production run for either simple batch systems or for complex reactors employing media feeds or recycle streams. In the latter case it is necessary that the q_p be calculated for the whole system, independent of internal streams, eg if the volume of conditioned medium leaving the system is V over a time period t and with a product concentration of c_p, then:

$$q_p = c_p . V. 10^6 / N.t \quad (\mu gms/10^6 cells/h)$$

For q_p averaged over the period t, and c_p in units of μgms/ml, V in mls, N total cells in the system and t in hours.

(1) "List of symbols with units recommended for use in Biotechnology"
Prepared for publication by R.C. Righelato, IUPAC Commission on Biotechnolgy, Pure and Appl. Chem., 54 (9), 1743-1749,1982.

Late paper

59 Hybridoma culture for the continuous production of monoclonal antibodies; a fractionalized factorial design experiment for process optimization

I. Wergeland, C. Wallberg and R. Ericsson

ABSTRACT

Monoclonal antibody production by cultured hybridoma cells is known to be influenced by several factors including pH,temperature, and dissolved oxygen.

In this work, a statistical method - reduced factorial design - has been applied to determine the statistically significant factors that affect cell growth and antibody production.A hybridoma cell line I.F6 was cultivated in DME supplemented with 4.5 grams of glucose/l and 0.5% FCS.The dissolved oxygen tension (DO), pH and dilution rate were manipulated systematically in two chemostats working in parallel during a 73 days period. The upper and lower factor levels were set at: pH;7.3 & 6.9, pO2;15% & 40%, dilution-rate; 0.62 & 0.48 reactor volum per day respectively. Temperature was controlled at 37°C. The concentrations of cells, glucose, lactate and monoclonal antibodies respectively were monitored off-line daily. Significant effects on these process values were observed as a result of changing pH, dissolved oxygen tension and dilution rate.Steady state cell concentrations were found to vary between $0.1*10^6$ and $1.2*10^6$ cells per ml.The corresponding antibody levels varied from a low 0 μg per ml to 20 μg per ml respectively.

We conclude that, within a 95% confidence interval,

1) pH and dilution rate were found to have a significant effect on cell productivity
2) changes of dissolved oxygen values between 15% and 40% had no significent effect on the cell productivity
3) monoclonal antibody productivity within the tested factor-levels was a function of the cell concentration only.

INTRODUCTION

The increasing interest for the production of pharmaceutical products in animal cells has focused attention on the need for a better understanding of the factors that affect culture parameters such as cell growth and product formation.This is particularly important for the design and scale up of cell culture to the production scale.

The primary objective of this study was to examine the effect of variation in pH, dissolved oxygen, and dilution rate on cell

productivity and monoclonal antibody formation.
In addition, we wanted to demonstrate the maintenance of a long term continuous culture (up to 3 months) without cell feedback.

Two important 'tools' are available for such studies.The first is the continuous culture method as described e.g. by Micael Tovey (1,2) who has applied the technique of continuous cultuvation on mammalian cells. The second is the method of statistical experimental design as outlined by G.Box et al (2), and J.Staeheli (3).

MATERIALS AND METHODS

GENERAL

A factorial design experiment is a systematic study of some factors believed to be important for the bio-process.(In our case pH, dissolved oxygen (DO) and dilution rate (D).)

Such a study is composed of a series of experiments, each performed with the chosen factors fixed at certain levels.
In this work hybridoma cells were cultivated in continuous culture of the chemostate type. Four experimental runs were necessary to determine the effects on the response parameters caused by the factors pH, DO, and dilution rate. The cell-, antibody-,glucose- and lactate concentrations were used as response parameters. The first run in the sequence was repeated at the end of the experiment, in order to calculate the standard deviation caused by instrumental drift. The four runs were performed in parallel by the use of two fermenters. As a result a total number of 10 runs were performed. The reason for performing experiments in parallel were especially two-fold; to increase the number of observations to obtain a more precise statistical evaluation of data and secondly to provide a 'back-up' in the event of a process failure in one of the fermenters.

CELL CULTURE

The hybridoma cell-line, IF6, producing a monoclonal antibody against Gonnococcus was cultivated in DME. The medium contained 4.5 gram glucose per ml and was further supplemented with 0.5% FCS, penicillin (100 U per ml) and streptomycin (100 μg per ml). The glutamine concentration was 0.58 gram per liter (4mM).

The two fermenters were each inoculated with $1*10^5$ cells per ml, and the culture volumes were kept at 700 ml. Dissolved oxygen was measured with a galvanometric electrode and pH by a standard Ingold electrode.

The gas exchange was performed by surface aeration and the pH was controlled with CO_2 and 1MNaOH. The fermenters were standard 2000 ml glass vessels with stainless steel top-plates. The temperature was regulated via a water jacket. Stirring was by a magnetically coupled flat blade impeller.

Medium was prepared from powder (Gibco) mixed in water of double destilled quality. The maximum storage time was one week (+4°C).

Samples were taken regularly - normally once a day - for cell counts, monoclonal antibody-, glucose- and lactate analysis as well as for pH calibration.

The temperature probes, pH probes as well as perfusion flow (or dilution rate) were checked/calibrated every day. The oxygen electrodes were checked twice in the experimental period and indicated a drift of less than ±5%.

EXPERIMENTAL DESIGN

Table 1 shows the 2^{3-1} fractional factorial design experiment where three selected factors (variables) were tested at two levels each in four runs (or sub-experiments).

The cells were inoculated at $1*10^5$ cells per ml in both fermenters and run II was started with DO=40% (or $P_{O2,1}$ = 0.8 atm), pH = 7.3 and D = 0.48 d^{-1} (48% of the culture volume continuously replaced with fresh medium every 24 hours). The steady state, as determined by constant glucose and lactate values, was normally established after 3-5 days. As soon as the steady state was established and observations terminated, factors were set for the next run. The experimental sequence was run II, III, IV, I and finally II again. The total experimental period was 70 days.

Our reasons for choosing these particular factor levels are breafly outlined below:

pH:

a) Litterature studies suggested pH 7.2-7.4 to be optimum.

For DME with 3.7 gram $NaHCO_3$ per liter in contact with a 5-8% CO_2 atmosphere the pH will stabilise at 7.3-7.4. Consequently we wanted to test one pH value in this range.

b) In closed batch-cultures the buffer capacity will normally be to low to compensate for carbondioxide and lactic acid formed in cell densities above $5*10^5$ cells per ml. As a result pH will be seen to decrease to a low 6.6-7.0 toward the end of the culture period. A pH value of 6.9 was consequently chosen as the second factor level.

Dissolved oxygen:

The concept of an 'optimum' oxygen tension for cell culture remains somewhat speculative, but values between 25% and 60% has been mentioned (5). We simply wanted to investigate if an optimum within the 15-40% range could be shown for the IF6 hybridom.

Dilution rate:

The only criteria for the factor settings were to keep them both well below a 'critical' dilution rate. 'Critical' referes to a dilution rate where wash-out occurs. This corresponds to the maximum cell doubling time observed for the cell line. For the cell line used in this experiment the maximum doubling time was 12-14 hours resulting in a critical dilution rate of about 1.1 per day.

Table I: Experimental design.

RUN (no)	FACTORS		
	DO(%)	pH	$D(d^{-1})$
I	15	6.9	.62
II	40	7.3	.62
III	40	6.9	.48
IV	15	7.3	.48

RESULTS AND DISCUSSION

RESPONSE PARAMETERS

As shown in Table II below, dramatic alterations of the response parameters were seen as a function of the different factor combinations. The data represent the average from the two chemostats working in parallel. The results from the second test of run II are indicated with brackets in all tables.
Glucose and lactate analysis was performed to determine when steady state was reached and those data are not reported futher.

Table II: Steady state response parameter values (Mean values of the two parallel fermenters.

RUN (no)	FACTORS			RESPONSE PARAMETERS AT STEADY STATE			
	DO(%)	pH	D(d-1)	Cell conc. (105/ml)	MAB conc. (mg/l)	Glucose (g/l)	Lactate (g/l)
I	15	6.9	.62	1	.75	4.2	.2
II	40	7.3	.62	6.3 (6.3	6 (8)	1.9 (2.5)	1.9 (1.8)
III	40	6.9	.48	5.3	9.0	3.5	0.8
IV	15	7.3	.48	12.5	14	.5	3.9

As shown in Table III, pH and dilution rate have significant effects on cell- and antibody concentrations, pH in particular.
As the antibody concentration is a function of cell concentration, those data only serve to point at the relation between the factors and the cell concentration at steady state.

Table III: Factor effects on the response parameters

FACTORS	EFFECTS ON RESPONSE PARAMETERS			
	Cell conc.	MAB conc.	Glucose	Lactate
DO (%) (15/40)	.5	.3	.5	-0.4
pH (6.9/7.3)	3.1	2.8	-1.2	1.2
D (d-1) (0.62/0.48)	2.6	3.9	-0.9	0.6
95% Conf.int.	±1.1	±2.2	±0.6	±0.6

PRODUCTION AND PRODUCTIVITY

In multiplying dilution rate with the corresponding response parameter values on specific cell- and antibody productivity data are obtained, Table IV. The effect on those response parameters are listed in Table V and indicates that:

a) pH is an important factor with 7.3 as the best alternative

b) the dilution rate is an important factor, with a change of 48% of the culture volume per day with fresh medium as the best alternative. (Thus, if one wants to reduce the medium consumption while keeping the total productivity constant, dilution rate could possibly be decreased further!)

Table IV: Cell and antibody productivity as a function of the four facto combinations

RUN (no)	FACTORS			PRODUCTIVITY (P) AT STEADY STATE	
				P_c=Productivity of cells	P_{mab}=Productivity of antibodies
	DO(%)	pH	$D(d^{-1})$	10^6 cells produced / liter and day	mg antibodies produced / liter and day
I	15	6.9	.62	62	.5
II	40	7.3	.62	400 (420)	3.9 (5.1)
III	40	6.9	.48	260	3.5
IV	15	7.3	.48	600	6.5

Table V: Effects on the cell and antibody productivity by the selected factors.

FACTORS		EFFECTS ON PRODUCTIVITY (P)	
		Pc	Pmab
DO (%)	(15/40)	-0.2	0.3
pH	(6.9/7.3)	1.7	1.5
D (d-1)	(0.62/0.48)	1.2	1.6
95% Conf.int.		±0.7	±1.4

Finally in Table VI data are recalculated to show the steady state antibody productivity per million cells and day - antibody volumetric productivity divided by the corresponding cell concentration. There are no significant indications that the factor levels affect antibody productivity on a per cell basis, see Table VII.

Table VI: Antibody formation per unit cell-time as a function of the four factor combinations.

RUN (no)	FACTORS			SPECIFIC MAB PRODUCTIVITY AT STEADY STATE
	DO(%)	pH	$D(d^{-1})$	microgram antibodies / million cells and day)
I	15	6.9	.62	4.7
II	40	7.3	.62	6.0 (7.6)
III	40	6.9	.48	8.6
IV	15	7.3	.48	5.4

Table VII: Effect on antibody formation per unit cell-time by the selected factors.

FACTORS		EFFECTS ON SPECIFIC ANTIBODY PRODUCTIVITY
DO (%)	(15/40)	0.13
pH	(6.9/7.3)	-0.03
D (d^{-1})	(0.62/0.48)	-0.07
95% Conf.int.		±0.15

GENERAL DISCUSSION

In this work we identify some factors of importance for optimal cell-mass and antibody production for one particular hybridoma cell line, IF6. We find, in this first screening experiment, pH and dilution rate to be of critical importance, and dissolved oxygen in the range tested to be of no importance.

To conserve time and energy in the otherwise demanding task of optimizing our process, we have chosen a combination of factorial design and chemostat culture. We find this to represent a powerful 'tool', which is recommende for process optimization.

As it is well known that hybridoma cells elicit individual characteristics, our findings are not necessarily directly applicable to other hybridoma. Also, as our data were obtained under chemostat growth rate control, caution should be exercised in applying these data to batch or fed-batch processes.
This because these cultures operate under a changing set of non-steady state condition e.g. with respect to nutrient and waste concentrations.

For the chemostate in operation the cellular growth rate must be under the control of one or several limiting nutrients. It is in this content interesting to note that in all factor combinations tested we find the steady state concentration of remaining glucose and lactate formed stoichometrically related to the glucose content of the continuous inflowing fresh medium (see Table II).

Furthermore, steady state concentrations of glucose are at all times well above limiting concentrations (6). Our data therefor suggest nutrients/factors other than glucose/lactate to be limiting. The nature of these limiting nutrients will be the object of further studies.

The fact that no effect on the IF6 hybridoma productivity could be shown by varying dissolved oxygen values in the range 15-40% is of particulare importance. This because oxygen availability is often the prime limiting factor in the scale-up of mammalian cell proceses. As the optimum for dissolved oxygen for this hybridoma must be lower than 15%, this should facilitate scale-up.

Finally, as we find monoclonal antibody productivity to be directely propotional to cellmass - under the experimental conditions - we see little need for an on-line monitoring of antibody concentration in our optimization studies. We will consequently be able to save time and effort by performing spot-checks to ensure this relation to hold true as our factorial design experiment continues towards a fully optimized process.

REFERENCES

1. Box,G.E.P.,Hunter,W.G. and Hunter,J.S.(1978).Statistics for experimenters, J.Wiley & Sons, New York.

2. Tovey,M. and Brouty-Boye,D.(1976).Characteristic of the chemostat culture of murine leukemia L1210 cells. Eksperimental cell research, 101, 346-354.

3. Tovey,M. and Brouty-Boye,D.(1976).The use of chemostate to study the relationship between cell growth rate, viability, and the

effect of interferon on L1210 cells. Eksperimental cell research, 118, 383-388.

4. Staeheli,J.(1987).Variable screening and optimization. Develop.biol.Standard., 66, 143-153.

5. Reuveny.S. et al (1985).Production of monoclonal antibodies in culture. Develop.biol.Standard., 60, 185-197.

6. Hu, Wei-Shu, 1986. Lecture at the course 'Mammalian cell reactor engineering', Minneapolis-St.Paul, US.

List of participants

ACKERMANN	Dr.	Mathias	Eidg. Vakzine-Institut Hagenaustr. 74 CH-4025 Basel SWITZERLAND
ALEXANDERSSON	Mr.	Sivert	Electrolux Fermentation 205 25 Malmo SWEDEN
ANDRE	Mr.	Guy	S.G.I. Setric Genie Industriel 15 Allees de Bellefontaine 31100 Toulouse FRANCE
ARATHOON	Dr.	W. Robert	460 Point San Bruno Avenue South San Francisco, CA 94080 USA
ARCHINARD	Dr.	Philippe	Bio Merieux Marcy l'Etoile 69752 Charbonnieres les Bains Cedex FRANCE
BAGNARELLI	Dr.	Patrizia	Institute of Microbiology Service of Virology Reg. Hosp. Torrette I-60100 Ancona ITALY
BAHNEMANN	Dr.	Hans	Pan American FMD Centre Caixa Postal 589 20001 Rio de Jeneiro BRAZIL
BAUD	Mr.	Henri	Roussel Uclaf 111 Route de Noisy 93230 Romainville FRANCE
BEAUPAIN	Dr.	Rene	Immunopharmacologie Experimentale (UA 579) Inst. Biomed. des Cordel. 15, rue de l'Ecole-de-Medecine 75270 Paris Cedex 06 FRANCE

BEBBINGTON	Dr.	C.	Celltech Ltd. 250, Bath Road Slough SL1 4DY UK
BEN-MICHAEL	Mr.	Abraham	c/o Savyon Diagnostics Ltd. 1, Jabontinsky Street Ramat Gan 52131 ISRAEL
BERG	Dr.	Guy	Charles River Biotechnical Services Manston Road Margate Kent CT9 4LT UK
BERTRAM	Dr.	Ken	Biotechnology Australia Pty. Ltd. P.O.B. 20 Roseville, NSW 2069 AUSTRALIA
BETHKE	Dr.	U.	Biotest Pharma GmbH Postfach 401109 D-6072 Dreieich FRG
BEUVERY	Dr.	E. Coen	National Institute for Public Health & Environmental Hygeine P.O. Box 1 3720 BA Bilthoven NETHERLANDS
BEY	Dr.	Elke	University of the Witwatersrand Dept. of Microbiology Johannesburg SOUTH AFRICA
BILLIG	Mr.	David	Pharmacia AB Bjorkgatan 30 S-75182 Uppsala SWEDEN
BOGE	Dr.	Armin	Impfstoffwerk Wellcome GmbH Postfach 1352 3006 Burgwedel 1 FRG
BOHAK	Dr.	Zvi	The Weizmann Institute of Science Department of Biophysics Rehovot 76100 ISRAEL

BOURCIER	Mr.	J-P	10, rue de Temara, P.B. 258 78104 St. Germain en Lave Cedex FRANCE
BRACHA	Dr.	Moshe	Inter-Yeda Ltd. Science-Based Industrial Park Kiryat Weizmann Ness-Ziona 76110 ISRAEL
BRAEUTIGAM	Mrs.	Sandra	Dechema Theodor-Heuss-Allee 25 6000 Frankfurt 97 FRG
BRATTLE	Mr.	Malcolm K.	Charles River Biotechnical Services Manston Road, Margate Kent CT9 4LT UK
BRO-JORGENSEN	Dr.	Klaus	Statens Seruminstitut Amager Boulevard 80 DK-2300 Copenhagen S DENMARK
BROWNE	Dr.	Jeffrey K.	Dept. of Molecular Cell Biology AmGen Inc. 1900 Oak Terrace Lane Thousand Oaks, CA 91320-1789 USA
BUNTEMEYER	Mr.	Heino	Gesellschaft fuer Biotechnologische Forschung mbH Mascheroder Weg 1 3300 Braunschweig FRG
BUTLER	Dr.	Michael	Manchester Polytechnic Biological Sciences Dept. Chester St., Manchester M15 GD UK
CANNAS	Dr.	Mario	Anatomia Umana Corso Massimo d'Azeglio 52 10126 Torino ITALY

CHARLIER	Dr.	Gerard	National Institute Veterinary Res. Groeselenberg 99 B-1180 Brussels BELGIUM
CIONI	Dr.	Costantino	Sclavo SPA. Viral Vaccines Dept.-Biopharm Div. Via Fiorentina 1 53100 Siena ITALY
CLARKE	Dr.	J.B.	Vrpl., Camr Porton Down Salisbury UK
CLEMENTI	Dr.	Massimo	Institute of Microbiology Service of Virology Regional Hospital of Torrette I-60100 Ancona ITALY
COLLINS	Mr.	W.D.	Commonwealth Serum Laboratories 45 Poplar Road Parkville Melbourne 3052 AUSTRALIA
COMER	Dr.	Michael	Boehringer Mannheim GmbH Forschungszentrum 8132 Tutzing FRG
CUTAYAR		J.	L'Air Liquide D.C.R.D. Centre de Recherche Claude Delorme B.P. 126 FRANCE
CYMBALISTA	Mr.	Samuel	Laboratoires Serono S.A. Aubonne, 1170 SWITZERLAND
D'OULTREMONT	Mr.	Philippe	Solvay & Cie S.A. Rue de Ransbeek 310 1120 Brussels BELGIUM
DE KORT	Dr.	Mari	Intervet International B.V. P.O.B. 31 NL-5830 AA Boxmeer NETHERLANDS

DELZER	Dr.	Juergen	BASF Aktiengesellshafen Abteilung ZHBIF - Geb. A30 D-6700 Ludwigshafen FRG
DRAGHI	Ms.	Enrica	Inter-Yeda Ltd. Science Based Industrial Park Kiryat Weizmann Ness Ziona 76 110 ISRAEL
DRAPER	Mrs.	Marian E.	Coopers Animal Health Ltd. Ash Road Pirbright Surrey GU24 ONQ UK
DUGRE	Dr.	Robert	Institut Armand-Frappier P.O. Box 100 Laval-des-Rapides Station Ville de Laval, Quebec H7N 4Z3 CANADA
DUNKER	Dr.	Reinhard	Biochemistry Research A4 E. Merck P.O. Box 4119 D-6100 Darmstadt FRG
EBERHARD	Dr.	Ulrich	Behringwerke AG Virologie Forschung Postfach 1140 355 Marburg FRG
EFRON	Mr.	Dov	Kibbutz Beit Haemek ISRAEL
EPSTEIN	Dr.	Nava	Israel Institute for Biological Research Dept. of Microbiology Ness-Ziona 70450 ISRAEL
ESPUNA	Dr.	Enric	Laboratorios Hipra Departamento de Virologia Amer (Girona) SPAIN

ESTEFANELL	Dr.	Jorge A.	Cooper-Zeltia, S.A. Apartado 16 Porrino Pontevedra 36400 SPAIN
EVENCHIK	Dr.	Zigmund	Israel Institute for Biological Research Ness-Ziona 70450 ISRAEL
EWART	Ms.	Kristiina	Orion Corporation Ltd. Orion Diagnostica P.O. Box 83 SF-02101 Espoo FINLAND
FABRY	Dr.	Luc	Smith Kline-R.I.T., S.A. Rue de l'Institut, 89 B-1330 Rixensart BELGIUM
FARRAM	Dr.	Eunice	5 Martin St. Hunters Hill NSW 2110 AUSTRALIA
FEDER	Dr.	Joseph	Monsanto Company Mail Zone R1B 800 North Lindbergh Blvd. St. Louis, Missouri 63167 USA
FIELD	Dr.	Raymond P.	Celltech Ltd. 244 Bath Road Slough Berks SL1 4DY UK
FORMAN	Mr.	Lawrence	Genentech Inc. 460 Pt. San Bruno Blvd. South San Francisco, CA 94080 USA
FOURNIER	Dr.	Pierre	Institut Merieux Marcy l'Etoile BP3 F-69752 Charbonnieres/Bains Cedex FRANCE
FRAUNE	Dr.	Elizabeth	B. Braun Melsungen AG Biotechnik Postfach 110 3508 Melsungen FRG

GABISON	Dr. David	Israel Institute for Biological Research Ness-Ziona 70450 ISRAEL
GALLILI	Dr. Gilad E.	Biological Laboratory Teva P.O. Box 27047 Jerusalem ISRAEL
GIEHRING	Dr. Horst	c/o Pfeifer & Langen Linnicher Strasse 48 D-5000 Koeln 41 FRG
GIVATI	Mr. Reuven	Israel Institute for Biological Research Ness-Ziona 70450 ISRAEL
GLUCK	Dr. Reinhard	Blinzernfeldweg 8 CH-3098 Koeniz SWITZERLAND
GOLDBERG	Dr. Alan M.	J. Hopkins School of Hygiene and Public Health 615 N. Wolfe St. Baltimore, MD 21205 USA
GOZES	Dr. Illana	The Weizmann Institute of Science Department of Hormone Research Rehovot ISRAEL
GOZES	Dr. Yehoshua	Israel Institute for Biological Research Ness-Ziona 70450 ISRAEL
GRAY	Dr. Peter	University of New South Wales Dept. of Biotechnology Sydney, AUSTRALIA
GREENFIELD	Prof Paul F.	University of Queensland Dept. of Chemical Engineering St. Lucia Queensland 4067 AUSTRALIA

GRIFFITHS	Prof J. Bryan	Phls Camr Porton Salisbury Wilts SP4 0JG UK
GROVES	Dr. David J.	University of Surrey Dept. of Biochemistry Guildford Surrey GU2 5XH UK
GUAZZONE	Mr. Bruno	c/o Chemap AG Hoelzliwiesenstr. 5 8604 Volketswil SWITZERLAND
GUTTER	Dr. Bezalel	Biological Lab. Teva P.O.B. 27047 Jerusalem ISRAEL
HAFFENDEN	Mr. Paul	Z.L. Bocknek Laboratories Inc. 165 Bethridge Road Rexdale, Ontario M9W 1N4 CANADA
HAGGSTROM	Dr. Margareta	Steridose System AB Datavagen 55 S-43600 Askim SWEDEN
HAMBURGER	Dr. David	IBT International Bio-Technologies Ltd. P.O. Box 151490 Jerusalem 93469 ISRAEL
HANDA	Dr. Anita	University of Surrey Dept. of Microbiology Guildford Surrey GU2 5XE UK
HANNINEN	Mr. Veli	Orion Corporation Ltd. Orion Diagnostica P.O. Box 83 SF-02101 Espoo FINLAND
HANSEN	Dr. Karen	Nordisk Gentofte A/S, R & D Dept. Niels Steensensvej 1 SK-2820 Gentofte DENMARK

HARDEFELT	Mr. Mats	SWEDEN
HASSELL	Mr. Thomas	Manchester Polytechnic Dept. of Biological Sciences Chester Street - J. Dalton Bldg. M15GD Manchester UK
HENNO	Dr. P.	C.N.T.S. Avenue des Tropiques B.P. 100 91943 Les Ulis Cedex FRANCE
HERMANN	Mrs. Wil H.	Duphar B.V. P.O. Box 2 1381 AA Weesp NETHERLANDS
HOFFMANN	Dr. Frieder	Brunswick Biotechnetics 4116 Sorrento Valley Blvd. San Diego, CA 92121 USA
HORAUD	Prof Florian	Institute Pasteur Unite de Virologie Medicale 25 rue de Dr. Roix 75724 Paris Cedex 15 FRANCE
HU	Prof Wie-Shou	University of Minnesota Dept. of Chem. Eng. & Materials Sci 421 Washington Ave., SE Mineapolis, MN 55455 USA
IANCONESCU	Prof Marius	Veterinary Institute P.O.B. 12 Beit Dagan 50250 ISRAEL
JALANKO	Mrs. Anu	Orion Genetic Engineering Lab. Valimotie 7 00380 Helsinki 38 FINLAND
JAMES	Dr. Keith	Universtiy of Edinburgh Dept. of Surgery, Univ. Med. School Teviot Place Edinburgh EH8 9AG - Scotland UK

KADOURI	Dr. Avinoam	The Weizmann Institute of Science Department of Biophysics Rehovot 76100 ISRAEL
KALYANPUR	Dr. Manohar	Millipore S.A. B.P. 307 78054 St. Quentin-en-Yvelines FRANCE
KATINGER	Prof Herman	Universitaet fuer Bodenkultur Institut fuer Mikrobiologie Peter Jordan Str. 82 A-1190 Wien AUSTRIA
KELLER	Mrs. Ingrid	c/o Pfeifer U. Langen Werk Dormagen Frankenstrasse 25 4047 Dormagen FRG
KILBURN	Prof Douglas G.	PHLS Centre for Applied Microbiolog Microbiology Research VRPL Porton Down Wilts SP4 OJG UK
KLEUSER	Dr. Beate	Sandoz Ltd Bldg. 506/304 4002 Basel SWITZERLAND
KLOFT	Dr. M.	Biotest Pharma GmbH, Dreiech Postfach 401109 6072 Dreiech FRG
KNUDSEN	Dr. Ida M.	Novo Industri A/S Novo Alle, DK-2880 Bagsvaerd DENMARK
KOEHL	Mr. Michael	Sanofi Elf Bio Recherches Labege Innopole B.P. 137 31328 Castanet-Tolosan FRANCE
KONGERSLEV	Dr. Leif	Nordisk Gentofte A/S R & D Dept. Niels Steensensvej 1 DK-2820 Gentofte DENMARK

LAGARDE	Ms.	Dominique	Pasteur Vaccins 1 Bd. R. Poincare 92430 Marnes-la-Coquette FRANCE
LANGJAHR	Mrs.	Renate	MBR Bio Reactor AG Werkstrasse 4 CH-8620 Wetzikon SWITZERLAND
LANGUET	Mr.	M.B.	Rhone Merieux Laboratoire IFFA 254 B.P. 7009 69342 Lyon Cedex 07 FRANCE
LAWRENCE	Mr.	Gordon C.	Glaxo Group Research Ltd. Fermentation Dept. Greenford Road Greenford, Middlesex UB6 OHE UK
LAZAR	Dr.	Arye	Israel Institute for Biological Research Ness-Ziona 70450 ISRAEL
LEBHERZ	Mr.	William B.	Program Resources Inc. NCI-Frederick Cancer Research Fac. P.O.B. B, Bldg. 434 Frederick, MD 21701 USA
LEHMANN	Dr.	Juergen	G.B.F. mbH Mascheroder Weg 1 D-3300 Braunschweig FRG
LEVY	Mr.	Reuven	Hadassah Medical School Dept. of Virology Jerusalem ISRAEL
LICK	Dr.	Carlo	BASF Abteilung ZHB1F - A30 D-6700 Ludwigshafen FRG
LINDNER	Ms.	Elisabeth	Kabigen AB, P3 S-112 87 Stockholm SWEDEN

LOBMANN	Mrs. Michele	c/o Smith Kline R/T 89, Rue de l'Institut 1330 Rixensart BELGIUM
LUNDGREN	Prof Erik	University of Umea Institute of Cell Molecular Biology S-90185 Umea SWEDEN
LUPKER	Dr. Jan	Sanofi Elf Bio Recherches B.P. 137, 31328 Castantet-Tolosan Cedex FRANCE
MAIORELLA	Dr. Brian L.	Cetus Corp. 1400 53rd St. Emeryville, CA 94608 USA
MARCUS	Mr. Dino	Israel Institute for Biological Research Ness-Ziona 70450 ISRAEL
MAULER	Dr. Rudolf	Behringwerke AG Virologie Forschung Postfach 1140 355 Marburg FRG
MAYER	Mr. Yitzhak	Ministry of Defense Tel Aviv ISRAEL
MEIGNIER	Dr. Bernard	Institut Merieux Marcy d'Etoile BP 3 F-69752 Charbonnieres-les-Bains FRANCE
MERTEN	Dr. Otto-W.	Institut Pasteur Technologie Cellulaire 28, rue du Docteur Roux F-75726 Paris-Cedex 15 FRANCE
MICHAEL	Dr. Amnon	Vineland Laboratories P.O. Box 561 Kfar Vitkin 40200 ISRAEL

MILLER	Prof Alain O.A.	Universite de l'Etat/Lab. de Biochemie Moleculaire-U. Biotech. 24, Avenue du Champ de Mars B-7000 Mons BELGIUM
MIZRAHI	Prof Avshalom	Israel Institute for Biological Research Dept. of Biotechnology Ness-Ziona 70450 ISRAEL
MOAV	Dr. Neomi	Inter-Yeda Ltd. Science Based Industrial Park Kiryat Weizmann Ness Ziona 76 110 ISRAEL
MORANDI	Dr. Maurizio	Sclavo SPA. R & D Biologics - Biopharm Division Via Fiorentina 1 53100 Siena ITALY
MORRIS	Dr. C.	Phls Camr Porton Down Salisbury, Wilts. SP4 0JG UK
MOTHERSILL	Dr. Carmel	St. Luke's Hospital Radiobiology Laboratory Highfield Road - Rathgar Dublin 6 IRELAND
MUNDT	Dr. Wolfgang	"IMMUNO" - Aktiengesellschaft fuer Chemisch-Medizinische Produkte Uferstrasse 15 A-2304 Orth/Donau AUSTRIA
MUNSTER	Dr. Michael	Boehringer Mannheim GmbH Bahnhofstrasse 9-15 D-8132 Tutzing FRG
MURAKAMI	Prof Hiroki	Kushu University 6-10-1, Hakozaki, Higashi-ku Fukuoka-shi Fukuoa-ken JAPAN

MURDIN	Dr.	Andrew	University of Surrey Department of Microbiology Guildford Surrey GU2 5XE UK
MUSGRAVE	Dr.	Stephen	University of Surrey Dept. of Microbiology Guildford, GU2 5EX UK
MacDONALD	Dr.	Caroline M	University of Strathclyde Immunology Division The Todd Centre G4 ONR, Glasgow, Scotland UK
NAGARANO	Mr.	Jacob	Ministry of Defense Tel Aviv ISRAEL
NEEMAN	Mrs.	Michal	The Weizmann Institute of Science Isotope Dept. Rehovot ISRAEL
NICHOLSON	Mr.	Ian	Celltech Limited 228 Bath Road Slough Berks SL1 4EN UK
NIEFFENEGGER	Mr.	Randolph	P.O. Box 1862 Woodland, CA 95695 USA
NILSSON	Dr.	Kjell	University of Lund Pure and Applied Biochemistry P.O.B. 124 S-221 00 Lund SWEDEN
NOBUO	Dr.	Fujiyoshi	Tokyo Research Laboratories Kyowa Hakko Kogyo Co. 3-6-6 Asahi-machi, Machida-shi Tokyo 194 JAPAN
NODA	Mr.	Koichi	Mochida Pharmaceutical Company 1-1-1, Kamiya Kita-ku Tokyo JAPAN

NOLL	Dr.	Gabriele	Behringwerke AG Postfach 1140 D-3550 Marburg 1 FRG
NORDENBERG	Dr.	Jardena	Beilinson Medical Center Rogoff Institute Petach Tikva 49100 ISRAEL
OHASHI	Dr.	Hideya	Kirin Brewery Co., Ltd. Pharmaceutical Lab 1-2-2, Soja, Soja-cho Maebashi-shi, Gunma-ken JAPAN
OLSEN	Mr.	Frank	Novo Industri A/S Novo Alle DK-2880 Bagsvaerd DENMARK
OMONO	Mr.	Soemon	Kubota Corporation 3-29-9, Hongo Bunkyo-ku Tokyo JAPAN
OTTESKOG	Ms.	Per	Kabigen AB, P5 S-112 87 Stockholm SWEDEN
OZERI	Dr.	Tali	Chaim Sheeba Medical Center Institute of Hematology Tel-Hashomer ISRAEL
PALASH	Mr.	Zvy	Inter-Yeda Ltd. Science-Based Industrial Park Kiryat Weizmann Ness-Ziona 76 110 ISRAEL
PANINA	Dr.	Gianfranco	Istituto Zooprofilattico Via Bianchi 7 25125 Brescia ITALY
PAPIER	Mr.	Yoel	Israel Institute for Biological Research Ness-Ziona 70450 ISRAEL

PARSONS	Mr.	David F.	Life Science Laboratories Ltd. Sedgewick Road Luton Beds LU4 9DT UK
PETERSEN	Dr.	Jorn M.	Novo Industri A/S Novo Alle DK-2880 Bagsvaerd DENMARK
PETRI	Dr.	Thomas	Schering AG Mikrobiol Biochemie Postfach 65 03 11 1000 Berlin 65 FRG
PETRICCIANI	Dr.	John C.	Biologicals World Health Organization 1211 Geneva 27 SWITZERLAND
PNIEL	Mr.	Rony	Israel Institute for Biological Research Ness-Ziona 70450 ISRAEL
PRICE	Mr.	Derek	UK
PRZYBOS	Dr.	Ewa	Polish Academy of Sciences Inst. of Systematic & Exper. Zoo. ul.Stawkowska 17 31-016 Krakow POLAND
PULLEN	Mr.	Ken	Wellcome Biotech Cell Culture Dept., Bldg. 112 Langley Court Beckenham, Kent BR3 3BS UK
RADOUX	Dr.	Daniel	Institut National des Radioelements (IRE) B-6220 Fleurus BELGIUM
REUVENY	Dr.	Shaul	Israel Institute for Biological Research Ness-Ziona 70450 ISRAEL

RUDGE	Dr.	John	Biochemical Recovery Group Dept. of Chemical Engineering P.O. Box 363 Edgbaston, Birmingham B15 2TT UK
RUETTEN	Mr.	Kurt Peter	MBR Bioreaktor AG Werkstrasse 3-4 CH-8620 Wetzikon SWITZERLAND
RUSHKIN	Ms.	Edna	The Weizmann Institute of Science Isotope Dept. Rehovot ISRAEL
SAYER	Ms.	Tania E.	Manchester Polytechnic Dept. of Biological Sciences Chester Street Manchester M1 5DG UK
SCHEIRER	Mr.	Winfried	Sandoz Forschunginstitut Brunner Strasse 59 A-1235 Vienna AUSTRIA
SCHLAEGER	Dr.	Ernst-J.	F. Hoffmann-La Roche & Co. AG ZFE/MB, 66/106 CH-4002 Basel SWITZERLAND
SCHOENFELD	Dr.	Richard	Integrated Genetics, Inc. 31 New York Avenue Framingham, MA 01701 USA
SCHONHERR	Dr.	Otto T.	Diosynth Kloosterstraat 6 5340 BH Oss NETHERLANDS
SCHULZ	Dr.	Rainer	Behringwerke AG P.O.B. 1140 D-3550 Marburg/L FRG
SCHWARZBARD	Dr.	Zivia	Inter-Yeda Ltd. Science-Based Industrial Park Kiryat Weizmann Ness-Ziona 76 110 ISRAEL

SENE	Dr.	Claude	IBF Biotechnics - Etablisement de Chaponost zi Les Troques BP 33 69630 Chaponost FRANCE
SEYMOUR	Dr.	Colin	St. Luke's Hospital Radiobiology Laboratory Highfield Road, Dublin 6 IRELAND
SHALITA	Dr.	Zamir P.	Israel Institute for Biological Research Ness-Ziona 70450 ISRAEL
SHALITIN	Dr.	Channa	Technion Dept. of Biology Haifa 32000 ISRAEL
SHER	Mrs.	Dalia	The Weizmann Institute of Science Department of Biophysics Rehovot 76100 ISRAEL
SHOMER	Mr.	Meir	Sheba Medical Center School of Medicine Embrology & Teratology Dept. Tel Hashomer 52621 ISRAEL
SHURTZ-SWIRSKI	Dr.	Revital	Israel Oceanographic & Limnological Research Ltd. Tel Shikmona, P.O.B. 8030 Haifa 31080 ISRAEL
SILBERKLANG	Dr.	Melvin	R-80W-207 P.O. Box 2000 Rahway, NJ 07065 USA
SILBERSTEIN	Mrs.	Lea	Israel Institute for Biological Research Ness-Ziona 70450 ISRAEL
SIMON	Dr.	Gad A.	Israel Institute for Biological Research Ness-Ziona 70450 ISRAEL

SOUVRAS	Dr. Michel	Institut Merieux Marcy l'Etoile BP 3 F-69752 Charbonnieres/Bains Cedex FRANCE
SPIER	Dr. Raymond	University of Surrey Dept. of Microbiology Guildford Surrey GU2 5XH UK
STEINER	Dr. Ulrich	Krutscheider Weg 55 5600 Wuppertal 11 FRG
STEPHENNE	Mr. Jean	Smith Kline - Rit Rue de L'Institut, 89 B-1330 Rixensart BELGIUM
SVENSSON	Ms. Karin	Monocarb AB S-223 57 Lund SWEDEN
TESTA	Prof Ivano	Istituto Clinica Medica Ospedale Torrette Ancona 60100 ITALY
THOMPSON	Mr. Keith J.	Bioscot Ltd. Kinos Buildings West Mains Road Edingburgh UK
THORPE	Ms. Jane	University of Surrey Dept. of Microbiology Guilford, Surrey GU2 5XE UK
TOLBERT	Dr. William R.	Invitron Corp. 8000 Maryland, # 860 Clayton, Missouri 63105 USA
TOMIBE	Dr. Katsuhiko	Teijin Ltd. Iino Bldg., 2-1-1, Uchisaiwaicho Chiyoda-ku Tokyo JAPAN

TROTEMANN	Dr.	Pierre	Institut Merieux Marcy l'Etoile BP 3 F-69752 Charbonnieres/Bains Cedex FRANCE
ULLRICH	Dr.	Axel	Genentech Inc. Dept. of Developmental Biology 460 Point San Bruno Blvd. South San Francisco, CA 94080 USA
URETZKY	Prof	Gideon	Hadassah Medical School Cardiac Surgery Research P.O. Box 12000 Jerusalem 91120 ISRAEL
V. GELDER	Mr.	P.T.J.A.	Diosynth Kloosterstraat 6 5340 BM Oss NETHERLANDS
V.D. VELDEN-DE GROOT	Mrs.	Tiny	National Institute for Public Health & Enviromental Hygiene P.O. Box 1 3720 BA Bilthoven NETHERLANDS
VAHERI	Mr.	Matti	Orion Corporation Ltd. Orion Diagnostica P.O. Box 83 SF-02101 Espoo FINLAND
VAN DER VLIET	Mr.	Marinus K.	Einsteinweg 101 2333 CR Leiden NETHERLANDS
VANDEST	Mr.	Patrick	Sanofi Recherche Montpellier FRANCE
VINAS	Dr.	Roger	Pasteur Vaccins Parc Industriel d'Incarville B.P. 101 27 100 Val de Reuil FRANCE
VINCENT-FALQUET	Dr.	J. Claude	Institut Merieux Marcy-L'etoile 69752 Charbonnieres les Bains FRANCE

WERENNE	Prof John	Universite Libre de Bruxelles Faculte des Sciences - CP 160 50, Av. F.D. Roosevelt B-1050 Bruxelles BELGIUM
WHELDON	Mr. Graham	Life Science Research, Eye, Suffolk, 1P23 7PX UK
WHITE	Mr. Michael E.	Biological Laboratories Teva Ltd. Industrial Zone Atarot P.O. Box 27047 Jerusalem 91270 ISRAEL
WHITE	Dr. Moshe D.	Israel Institute for Biological Research Ness-Ziona 70450 ISRAEL
WHITESIDE	Dr. Peter	University of Surrey Dept. of Microbiology Guildford Surry GU2 5XE UK
WIKSTROM	Mr. Einar	Kabivitrum AB S-152 00 Strangnas SWEDEN
WINTER	Dr. Peter	Dechema-Institut Theod-Heuss-Allee 25 D-6000 Frankfurt/Main FRG
WIRTH	Dr. Manfred	Gesellschaft fuer Biotechnologische Forschung mbH (GBF) Mascheroder Weg 1 3300 Braunschweig FRG
WITTEK	Mr. Bernd H.	Dr. Mueller-Lierheim KG Behringstrasse 6 8033 Planegg FRG
WUDTKE	Mrs. Marita	Institut fuer Technische Chemie Callinstrasse 3 3000 Hannover 1 FRG

WURM	Dr.	Florian M.	Genetech Inc.- Dept. Cell Culture Research and Development 460 Point San Bruno Blvd. South San Fracnsico, CA 94080 USA
YADIN	Mr.	Hagai	Ministry of Agriculture Kiryat Tel Aviv ISRAEL
YAMANE	Prof	Isao	Yagiyama-Minami, 6-6-20 Sendai 982 JAPAN
YARKONI	Dr.	Eliyahu	Vineland Laboratories P.O. Box 561 Kfar Vitkin 40200 ISRAEL
ZAHN	Dr.	Gabriele	Dr. Karl Thomae GmbH Dept. of Biochemistry Postfach 17 55 D-7950 Biberach/Riss FRG
ZAKAY-RONES	Prof	Zichria	The Hebrew University-Hadassah Medical School-Virology Dept. P.O.B. 1172 Jerusalem 91010 ISRAEL
ZEICHER	Dr.	Marc	Ire Cell Targ 30 Clos Chapelle aux Champs 1200 Brussels BELGIUM
ZILIOTTO	Dr.	Raul	c/o Farmitalia Carlo Erba Via Carlo Imbonati 24 20159 Milano ITALY
ZOLETTO	Dr.	Renzo	Istituto Zooprofilattico Sperimentale delle Venezie Via G. Orus 2 35100 Padova ITALY

List of exhibitors

AMIT TECHNOLOGY
P.O.B. 349
Givatayim 53102, Israel
Tel: (02)810452 Tlx: 26127

Representing:

LH Fermentation, UK -	Fermentor systems, direct drive and air lift systems, magnetic stirrers, shakers, and computerized control attachments.

B. BRAUN MELSUNGEN AG
P.O.B. 110
3508 Melsungen, FRG
Tel: (05661)713921 Tlx:999 44360

BIOSTAT EC - Fermentor for the cultivation of animal tissue cells.
BIOPEM sampling device for continuous sampling in biotechnological processes.

BIO-INSTRUMENTS LTD.
P.O.B. 9519
Jerusalem 91094, Israel
Tel: (02)817380, 825384 Fax: (201)2774222

Representing:

New Brunswick Scientific, USA -	CELLIGEN - the most advanced cell culture system for microcarriers and suspended cells; tissue culture apparatus; incubators; full range of equipment for upstream and downstream processes.
Solohill Engineering, USA -	Glass and plastic microcarrier beads.

Z.L. BOCKNEK LABORATORIES INC.
165 Bethridge Road
Rexdale, Ontario, Canada M9W 1N4
Tel: (416)7450796 Tlx: 6989319

Bocknek Labs specializes in low endotoxin animal sera and liquid tissue culture media.

CELLTECH LTD.
228 Bath Road
Slough, Berks SL1 4EN, UK
Tel: (44753)77866 Tlx: 846701

Contract production of monoclonal antibodies (g-Kg scale) to international regulatory standards for therapeutic proteins, contract research and pharmaceutical product development, immunopurification and assay reagents for industrial scale production of interferons and interleukin-2.

DEXMOR LTD.
P.O.B. 13141
Tel Aviv 61131, Israel
Tel: (03)498503 Tlx: 341486 Fax: (972/3)498505

Representing:

Pharmacia, Sweden -	Separation products, large scale equipment.

ELDAN ELECTRONIC INSTRUMENT CO. LTD.
P.O.B. 1270
Jerusalem 91012, Israel
Tel: (02)528211 Tlx:25231 Fax:(02)534289

Reperesenting:

Biolafitte, France -	Fermentors for the growth of bacteria and cell cultures
Endotronics, USA -	Systems for growth of animal cells using hollow fiber method.
Jouan, France -	Cooled centrifuge, CO_2 incubator with infra-red control, dry incubator for heating purposes.
LKB, Sweden -	POLESPOR electrophoresis - system for separating large pieces of DNA; Fraction collector and recorder - system for liquid chromatorgaphy.

ISRAEL FINE CHEMICALS
P.O.B. 33506
Tel Aviv 61334, Israel
Tel: (03)7510170 Tlx: 35553

Representing:

Costar, Holland -	Tissue culture tubes, flasks, dishes, plates, scrapers, clusters, EIA/RIA plates & strips, disposable pipettes, freeze-vials, media, media supplements, hybridoma growth factors.
Drummond, USA -	Pipet-Aid, microdispensers and accesories.
Elkay, UK -	Plasticsware: cuvettes, microcentrifuge and centrifuge tubes, testtubes, sample holders and cups, tubing, tips, timers, biohazard bags, identity tape.
Flow, UK -	Cell and plant tissue culture ware (glass and plastic), microbiology diagnostic reagents, laminar air flow and filtration equipment, Titertek.
Meloy, USA -	Antisera to human and mouse serum proteins, fluorescein-conjugated antisera, immunodiffusion kits.
Nichiryo, Japan -	Repetivitve syringe dispenser, bottle dispensers, dilutor-dispensers, pipettes.
Northumbria Biologicals, UK -	Restriciton enzymes, DNA sequencing reagents, cloning systems, electrophoresis apparatus.
Tecnomara, Switzerland, -	Bunsen burner and pipetboy.

ISRAEL INSTITUTE FOR BIOLOGICAL RESEARCH
P.O.B. 19
Ness Ziona 70450, Israel
Tel: (08)481656 Tlx: 361933 IIBR Fax: (08)475594

Applied research and high technology R&D contracts in biotechnology and drug design.
Tailored scientific services in genetic engineering, fermentation, pharmacological screening and toxicology studies.
High quality products: lymphokines, monoclonal antibodies, CEA antigen and fine chemicals.

LEVANT X-RAY LTD.
P.O.B. 400
Herzliya 46103, Israel
Tel: (052)521666 Tlx: 341954 Fax: (052)543654

Representing:

DuPont/Sorvall, USA -	Laboratory centrifuges.

LANDSEAS LTD.
P.O.B. 23011
Tel Aviv 61230, Israel
Tel: (03)299091 Tlx: 342216

Representing:

Millipore Intertech, France -	Millipore offer tangential flow systems for rapid concentration of extra or intracellular mammalian cell products.
Setric Genie Industriel, (SGI), France -	Cell culture fermentation systems, range 10-500 liters.

M.B.R. BIO REACTOR AG
Werkstrasse 4
CH-8620 Wetzikon, Switzerland
Tel: (01)9307806 Tlx: 875133

Bioreactors; downstream equipment; process control systems.

MODCON LTD.
Process Control Instrumentation Supply & Installation
P.O.B. 15
Kiryat Motzkin 26100, Israel
Tel: (04)710642, 731138 Tlx: 471531

Representing:

Dr. W. Ingold, Switzerland -	Electrodes, redox and pH meters for laboratory and industry.
IKA Laboratories, FRG -	Laboratory equipment.

ORGENICS LTD.
P.O.B. 360
Yavne 70650, Israel
Tel: (08)438754 Tlx:341982 Fax: (08)439901

Representing:
Charles River Biotechnical Services Ltd. (CRBS), UK - OPTICELL SYSTEMS
- computer controlled immobilised cell culture bioreactors. The latest 5200 bench - top unit will be displayed..
Contract production of biomolecules - in vivo and in vitro services are available.
Safety testing of biologicals - full range of tests are available to meet government regulations.

PFEIFER & LANGEN
Frankenstrasse 25
D-4047 Dormagen, FRG
Tel: (221)4980375 Tlx: 8882818

DORMACELL - microcarriers for animal cell cultures.
DORMAGEL - separation media for bioactive substances.
DEXTRANS PS - for aqueous phase separation systems.

STERILIN LTD.
Sterilin House, Clockhouse Lane
Feltham, Middlesex TW18 8QS, UK
Tel: (07842)51155
Tlx: 8812910

CEL-CULT tissue culture plastics (Represented in Israel by Maagar Ltd.)
BELLCO biotechnology equipment (Represented in Israel by Isramex Ltd.)

TZAMAL MODERN MEDICAL EQUPMENT LTD.
P.O.B. 88
Ramat Gan 52100, Israel
Tel: (03)700650 Tlx: 341449

Representing:

GIBCO - B.R.L., USA, UK -	Products for tissue culture - sera and media.
Nunc, Denmark -	Disposable plasticsware for tissue culture and biological research.

ALPHABETICAL LIST OF COMPANIES PARTICIPATING AND/OR REPRESENTED AT EXHIBITION

Amit Technology, Israel
B. Braun Melsungen Biotechnics Division, FRG
Bio-Instruments Ltd., Israel
Biolafitte, France
Bocknek Laboratories, Inc., Canada
Celltech Ltd., UK
Charles River Biotechnical Services, UK
Costar, Netherlands
Dexmor Ltd., Israel
Drummond, USA
DuPont/Sorvall, USA
Eldan Ltd., Israel
Elkay, UK
Endotronics, USA
Flow, UK
GIBCO - B.R.L., USA
IKA Laboratories, FRG
Ingold Dr. W., Switzerland
Israel Fine Chemicals Ltd., Israel
Israel Institute for Biological Research, Israel
Jouan France
Landseas Ltd., Israel
Levant X-Ray Ltd., Israel
L.H. Fermentation, UK
L.K.B., Sweden
M.B.R. Bioreactor AG, Switzerland
Meloy, USA
Millipore Intertech, France
Modcon Ltd., Israel
New Brunswick Scientific, USA
Nichiryo, Japan
Northumbria Biologicals, UK
Nunc A/S, Denmark
Orgenics Ltd., Israel
Pfeifer & Langen, FRG
Pharmacia, Sweden
Setric Genie Industriel, France
Solohill Engineering, USA
Sterilin Ltd., UK
Tecnomara, Switzerland
Tzamal Ltd., Israel